Calcium Signaling in Biology and Medicine

Calcium Signaling in Biology and Medicine

Editor: Mylo Sanders

AMERICAN
MEDICAL PUBLISHERS
www.americanmedicalpublishers.com

Cataloging-in-Publication Data

Calcium signaling in biology and medicine / edited by Mylo Sanders.
    p. cm.
Includes bibliographical references and index.
ISBN 978-1-63927-611-0
1. Calcium--Physiological effect. 2. Cellular signal transduction. 3. Intracellular calcium.
4. Calcium in the body. 5. Cell physiology. 6. Cytology. I. Sanders, Mylo.
QP535.C2 C35 2023
572.516--dc23

American Medical Publishers,
41 Flatbush Avenue,
1st Floor, New York,
NY 11217, USA

ISBN 978-1-63927-611-0 (Hardback)

Contents

Permissions

List of Contributors

Index

Preface

In my initial years as a student, I used to run to the library at every possible instance to grab a book and learn something new. Books were my primary source of knowledge and I would not have come such a long way without all that I learnt from them. Thus, when I was approached to edit this book; I became understandably nostalgic. It was an absolute honor to be considered worthy of guiding the current generation as well as those to come. I put all my knowledge and hard work into making this book most beneficial for its readers.

The process that increases low cytoplasmic concentrations in order to activate specific, essential events required by the cell is defined as calcium signaling. It uses calcium ions (Ca2+) to communicate and carry out intracellular processes often as a step in signal transduction. Calcium signaling plays a crucial role in the regulation of a diverse set of essential functions within the human body. The management of various neurodegenerative disorders such as Alzheimer's disease (AD), Huntington's disease (HD), and Parkinson's disease (PD) requires calcium signaling. It has been observed that disruption in calcium signaling may lead to synaptic deficits which further leads to accumulation of Aβ plaques and neurofibrillary tangles. Calcium plays an important role in various biological processes and calcium signaling has been linked to various diseases. Therefore, it is currently being explored as a therapeutic target for various types of cancer. This book elucidates the concepts and innovative models around prospective developments with respect to the importance of calcium signaling in biology and medicine. It will serve as a valuable source of reference for those interested in this field of study.

I wish to thank my publisher for supporting me at every step. I would also like to thank all the authors who have contributed their researches in this book. I hope this book will be a valuable contribution to the progress of the field.

Editor

Various Aspects of Calcium Signaling in the Regulation of Apoptosis, Autophagy, Cell Proliferation and Cancer

Simone Patergnani [1](ID), Alberto Danese [1], Esmaa Bouhamida [1], Gianluca Aguiari [2], Maurizio Previati [3](ID), Paolo Pinton [1,*](ID) and Carlotta Giorgi [1,*]

[1]   Department of Medical Sciences, Laboratory for Technologies of Advanced Therapies, University of Ferrara, 44121 Ferrara, Italy; simone.patergnani@unife.it (S.P.); alberto.danese@unife.it (A.D.); esmaa.bouhamida@unife.it (E.B.)

[2]   Department of Biomedical and Surgical Specialty Sciences, University of Ferrara, 44121 Ferrara, Italy; gianluca.aguiari@unife.it

[3]   Department of Morphology, Surgery and Experimental Medicine, Section of Human Anatomy and Histology, Laboratory for Technologies of Advanced Therapies (LTTA), University of Ferrara, 44121 Ferrara, Italy; maurizio.previati@unife.it

*   Correspondence: paolo.pinton@unife.it (P.P.); carlotta.giorgi@unife.it (C.G.)

Abstract: Calcium (Ca^{2+}) is a major second messenger in cells and is essential for the fate and survival of all higher organisms. Different Ca^{2+} channels, pumps, or exchangers regulate variations in the duration and levels of intracellular Ca^{2+}, which may be transient or sustained. These changes are then decoded by an elaborate toolkit of Ca^{2+}-sensors, which translate Ca^{2+} signal to intracellular operational cell machinery, thereby regulating numerous Ca^{2+}-dependent physiological processes. Alterations to Ca^{2+} homoeostasis and signaling are often deleterious and are associated with certain pathological states, including cancer. Altered Ca^{2+} transmission has been implicated in a variety of processes fundamental for the uncontrolled proliferation and invasiveness of tumor cells and other processes important for cancer progression, such as the development of resistance to cancer therapies. Here, we review what is known about Ca^{2+} signaling and how this fundamental second messenger regulates life and death decisions in the context of cancer, with particular attention directed to cell proliferation, apoptosis, and autophagy. We also explore the intersections of Ca^{2+} and the therapeutic targeting of cancer cells, summarizing the therapeutic opportunities for Ca^{2+} signal modulators to improve the effectiveness of current anticancer therapies.

Keywords: calcium; cancer; apoptosis; autophagy; cell cycle; therapy; chemotherapy

1. Introduction: A General Overview of Ca^{2+} Signaling

In resting cells, the intracellular free Ca^{2+} concentration ($[Ca^{2+}]_i$) is maintained at lower levels than extracellular fluid. Indeed, there is a 20,000-fold gradient between outside (about 1.2 mM) and inside (approximately 10–100 nM) of cells. Moreover, in the mitochondria and in the nucleus, the concentrations of Ca^{2+} are similar to those in the cytoplasm. In the endoplasmic reticulum (ER), considered the main intracellular Ca^{2+} store, the $[Ca^{2+}]$ ranges between 100 and 800 µM [1]. In addition, direct measurements of Ca^{2+} levels show that lysosomes present an internal $[Ca^{2+}]$ of about ≈500 µM [2]. Therefore, it exists an elaborate system of Ca^{2+}-transporters, -channels, -exchangers, -binding/buffering proteins, and -pumps that finely regulate Ca^{2+} flow inside and outside of cells and among intracellular organelles [3]. This network permits preservation of a low resting $[Ca^{2+}]$ and regulates the propagation of intracellular Ca^{2+} changes that are fundamental to intracellularly transmitted biological information and important physiologic processes, including metabolism, cell proliferation and death, protein phosphorylation, gene transcription, neurotransmission, contraction, and secretion [4,5]. During cell stimulation the $[Ca^{2+}]_i$ can increase more than twofold at the micromolar level. Different channels situated in the plasma membrane (PM) induce the influx of extracellular Ca^{2+} into the cells. Among these channels, the most important are transient receptor potential channels (TRPC) [6], store-operated Ca^{2+} entry (SOCE) channels such as ORAI and STIM [7], voltage-gated Ca^{2+} channels (VGCC) in excitable cells [8], receptor-operated Ca^{2+} channels such as the N-methyl-d-aspartate receptor (NMDA) [9] and purinergic P2 receptors [10], whose activation determines cytosolic Ca^{2+} influx. Intracellular Ca^{2+} increases may be also due to Ca^{2+} release from internal stores, mainly via inositol 1,4,5-triphosphate receptors (IP3Rs) situated on the ER [11,12]. IP3Rs are large-conductance cation channels that are activated in response to the activation of cell surface receptors [13]. Despite different physiological and pharmacological profiles, ryanodine receptors (RyRs) have an approximatively 40% homology with IP3Rs and are the Ca^{2+} release channels on the sarcoplasmic reticulum of muscle cells [14]. A prolonged elevation of $[Ca^{2+}]_i$ has adverse effects for the cells. Therefore, different channels, pumps, and buffering systems reestablish low $[Ca^{2+}]_i$. The reuptake of Ca^{2+} into the ER lumen is allowed by the activity of sarcoendoplasmic reticulum Ca^{2+}-ATPase (SERCA), which pumps Ca^{2+} into the ER with a stoichiometry of 2:1 Ca^{2+}/ATP and by the secretory protein calcium ATPase (SPCA), which transports Ca^{2+} into the Golgi apparatus [15]. Plasma membrane Ca^{2+} transport ATPase (PMCA) and Na^+/Ca^{2+} exchanger (NCX) are the two mechanisms situated on the PM responsible for Ca^{2+} extrusion. PMCA is a pump that belongs to the class of P-type ATPases that pump Ca^{2+} across the PM out of the cell at the expense of ATP [16,17]. NCX permits Ca^{2+} extrusion against its gradient without energy consumption by using the electrochemical gradient of Na^+. For each Ca^{2+} ion extruded, three Na^+ ions enter the cell [18]. Additionally, mitochondria significantly contribute to the signaling pattern of released intracellular Ca^{2+}. Indeed, these organelles may act as Ca^{2+} buffers [19]. It is widely accepted that Ca^{2+} entry into mitochondria is mediated by the activity of the mitochondrial calcium uniporter (MCU) complex, composed of the pore-forming subunit of the MCU channel together with several regulatory proteins (MICU1, MICU2, MICU3, MCUR1, MCUb, and EMRE) [20]. Advances in the studies regarding Ca^{2+} dynamics have revealed that a network of membrane contact sites has a determinant role in Ca^{2+} signaling. These contacts create microdomains that permit the exchange of metabolites and signals between membranes of different compartments. The structural and functional interactions between the ER and mitochondria (the mitochondria associated membranes, MAMs) represent the main central hub for controlling Ca^{2+} exchange between these two compartments [21]. Disruption of MAMs result in the suppression of ER Ca^{2+}-release and alters mitochondrial Ca^{2+} accumulation (Figure 1). ER membranes are also interconnected with the membranes of lysosomes to form the ER-lysosome membrane contact sites. It has been proposed that the IP3R-mediated ER release of Ca^{2+} is a mechanism for mediating the reestablishment of Ca^{2+} levels in lysosomes [22]. However, the Ca^{2+} transporter mediating this Ca^{2+} transmission remains unidentified. In contrast, the identity of channels regulating lysosomal Ca^{2+} release has been established. Several channels mediate this Ca^{2+} transport. Among these channels, the mucolipin subgroup of the TRP ion channel family, in particular the isoform TRPML1, represents the most well-established lysosomal Ca^{2+} release channels [23].

Overall, all these mechanisms preserve the correct Ca^{2+} homeostasis of the cell and regulate the spatiotemporal patterning of the Ca^{2+} signal. Any alterations to this highly connected network of Ca^{2+} transporters, channels, exchangers, binding/buffering proteins, and pumps determine the unregulated Ca^{2+} dynamics that affect almost every aspect of cell function, such as proliferation, gene expression, cell death, and protein phosphorylation and dephosphorylation [24]. There is evidence that cancer cells have disrupted Ca^{2+} signaling, where the expression of Ca^{2+} channels/pumps and Ca^{2+}-regulating proteins is altered [3]. Therefore, remodeling of these derailed Ca^{2+} features may be a potential target for cancer therapies. In view of this possibility, we outline the contributions of Ca^{2+} signaling to the cell cycle and cell proliferation, apoptosis, and autophagy with particular attention to the cancer context. We also focus on the potential impact of Ca^{2+} signal modulation in cancer therapy.

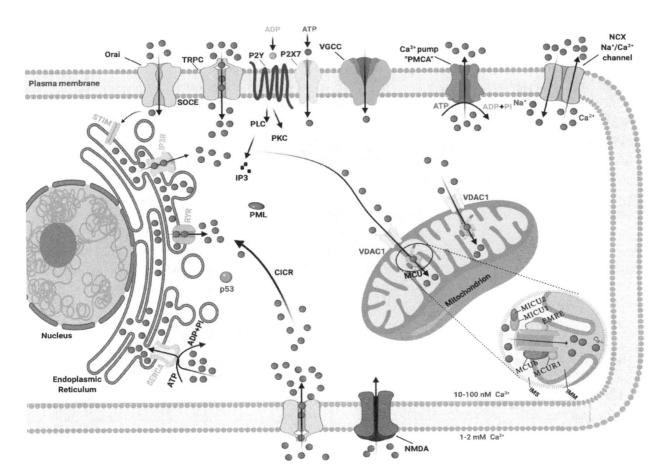

Figure 1. The intracellular Calcium (Ca^{2+}) signaling. Different Ca^{2+} transporters, channels, exchangers, binding/buffering proteins and pumps mediate the regulation of cytosolic Ca^{2+} concentration. In the plasma membrane (PM), PM Ca^{2+}-ATPases (PMCA) pumps, transient receptor potential channels (TRPC), voltage-gated Ca^{2+} channels (VGCC), Na^+/Ca^{2+} exchanger (NCX), and purinergic P2 receptors regulate the transport of Ca^{2+} ions inside and outside cells. Inositol 1,4,5-triphosphate receptors (IP3R), ryanodine receptors (RyR), and sarcoendoplasmic reticulum Ca^{2+}-ATPase (SERCA) pumps control the storage of Ca^{2+} in the endoplasmic reticulum. Finally, voltage-dependent anion channels (VDAC) and members of the mitochondrial Ca^{2+} uniporter family are critical for controlling the mitochondrial Ca^{2+} uptake. Created with BioRender.com.

2. Ca^{2+} Homeostasis during Cell Cycle and Tumor Growth

In recent years, the importance of cell cycle progression regulating by Ca^{2+} signals has been recognized, especially upon the development of probes that allow a very sensible and direct visualization of Ca^{2+} transients. Spontaneous Ca^{2+} oscillations at the three major cell cycle checkpoints have been described. For example, transient $[Ca^{2+}]_i$ increases during the G1/S phase transition [25], the G2/M transition [26], and the metaphase to anaphase transition [27]. Ca^{2+} is also required in the early G1 phase, when cells re-enter the cell cycle, to promote the activation of c-AMP-responsive element binding protein, AP1 (FOS and JUN) transcription factors, and the nuclear factor of activated T-cell (NFAT) [28]. The cell cycle is principally controlled by the expression of protein complexes organized around cyclin-dependent protein kinases (CDKs), which coordinate the entry into the next phase of the cell cycle only when bound to a cyclin. The essential bridge between Ca^{2+} ions and CDK/cycline complexes is undoubtedly represented by the Ca^{2+}-sensors calmodulin (CAM) and calcineurin (CaN). These Ca^{2+}-binding proteins, and intermediary proteins such as Ca^{2+}/calmodulin-dependent protein kinases (CAMKI, CAMKII, and CAMKIII), interact with CDK/cycline complexes regulating crucial cell cycle events, including DNA synthesis (i.e., by cyclin D1-CDK4 regulation through CAMKI) [29], microtubule stability regulation, e.g., by decreasing the amount of Ca^{2+} required for microtubule depolymerization [30] and by interacting with nucleoporin p62 [31] and for cytokinesis completion [32]. The essential role played by Ca^{2+} as a regulator of cell cycle progression has been extensively presented in publications based on the use of CAM activity inhibitors. In particular, treatment with W-7 and W-13 CAM antagonists induced G1 phase cell cycle arrest by downregulating cyclins and upregulating p21 [33]. Moreover, the microinjection of monoclonal antibodies against CAM inhibited the synthesis of DNA in a dose-dependent manner [34]. One of the most fascinating aspects of studying these cell cycle progression regulation mechanisms is undoubtedly the detection of new potential targets in the fight against cancer. Indeed, the most important characteristic of cancer cells is certainly their ability to undergo biological changes that sustain their unlimited replicative capacities. The behavioral study of Ca^{2+} channels and pumps in relation to the cell cycle and proliferation has turned out to be very important, especially in recent years. Some examples of how a perturbation of Ca^{2+} signaling can lead to cell cycle dysregulation with consequent repercussions on tumor pathologies are worthy of description. Cytosolic Ca^{2+} levels modulate guanosine exchange factor and GTPase activating protein, which are a RAS stimulator and a RAS inhibitor, respectively. RAS, in turn, stimulates the proliferative mitogen-activated protein kinase (MAPK) pathway, which initiates the cells transition into the S phase because of phosphorylation of the tumor suppressor RB1 upon cytoplasmic cyclin D1 upregulation. Constitutively high cytosolic Ca^{2+} levels in cancer cells can lead to uncontrolled growth through the removal of the G1/S transition checkpoint [35]. ORAI3, a SOCE component, is overexpressed in breast cancer biopsy samples and is involved in breast cancer cell proliferation and cell cycle progression by modulating the G1 phase and G1/S transition regulator protein activity [36]. ORAI3 is an upstream regulator of c-myc that controls the cell cycle and proliferation in breast cancer by modulating the expression of cyclins D1 and E, CDKs 4 and 2, cyclin-dependent kinase inhibitor p21, and tumor-suppressing protein p53 [37]. A large number of studies have indicated key roles for cyclin D and cyclin E expression in breast cancer cell cycle deregulation; in fact, cyclins D1 and E proteins are overexpressed in more than 50% of breast tumors [38–40]. Additionally, VGCCs are associated with cell proliferation regulation. Specific VGCCs family genes were downregulated in breast, kidney, brain, and lung cancers, showing that these Ca^{2+} channels play roles as tumor suppressor genes [41]. On the other hand, members of the VGCCs family are expressed at detectable levels in melanoma cells but not in untransformed melanocytes, and the use of T-type channel inhibitors induces cell cycle arrest with a significant increase of the percentage of cells in the G1 phase and a reduction of cells in the S phase [42].

Changes in the expression of TRPCs have been implicated in prostate cancer. In particular, transient receptor potential vanilloid subfamily member 6 (TRPV6) in prostate cancer reduces the activation of NFAT and decreases cell accumulation in the S phase of the cell cycle [43].

It has been reported that CAMKs expression alterations have repercussions on cell cycle progression in several tumor pathologies. Parmer et al. described CAMKIII as a potential pharmacological target against glioma because of its important link to cell proliferation, viability, and malignancy [44]. Chemical inhibition of CAMKIII resulted in the reduction of growth of glioma cells line, which was mirrored by a blocked G1 phase transition in the cell cycle. In breast cancer, CAMKIII activation leads to the phosphorylation of elongation factor-2 and transient inhibition of protein synthesis. These events are controlled by mitogens and are predominant in the S phase of the cell cycle [45]. CAMKII has been described to be crucial in T cell lymphoma cell proliferation; its genetic ablation drives a significant increase in the percentage of G2/M phase cells and a decrease in the percentage of S phase cells, outcomes that are consistent with the inhibition of cell proliferation [46].

3. Role of Ca^{2+} in Apoptosis and Cancer

A main hallmark of cancer cells is evasion of programmed cell death (PCD) [47]. PCD is a genetically determined cell routine in which cells undergo an unexpected decline in homeostasis and functionality, triggering several intracellular pathways and ultimately cell death. Different types of PCD exists. During viral or microbial infections, PCD is a part of the host immune response, with traits similar to those of apoptosis and necrosis in a process referred to as pyroptosis or necroptosis.

Pyroptosis is a caspase-dependent form of PCD that leads to membrane permeabilization and cell swelling through gasdermin D activation. Pyroptosis is triggered via inflammasome that was induced when the cell senses changes after viral or microbial invasion and is linked to atherosclerosis, metabolic disease, and neuroinflammatory disorders [48,49]. Another type of PCD is necroptosis, an inflammation-dependent form of PCD, which is similar to necrosis in the absence of caspase involvement and eventual membrane permeabilization, cell swelling, and lysis, with the subsequent leakage of a plethora of proinflammatory molecules. [50]. However, the best-characterized form of PCD is apoptosis. Apoptosis is a strictly controlled phenomenon, typically manifested by chromatin condensation, DNA and nuclear fragmentation, mitochondrial failure, proteolytic enzyme activation and membrane blebbing, even with cell membrane integrity. The cell membrane forms the interface with cell-impermeant stimuli, which interact with membrane receptors triggering the so-called intrinsic pathways. Thus, membrane-induced apoptosis depends upon extracellular ligands, such as tumor necrosis factor-α (TNFα) and first apoptosis signal (FAS) ligand, and their receptors, TNFRs and FAS. Employing several adaptors, some of which carry the death effector domain, these activated receptors induce the formation of the death-inducing signaling complex (DISC). DISC stimulates the autoproteolytic cleavage of initiator caspase-8, which in turn activates the executioner caspase-3, -6, and -7. This proteolytic cascade strongly amplifies the initial signal and initiates the cleavage of hundreds of cellular targets and is thus crucial for the main morphological features of apoptosis [51] (Figure 2, upper panel).

In addition to the extracellular-driven intrinsic pathway, an intracellular, mitochondria-centered pathway is activated in apoptosis, with the function of coupling insurmountable mitochondrial stress to cell death. Nevertheless, a wide number of stress conditions, including hypoxia, alteration, or poisoning of the electron transfer chain, unbuffered ROS production, and imbalanced mitochondrial protein homeostasis, can initiate mitochondrial permeability transition (MPT) [52,53]. MPT, followed by mitochondrial osmotic imbalance and mitochondrial outer membrane permeabilization, allows the release of several mitochondrial proteins, such as cytochrome C (cyt-C). In particular, cytoplasmic cyt-C binds to apoptotic protease activating factor 1 to form a multiprotein complex able to recruit and activate the initiator caspase-9 via the caspase recruitment domain. Caspase-9 in turn cleaves and activates the other executioners, namely, caspase-3, -6, and -7 [51]. An important regulatory mechanism of the intrinsic pathway is the mitochondrial Ca^{2+} load. Ca^{2+} is an important regulator of Krebs cycle dehydrogenases [54] and normally accumulated in the mitochondrial matrix at concentrations 10-fold higher than those measured in the cytosol. Under specific conditions, Ca^{2+} overload can trigger MTP by opening the permeability transition pore with the consequent release of apoptogenic factors [52,55] (Figure 2, upper panel). During carcinogenesis, cancer cells use different machinery to circumvent apoptosis and acquire a profound survival and proliferative advantage. They accumulate genetic alterations that increase or decrease the expression of pro- and/or antiapoptotic genes. Moreover, cancer cells can prevent apoptosis through post-translation modification, such as phosphorylation/dephosphorylation. In contrast, cancer cells may also evade apoptosis by reducing the Ca^{2+} signaling necessary to prompt the apoptotic machinery. The latest evidence shows that Ca^{2+} release from the ER is the main mechanism regulating the mitochondrial Ca^{2+} remodeling and apoptosis [56]. The first observation was obtained by studying B-cell lymphoma-2 (BCL-2) proteins. These proteins are classified into antiapoptotic category (BCL-2, BCL-xL, and Mcl-1) and a pro-apoptotic category (like Bax, Bak, Bim, Bid, etc.). Evidence demonstrates that antiapoptotic BCL-2 proteins regulate the apoptotic program by controlling ER-mitochondrial Ca^{2+} transfer in both organelles, and in particular, recent studies indicate that these proteins also exert antiapoptotic functions at MAMs levels [57]. Overexpression of pro-apoptotic BCL-2 proteins was found to reduce both ER-Ca^{2+} release either by direct control of IP3R3-induced pore opening or by lowering the Ca^{2+} content of the ER [58,59]. As a consequence, Ca^{2+}-induced MPT is prevented, and the apoptotic program is abolished. Furthermore, it has also been demonstrated that BCL-2, BCL-XL, and Mcl-1 determine pro-survival IP3Rs-mediated Ca^{2+} oscillations that are necessary to increase mitochondrial energy production and stimulate cell proliferation [60]. Overall, these proteins impact three important aspects of cancer development: cell death, survival, and energy production. Consistently, upregulation of pro-apoptotic BCL-2 members was found in different human cancer samples and was associated with the invasion and metastasis of colon, breast, and gastric cancer [61]. BCL-2 members are not the only proteins that regulate apoptosis and cell proliferation by modulating the ER-Ca^{2+} release into mitochondria. The oncogene RAS plays a pivotal role in tumor growth and maintenance of the tumor environment [62]. To exert this function, RAS deregulates ER Ca^{2+} dynamics with the consequent inhibition of apoptosis, impairment to mitochondrial metabolism, and promotion of malignant cell survival [63]. Additionally, the oncogene AKT phosphorylates and inactivates several proteins (such as Bad, Bax, and hexokinase-2) that normally work to promote the Ca^{2+}-dependent apoptotic response. Furthermore, AKT inhibits the apoptotic process by exerting a direct control of IP3R3 opening, thus avoiding the Ca^{2+} overload necessary to activate the intrinsic apoptosis [64]. If oncogene proteins promote cell survival and proliferation by blocking Ca^{2+}-mediated apoptosis, it is not surprising that tumor suppressors activate the same mechanism. Protein phosphatase and tensin homolog (PTEN), which is frequently lost or mutated in several cancers, counteracts the activity of AKT and restores Ca^{2+} transfer and reestablishes subsequent cell death [65]. In addition, PTEN was also recently found to block the proteosomal degradation of IP3R3 provoked by the F-box protein FBXL2 [66]. The activity of AKT is also balanced by the tumor suppressor promyelocytic leukemia protein (PML), which, together with IP3R3, AKT, and the phosphatase PP2a, creates a complex that rules ER-mitochondria Ca^{2+} transfer [67]. BRCA1-associated protein 1 (BAP1) is a tumor suppressor frequently mutated in diverse malignancies, especially in mesotheliomas, for which alteration of Ca^{2+} dynamics had previously described [68]. BAP1 works as a deubiquitinating enzyme and is involved in different processes, such as DNA repair and transcription. Recently, it has been demonstrated that BAP1 also deubiquitylates and stabilizes IP3R3. Therefore, following DNA damage exposure, cells can undergo to apoptosis by activating ER-mitochondria Ca^{2+} transfer [69]. Additionally, p53 regulates tumorigenesis by modulating ER-mitochondria Ca^{2+} flux. In this case, the tumor suppressor was found to improve intracellular Ca^{2+} accumulation by increasing SERCA pump activities [70] (Figure 2, lower panel). Apart from the well-established roles for ER-Ca^{2+} dynamics in cancer, recent investigations suggest that impairments in lysosomal Ca^{2+} processes are also important in driving tumorigenesis. Consistent with this finding, cancers of the bladder, head and neck region, and thyroid exhibit increased expression of the gene mucolipin 1 [71], which encodes the lysosomal Ca^{2+} release channel (transient receptor potential mucolipin 1, TRPML1). Consistent with this, TRPML1 inhibition reduces the proliferation of cancer cells [71]. Additionally, the expression of TRPML2 isoforms has been found to be highly expressed in glioma tissues [72]. The transcription factor EB (TFEB) is a master regulator of lysosome function. Altered expression and/or activity of TFEB has been found in pancreatic, kidney, and non-small cell lung cancers [73–75] and is associated with aggressive clinical features in colorectal cancer [76]. Interestingly, it has been demonstrated that TFEB activities are highly modulated by a Ca^{2+}-enriched microenvironment that is created following lysosomal Ca^{2+} release mediated by TRPML1 channels [77] and that TFEB itself modulates the lysosomal Ca^{2+} buffering capacity [78], thereby suggesting a primary role of lysosomal Ca^{2+} in TFEB-associated cancers. Recent advances in RNA research have revealed that the levels of microRNAs (miRs), a class of small noncoding RNAs that regulate various target genes leading to a decrease in target protein levels, are associated with a variety of human diseases, including cancer. In this context, miRs not only regulate the functions of several oncogenes and tumor suppressors but also target genes that control intracellular Ca^{2+} dynamics. Among these miRNAs, oncogenic miR-25 provokes the downregulation of MCU with subsequent decreases in mitochondrial Ca^{2+} uptake and a reduction in the apoptotic process. Accordingly, prostate and colon cancer cells express increased miR-25 levels and present reduced MCU levels [79]. miR-25-dependent MCU downregulation has also been observed in pulmonary artery smooth muscle cells, where decreases in mitochondrial Ca^{2+} levels cause the activation of a cancer-like phenotype characterized by increased cellular proliferation, migration, and apoptotic resistance. In addition to miR-25, the authors also identified miR-138 as a regulator of MCU expression and demonstrated that nebulizing anti-miR-25 and miR-138 restored MCU expression and abolished the cancer-like phenotype [80]. Another miR involved in cancer is miR-34, whose aberrant expression has been detected in the T lymphocytes of cancer patients [81,82]. It has been observed that miR-34 is also a regulator of SOCE by targeting the expression of IP3R2, STIM1, and ORAI3 in immune cells. These results suggest that miR-34 may control the activities of pro- and antiapoptotic genes by regulating Ca^{2+} signaling, thereby controlling the activation and proliferation of T cells and inducing the inhibition of the antitumor immune response. Finally, several other miRs have been suggested to regulate Ca^{2+} homeostasis and apoptosis. Despite this, a direct correlation between these miRs, Ca^{2+}, and cancers has not been demonstrated. Only to cite a few, miR-132 influences Ca^{2+} levels by regulating the expression of the exchanger NCX [83]. MiR-7 reduces voltage-dependent anion channels 1 (VDAC1) expression and diminishes the efflux of Ca^{2+} from mitochondria [84]. MiR-1 regulates the expression of MCU and protects mitochondria from Ca^{2+} overload in cardiac myocytes during development [85].

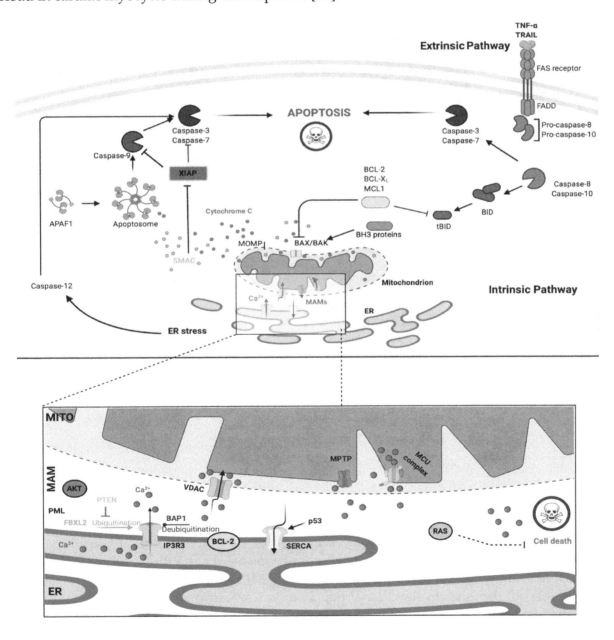

Figure 2. Apoptosis and Calcium (Ca^{2+}) dynamics in cancer. Apoptosis is the best-characterized and

studied programmed cell death. In the extrinsic pathway, extracellular ligands determine the formation of the death-inducing signaling complex that activates the caspases cascade. The intrinsic apoptotic pathway is characterized by permeabilization of the mitochondria that allows the release of cytochrome c (cyt-c) and other apoptogenic factors in the cytosol. Once released, these factors bind apoptotic protease activating factor 1 (APAF1) and form a multiprotein complex called the apoptsome that recruits and activates the caspases. Ca^{2+} has a major role during intrinsic apoptosis, and excessive mitochondrial Ca^{2+} accumulation may trigger apoptosis. Different proteins were found to control apoptotic machinery by regulating Ca^{2+} flux between endoplasmic reticulum (ER) and mitochondria. Antiapoptotic B-cell lymphoma-2 (BCL-2) members block apoptotic program by lowering Ca^{2+} levels in the ER, thereby attenuating subsequent Ca^{2+} release. p53 localizes at the ER–mitochondria interface to improve Ca^{2+} dynamics and apoptosis by increasing sarcoendoplasmic reticulum Ca^{2+}-ATPase (SERCA) pumps activities. Additionally, the tumor suppressors promyelocytic leukemia protein (PML), BRCA1-associated protein 1 (BAP1), and phosphatase and tensin homolog (PTEN) move to the mitochondria associated membranes (MAMs) to regulate Ca^{2+}-dependent apoptosis. They determine the activation of Ca^{2+} release from the ER by modulating the activity of inositol 1,4,5-triphosphate receptor 3 (IP3R3). Mutations or loss of these tumor suppressors are frequently found in diverse human tumor samples, where they lead to a reduction in Ca^{2+} homeostasis and the apoptosis rate, favoring cellular proliferation, tumor growth, maintenance, and metastasis. Created with BioRender.com.

4. The Regulation of Autophagy by Calcium Signals and Its Involvement in Cancer

Autophagy is an intracellular catabolic process that targets and isolates cytoplasmic components, ranging from low-dimension biological macromolecules to whole organelles, and successively enables their delivery to lysosomes for degradation. As a whole, autophagy plays two main roles. First, it is a homeostatic mechanism, ensuring the removal of damaged proteins and organelles. Selective forms of autophagy can specifically target mitochondria (mitophagy), the endoplasmic reticulum (reticulophagy), peroxisomes (pexophagy), lipid droplets (lipophagy), or invading pathogens (xenophagy) [86]. Second, the degraded material is a source of amino acids and lipids for the subsequent de novo synthesis of proteins and lipids. This recycling is of particular importance in the presence of conditions limiting the availability of amino acids, such as during starvation, when the presence of the whole amino acid pool can be guaranteed only through the demolition of cellular proteins, which serve as a reservoir of amino acids. For this reason, autophagy is mainly regarded as a survival mechanism executed during shortage conditions and also during similar stressful circumstances, such as hypoxia or pathogen invasion [87]. Autophagy is initiated with the formation of double-membrane lined vesicles, which gather and fuse, engulfing portions of the cytoplasm. The resulting double-membrane vacuoles are called autophagosomes (APs), which can fuse with vesicles in the endocytic pathway at different stages of maturation or directly with the lysosome, becoming an autolysosome. At this point, the acidic hydrolases break down the macromolecules into smaller constituents that are released back into the cytosol by lysosomal transporters and permeases.

In APs formation, unc-51 like autophagy activating kinase 1-2/autophagy-related 13/200-kDa focal adhesion kinase family-interacting protein (ULK/ATG13/FIP200) complex is the upstream regulator [88]. The ULK1 complex and related adaptor proteins are controlled through the action of kinases such as mammalian target of rapamycin (mTORC) and 5′ adenosine monophosphate-activated protein kinase (AMPK). mTORC is a protein complex that integrates different stimuli involved in nutritional status and oxygen levels to regulate several cellular processes, in particular inhibiting autophagy via direct phosphorylation of ULK1. On the other hand, AMPK stimulates autophagy through ULK1 phosphorylation in response to nutritional deprivation, oxygen unavailability, and mitochondrial dysfunction [89]. Activated ULK1 and 2 proteins, in turn, not only inhibit mTORC and AMPK but also phosphorylate and activate coiled-coil, moesin-like BCL2 interacting protein (BECN1). BECN1 is part of a complex that includes class III phosphatidylinositol 3-kinase (PI3K) and its regulatory proteins, which, when activated, are involved in the nucleation and elongation of the phagophore upon activation.

The first step is the synthesis of phosphatidylinositol-3-phosphate (PI3P) by phosphorylation of phosphatidylinositol in the membrane of the ER, mitochondria, Golgi complex, endosomes, or PM [90]. PI3P recruits several adaptor proteins involved in the elongation of a sack-like, omega-shaped structure, which grows and closes around and binding the material to be digested. The BECN1 interactome, formed by BECN1 and various interacting proteins, regulates the nucleation and elongation of phagophore [91]. While PI3P behaves as a positive regulator of autophagy and is involved in the recruitment of various adaptor proteins, BECN1 can be negatively regulated by the antiapoptotic proteins BCL-2, BCL-XL and other members of BCL-2 family. On the one hand, these proteins bind to BECN1 through the BCL-2-homology-3 (BH3) domain and inhibit autophagy by disrupting the interaction between BECN1 and the class III PI3K complex. On the other hand, BCL-2 phosphorylation can attenuate BECN1 sequestration and autophagy inhibition [91]. Two systems, the ATG12-ATG5-ATG16L1 and microtubule-associated proteins 1A/1B light chain 3 (LC3)-phosphatidylethanolamine (PE) complexes, seem to be essential for the growth and closure of the APs. After successive protein interactions, the resulting protein complex joins LC3 at membrane lipid PE. The lapidated complex, bound to the autophagosomal membrane, recruits other adaptor proteins to recognize cargo material, elongate and close the vesicle. The last stage is fusion with lysosomal membrane, followed by lysosomal compartment acidification, decomposition of macromolecules by hydrolases and lipases, and recycling of base constituents (Figure 3). Lysosomes not only represent degradative mediators of AVs but are also signaling scaffolds for AMPK and mTOR autophagy-related activities. For example, the kinase activities of mTOR are regulated by the GTPase RAS homolog enriched in brain (RHEB), which is situated on the lysosome surface, and mTOR itself has been found localized on the lysosomal compartment [92,93]. Similarly, recent investigations demonstrated that AMPK is a resident protein of lysosomes and that the lysosomal protein complex composed of vacuolar-type ATPase (V-ATPase) and Ragulator (required for the mTOR signaling pathway) is essential to phosphorylate and activate AMPK in response to nutrient starvation [94,95]. Autophagy was first described more than 50 years ago. Nevertheless, only in the last two decades have the functions of this catabolic process been elucidated, and currently, autophagy is considered one of the main mechanisms regulating the pathophysiology of many human diseases [96–98]. In particular, defects and alteration in the autophagic process have been associated with tumor growth, tumor suppression, cancer-drug resistance, and metastasis [99,100]. Autophagy may preserve the genomic stability, remove damaged organelles and their defective proteins after cell injury, thus battling and counteracting cancer development. Defective levels of the autophagy gene BECN1 were indeed found in human hepatocellular carcinoma and prostate, breast, and ovarian cancers [101]. Similarly, mutations in different autophagy related genes (ATG5, ATG2B, ATG16L1, and ATG9B) were observed in hepatocellular carcinoma and gastric and colorectal cancers [102]. Autophagy also prevents tumor formation by counteracting the chronic inflammatory condition typical of the tumor environment. For example, oncogenic transformation in lung cell carcinoma was correlated with increased activation of IL-6 and reduced autophagy [103]. Deficiency of ATG16L1 can provoke activation of IL-1β and IL-18 and is associated with an elevated risk of colorectal cancers [104,105].

On the other hand, autophagy is a defense mechanism that sustains the tumor metabolism and promotes tumor development and metastasis. Consistent with this idea, different studies describe increased autophagy activities in different cancer types [106–109] and correlated augmented autophagic marker levels with more aggressive tumor phenotypes [110]. Several reports show that genetic inhibition of ATG genes, such as ATG7, prevents tumor formation and the progression of colorectal, lung, and prostate cancers and glioblastoma [111–113]. Autophagy is also used by cancer cells to evade several cancer treatments, including radiation therapy and chemotherapy, and typical cellular stress conditions of cancer cells (hypoxia, nutrient deprivation, and metabolic stress) induce cytoprotective and pro-survival autophagy.

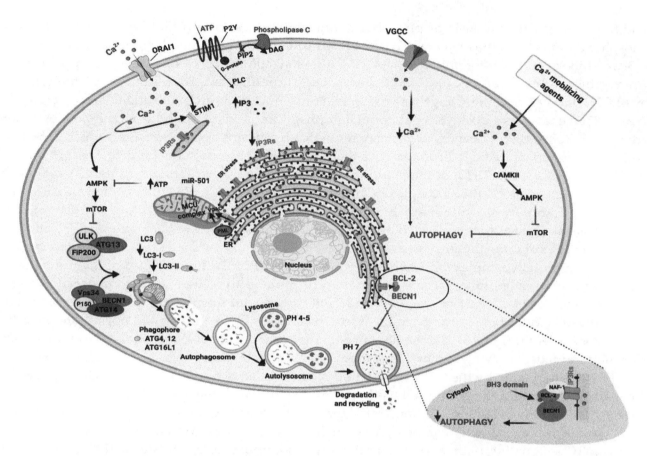

Figure 3. Autophagy, Ca^{2+} and cancer. Autophagy is a key process necessary for the maintenance of the correct cell homeostasis. The unc-51 like autophagy activating kinase 1-2/autophagy-related 13/200-kDa focal adhesion kinase family-interacting protein (ULK/ATG13/FIP200) complex together with other proteins, such as coiled-coil, moesin-like BCL2 interacting protein (BECN1), controls the formation and elongation of autophagosome vesicles. The activity of BECN1 is also regulated by the portion of B-cell lymphoma-2 (BCL-2) pool that is localized in the endoplasmic reticulum (ER). A series of autophagy-related genes (ATG) is essential to the growth and closure of the autophagosome. Additionally, Ca^{2+} signal intervenes to modulate the autophagic machinery. A correct Ca^{2+} transfer between the ER and mitochondria permits optimal mitochondrial Ca^{2+} uptake and consequent ATP production. This signal downregulates the activation of the energy sensor of ATP/AMP ratio, 5' adenosine monophosphate-activated protein kinase (AMPK), which is the most investigated positive regulator of autophagic induction mechanism, AMPK-ULK1-mammalian target of rapamycin (mTOR). When Ca^{2+} transfer from the ER to mitochondria and/or Ca^{2+} is imported into mitochondria is compromised, AMPK is activated, and survival autophagy is induced. For example, tumors characterized by loss of the tumor suppressor promyelocytic leukemia protein (PML) have enhanced autophagy levels and attenuated Ca^{2+} dynamics. In renal carcinoma, miR-501 decreases the activity of mitochondrial Ca^{2+} uniporter (MCU) channel, provoking a reduction in ATP production and recruitment of AMPK-ULK-mTOR pathway. In contrast, it has been observed that reduced Ca^{2+} dynamics may also activate autophagic cell death. Cancers have increased voltage-gated Ca^{2+} channels (VGCC): their inhibition reduces Ca^{2+} entry and activates autophagy to reduce cell proliferation. A decrease in ORAI1 delays cytoplasmic Ca^{2+} clearance and activates autophagy. Diverse Ca^{2+} mobilizing agents increase in intracellular Ca^{2+} levels and activate pro-survival autophagy by activating Ca^{2+}/calmodulin-dependent protein kinases 2 (CAMKII). Created with BioRender.com.

Autophagic dynamics in cancer are modulated by Ca^{2+} signaling. The inhibition of VGCCs induces the activation of autophagy in human adenocarcinoma and endometrial carcinoma. This process is accompanied by both increases in the apoptosis rate and decreased proliferation and migration, suggesting that intracellular Ca^{2+} is necessary for the protection of cancer cells against autophagic death [114]. ORAI1 is another protein involved in the regulation of Ca^{2+}-mediated autophagy in cancer cells. It has been shown that ORAI1 downregulation delays cytoplasmic Ca^{2+} clearance, thereby promoting the activation of several Ca^{2+}-dependent kinases. This activation signal impacts the expression of the cyclin-dependent kinase inhibitor p21, which results in activation of autophagy, cell growth arrest, and increased cell survival [115]. Additionally, diverse Ca^{2+}-mobilizing agents (thapsigargin, ATP, ionomycin, and chemotherapic agents) [116,117] and nutrient withdrawal [118] provoke increases in $[Ca^{2+}]_c$ levels and the simultaneous activation of pro-survival autophagy in cancer cells. Consistently, addition of intracellular Ca^{2+} chelators prevented autophagic activation and induced cell death. Different downstream effectors were proposed to regulate Ca^{2+}-dependent autophagy. For example, it was suggested that increases in $[Ca^{2+}]_c$ determined activation and phosphorylation of protein kinase $C\theta$ stimulating LC3-II conversion and autophagy [119]. Rapid elevation in $[Ca^{2+}]_c$ was also associated to activation of the ERK pathway, which activates mitochondrial depolarization, autophagy, and apoptosis [120]. However, the primary key regulating factor of autophagy following intracellular Ca^{2+} elevation is likely CAMK2. Indeed, an increase in $[Ca^{2+}]_c$ promotes activation of CAMK2, which, in turn, activates autophagy through the regulation of the AMPK/mTOR pathway [116]. Importantly, it has also been demonstrated that $[Ca^{2+}]_c$ dynamics are regulated at ER levels by BCL-2 protein, which lower $[Ca^{2+}]_{ER}$ levels, thus reducing the Ca^{2+} leak from ER [58].

In contrast, inhibition of Ca^{2+} mobilization from the ER can also increase autophagic flux. In this scenario, IP3Rs are the primary regulator. Lithium and L-690330 stimulate autophagy by reducing the levels of IP3 and inositol and consequent IP3Rs activities [121]. Consistent with this process, the downregulation of IP3R3 levels reduces Ca^{2+} dynamics and promotes autophagy [122]. Interestingly, all these dynamics have been attributed to Ca^{2+}-dependent mitochondrial activation. Correct IP3R-mediated Ca^{2+} transfer between ER and mitochondria supports the tricarboxylic acid cycle and consequent ATP production. This signal determines the inhibition of the energy sensor of the ATP/AMP ratio, AMPK, which is also the most investigated positive regulator of autophagic induction mechanism: AMPK-ULK1-mTOR [122]. Recently, it was demonstrated that the acquisition of tumor-promoting behaviors in tumors lacking the tumor suppressor protein PML relies on the IP3R3-dependent regulation of autophagy. Indeed, PML loss decreases the transfer of Ca^{2+} from the ER to mitochondria with a subsequent decrease in ATP production, which determines the activation of the AMPK pathway, thereby promoting pro-survival autophagy [100]. Similar effects were also found in renal cell carcinoma, where upregulated expression of miR-501 increases autophagy by activating AMPK. In this study, the authors demonstrated that the miR-501 decreases the activity of the MCU channel, provoking a reduction in mitochondrial activities and resulting in reduced ATP production and activation of the AMPK/ULK1 pathway [123] (Figure 3).

5. New Strategies for Ca^{2+} Signaling in Cancer Therapy: Ca^{2+} Channels and Pumps as Targets

Ca^{2+} is an important second messenger that regulates different cellular processes linked to cancer such as cell proliferation, apoptosis, and autophagy. Dysregulation of Ca^{2+} signaling may contribute to cancer development and expansion; therefore, targeting Ca^{2+} signaling pathways may be a good option for cancer treatment. Alterations in Ca^{2+} homeostasis may occur by impaired Ca^{2+} channel/pump expression, mutation or protein mislocalization that lead to the remodeling of various signaling pathways contributing to carcinogenesis [124]. The involvement and pharmacological targeting of Ca^{2+}-related proteins associated with cancer are discussed below.

5.1. TRP Channels

TRPCs are classified into different subfamilies (TRPA, TRPC, TRPM, TRPML, TRPP, TRPN, and TRPV) and are implicated in different diseases, including cancer [125]. Alterations of the TRPCs (canonical) subgroup, in particular TRPC1, TRPC3, and TRPC6, are associated with a variety of cancer types, including breast, pancreatic, glioblastoma, lung, hepatic, myeloma, and thyroid cancers [126]. The dysfunction of the TRPM (melastatin) subfamily is also involved in different cancers.

In particular, TRPM1 was found to be decreased in melanoma; TRPM2 is overexpressed in prostate, breast, and pancreatic cancer; and TRPM4 and TRPM5 are upregulated in prostate and lung cancers, respectively. TRPM7 is increased in breast and pancreatic cancers, while TRPM8 is markedly increased in prostate cancers and in pancreatic carcinoma [127]. Furthermore, the dysregulation of TRPV (vanilloid) channels is mostly associated with prostate cancers but is involved also in other tumors. TRPV1 and TRPV2 expression affects bladder and prostate cancer. TRPV4 is downregulated in prostate, skin, and breast cancers, while TRPV6 is boosted in various tumors, including prostate and breast cancers [125,128]. Most of these TRP channels can be targeted for cancer therapy. In this regard, the treatment of different colon cancer cell lines with 20-GPPD, a metabolite of ginseng, induces apoptosis by intracellular Ca^{2+} elevation through the activation of TRPCs. Moreover, the inhibition of TRPCs by the SKF96365 compound caused cell cycle arrest in glioblastoma cells [125,129]. The inhibition of TRPM7 channels by carvacrol treatment reduced the viability, migration, and invasion of U87 glioma cells by inactivating the RAS/MEK/MAPK and PI3K/AKT signaling pathways [130]. In addition, treatment with D-3263, an activator of TRPM8, induces the apoptosis in different cancer cell lines, decreases mice prostate hyperplasia and has been used in phase I clinical trial (https://clinicaltrials.gov/ct2/show/NCT00839631) for the treatment of various solid tumors [125]. This clinical trial not only evaluated the safety and pharmacokinetic profile of D-3263 hydrochloride in a group of patients with advanced solid tumors refractory to conventional therapy but also assessed its antitumor activity. Despite preliminary results showing disease stabilization in persons with prostate cancer [131], no recent clinical results have been reported. The activation of TRPV1 by capsaicin and TRPV2 by cannabidiol generates a continuous influx of intracellular Ca^{2+}, inducing the apoptosis in prostate and bladder cancer cells, respectively [129,132]. Finally, the pharmacological targeting of TRPV6 by the peptide SOR-C13 led to the inhibition of cell growth in cellular and animal models for ovarian and prostate cancers [128]. Interestingly, SOR-C13 has been successful in a phase I study. SOR-C13 was found to be safe, well tolerated, and displayed anticancer activity in 12 of the 22 evaluable patients affected by advanced solid epithelial tumors [133].

5.2. VGCCs and Purinergic P2 Receptors

VGGCs are classified in high voltage activated (HVA) and low voltage activated (LVA) channels according to their pharmacological and electrophysiological profiles. They regulate various Ca^{2+}-dependent cellular processes, including cell proliferation, survival, and differentiation [134]. LVA are also known as T-type Ca^{2+} channels and are frequently altered in different cancer types. The upregulation of T-type Ca^{2+} channels has been mainly observed in prostate, breast, and ovarian cancers; however, it has also been found in melanoma, retinoblastoma, glioma, glioblastoma, hepatocellular, colon, and esophageal cancers cells [134]. The pharmacological inhibition of T-type Ca^{2+} channels by using the channel blocker mibefradil reduced esophageal and colon cancer cell proliferation by upregulating p53. Moreover, this inhibitor induced the apoptosis in glioblastoma cells and ovarian cancer cells. Furthermore, the administration of mibefradil or NNC-55-096, another T-type Ca^{2+} channel blocker, decreased tumor growth in xenograft models of glioblastoma and ovarian cancers [134].

Purinergic P2 receptors are classified in two subfamilies named P2X and P2Y. P2X receptors, activated by ATP, are ligand-gated nonselective cation channels formed by homotrimeric or heterotrimeric complexes of seven different subunits (P2X1–7). The P2Y receptor category, comprising eight members (P2Y1, P2Y2, P2Y4, P2Y6, and P2Y11–14), may be activated by ATP, ADP, UTP, UDP, and UDP-glucose. In particular, P2Y2, P2Y4, and P2Y6 receptors are coupled to Gq proteins and their stimulation leads to Ca^{2+} mobilization by the activation of IP3Rs and SOCE channels [10]. In cancer cells, P2X7R dysfunction impairs the ability of this receptor to open the macropore in response to high extracellular ATP concentration present in the tumor microenvironment, preventing prolonged plasma membrane depolarization and cell death. Moreover, many tumor types including prostate, lung, kidney, colorectal, gastric, and breast cancers, express mutated forms of P2X7R, which are associated with tumor development, survival, and metastasis [135]. Furthermore, P2X3R and P2X5R were found to be overexpressed in hepatocellular carcinoma and squamous cell carcinoma, respectively [10]. Among P2Y family members, P2Y2 is upregulated in breast, hepatoma, pancreatic adenocarcinoma, and colon cancers while P2Y4 is overexpressed in colon cancers [125]. Many anti-P2X7R molecules have been developed in order to treat different diseases, including cancer. Among them, BIL010t and BIL06v, which have been tested in basal cell carcinoma (BCC) and other solid tumors, seem to be the most promising therapeutics [135]. Consistently, a phase I clinical trial demonstrates that BIL010t is safe, tolerable, and reduces primary lesions of BCC [136], meanwhile the phase I study for BIL06v in advanced tumors is ongoing (registration number: ACTRN12618000838213). The pharmacologic inhibition of P2RY2 by using its selective antagonist AR-C118925XX reduces tumor cell growth in xenograft models of pancreatic ductal adenocarcinoma [137].

5.3. SOCE Machinery Proteins (ORAI and STIM)

ORAI channels and STIM Ca^{2+}-sensors are molecular SOCE components. Currently, in mammalian cells, three isoforms of ORAI (ORAI1, ORAI2, and ORAI3) and two isoforms of STIM (STIM1 and STIM2) have been identified. Remodeling of Ca^{2+} signals due to SOCE dysregulation may cause various diseases, including cancer. In fact, activating ORAI1 mutations were found in different types of cancer, including colorectal, stomach, and uterine cancers [138]. Moreover, increased expression of ORAI1 and STIM1 is involved in glioblastoma, pancreatic adenocarcinoma, and breast, prostate, liver, and kidney cancers [125]. Altered expression of STIM2 was found in melanoma and colorectal cancers, while high levels of ORAI2 were observed in acute myeloid leukemia cell lines [138]. Nevertheless, the channel most involved in carcinogenesis is ORAI3, which is expressed in mammalian cells only. Increased levels of ORAI3 form an SOC channel that drives tumorigenesis in estrogen receptor-positive breast cancer as well as in lung adenocarcinoma. Moreover, the interaction between ORAI3 and ORAI1 leads to the generation of arachidonic/leukotriene-regulated heteromeric Ca^{2+} channels expressed in prostate and colorectal cancers but not in healthy tissue [139,140].

The first described inhibitor for ORAI1 channel was SKF-96365, which is able to reduce the growth and migration of breast cancer cells [139]. SOCE channels are also inhibited by trivalent ions such as La^{3+} and Gd^{3+}; however, these channel blockers, as well as SKF-96365, are not SOCE-specific inhibitors; therefore, treatment with these compounds may cause side effects. In this regard, DPB-162AE and DPB-163AE, derivatives of 2-APB, have been developed and are potent SOCE inhibitors capable of inhibiting SOC channels without affecting IP3Rs activity. In addition, RO2959, which inhibits the ORAI1-mediated current, may represent an important therapeutic tool since it is able to selectively increase the ORAI1 channel [141]. Another promising molecule for cancer treatment is ML-9, which inhibits SOCE, blocking STIM1 plasma membrane translocation. This compound administered alone or in combination with other drugs induces prostate cancer cell death [125]. As these compounds have been tested only in cellular models, future studies will be needed to corroborate their effectiveness in cancer therapy. Some drugs with anticancer properties including rapamycin and its analogs, are able to inhibit STIM1- and ORAI1-dependent Ca^{2+} influx. These mTOR inhibitors are being tested in different clinical trials as anticancer therapy [141]. ORAI1-dependent Ca^{2+} influx was also found to be crucial for activating the cell death induced by the anti-CD20 monoclonal antibody GA101/obinutuzumab in non-Hodgkin lymphoma and primary B-cell chronic lymphocytic leukemia cells. Moreover, in addition to ORAI1-dependent Ca^{2+} influx, in this study, it was demonstrated that GA101 determines intracellular Ca^{2+} elevation by provoking Ca^{2+} release from lysosomes [142].

5.4. IP3Rs and Ca^{2+}-ATPases

IP3Rs consist of three isoforms (IP3R1, IP3R2, and IP3R3) that are activated by the generation of intracellular IP3. The most isoform of IP3 receptor isoforms involved in carcinogenesis is IP3R3. The dysfunction of this receptor was found in clear cell renal cell carcinoma cells and in colorectal and ovarian cancer cell lines, where this receptor exerted proliferative and antiapoptotic effects [143]. Furthermore, the upregulation of IP3R2 seems to be associated with the growth of chronic lymphocytic leukemia cells [144]. Only a few IP3Rs inhibitors have been tested in cancer models. However, the caffeine treatment of "in vitro" and "in vivo" models of glioblastoma inhibited cell migration, and it increased the survival of a mouse xenograft model of glioblastoma by inhibiting the IP3R3 receptor channel [145].

Altered expression or mutations of SERCA isoforms (SERCA2 and SERCA3) was observed in several cancer types, including colon, gastric, lung, and prostate carcinoma [125,146,147]. The upregulation of SPCA1 and the translocation of SPCA2 to the plasma membrane were found in breast cancers, where these pump types seem to promote Ca^{2+}-dependent cell proliferation [148]. In addition, PMCA isoforms are dysregulated in various cancer types. In particular, PMCA2 is overexpressed in different breast cancer cell lines, while PMCA1 is upregulated in colon cancer cells but downregulated in oral squamous cell carcinoma cell lines. Moreover, the expression of PMCA4 was found to be reduced in colon cancer cells and in breast and colon cancers tissues [149,150]. In contrast, this pump was found to be overexpressed in different pancreatic ductal adenocarcinoma tumors, where it correlates with poor patient survival [151,152]. However, the different expression levels of PMCA ATPases observed in various cancers suggests that these pumps can function in different ways depending on the tumor type. Nevertheless, the contribution of these molecules to cancer development and progression remains unclear and needs further investigation.

In the last years, many compounds able to inhibit SERCA for cancer treatment have been produced. Mipsagargin (G-202), a thapsigargin derivate, has been tested on different solid tumors, including prostate cancers, glioblastoma, kidney, and hepatocellular carcinoma, in phase I and II clinical trials. The results obtained in phase I demonstrate that mipsagargin displays a favorable pharmacokinetic profile and acceptable tolerability. Furthermore, significant disease stabilization was observed, suggesting possible antitumor activity [153]. The results obtained in phase II supported the hypothesis of antitumor activity and demonstrated that mipsagargin induces prolonged disease stabilization in patients affected by hepatocellular carcinoma and may represent an effective therapeutic treatment for advanced tumors [154]. Treatment with curcumin, another SERCA inhibitor, promotes apoptosis of cells derived from various tumors, such as breast, lung, ovarian, and colon cancers [155]. In addition, PMCA inhibitors were developed for anticancer therapy. In fact, treatment with the selective PMCA inhibitor [Pt(O,O0-acac)(γ-acac)(DMS)] induced the apoptosis of MCF7 breast cancer cells by elevating cytosolic Ca^{2+} levels [156]. Furthermore, resveratrol and its derivatives reduce cell viability through the increase of intracellular Ca^{2+} levels by inhibiting PMCA in prostate cancer cells. Unfortunately, the function of the latter compounds is exerted by the activation of IP3Rs; therefore, they cannot be considered PMCA-specific inhibitors [125,157].

5.5. MCU and VDAC

Alterations in the MCU complex expression/function were found in different cancer types. High expression levels of MCU were detected in colorectal, ovarian, pancreatic, stomach, and prostate cancers, while genetic mutations were observed mainly in prostate, breast, and uterine cancers. Genetic modifications linked to cancer development were also detected in the other components of the MCU complex [158]. Based on these observations, the targeting of MCU channel for cancer therapy may be an intriguing option, especially for cancer patients who overexpress the MCU protein. However, compounds able to inhibit this channel, including ruthenium red and its derivative ruthenium 360, are nonspecific and lead to different side effects. Recently, a new membrane-permeant MCU complex inhibitor named DS16570511 was identified, but its anticancer properties need further investigation [158].

Another important mitochondrial protein is VDAC; this pore is a nonselective Ca^{2+}-permeable pore located on the outer membrane of mitochondrion, where it regulates the flux of ions and metabolites from cytosol to mitochondria and vice versa. Three isoforms for this channel, VDAC1, VDAC2, and VDAC3, have been identified in mammalian cells [159]. Based on the assumption that VDAC pores regulate mitochondrial Ca^{2+} fluxes, it is speculated that VDACs may be involved in the control of cell proliferation and apoptosis; therefore, these channels may affect the fate of cancer cells. VDAC1 is upregulated in a variety of human cancer cell lines, while VDAC2 is overexpressed in melanoma, mesothelioma, and thyroid cancer cells [159]. The targeting of VDAC isoforms may be an important option for cancer treatment. In fact, the administration of R-Tf-D-LP4, a VDAC-based peptide, in xenograft mouse models of glioblastoma, lung, and breast cancer inhibited tumor growth, causing massive cancer cell death [160].

6. Plan of Action in Cancer Therapy: Intersection between Cancer Therapies with Ca^{2+} Signaling

Proliferation, invasiveness, cell death, neovascularization, gene transcription, protein production, and phosphorylation/dephosphorylation events are some of the numerous targets of anticancer compounds. Given that Ca^{2+} signaling is extensively involved in these molecular processes, it is not surprising that an anticancer agent may indirectly modulate Ca^{2+} dynamics in cancer cells. Various studies demonstrated that chemotherapeutic agents modulate intracellular Ca^{2+} levels. For example, 5-fluorouracil (5FU) is an approved anticancer treatment for several cancer types. It has been observed that 5FU mediates in hepatocarcinoma cell death by diminishing Ca^{2+} influx. Indeed, 5FU administration decreased ORAI1 levels and induced autophagic cell death by inhibiting PI3K/AKT/mTOR pathway [161]. In contrast, in colon carcinoma cells, 5FU mediated its cytotoxic effects by increasing intracellular Ca^{2+} amounts to a level necessary to activate calmodulin, which, in turn, phosphorylated p53 to trigger apoptosis [162]. Similar effects were also found with the clinical chemotherapeutic agent cisplatin, which initiated ER stress, the unfolded protein response and Ca^{2+}-mediated apoptosis [163]. Additionally, dexamethasone and other glucocorticoid hormones used for the treatment of lymphoid malignancies increased intracellular Ca^{2+} transport. However, in this case, Ca^{2+} dynamics were associated with chemoresistance. Indeed, both inhibition of TRPCs and Ca^{2+} chelation increased the sensitivity of human leukemia cells to dexamethasone [164,165]. Studies on the chemotherapeutic drugs doxorubicin and simvastin (belonging to the anthracyclines family) reported a direct effect on Ca^{2+} signaling. Accordingly, these drugs induced the persistent release of Ca^{2+} from intracellular stores, provoking mitochondrial Ca^{2+} accumulation and apoptosis. Interestingly, doxorubicin also promoted the binding of p53 to SERCA in the ER. In this state, p53 increased Ca^{2+} transmission between the ER and mitochondria to induce apoptosis [166]. Taxane paclitaxel is widely used in clinical practice for ovarian, breast, neck and head cancers. Paclitaxel induces cytosolic Ca^{2+} oscillations that affect neuronal Ca^{2+} sensor 1 proteins, leading to Ca^{2+} release from the ER via an IP3R3-dependent pathway [167]. Photodynamic therapy (PTD) refers to the use of photosensitizing agents to kill cancerous cells by generating oxidative stress capable of causing damage to cell membranes, proteins, and/or DNA. PTD may promote its anticancer effects by increasing intracellular Ca^{2+} concentration and activating the apoptotic pathway in a p53-dependent pathway [168]. Finally, there is mounting preclinical evidence showing that modulating the autophagic response may improve the efficacy of conventional anticancer drugs for late-stage tumors. Intriguingly, a growing body of evidence highlights that a series of autophagic inhibitors modulate Ca^{2+} signaling in tumors, particularly 4-aminoquinoline antimalarial compounds chloroquine (CQ) and hydroxychloroquine (HCQ). Currently, more than 30 clinical studies are evaluating the antitumor efficacy of CQ and HCQ. Publications reporting the clinical trial results are encouraging. Indeed, most of these investigations describe positive and/or partial effects of CQ and HCQ in reducing tumor growth alone or in combination with the conventional therapies used for several cancer types [169]. It has been demonstrated that CQ decreased the intracellular Ca^{2+} accumulation by inhibiting the IP3Rs-dependent ER Ca^{2+} release and the Ca^{2+} influx mediated by TRPCs, ORAI, and STIM channels [170]. This effect was found in primary B lymphocytes, suggesting that CQ cooperate with Ca^{2+} signaling to modulate the immunological response [171]. This hypothesis was confirmed in a recent study showing that CQ drives the switch of tumor-associated macrophagy (TAM) from the M2 phenotype to the tumor-killing M1 phenotype. In this scenario, CQ increased the intracellular Ca^{2+} levels that were necessary to activate p38, NF-kB, and TFEB to reprogram the TAM phenotype [171]. CQ exerts its anticancer effects by modulating Ca^{2+} homeostasis also in solid tumors associated with PML absence or downregulation. Loss of PML conferred resistance to chemotherapies due to a reduction in ER-mitochondria Ca^{2+}

transmission that activates autophagy and establishes a metabolic advantage for the cancer cells. As a consequence of blocking autophagy with the specific inhibitors (CQ, 3-methyladenine or siRNA BECN1), the apoptotic process was rescued in vitro and in vivo [100]. Similar results were obtained in glioblastoma cells, where CQ promoted impairment in protein folding, ER stress, subsequent Ca^{2+} release, and activation of apoptosis. Interestingly, specific MCU inhibitors or MCU silencing abrogated CQ-dependent effects, thus confirming the importance of mitochondrial Ca^{2+} overload for cell death induction [172]. A reduction in tumor growth and the activation of cell death was observed upon the inhibition of essential autophagy-related genes. Consistent with this finding, knocking down ATG5 led to recovered Ca^{2+} mobilization in glioma cells that had previously been rendered sensitive to anticancer therapy [173]; in addition, the genetic ablation of ATG7 in renal cell carcinoma counteracts the excessive autophagic level caused by a reduction in mitochondrial Ca^{2+} uptake and ATP production and diminishing cancer cell proliferation and migration [123]. Another emerging approach to counteract tumor-promoting conditions is cancer immunotherapy, which improves the cancer-killing efficiency of tumor-infiltrating T lymphocytes (TILs). A characteristic of TILs is the expression of the cell surface receptor programmed death-1 (PD-1) [174]. PD-1 binding to its ligands, PD-L1 or PD-L2, inhibits the activation of T cells. Antibodies blocking the PD-1/PD-L1 signaling pathway reactivate the T-cell-mediated immune response and are employed for the treatment of patients with cancer. Unfortunately, some patients initially respond to immunotherapy but then suffer rapid disease progression. These antibodies, such as pembrolizumab, also modulate intracellular Ca^{2+} signaling, which improves the chemotaxis of T cells by increasing intracellular Ca^{2+} influx [175]. Furthermore, it has been shown that Ca^{2+} signaling inhibits PDL1 and PDL2 expression [176] and that Ca^{2+} flux is abolished when T cells express high levels of PD-1 [177]. Therefore, cancer immunotherapy may be further improved by coupling commonly used antibodies blocking PD-1/PD-L1 signaling and regulators of Ca^{2+} transmission. However, a greater understanding of the steps involving Ca^{2+} signaling during cancer immunotherapy is required.

7. Discussion

Cellular Ca^{2+} is a ubiquitous signal that contributes to the control of diverse cellular functions. Uncontrolled remodeling of Ca^{2+} flux contributes to severe pathophysiology processes and often intersects key aspects of cancer progression, such as tumor proliferation, malignant transformation, escape from cell death, and resistance to anticancer agents. Accumulating preclinical and clinical evidence supports the relationship between Ca^{2+} and cancer, indicating that Ca^{2+} signaling is a reliable target for novel anticancer treatments. As summarized in this review, defects in Ca^{2+} channels/transporters/pumps are typical features of cancerous cells and confer low sensitivity to cell death inducers, thus sustaining the tumor growth and metastasis. Hence, pharmacological modulation of these proteins may be a reliable approach to restore the effectiveness of current cancer treatment regimes (Table 1). However, before thinking about an effective therapeutic intervention based on pharmacological modulation of Ca^{2+}-regulators, it is important to consider other critical aspects. Targeting these processes is difficult, and most importantly, the tumor environment presents substantial cellular heterogeneity in which only a subset of cancer cells should be targeted. Recent investigations have made progress in overcoming these problems. For example, by encapsulating pumps and channel agonists in lipid nanocapsules, it is possible to efficiently modulate the activities of these pumps and channels and the related cellular processes [178]. In addition, by coupling these agents to peptides to create a prodrug that is activated only by a cancer-specific protease, the cytotoxic effects of Ca^{2+} modulators can be solely directed to the cancer cell population [179,180]. Further studies are needed to verify the effective toxicity and pharmacokinetic of these modulators prior to performing clinical testing. Autophagy has also attracted attention in the cancer context. It supplies nutrients to the tumor, suppresses the immune response, and helps cancer cells evade cell death and conventional chemotherapy. Despite the interconnections between autophagy and Ca^{2+} in cancer, this area of study is still in its infancy, with a number of studies starting to explore and highlight the importance of these interconnections. Removing the remaining gap in our knowledge on the intersections between Ca^{2+} and cancer will help researchers better understand the multiple molecular mechanisms that affect tumor development, maintenance, and metastasis and help clinicians design and develop new-generation drugs with the final aim of breaking all the defense barriers of cancer.

Table 1. Summary of the main compounds targeting Ca^{2+} channels/transporters/pumps.

Channel/Transporter/Pump	Compound	Cancer
TRPCs	20-GPPD	Colorectal
	SKF96365	Glioblastoma
	Carvacrol	Glioma
	D-3263	Prostate, colon, breast, lung, pancreas, leiomyosarcoma, and Kaposi's sarcoma
	Capsaicin	Prostate
	Cannabidiol	Bladder
	SOR-C13	Ovarian and prostate
	Dexamethasone	Leukemia
VGGCs	Mibefradil	Esophageal, colon, glioblastoma, and ovarian
	NNC-55-096	Glioblastoma and ovarian
Purinergic P2 receptors	BIL010t	Basal cell carcinoma
	BIL06v	Advanced or metastatic solid tumors
	AR-C118925XX	Pancreatic ductal adenocarcinoma
ORAI and STIM	SKF96365	Breast
	DPB-162AE/-163AE	Colon and glioma
	ML-9	Prostate
	GA101/obinutuzumab	Non-Hodgkin lymphoma and leukemia
	5-Fluorouracil	Hepatocarcinoma
SERCA	Mipsargargin	Prostate cancers, glioblastoma, kidney, and hepatocellular carcinoma
	Curcumin	Breast, lung, ovarian, and colon
PMCA	Pt(O,O0-acac)(γ-acac)(DMS)	Breast
	Resveratrol	Prostate
IP3R3	Paclitaxel	Ovarian, breast, neck and head
VDAC	R-Tf-D-LP4	Glioblastoma, lung, and breast

Author Contributions: S.P., C.G. and P.P. conceived the article; S.P., A.D., G.A., M.P. and E.B. wrote the first version of the manuscript with constructive input from C.G. and P.P.; E.B. prepared display items (with https://biorender.com) under the supervision of C.G. and P.P. Figures are original and have not been published before. S.P., P.P. and C.G. reviewed and edited the manuscript before submission. All authors have read and agreed to the published version of the manuscript.

Abbreviations

AMPK	5′ adenosine monophosphate-activated protein kinase
APs	autophagosomes

ATG	autophagy-related
BAP1	BRCA1-associated protein 1
BCL-2	B-cell lymphoma 2
BECN1	coiled-coil, moesin-like BCL2 interacting protein
Ca^{2+}	calcium
Cyt-C	cytochrome C
CAM	calmodulin
CAMK	Ca^{2+}/Calmodulin Dependent Protein Kinases
CaN	calcineurin
CDK	cyclin-dependent protein kinases
DISC	death-inducing signaling complex
ER	endoplasmic reticulum
HVA	high voltage activated
IP3Rs	inositol 1,4,5-triphosphate receptors
LC3	microtubule-associated proteins 1A/1B light chain 3
LVA	low voltage activated
MAMs	mitochondria associated membranes
MAPK	mitogen-activated protein kinase
MCU	mitochondrial calcium uniporter
MPT	mitochondrial permeability transition
mTOR	mammalian target of rapamycin
NCX	Na^+/Ca^{2+} exchanger
NFAT	nuclear factor of activated T-cell
NMDA	N-methyl-d-aspartate receptor
PCD	programmed cell death
PD-1	programmed death-1
PI3K	phosphatidylinositol 3-kinase
PI3P	phosphatidylinositol-3-phosphate
PM	plasma membrane
PMCA	plasma membrane Ca^{2+} transport ATPase
PML	promyelocytic leukemia protein
PTEN	protein phosphatase and tensin homolog
RHEB	RAS homolog enriched in brain
RIPK	receptor-interacting serine/threonine-protein kinase
RyRs	ryanodine receptors
SERCA	sarcoendoplasmic reticulum Ca^{2+}-ATPase
SOCE	store-operated Ca^{2+} entry
SPCA	secretory protein calcium ATPase
TFEB	transcription factor EB
TIL	tumor-infiltrating T lymphocytes
TRPML1	transient receptor potential mucolipin 1
TRPC	transient receptor channels
ULK	Unc-51 like autophagy activating kinase
V-ATPase	vacuolar-type ATPase
VDAC	Voltage-dependent anion channels
VGCC	voltage-gated Ca^{2+} channels

References

1. Raffaello, A.; Mammucari, C.; Gherardi, G.; Rizzuto, R. Calcium at the Center of Cell Signaling: Interplay between Endoplasmic Reticulum, Mitochondria, and Lysosomes. *Trends Biochem. Sci.* **2016**, *41*, 1035–1049. [CrossRef] [PubMed]

2. Christensen, K.A.; Myers, J.T.; Swanson, J.A. pH-dependent regulation of lysosomal calcium in macrophages. *J. Cell Sci.* **2002**, *115*, 599–607. [PubMed]

3. Marchi, S.; Giorgi, C.; Galluzzi, L.; Pinton, P. Ca^{2+} Fluxes and Cancer. *Mol. Cell* **2020**, *78*, 1055–1069. [CrossRef] [PubMed]

4. Bootman, M.D.; Bultynck, G. Fundamentals of Cellular Calcium Signaling: A Primer. *Cold Spring Harb. Perspect. Biol.* **2020**, *12*, a038802. [CrossRef] [PubMed]

5. Missiroli, S.; Perrone, M.; Genovese, I.; Pinton, P.; Giorgi, C. Cancer metabolism and mitochondria: Finding novel mechanisms to fight tumours. *EBioMedicine* **2020**, *59*, 102943. [CrossRef] [PubMed]

6. Venkatachalam, K.; Montell, C. TRP channels. *Annu. Rev. Biochem.* **2007**, *76*, 387–417. [CrossRef] [PubMed]

7. Hogan, P.G.; Rao, A. Store-operated calcium entry: Mechanisms and modulation. *Biochem. Biophys. Res. Commun.* **2015**, *460*, 40–49. [CrossRef]

8. Catterall, W.A. Voltage-gated calcium channels. *Cold Spring Harb. Perspect. Biol.* **2011**, *3*, a003947. [CrossRef]

9. Kavalali, E.T. Neuronal Ca^{2+} signalling at rest and during spontaneous neurotransmission. *J. Physiol.* **2020**, *598*, 1649–1654. [CrossRef] [PubMed]

10. Di Virgilio, F.; Adinolfi, E. Extracellular purines, purinergic receptors and tumor growth. *Oncogene* **2017**, *36*, 293–303. [CrossRef]

11. Berridge, M.J. The Inositol Trisphosphate/Calcium Signaling Pathway in Health and Disease. *Physiol. Rev.* **2016**, *96*, 1261–1296. [CrossRef]

12. Vallese, F.; Barazzuol, L.; Maso, L.; Brini, M.; Cali, T. ER-Mitochondria Calcium Transfer, Organelle Contacts and Neurodegenerative Diseases. *Adv. Exp. Med. Biol.* **2020**, *1131*, 719–746. [CrossRef] [PubMed]

13. Gaspers, L.D.; Bartlett, P.J.; Politi, A.; Burnett, P.; Metzger, W.; Johnston, J.; Joseph, S.K.; Hofer, T.; Thomas, A.P. Hormone-induced calcium oscillations depend on cross-coupling with inositol 1,4,5-trisphosphate oscillations. *Cell Rep.* **2014**, *9*, 1209–1218. [CrossRef]

14. Zalk, R.; Clarke, O.B.; des Georges, A.; Grassucci, R.A.; Reiken, S.; Mancia, F.; Hendrickson, W.A.; Frank, J.; Marks, A.R. Structure of a mammalian ryanodine receptor. *Nature* **2015**, *517*, 44–49. [CrossRef] [PubMed]

15. Vandecaetsbeek, I.; Vangheluwe, P.; Raeymaekers, L.; Wuytack, F.; Vanoevelen, J. The Ca^{2+} pumps of the endoplasmic reticulum and Golgi apparatus. *Cold Spring Harb. Perspect. Biol.* **2011**, *3*, a004184. [CrossRef]

16. Bruce, J.I.E. Metabolic regulation of the PMCA: Role in cell death and survival. *Cell Calcium* **2018**, *69*, 28–36. [CrossRef]

17. Cali, T.; Brini, M.; Carafoli, E. The PMCA pumps in genetically determined neuronal pathologies. *Neurosci. Lett.* **2018**, *663*, 2–11. [CrossRef]

18. Verkhratsky, A.; Trebak, M.; Perocchi, F.; Khananshvili, D.; Sekler, I. Crosslink between calcium and sodium signalling. *Exp. Physiol.* **2018**, *103*, 157–169. [CrossRef] [PubMed]

19. Giorgi, C.; Marchi, S.; Pinton, P. The machineries, regulation and cellular functions of mitochondrial calcium. *Nat. Rev. Mol. Cell Biol.* **2018**, *19*, 713–730. [CrossRef]

20. Kamer, K.J.; Mootha, V.K. The molecular era of the mitochondrial calcium uniporter. *Nat. Rev. Mol. Cell Biol.* **2015**, *16*, 545–553. [CrossRef]

21. Filadi, R.; Greotti, E.; Pizzo, P. Highlighting the endoplasmic reticulum-mitochondria connection: Focus on Mitofusin 2. *Pharmacol. Res.* **2018**, *128*, 42–51. [CrossRef]

22. Garrity, A.G.; Wang, W.; Collier, C.M.; Levey, S.A.; Gao, Q.; Xu, H. The endoplasmic reticulum, not the pH gradient, drives calcium refilling of lysosomes. *eLife* **2016**, *5*, e15887. [CrossRef]

23. Lloyd-Evans, E.; Waller-Evans, H. Lysosomal Ca^{2+} Homeostasis and Signaling in Health and Disease. *Cold Spring Harb. Perspect. Biol.* **2020**, *12*, a035311. [CrossRef]

24. Giorgi, C.; Danese, A.; Missiroli, S.; Patergnani, S.; Pinton, P. Calcium Dynamics as a Machine for Decoding Signals. *Trends Cell Biol.* **2018**, *28*, 258–273. [CrossRef]

25. Russa, A.D.; Maesawa, C.; Satoh, Y. Spontaneous $[Ca^{2+}]_i$ oscillations in G1/S phase-synchronized cells. *J. Electron Microsc.* **2009**, *58*, 321–329. [CrossRef]

26. Patel, R.; Holt, M.; Philipova, R.; Moss, S.; Schulman, H.; Hidaka, H.; Whitaker, M. Calcium/calmodulin-dependent phosphorylation and activation of human Cdc25-C at the G2/M phase transition in HeLa cells. *J. Biol. Chem.* **1999**, *274*, 7958–7968. [CrossRef] [PubMed]

27. Heim, A.; Tischer, T.; Mayer, T.U. Calcineurin promotes APC/C activation at meiotic exit by acting on both XErp1 and Cdc20. *EMBO Rep.* **2018**, *19*, e46433. [CrossRef]

28. Rao, A. Signaling to gene expression: Calcium, calcineurin and NFAT. *Nat. Immunol.* **2009**, *10*, 3–5. [CrossRef] [PubMed]

29. Kahl, C.R.; Means, A.R. Calcineurin regulates cyclin D1 accumulation in growth-stimulated fibroblasts. *Mol. Biol. Cell* **2004**, *15*, 1833–1842. [CrossRef]

30. Keith, C.; DiPaola, M.; Maxfield, F.R.; Shelanski, M.L. Microinjection of Ca++-calmodulin causes a localized depolymerization of microtubules. *J. Cell Biol.* **1983**, *97*, 1918–1924. [CrossRef] [PubMed]

31. Dinsmore, J.H.; Sloboda, R.D. Calcium and calmodulin-dependent phosphorylation of a 62 kd protein induces microtubule depolymerization in sea urchin mitotic apparatuses. *Cell* **1988**, *53*, 769–780. [CrossRef]

32. Chircop, M.; Malladi, C.S.; Lian, A.T.; Page, S.L.; Zavortink, M.; Gordon, C.P.; McCluskey, A.; Robinson, P.J. Calcineurin activity is required for the completion of cytokinesis. *Cell. Mol. Life Sci.* **2010**, *67*, 3725–3737. [CrossRef]

33. Yokokura, S.; Yurimoto, S.; Matsuoka, A.; Imataki, O.; Dobashi, H.; Bandoh, S.; Matsunaga, T. Calmodulin antagonists induce cell cycle arrest and apoptosis in vitro and inhibit tumor growth in vivo in human multiple myeloma. *BMC Cancer* **2014**, *14*, 882. [CrossRef]

34. Machaca, K. Ca^{2+} signaling, genes and the cell cycle. *Cell Calcium* **2011**, *49*, 323–330. [CrossRef]

35. Xu, M.; Seas, A.; Kiyani, M.; Ji, K.S.Y.; Bell, H.N. A temporal examination of calcium signaling in cancer-from tumorigenesis, to immune evasion, and metastasis. *Cell Biosci.* **2018**, *8*, 25. [CrossRef] [PubMed]

36. Faouzi, M.; Hague, F.; Potier, M.; Ahidouch, A.; Sevestre, H.; Ouadid-Ahidouch, H. Down-regulation of Orai3 arrests cell-cycle progression and induces apoptosis in breast cancer cells but not in normal breast epithelial cells. *J. Cell. Physiol.* **2011**, *226*, 542–551. [CrossRef] [PubMed]

37. Faouzi, M.; Kischel, P.; Hague, F.; Ahidouch, A.; Benzerdjeb, N.; Sevestre, H.; Penner, R.; Ouadid-Ahidouch, H. ORAI3 silencing alters cell proliferation and cell cycle progression via c-myc pathway in breast cancer cells. *Biochim. Biophys. Acta* **2013**, *1833*, 752–760. [CrossRef]

38. Ahlin, C.; Lundgren, C.; Embretsen-Varro, E.; Jirstrom, K.; Blomqvist, C.; Fjallskog, M. High expression of cyclin D1 is associated to high proliferation rate and increased risk of mortality in women with ER-positive but not in ER-negative breast cancers. *Breast Cancer Res. Treat.* **2017**, *164*, 667–678. [CrossRef]

39. Arnold, A.; Papanikolaou, A. Cyclin D1 in breast cancer pathogenesis. *J. Clin. Oncol. Off. J. Am. Soc. Clin. Oncol.* **2005**, *23*, 4215–4224. [CrossRef]

40. Reis-Filho, J.S.; Savage, K.; Lambros, M.B.; James, M.; Steele, D.; Jones, R.L.; Dowsett, M. Cyclin D1 protein overexpression and CCND1 amplification in breast carcinomas: An immunohistochemical and chromogenic in situ hybridisation analysis. *Mod. Pathol.* **2006**, *19*, 999–1009. [CrossRef]

41. Phan, N.N.; Wang, C.Y.; Chen, C.F.; Sun, Z.; Lai, M.D.; Lin, Y.C. Voltage-gated calcium channels: Novel targets for cancer therapy. *Oncol. Lett.* **2017**, *14*, 2059–2074. [CrossRef] [PubMed]

42. Das, A.; Pushparaj, C.; Bahi, N.; Sorolla, A.; Herreros, J.; Pamplona, R.; Vilella, R.; Matias-Guiu, X.; Marti, R.M.; Canti, C. Functional expression of voltage-gated calcium channels in human melanoma. *Pigment Cell Melanoma Res.* **2012**, *25*, 200–212. [CrossRef]

43. Lehen'kyi, V.; Flourakis, M.; Skryma, R.; Prevarskaya, N. TRPV6 channel controls prostate cancer cell proliferation via Ca^{2+}/NFAT-dependent pathways. *Oncogene* **2007**, *26*, 7380–7385. [CrossRef]

44. Parmer, T.G.; Ward, M.D.; Hait, W.N. Effects of rottlerin, an inhibitor of calmodulin-dependent protein kinase III, on cellular proliferation, viability, and cell cycle distribution in malignant glioma cells. *Cell Growth Differ. Mol. Biol. J. Am. Assoc. Cancer Res.* **1997**, *8*, 327–334.

45. Parmer, T.G.; Ward, M.D.; Yurkow, E.J.; Vyas, V.H.; Kearney, T.J.; Hait, W.N. Activity and regulation by growth factors of calmodulin-dependent protein kinase III (elongation factor 2-kinase) in human breast cancer. *Br. J. Cancer* **1999**, *79*, 59–64. [CrossRef]

46. Gu, Y.; Zhang, J.; Ma, X.; Kim, B.W.; Wang, H.; Li, J.; Pan, Y.; Xu, Y.; Ding, L.; Yang, L.; et al. Stabilization of the c-Myc Protein by CAMKIIgamma Promotes T Cell Lymphoma. *Cancer Cell* **2017**, *32*, 115–128.e117. [CrossRef]

47. Fouad, Y.A.; Aanei, C. Revisiting the hallmarks of cancer. *Am. J. Cancer Res.* **2017**, *7*, 1016–1036.

48. Malik, A.; Kanneganti, T.D. Inflammasome activation and assembly at a glance. *J. Cell Sci.* **2017**, *130*, 3955–3963. [CrossRef] [PubMed]

49. Missiroli, S.; Patergnani, S.; Caroccia, N.; Pedriali, G.; Perrone, M.; Previati, M.; Wieckowski, M.R.; Giorgi, C. Mitochondria-associated membranes (MAMs) and inflammation. *Cell Death Dis.* **2018**, *9*, 329. [CrossRef]

50. Cai, Z.; Jitkaew, S.; Zhao, J.; Chiang, H.C.; Choksi, S.; Liu, J.; Ward, Y.; Wu, L.G.; Liu, Z.G. Plasma membrane translocation of trimerized MLKL protein is required for TNF-induced necroptosis. *Nat. Cell Biol.* **2014**, *16*, 55–65. [CrossRef]

51. Nagata, S. Apoptosis and Clearance of Apoptotic Cells. *Annu. Rev. Immunol.* **2018**, *36*, 489–517. [CrossRef] [PubMed]

52. Bonora, M.; Patergnani, S.; Ramaccini, D.; Morciano, G.; Pedriali, G.; Kahsay, A.E.; Bouhamida, E.; Giorgi, C.; Wieckowski, M.R.; Pinton, P. Physiopathology of the Permeability Transition Pore: Molecular Mechanisms in Human Pathology. *Biomolecules* **2020**, *10*, 998. [CrossRef]

53. Danese, A.; Marchi, S.; Vitto, V.A.M.; Modesti, L.; Leo, S.; Wieckowski, M.R.; Giorgi, C.; Pinton, P. Cancer-Related Increases and Decreases in Calcium Signaling at the Endoplasmic Reticulum-Mitochondria Interface (MAMs). *Rev. Physiol. Biochem. Pharmacol.* **2020**. [CrossRef]

54. Patergnani, S.; Baldassari, F.; De Marchi, E.; Karkucinska-Wieckowska, A.; Wieckowski, M.R.; Pinton, P. Methods to monitor and compare mitochondrial and glycolytic ATP production. *Methods Enzymol.* **2014**, *542*, 313–332. [CrossRef]

55. Bonora, M.; Morganti, C.; Morciano, G.; Pedriali, G.; Lebiedzinska-Arciszewska, M.; Aquila, G.; Giorgi, C.; Rizzo, P.; Campo, G.; Ferrari, R.; et al. Mitochondrial permeability transition involves dissociation of F_1F_O ATP synthase dimers and C-ring conformation. *EMBO Rep.* **2017**, *18*, 1077–1089. [CrossRef] [PubMed]

56. Kerkhofs, M.; Bittremieux, M.; Morciano, G.; Giorgi, C.; Pinton, P.; Parys, J.B.; Bultynck, G. Emerging molecular mechanisms in chemotherapy: Ca^{2+} signaling at the mitochondria-associated endoplasmic reticulum membranes. *Cell Death Dis.* **2018**, *9*, 334. [CrossRef]

57. Vervliet, T.; Parys, J.B.; Bultynck, G. Bcl-2 proteins and calcium signaling: Complexity beneath the surface. *Oncogene* **2016**, *35*, 5079–5092. [CrossRef]

58. Pinton, P.; Ferrari, D.; Magalhaes, P.; Schulze-Osthoff, K.; Di Virgilio, F.; Pozzan, T.; Rizzuto, R. Reduced loading of intracellular Ca^{2+} stores and downregulation of capacitative Ca^{2+} influx in Bcl-2-overexpressing cells. *J. Cell Biol.* **2000**, *148*, 857–862. [CrossRef]

59. Rong, Y.P.; Aromolaran, A.S.; Bultynck, G.; Zhong, F.; Li, X.; McColl, K.; Matsuyama, S.; Herlitze, S.; Roderick, H.L.; Bootman, M.D.; et al. Targeting Bcl-2-IP3 receptor interaction to reverse Bcl-2's inhibition of apoptotic calcium signals. *Mol. Cell* **2008**, *31*, 255–265. [CrossRef]

60. Pihan, P.; Carreras-Sureda, A.; Hetz, C. BCL-2 family: Integrating stress responses at the ER to control cell demise. *Cell Death Differ.* **2017**, *24*, 1478–1487. [CrossRef]

61. Um, H.D. Bcl-2 family proteins as regulators of cancer cell invasion and metastasis: A review focusing on mitochondrial respiration and reactive oxygen species. *Oncotarget* **2016**, *7*, 5193–5203. [CrossRef]

62. Malumbres, M.; Barbacid, M. RAS oncogenes: The first 30 years. *Nat. Rev. Cancer* **2003**, *3*, 459–465. [CrossRef]

63. Rimessi, A.; Marchi, S.; Patergnani, S.; Pinton, P. H-Ras-driven tumoral maintenance is sustained through caveolin-1-dependent alterations in calcium signaling. *Oncogene* **2014**, *33*, 2329–2340. [CrossRef] [PubMed]

64. Marchi, S.; Marinello, M.; Bononi, A.; Bonora, M.; Giorgi, C.; Rimessi, A.; Pinton, P. Selective modulation of subtype III IP$_3$R by Akt regulates ER Ca^{2+} release and apoptosis. *Cell Death Dis.* **2012**, *3*, e304. [CrossRef]

65. Bononi, A.; Bonora, M.; Marchi, S.; Missiroli, S.; Poletti, F.; Giorgi, C.; Pandolfi, P.P.; Pinton, P. Identification of PTEN at the ER and MAMs and its regulation of Ca^{2+} signaling and apoptosis in a protein phosphatase-dependent manner. *Cell Death Differ.* **2013**, *20*, 1631–1643. [CrossRef]

66. Kuchay, S.; Giorgi, C.; Simoneschi, D.; Pagan, J.; Missiroli, S.; Saraf, A.; Florens, L.; Washburn, M.P.; Collazo-Lorduy, A.; Castillo-Martin, M.; et al. PTEN counteracts FBXL2 to promote IP3R3- and Ca^{2+}-mediated apoptosis limiting tumour growth. *Nature* **2017**, *546*, 554–558. [CrossRef] [PubMed]

67. Pinton, P.; Giorgi, C.; Pandolfi, P.P. The role of PML in the control of apoptotic cell fate: A new key player at ER-mitochondria sites. *Cell Death Differ.* **2011**, *18*, 1450–1456. [CrossRef] [PubMed]

68. Patergnani, S.; Giorgi, C.; Maniero, S.; Missiroli, S.; Maniscalco, P.; Bononi, I.; Martini, F.; Cavallesco, G.; Tognon, M.; Pinton, P. The endoplasmic reticulum mitochondrial calcium cross talk is downregulated in malignant pleural mesothelioma cells and plays a critical role in apoptosis inhibition. *Oncotarget* **2015**, *6*, 23427–23444. [CrossRef]

69. Bononi, A.; Giorgi, C.; Patergnani, S.; Larson, D.; Verbruggen, K.; Tanji, M.; Pellegrini, L.; Signorato, V.; Olivetto, F.; Pastorino, S.; et al. BAP1 regulates IP3R3-mediated Ca^{2+} flux to mitochondria suppressing cell transformation. *Nature* **2017**, *546*, 549–553. [CrossRef]

70. Giorgi, C.; Bonora, M.; Pinton, P. Inside the tumor: p53 modulates calcium homeostasis. *Cell Cycle* **2015**, *14*, 933–934. [CrossRef]

71. Jung, J.; Cho, K.J.; Naji, A.K.; Clemons, K.N.; Wong, C.O.; Villanueva, M.; Gregory, S.; Karagas, N.E.; Tan, L.; Liang, H.; et al. HRAS-driven cancer cells are vulnerable to TRPML1 inhibition. *EMBO Rep.* **2019**, *20*, e46685. [CrossRef] [PubMed]

72. Morelli, M.B.; Nabissi, M.; Amantini, C.; Tomassoni, D.; Rossi, F.; Cardinali, C.; Santoni, M.; Arcella, A.; Oliva, M.A.; Santoni, A.; et al. Overexpression of transient receptor potential mucolipin-2 ion channels in gliomas: Role in tumor growth and progression. *Oncotarget* **2016**, *7*, 43654–43668. [CrossRef]

73. Giatromanolaki, A.; Kalamida, D.; Sivridis, E.; Karagounis, I.V.; Gatter, K.C.; Harris, A.L.; Koukourakis, M.I. Increased expression of transcription factor EB (TFEB) is associated with autophagy, migratory phenotype and poor prognosis in non-small cell lung cancer. *Lung Cancer* **2015**, *90*, 98–105. [CrossRef]

74. Kauffman, E.C.; Ricketts, C.J.; Rais-Bahrami, S.; Yang, Y.; Merino, M.J.; Bottaro, D.P.; Srinivasan, R.; Linehan, W.M. Molecular genetics and cellular features of TFE3 and TFEB fusion kidney cancers. *Nat. Rev. Urol.* **2014**, *11*, 465–475. [CrossRef]

75. Marchand, B.; Arsenault, D.; Raymond-Fleury, A.; Boisvert, F.M.; Boucher, M.J. Glycogen synthase kinase-3 (GSK3) inhibition induces prosurvival autophagic signals in human pancreatic cancer cells. *J. Biol. Chem.* **2015**, *290*, 5592–5605. [CrossRef]

76. Liang, J.; Jia, X.; Wang, K.; Zhao, N. High expression of TFEB is associated with aggressive clinical features in colorectal cancer. *Oncotargets Ther.* **2018**, *11*, 8089–8098. [CrossRef]

77. Medina, D.L.; Di Paola, S.; Peluso, I.; Armani, A.; De Stefani, D.; Venditti, R.; Montefusco, S.; Scotto-Rosato, A.; Prezioso, C.; Forrester, A.; et al. Lysosomal calcium signalling regulates autophagy through calcineurin and TFEB. *Nat. Cell Biol.* **2015**, *17*, 288–299. [CrossRef]

78. Sbano, L.; Bonora, M.; Marchi, S.; Baldassari, F.; Medina, D.L.; Ballabio, A.; Giorgi, C.; Pinton, P. TFEB-mediated increase in peripheral lysosomes regulates store-operated calcium entry. *Sci. Rep.* **2017**, *7*, 40797. [CrossRef]

79. Marchi, S.; Lupini, L.; Patergnani, S.; Rimessi, A.; Missiroli, S.; Bonora, M.; Bononi, A.; Corra, F.; Giorgi, C.; De Marchi, E.; et al. Downregulation of the mitochondrial calcium uniporter by cancer-related miR-25. *Curr. Biol.* **2013**, *23*, 58–63. [CrossRef] [PubMed]

80. Hong, Z.; Chen, K.H.; DasGupta, A.; Potus, F.; Dunham-Snary, K.; Bonnet, S.; Tian, L.; Fu, J.; Breuils-Bonnet, S.; Provencher, S.; et al. MicroRNA-138 and MicroRNA-25 Down-regulate Mitochondrial Calcium Uniporter, Causing the Pulmonary Arterial Hypertension Cancer Phenotype. *Am. J. Respir. Crit. Care Med.* **2017**, *195*, 515–529. [CrossRef]

81. Leidinger, P.; Backes, C.; Dahmke, I.N.; Galata, V.; Huwer, H.; Stehle, I.; Bals, R.; Keller, A.; Meese, E. What makes a blood cell based miRNA expression pattern disease specific?—A miRNome analysis of blood cell subsets in lung cancer patients and healthy controls. *Oncotarget* **2014**, *5*, 9484–9497. [CrossRef]

82. Saito, Y.; Nakaoka, T.; Saito, H. microRNA-34a as a Therapeutic Agent against Human Cancer. *J. Clin. Med.* **2015**, *4*, 1951–1959. [CrossRef] [PubMed]

83. Hong, S.; Lee, J.; Seo, H.H.; Lee, C.Y.; Yoo, K.J.; Kim, S.M.; Lee, S.; Hwang, K.C.; Choi, E. Na$^+$-Ca^{2+} exchanger targeting miR-132 prevents apoptosis of cardiomyocytes under hypoxic condition by suppressing Ca^{2+} overload. *Biochem. Biophys. Res. Commun.* **2015**, *460*, 931–937. [CrossRef]

84. Chaudhuri, A.D.; Choi, D.C.; Kabaria, S.; Tran, A.; Junn, E. MicroRNA-7 Regulates the Function of Mitochondrial Permeability Transition Pore by Targeting VDAC1 Expression. *J. Biol. Chem.* **2016**, *291*, 6483–6493. [CrossRef]

85. Zaglia, T.; Ceriotti, P.; Campo, A.; Borile, G.; Armani, A.; Carullo, P.; Prando, V.; Coppini, R.; Vida, V.; Stolen, T.O.; et al. Content of mitochondrial calcium uniporter (MCU) in cardiomyocytes is regulated by microRNA-1 in physiologic and pathologic hypertrophy. *Proc. Natl. Acad. Sci. USA* **2017**, *114*, E9006–E9015. [CrossRef] [PubMed]

86. Patergnani, S.; Pinton, P. Mitophagy and mitochondrial balance. *Methods Mol. Biol.* **2015**, *1241*, 181–194. [CrossRef]

87. Dikic, I.; Elazar, Z. Mechanism and medical implications of mammalian autophagy. *Nat. Rev. Mol. Cell Biol.* **2018**, *19*, 349–364. [CrossRef]

88. Hosokawa, N.; Hara, T.; Kaizuka, T.; Kishi, C.; Takamura, A.; Miura, Y.; Iemura, S.; Natsume, T.; Takehana, K.; Yamada, N.; et al. Nutrient-dependent mTORC1 association with the ULK1-Atg13-FIP200 complex required for autophagy. *Mol. Biol. Cell* **2009**, *20*, 1981–1991. [CrossRef]

89. Kim, J.; Kundu, M.; Viollet, B.; Guan, K.L. AMPK and mTOR regulate autophagy through direct phosphorylation of Ulk1. *Nat. Cell Biol.* **2011**, *13*, 132–141. [CrossRef]

90. Tooze, S.A.; Yoshimori, T. The origin of the autophagosomal membrane. *Nat. Cell Biol.* **2010**, *12*, 831–835. [CrossRef]

91. Xu, H.D.; Qin, Z.H. Beclin 1, Bcl-2 and Autophagy. *Adv. Exp. Med. Biol.* **2019**, *1206*, 109–126. [CrossRef] [PubMed]

92. Long, X.; Lin, Y.; Ortiz-Vega, S.; Yonezawa, K.; Avruch, J. Rheb binds and regulates the mTOR kinase. *Curr. Biol.* **2005**, *15*, 702–713. [CrossRef]

93. Sancak, Y.; Bar-Peled, L.; Zoncu, R.; Markhard, A.L.; Nada, S.; Sabatini, D.M. Ragulator-Rag complex targets mTORC1 to the lysosomal surface and is necessary for its activation by amino acids. *Cell* **2010**, *141*, 290–303. [CrossRef]

94. Zhang, C.S.; Jiang, B.; Li, M.; Zhu, M.; Peng, Y.; Zhang, Y.L.; Wu, Y.Q.; Li, T.Y.; Liang, Y.; Lu, Z.; et al. The lysosomal v-ATPase-Ragulator complex is a common activator for AMPK and mTORC1, acting as a switch between catabolism and anabolism. *Cell Metab.* **2014**, *20*, 526–540. [CrossRef]

95. Chapel, A.; Kieffer-Jaquinod, S.; Sagne, C.; Verdon, Q.; Ivaldi, C.; Mellal, M.; Thirion, J.; Jadot, M.; Bruley, C.; Garin, J.; et al. An extended proteome map of the lysosomal membrane reveals novel potential transporters. *Mol. Cell. Proteom.* **2013**, *12*, 1572–1588. [CrossRef]

96. Castellazzi, M.; Patergnani, S.; Donadio, M.; Giorgi, C.; Bonora, M.; Fainardi, E.; Casetta, I.; Granieri, E.; Pugliatti, M.; Pinton, P. Correlation between auto/mitophagic processes and magnetic resonance imaging activity in multiple sclerosis patients. *J. Neuroinflamm.* **2019**, *16*, 131. [CrossRef]

97. Patergnani, S.; Castellazzi, M.; Bonora, M.; Marchi, S.; Casetta, I.; Pugliatti, M.; Giorgi, C.; Granieri, E.; Pinton, P. Autophagy and mitophagy elements are increased in body fluids of multiple sclerosis-affected individuals. *J. Neurol. Neurosurg. Psychiatry* **2018**, *89*, 439–441. [CrossRef] [PubMed]

98. Saha, S.; Panigrahi, D.P.; Patil, S.; Bhutia, S.K. Autophagy in health and disease: A comprehensive review. *Biomed. Pharmacother.* **2018**, *104*, 485–495. [CrossRef]

99. Xue, J.; Patergnani, S.; Giorgi, C.; Suarez, J.; Goto, K.; Bononi, A.; Tanji, M.; Novelli, F.; Pastorino, S.; Xu, R.; et al. Asbestos induces mesothelial cell transformation via HMGB1-driven autophagy. *Proc. Natl. Acad. Sci. USA* **2020**, *117*, 25543–25552. [CrossRef]

100. Missiroli, S.; Bonora, M.; Patergnani, S.; Poletti, F.; Perrone, M.; Gafa, R.; Magri, E.; Raimondi, A.; Lanza, G.; Tacchetti, C.; et al. PML at Mitochondria-Associated Membranes Is Critical for the Repression of Autophagy and Cancer Development. *Cell Rep.* **2016**, *16*, 2415–2427. [CrossRef]

101. Vega-Rubin-de-Celis, S. The Role of Beclin 1-Dependent Autophagy in Cancer. *Biology* **2019**, *9*, 4. [CrossRef]

102. Li, X.; He, S.; Ma, B. Autophagy and autophagy-related proteins in cancer. *Mol. Cancer* **2020**, *19*, 12. [CrossRef]

103. Qi, Y.; Zhang, M.; Li, H.; Frank, J.A.; Dai, L.; Liu, H.; Zhang, Z.; Wang, C.; Chen, G. Autophagy inhibition by sustained overproduction of IL6 contributes to arsenic carcinogenesis. *Cancer Res.* **2014**, *74*, 3740–3752. [CrossRef]

104. Hampe, J.; Franke, A.; Rosenstiel, P.; Till, A.; Teuber, M.; Huse, K.; Albrecht, M.; Mayr, G.; De La Vega, F.M.; Briggs, J.; et al. A genome-wide association scan of nonsynonymous SNPs identifies a susceptibility variant for Crohn disease in ATG16L1. *Nat. Genet.* **2007**, *39*, 207–211. [CrossRef]

105. Saitoh, T.; Fujita, N.; Jang, M.H.; Uematsu, S.; Yang, B.G.; Satoh, T.; Omori, H.; Noda, T.; Yamamoto, N.; Komatsu, M.; et al. Loss of the autophagy protein Atg16L1 enhances endotoxin-induced IL-1beta production. *Nature* **2008**, *456*, 264–268. [CrossRef]

106. Altman, B.J.; Jacobs, S.R.; Mason, E.F.; Michalek, R.D.; MacIntyre, A.N.; Coloff, J.L.; Ilkayeva, O.; Jia, W.; He, Y.W.; Rathmell, J.C. Autophagy is essential to suppress cell stress and to allow BCR-Abl-mediated leukemogenesis. *Oncogene* **2011**, *30*, 1855–1867. [CrossRef]

107. Folkerts, H.; Hilgendorf, S.; Wierenga, A.T.J.; Jaques, J.; Mulder, A.B.; Coffer, P.J.; Schuringa, J.J.; Vellenga, E. Inhibition of autophagy as a treatment strategy for p53 wild-type acute myeloid leukemia. *Cell Death Dis.* **2017**, *8*, e2927. [CrossRef]

108. Umemura, A.; He, F.; Taniguchi, K.; Nakagawa, H.; Yamachika, S.; Font-Burgada, J.; Zhong, Z.; Subramaniam, S.; Raghunandan, S.; Duran, A.; et al. p62, Upregulated during Preneoplasia, Induces Hepatocellular Carcinogenesis by Maintaining Survival of Stressed HCC-Initiating Cells. *Cancer Cell* **2016**, *29*, 935–948. [CrossRef]

109. Ma, X.H.; Piao, S.; Wang, D.; McAfee, Q.W.; Nathanson, K.L.; Lum, J.J.; Li, L.Z.; Amaravadi, R.K. Measurements of tumor cell autophagy predict invasiveness, resistance to chemotherapy, and survival in melanoma. *Clin. Cancer Res. Off. J. Am. Assoc. Cancer Res.* **2011**, *17*, 3478–3489. [CrossRef]

110. Lazova, R.; Camp, R.L.; Klump, V.; Siddiqui, S.F.; Amaravadi, R.K.; Pawelek, J.M. Punctate LC3B expression is a common feature of solid tumors and associated with proliferation, metastasis, and poor outcome. *Clin. Cancer Res. Off. J. Am. Assoc. Cancer Res.* **2012**, *18*, 370–379. [CrossRef] [PubMed]

111. Santanam, U.; Banach-Petrosky, W.; Abate-Shen, C.; Shen, M.M.; White, E.; DiPaola, R.S. Atg7 cooperates with Pten loss to drive prostate cancer tumor growth. *Genes Dev.* **2016**, *30*, 399–407. [CrossRef]

112. Gammoh, N.; Fraser, J.; Puente, C.; Syred, H.M.; Kang, H.; Ozawa, T.; Lam, D.; Acosta, J.C.; Finch, A.J.; Holland, E.; et al. Suppression of autophagy impedes glioblastoma development and induces senescence. *Autophagy* **2016**, *12*, 1431–1439. [CrossRef] [PubMed]

113. Levy, J.M.; Thorburn, A. Modulation of pediatric brain tumor autophagy and chemosensitivity. *J. Neuro Oncol.* **2012**, *106*, 281–290. [CrossRef]

114. Buchanan, P.J.; McCloskey, K.D. CaV channels and cancer: Canonical functions indicate benefits of repurposed drugs as cancer therapeutics. *Eur. Biophys. J.* **2016**, *45*, 621–633. [CrossRef] [PubMed]

115. Abdelmohsen, K.; Srikantan, S.; Tominaga, K.; Kang, M.J.; Yaniv, Y.; Martindale, J.L.; Yang, X.; Park, S.S.; Becker, K.G.; Subramanian, M.; et al. Growth inhibition by miR-519 via multiple p21-inducing pathways. *Mol. Cell. Biol.* **2012**, *32*, 2530–2548. [CrossRef] [PubMed]

116. Hoyer-Hansen, M.; Bastholm, L.; Szyniarowski, P.; Campanella, M.; Szabadkai, G.; Farkas, T.; Bianchi, K.; Fehrenbacher, N.; Elling, F.; Rizzuto, R.; et al. Control of macroautophagy by calcium, calmodulin-dependent kinase kinase-beta, and Bcl-2. *Mol. Cell* **2007**, *25*, 193–205. [CrossRef]

117. Mathiasen, I.S.; Sergeev, I.N.; Bastholm, L.; Elling, F.; Norman, A.W.; Jaattela, M. Calcium and calpain as key mediators of apoptosis-like death induced by vitamin D compounds in breast cancer cells. *J. Biol. Chem.* **2002**, *277*, 30738–30745. [CrossRef]

118. Decuypere, J.P.; Welkenhuyzen, K.; Luyten, T.; Ponsaerts, R.; Dewaele, M.; Molgo, J.; Agostinis, P.; Missiaen, L.; De Smedt, H.; Parys, J.B.; et al. Ins(1,4,5)P_3 receptor-mediated Ca^{2+} signaling and autophagy induction are interrelated. *Autophagy* **2011**, *7*, 1472–1489. [CrossRef]

119. Sakaki, K.; Wu, J.; Kaufman, R.J. Protein kinase Ctheta is required for autophagy in response to stress in the endoplasmic reticulum. *J. Biol. Chem.* **2008**, *283*, 15370–15380. [CrossRef]

120. Wang, S.H.; Shih, Y.L.; Ko, W.C.; Wei, Y.H.; Shih, C.M. Cadmium-induced autophagy and apoptosis are mediated by a calcium signaling pathway. *Cell. Mol. Life Sci.* **2008**, *65*, 3640–3652. [CrossRef]

121. Sarkar, S.; Floto, R.A.; Berger, Z.; Imarisio, S.; Cordenier, A.; Pasco, M.; Cook, L.J.; Rubinsztein, D.C. Lithium induces autophagy by inhibiting inositol monophosphatase. *J. Cell Biol.* **2005**, *170*, 1101–1111. [CrossRef]

122. Cardenas, C.; Miller, R.A.; Smith, I.; Bui, T.; Molgo, J.; Muller, M.; Vais, H.; Cheung, K.H.; Yang, J.; Parker, I.; et al. Essential regulation of cell bioenergetics by constitutive InsP3 receptor Ca^{2+} transfer to mitochondria. *Cell* **2010**, *142*, 270–283. [CrossRef]

123. Patergnani, S.; Guzzo, S.; Mangolini, A.; dell'Atti, L.; Pinton, P.; Aguiari, G. The induction of AMPK-dependent autophagy leads to P53 degradation and affects cell growth and migration in kidney cancer cells. *Exp. Cell Res.* **2020**, *395*, 112190. [CrossRef]

124. Bonora, M.; Giorgi, C.; Pinton, P. Novel frontiers in calcium signaling: A possible target for chemotherapy. *Pharmacol. Res.* **2015**, *99*, 82–85. [CrossRef]

125. Cui, C.; Merritt, R.; Fu, L.; Pan, Z. Targeting calcium signaling in cancer therapy. *Acta Pharm. Sin. B* **2017**, *7*, 3–17. [CrossRef] [PubMed]

126. Elzamzamy, O.M.; Penner, R.; Hazlehurst, L.A. The Role of TRPC1 in Modulating Cancer Progression. *Cells* **2020**, *9*, 388. [CrossRef]

127. Hantute-Ghesquier, A.; Haustrate, A.; Prevarskaya, N.; Lehen'kyi, V. TRPM Family Channels in Cancer. *Pharmaceuticals* **2018**, *11*, 58. [CrossRef]

128. Stewart, J.M. TRPV6 as a Target for Cancer Therapy. *J. Cancer* **2020**, *11*, 374–387. [CrossRef]

129. Santoni, G.; Maggi, F.; Morelli, M.B.; Santoni, M.; Marinelli, O. Transient Receptor Potential Cation Channels in Cancer Therapy. *Med. Sci.* **2019**, *7*, 108. [CrossRef]

130. Chen, W.L.; Barszczyk, A.; Turlova, E.; Deurloo, M.; Liu, B.; Yang, B.B.; Rutka, J.T.; Feng, Z.P.; Sun, H.S. Inhibition of TRPM7 by carvacrol suppresses glioblastoma cell proliferation, migration and invasion. *Oncotarget* **2015**, *6*, 16321–16340. [CrossRef] [PubMed]

131. Tolcher, A.; Patnaik, A.; Papadopoulos, K.; Mays, T.; Stephan, T.; Humble, D.; Frohlich, M.; Sims, R. 376 Preliminary results from a Phase 1 study of D-3263 HCl, a TRPM8 calcium channel agonist, in patients with advanced cancer. *EJC Suppl.* **2010**, *8*, 119. [CrossRef]

132. Yamada, T.; Ueda, T.; Shibata, Y.; Ikegami, Y.; Saito, M.; Ishida, Y.; Ugawa, S.; Kohri, K.; Shimada, S. TRPV2 activation induces apoptotic cell death in human T24 bladder cancer cells: A potential therapeutic target for bladder cancer. *Urology* **2010**, *76*, 509.e1–509.e7. [CrossRef]

133. Fu, S.; Hirte, H.; Welch, S.; Ilenchuk, T.T.; Lutes, T.; Rice, C.; Fields, N.; Nemet, A.; Dugourd, D.; Piha-Paul, S.; et al. First-in-human phase I study of SOR-C13, a TRPV6 calcium channel inhibitor, in patients with advanced solid tumors. *Investig. New Drugs* **2017**, *35*, 324–333. [CrossRef]

134. Antal, L.; Martin-Caraballo, M. T-type Calcium Channels in Cancer. *Cancers* **2019**, *11*, 134. [CrossRef]

135. Lara, R.; Adinolfi, E.; Harwood, C.A.; Philpott, M.; Barden, J.A.; Di Virgilio, F.; McNulty, S. P2X7 in Cancer: From Molecular Mechanisms to Therapeutics. *Front. Pharmacol.* **2020**, *11*, 793. [CrossRef]

136. Gilbert, S.M.; Gidley Baird, A.; Glazer, S.; Barden, J.A.; Glazer, A.; Teh, L.C.; King, J. A phase I clinical trial demonstrates that nfP2X7 -targeted antibodies provide a novel, safe and tolerable topical therapy for basal cell carcinoma. *Br. J. Dermatol.* **2017**, *177*, 117–124. [CrossRef]

137. Hu, L.P.; Zhang, X.X.; Jiang, S.H.; Tao, L.Y.; Li, Q.; Zhu, L.L.; Yang, M.W.; Huo, Y.M.; Jiang, Y.S.; Tian, G.A.; et al. Targeting Purinergic Receptor P2Y2 Prevents the Growth of Pancreatic Ductal Adenocarcinoma by Inhibiting Cancer Cell Glycolysis. *Clin. Cancer Res. Off. J. Am. Assoc. Cancer Res.* **2019**, *25*, 1318–1330. [CrossRef]

138. Chalmers, S.B.; Monteith, G.R. ORAI channels and cancer. *Cell Calcium* **2018**, *74*, 160–167. [CrossRef] [PubMed]

139. Tanwar, J.; Arora, S.; Motiani, R.K. Orai3: Oncochannel with therapeutic potential. *Cell Calcium* **2020**, *90*, 102247. [CrossRef] [PubMed]

140. Dubois, C.; Vanden Abeele, F.; Lehen'kyi, V.; Gkika, D.; Guarmit, B.; Lepage, G.; Slomianny, C.; Borowiec, A.S.; Bidaux, G.; Benahmed, M.; et al. Remodeling of channel-forming ORAI proteins determines an oncogenic switch in prostate cancer. *Cancer Cell* **2014**, *26*, 19–32. [CrossRef]

141. Moccia, F.; Zuccolo, E.; Poletto, V.; Turin, I.; Guerra, G.; Pedrazzoli, P.; Rosti, V.; Porta, C.; Montagna, D. Targeting Stim and Orai Proteins as an Alternative Approach in Anticancer Therapy. *Curr. Med. Chem.* **2016**, *23*, 3450–3480. [CrossRef]

142. Latour, S.; Zanese, M.; Le Morvan, V.; Vacher, A.M.; Menard, N.; Bijou, F.; Durrieu, F.; Soubeyran, P.; Savina, A.; Vacher, P.; et al. Role of Calcium Signaling in GA101-Induced Cell Death in Malignant Human B Cells. *Cancers* **2019**, *11*, 291. [CrossRef]

143. Rezuchova, I.; Hudecova, S.; Soltysova, A.; Matuskova, M.; Durinikova, E.; Chovancova, B.; Zuzcak, M.; Cihova, M.; Burikova, M.; Penesova, A.; et al. Type 3 inositol 1,4,5-trisphosphate receptor has antiapoptotic and proliferative role in cancer cells. *Cell Death Dis.* **2019**, *10*, 186. [CrossRef]

144. Akl, H.; Monaco, G.; La Rovere, R.; Welkenhuyzen, K.; Kiviluoto, S.; Vervliet, T.; Molgo, J.; Distelhorst, C.W.; Missiaen, L.; Mikoshiba, K.; et al. IP3R2 levels dictate the apoptotic sensitivity of diffuse large B-cell lymphoma cells to an IP3R-derived peptide targeting the BH4 domain of Bcl-2. *Cell Death Dis.* **2013**, *4*, e632. [CrossRef]

145. Kang, S.S.; Han, K.S.; Ku, B.M.; Lee, Y.K.; Hong, J.; Shin, H.Y.; Almonte, A.G.; Woo, D.H.; Brat, D.J.; Hwang, E.M.; et al. Caffeine-mediated inhibition of calcium release channel inositol 1,4,5-trisphosphate receptor subtype 3 blocks glioblastoma invasion and extends survival. *Cancer Res.* **2010**, *70*, 1173–1183. [CrossRef]

146. Gelebart, P.; Kovacs, T.; Brouland, J.P.; van Gorp, R.; Grossmann, J.; Rivard, N.; Panis, Y.; Martin, V.; Bredoux, R.; Enouf, J.; et al. Expression of endomembrane calcium pumps in colon and gastric cancer cells. Induction of SERCA3 expression during differentiation. *J. Biol. Chem.* **2002**, *277*, 26310–26320. [CrossRef]

147. Fan, L.; Li, A.; Li, W.; Cai, P.; Yang, B.; Zhang, M.; Gu, Y.; Shu, Y.; Sun, Y.; Shen, Y.; et al. Novel role of Sarco/endoplasmic reticulum calcium ATPase 2 in development of colorectal cancer and its regulation by F36, a curcumin analog. *Biomed. Pharmacother.* **2014**, *68*, 1141–1148. [CrossRef]

148. Bruce, J.I.E.; James, A.D. Targeting the Calcium Signalling Machinery in Cancer. *Cancers* **2020**, *12*, 2351. [CrossRef]
149. Varga, K.; Hollosi, A.; Paszty, K.; Hegedus, L.; Szakacs, G.; Timar, J.; Papp, B.; Enyedi, A.; Padanyi, R. Expression of calcium pumps is differentially regulated by histone deacetylase inhibitors and estrogen receptor alpha in breast cancer cells. *BMC Cancer* **2018**, *18*, 1029. [CrossRef] [PubMed]
150. Ruschoff, J.H.; Brandenburger, T.; Strehler, E.E.; Filoteo, A.G.; Heinmoller, E.; Aumuller, G.; Wilhelm, B. Plasma membrane calcium ATPase expression in human colon multistep carcinogenesis. *Cancer Investig.* **2012**, *30*, 251–257. [CrossRef] [PubMed]
151. Curry, M.C.; Roberts-Thomson, S.J.; Monteith, G.R. Plasma membrane calcium ATPases and cancer. *BioFactors* **2011**, *37*, 132–138. [CrossRef]
152. Sritangos, P.; Pena Alarcon, E.; James, A.D.; Sultan, A.; Richardson, D.A.; Bruce, J.I.E. Plasma Membrane Ca^{2+} ATPase Isoform 4 (PMCA4) Has an Important Role in Numerous Hallmarks of Pancreatic Cancer. *Cancers* **2020**, *12*, 218. [CrossRef]
153. Mahalingam, D.; Wilding, G.; Denmeade, S.; Sarantopoulas, J.; Cosgrove, D.; Cetnar, J.; Azad, N.; Bruce, J.; Kurman, M.; Allgood, V.E.; et al. Mipsagargin, a novel thapsigargin-based PSMA-activated prodrug: Results of a first-in-man phase I clinical trial in patients with refractory, advanced or metastatic solid tumours. *Br. J. Cancer* **2016**, *114*, 986–994. [CrossRef]
154. Mahalingam, D.; Peguero, J.; Cen, P.; Arora, S.P.; Sarantopoulos, J.; Rowe, J.; Allgood, V.; Tubb, B.; Campos, L. A Phase II, Multicenter, Single-Arm Study of Mipsagargin (G-202) as a Second-Line Therapy Following Sorafenib for Adult Patients with Progressive Advanced Hepatocellular Carcinoma. *Cancers* **2019**, *11*, 833. [CrossRef]
155. Peterkova, L.; Kmonickova, E.; Ruml, T.; Rimpelova, S. Sarco/Endoplasmic Reticulum Calcium ATPase Inhibitors: Beyond Anticancer Perspective. *J. Med. Chem.* **2020**, *63*, 1937–1963. [CrossRef]
156. Muscella, A.; Calabriso, N.; Vetrugno, C.; Fanizzi, F.P.; De Pascali, S.A.; Storelli, C.; Marsigliante, S. The platinum (II) complex [Pt(O,O'-acac)(gamma-acac)(DMS)] alters the intracellular calcium homeostasis in MCF-7 breast cancer cells. *Biochem. Pharmacol.* **2011**, *81*, 91–103. [CrossRef] [PubMed]
157. Peterson, J.A.; Crowther, C.M.; Andrus, M.B.; Kenealey, J.D. Resveratrol derivatives increase cytosolic calcium by inhibiting plasma membrane ATPase and inducing calcium release from the endoplasmic reticulum in prostate cancer cells. *Biochem. Biophys. Rep.* **2019**, *19*, 100667. [CrossRef]
158. Vultur, A.; Gibhardt, C.S.; Stanisz, H.; Bogeski, I. The role of the mitochondrial calcium uniporter (MCU) complex in cancer. *Pflug. Arch. Eur. J. Physiol.* **2018**, *470*, 1149–1163. [CrossRef]
159. Mazure, N.M. VDAC in cancer. *Biochim. Biophys. Acta. Bioenerg.* **2017**, *1858*, 665–673. [CrossRef] [PubMed]
160. Shteinfer-Kuzmine, A.; Amsalem, Z.; Arif, T.; Zooravlov, A.; Shoshan-Barmatz, V. Selective induction of cancer cell death by VDAC1-based peptides and their potential use in cancer therapy. *Mol. Oncol.* **2018**, *12*, 1077–1103. [CrossRef] [PubMed]
161. Tang, B.D.; Xia, X.; Lv, X.F.; Yu, B.X.; Yuan, J.N.; Mai, X.Y.; Shang, J.Y.; Zhou, J.G.; Liang, S.J.; Pang, R.P. Inhibition of Orai1-mediated Ca^{2+} entry enhances chemosensitivity of HepG2 hepatocarcinoma cells to 5-fluorouracil. *J. Cell. Mol. Med.* **2017**, *21*, 904–915. [CrossRef]
162. Can, G.; Akpinar, B.; Baran, Y.; Zhivotovsky, B.; Olsson, M. 5-Fluorouracil signaling through a calcium-calmodulin-dependent pathway is required for p53 activation and apoptosis in colon carcinoma cells. *Oncogene* **2013**, *32*, 4529–4538. [CrossRef]
163. Shen, L.; Wen, N.; Xia, M.; Zhang, Y.U.; Liu, W.; Xu, Y.E.; Sun, L. Calcium efflux from the endoplasmic reticulum regulates cisplatin-induced apoptosis in human cervical cancer HeLa cells. *Oncol. Lett.* **2016**, *11*, 2411–2419. [CrossRef] [PubMed]
164. Abdoul-Azize, S.; Dubus, I.; Vannier, J.P. Improvement of dexamethasone sensitivity by chelation of intracellular Ca^{2+} in pediatric acute lymphoblastic leukemia cells through the prosurvival kinase ERK1/2 deactivation. *Oncotarget* **2017**, *8*, 27339–27352. [CrossRef] [PubMed]
165. Abdoul-Azize, S.; Buquet, C.; Vannier, J.P.; Dubus, I. Pyr3, a TRPC3 channel blocker, potentiates dexamethasone sensitivity and apoptosis in acute lymphoblastic leukemia cells by disturbing Ca^{2+} signaling, mitochondrial membrane potential changes and reactive oxygen species production. *Eur. J. Pharmacol.* **2016**, *784*, 90–98. [CrossRef]

166. Giorgi, C.; Bonora, M.; Sorrentino, G.; Missiroli, S.; Poletti, F.; Suski, J.M.; Galindo Ramirez, F.; Rizzuto, R.; Di Virgilio, F.; Zito, E.; et al. p53 at the endoplasmic reticulum regulates apoptosis in a Ca^{2+}-dependent manner. *Proc. Natl. Acad. Sci. USA* **2015**, *112*, 1779–1784. [CrossRef]

167. Boehmerle, W.; Splittgerber, U.; Lazarus, M.B.; McKenzie, K.M.; Johnston, D.G.; Austin, D.J.; Ehrlich, B.E. Paclitaxel induces calcium oscillations via an inositol 1,4,5-trisphosphate receptor and neuronal calcium sensor 1-dependent mechanism. *Proc. Natl. Acad. Sci. USA* **2006**, *103*, 18356–18361. [CrossRef] [PubMed]

168. Giorgi, C.; Bonora, M.; Missiroli, S.; Poletti, F.; Ramirez, F.G.; Morciano, G.; Morganti, C.; Pandolfi, P.P.; Mammano, F.; Pinton, P. Intravital imaging reveals p53-dependent cancer cell death induced by phototherapy via calcium signaling. *Oncotarget* **2015**, *6*, 1435–1445. [CrossRef]

169. Verbaanderd, C.; Maes, H.; Schaaf, M.B.; Sukhatme, V.P.; Pantziarka, P.; Sukhatme, V.; Agostinis, P.; Bouche, G. Repurposing Drugs in Oncology (ReDO)-chloroquine and hydroxychloroquine as anti-cancer agents. *Ecancermedicalscience* **2017**, *11*, 781. [CrossRef]

170. Wu, Y.F.; Zhao, P.; Luo, X.; Xu, J.C.; Xue, L.; Zhou, Q.; Xiong, M.; Shen, J.; Peng, Y.B.; Yu, M.F.; et al. Chloroquine inhibits Ca^{2+} permeable ion channels-mediated Ca^{2+} signaling in primary B lymphocytes. *Cell Biosci.* **2017**, *7*, 28. [CrossRef]

171. Chen, D.; Xie, J.; Fiskesund, R.; Dong, W.; Liang, X.; Lv, J.; Jin, X.; Liu, J.; Mo, S.; Zhang, T.; et al. Chloroquine modulates antitumor immune response by resetting tumor-associated macrophages toward M1 phenotype. *Nat. Commun.* **2018**, *9*, 873. [CrossRef]

172. Jang, E.; Kim, I.Y.; Kim, H.; Lee, D.M.; Seo, D.Y.; Lee, J.A.; Choi, K.S.; Kim, E. Quercetin and chloroquine synergistically kill glioma cells by inducing organelle stress and disrupting Ca^{2+} homeostasis. *Biochem. Pharmacol.* **2020**, *178*, 114098. [CrossRef]

173. Vu, H.T.; Kobayashi, M.; Hegazy, A.M.; Tadokoro, Y.; Ueno, M.; Kasahara, A.; Takase, Y.; Nomura, N.; Peng, H.; Ito, C.; et al. Autophagy inhibition synergizes with calcium mobilization to achieve efficient therapy of malignant gliomas. *Cancer Sci.* **2018**, *109*, 2497–2508. [CrossRef]

174. Chen, X.; Pan, X.; Zhang, W.; Guo, H.; Cheng, S.; He, Q.; Yang, B.; Ding, L. Epigenetic strategies synergize with PD-L1/PD-1 targeted cancer immunotherapies to enhance antitumor responses. *Acta Pharm. Sin. B* **2020**, *10*, 723–733. [CrossRef]

175. Newton, H.S.; Gawali, V.S.; Chimote, A.A.; Lehn, M.A.; Palackdharry, S.M.; Hinrichs, B.H.; Jandarov, R.; Hildeman, D.; Janssen, E.M.; Wise-Draper, T.M.; et al. PD1 blockade enhances K^+ channel activity, Ca^{2+} signaling, and migratory ability in cytotoxic T lymphocytes of patients with head and neck cancer. *J. Immunother. Cancer* **2020**, *8*, e000844. [CrossRef] [PubMed]

176. Ou, X.; Xu, S.; Li, Y.F.; Lam, K.P. Adaptor protein DOK3 promotes plasma cell differentiation by regulating the expression of programmed cell death 1 ligands. *Proc. Natl. Acad. Sci. USA* **2014**, *111*, 11431–11436. [CrossRef]

177. Wei, F.; Zhong, S.; Ma, Z.; Kong, H.; Medvec, A.; Ahmed, R.; Freeman, G.J.; Krogsgaard, M.; Riley, J.L. Strength of PD-1 signaling differentially affects T-cell effector functions. *Proc. Natl. Acad. Sci. USA* **2013**, *110*, E2480–E2489. [CrossRef] [PubMed]

178. Grolez, G.P.; Hammadi, M.; Barras, A.; Gordienko, D.; Slomianny, C.; Volkel, P.; Angrand, P.O.; Pinault, M.; Guimaraes, C.; Potier-Cartereau, M.; et al. Encapsulation of a TRPM8 Agonist, WS12, in Lipid Nanocapsules Potentiates PC3 Prostate Cancer Cell Migration Inhibition through Channel Activation. *Sci. Rep.* **2019**, *9*, 7926. [CrossRef]

179. Dubois, C.; Vanden Abeele, F.; Sehgal, P.; Olesen, C.; Junker, S.; Christensen, S.B.; Prevarskaya, N.; Moller, J.V. Differential effects of thapsigargin analogues on apoptosis of prostate cancer cells: Complex regulation by intracellular calcium. *FEBS J.* **2013**, *280*, 5430–5440. [CrossRef]

180. Denmeade, S.R.; Isaacs, J.T. The SERCA pump as a therapeutic target: Making a "smart bomb" for prostate cancer. *Cancer Biol. Ther.* **2005**, *4*, 14–22. [CrossRef]

2

Automated Intracellular Calcium Profiles Extraction from Endothelial Cells Using Digital Fluorescence Images

Marcial Sanchez-Tecuatl [1,*], **Ajelet Vargaz-Guadarrama** [2], **Juan Manuel Ramirez-Cortes** [1,*], **Pilar Gomez-Gil** [3], **Francesco Moccia** [4] **and Roberto Berra-Romani** [2,*]

[1] Electronics Department, National Institute of Astrophysics, Optics and Electronics, 72840 Puebla, Mexico
[2] Biomedicine School, Benemérita Universidad Autónoma de Puebla, 72410 Puebla, Mexico; ajeletvargaz7@hotmail.com
[3] Computer Science Department, National Institute of Astrophysics, Optics and Electronics, 72840 Puebla, Mexico; pgomez@inaoep.mx
[4] Department of Biology and Biotechnology "Lazzaro Spallanzani", University of Pavia, 27100 Pavia, Italy; francesco.moccia@unipv.it
[*] Correspondence: marcial_st@hotmail.com (M.S.-T.); jmram@inaoep.mx (J.M.R.-C.); rberra001@hotmail.com (R.B.-R.)

Abstract: Endothelial cells perform a wide variety of fundamental functions for the cardiovascular system, their proliferation and migration being strongly regulated by their intracellular calcium concentration. Hence it is extremely important to carefully measure endothelial calcium signals under different stimuli. A proposal to automate the intracellular calcium profiles extraction from fluorescence image sequences is presented. Digital image processing techniques were combined with a multi-target tracking approach supported by Kalman estimation. The system was tested with image sequences from two different stimuli. The first one was a chemical stimulus, that is, ATP, which caused small movements in the cells trajectories, thereby suggesting that the bath application of the agonist does not generate significant artifacts. The second one was a mechanical stimulus delivered by a glass microelectrode, which caused major changes in cell trajectories. The importance of the tracking block is evidenced since more accurate profiles were extracted, mainly for cells closest to the stimulated area. Two important contributions of this work are the automatic relocation of the region of interest assigned to the cells and the possibility of data extraction from big image sets in efficient and expedite way. The system may adapt to different kind of cell images and may allow the extraction of other useful features.

Keywords: intracellular calcium; endothelium; cell tracking; multi-target tracking; Kalman

1. Introduction

Vascular endothelium has long been considered as a homogeneous population of inert cells that forms a permeable barrier between flowing blood and surrounding tissues, regulating the exchange of nutrients, oxygen and catabolic waste at capillary level. This rather conventional view has recently given the way to a more realistic interpretation of the multifaceted role played by endothelial cells (ECs) within the cardiovascular system. The endothelium constitutes a multifunctional signal transduction surface that is now regarded as the largest endocrine organ of the body by virtue of its extension (about 350 m^2) and ability to release a broad spectrum of paracrine mediators [1]. For instance, ECs control the vascular tone, platelet activation and aggregation and angiogenesis. Endothelial dysfunction ultimately results in severe cardiovascular disorders, such as hypertension, atherosclerosis, coronary artery disease and stroke, which may cause patient's death [2,3].

An increase in intracellular Ca^{2+} concentration ($[Ca^{2+}]_i$) represents one of the most important signaling pathways whereby ECs exert their sophisticated control of cardiovascular homeostasis [4–6]. For instance, extracellular autacoids, such as adenosine trisphosphate (ATP) and acetylcholine, cause a robust increase in endothelial $[Ca^{2+}]_i$ to recruit the Ca^{2+}/Calmodulin-dependent endothelial nitric oxide (NO) synthase (eNOS), thereby inducing NO release and vasodilation [7]. Likewise, endothelial Ca^{2+} signals drive the production of additional vasorelaxing mediators, such as prostacyclin I2 and hydrogen sulphide [8] and engage in Ca^{2+}-dependent intermediate and small-conductance K^+ channels to activate endothelial-dependent hyperpolarization (EDH) [9].

The intact endothelium of excised rat aorta provides an ideal preparation to investigate endothelial Ca^{2+} handling as the intracellular Ca^{2+} toolkit of native endothelium may differ from that described in cultured ECs [10–12]. This preparation, therefore, is suitable for both studying the Ca^{2+} response to physiological agonists, including ATP and acetylcholine [11,13,14] and investigating the Ca^{2+} response to mechanical injury that ultimately results in the process of endothelial wound healing [6,12,15]. Unraveling the signal transduction pathways engaged by physiological agonists and mechanical injury is essential to design novel therapeutic treatments aimed at restoring endothelial integrity and limiting cardiovascular diseases. Nevertheless, tracking endothelial Ca^{2+} signaling in their native microenvironment is a rather challenging task.

The ratiometric Ca^{2+}-sensitive fluorophore Fura-2 is the most suitable fluorescent dye to obtain quantitative data on $[Ca^{2+}]_i$ and to carry out long-lasting recording due to its little photobleaching [15,16]. It has been widely used to measure intracellular Ca^{2+} signals in intact endothelium of excised rat aorta [6,11–15] as well as of other vascular segments [17,18].

The term calcium profile (CP) indicates a time-dependent change in the mean fluorescence emitted by a Ca^{2+} indicator, such as Fura-2, within a user-defined area, also termed region of interest (ROI), manually drawn around a single cell. The CP extraction procedure consists in collecting the mean intensity within each ROI for every frame throughout a ratio image sequence. This procedure provides a meaningful profile of the Ca^{2+} signal as a function of time when the ROI surrounds only one cell nucleus, which appears brighter as it is thicker cell area. For a conventional experiment, the CP of multiple cells is extracted from ratio image sequences acquired for hundreds or even thousands of frames (acquisition frequency 0.3 Hz): manual ROI analysis is time consuming and labor intensive and may introduce artificial signals as the position and plane of focus of the cells may change radically during an experiment. This leads the users to manually relocate the ROIs frame per frame and one by one if needed. Consequently, the process becomes tedious, impractical and even more time consuming.

The Ca^{2+} response to ATP or acetylcholine in the intact endothelium of excised rat aorta consists in a biphasic increase in $[Ca^{2+}]_i$, which comprises a rapid Ca^{2+} transient followed by a plateau phase of intermediate amplitude between the resting Ca^{2+} levels and the initial Ca^{2+} response [11,13,14]. Similarly, the Ca^{2+} signal induced by mechanical injury consists in a rapid Ca^{2+} peak followed by a long-lasting decline towards the baseline in the majority ECs adjacent to the lesion site. The remaining fraction of cells in the first row also displayed periodic Ca^{2+} oscillations which overlapped the decay phase and were initiated after the initial Ca^{2+} spike. Furthermore, repetitive Ca^{2+} spikes and heterogeneous Ca^{2+} signals were detected also in ECs further back from the wound edge (see Supplementary Material in Reference [15]). As discussed elsewhere [6], injury-induced intra- and intercellular oscillations in $[Ca^{2+}]_i$ might play a key role in healing damaged endothelium by inducing surviving ECs to proliferate and migrate in order to cover the denuded membrane basement.

Despite the obvious importance of recording the Ca^{2+} signal in all the cells that lie beyond the injury edge, several technical difficulties prevented us from successfully accomplishing this task. The main technical difficulties were: (1) cell selection was performed by manual ROI drawing around the cells: in the typical microscope visual field of intact endothelium at $40\times$ magnification, around 100–150 cells can be visualized, this resulted in a time consuming cell selection; (2) as endothelial borders on in situ endothelium were not clearly identifiable, a ROI might include part of the cytoplasm of adjacent cells, so a special attention should be taken in order to draw ROI very close to the nucleus area, increasing the ROI time selection and error possibilities; (3) as ECs of intact vessels are located over a thick layer of vascular smooth muscle that contract or relax in response to vasoactive substances released by the chemical or mechanical stimulated endothelium, this results in the movement of ECs during the experiment protocol in such a way that ROI must be repositioned manually throughout the experiment and according to the specific trajectory of each endothelial cell; (4) under the mechanical injury procedure, cells that are damaged from the glass micro-pipette tip (see Section 4. Materials and Methods), immediately loose Fura-2 fluorescence and disappear from the microscope field (here defined as OFF-LINE cells, in contrast to ON-LINE which are defined as the cells present); in addition, as shown in Reference [15], they are able to incorporate ethidium bromide, a fluorescent molecule unable to cross an intact plasma membrane and therefore indicative of damaged/dead cells. A wrong interpretation of data will occur if the CPs from ROIs that enclose those damaged/dead cells are mistakenly taken in to account. It should also point out that the simple response to Ca^{2+}-releasing agonists, such as ATP and acetylcholine, is hampered by the fact that bath application of these agonists stimulates the underlying layer of vascular smooth muscle cells (VMSCs). The following contraction of the muscular component of the artery wall causes many ECs to transiently disappear during some frames because of movement artefacts, thereby emitting less fluorescence and introducing a false negative in the Ca^{2+} signal.

In this work, a proposal to automate the CP extraction process is presented. Taking advantage from the slow cell movement, a tracking scheme based on intersections was implemented, which proved to be a simple and fast solution to avoid the movement artefact of the endothelial monolayer in situ caused by either physiological (due to VSMC contraction) of mechanical (due to the direct damage of the micro-pipette tip) stimulation. Although overlap-based tracking is a straightforward solution in many cases [19,20], problems related to partially missing data require the use of complementary approaches for robustness improvement. In this work, a Kalman filter is incorporated into the proposed system as a computational tool to keep track of cells in movement even when they momentarily vanish. Kalman estimation has been successfully used for object tracking in several applications [21–23]. The position, velocity and acceleration of the cells are estimated in every iteration to be able to predict the position of the current cell in the case it has disappeared.

2. Results

The intracellular CPs are extracted with the aim of modeling the endothelium behavior under different stimuli. With that purpose in mind, the application was tested with several image sequences obtained from two principal types of experiments: Endothelium under chemical (ATP) stimulus and Endothelium under mechanical stimulus. First, we will illustrate the procedure that we exploited to automate CP extraction in in situ ECs and then get into the details of this novel algorithm.

Online Supplementary Video 1 shows the typical experimental procedure to automate CP extraction in in situ ECs that undergo a mechanical injury imposed by the tip of a glass micro-pipette, as described in Materials and Methods. The upper left movie shows the inner wall of the aortic rings loaded with the Ca^{2+}-sensitive fluorophore, Fura-2, which was fully covered by fluorescent ECs. At frame 0, an example of the adaptive segmentation outcome is shown, bordering all ECs visualized in the microscope field. For illustrative proposes, a single EC was selected and a orange ROI was automatically drawn around it in order to extract its CP and tracking its frame to frame position during the experiment (see Video 1 at frame 3). The CP was simultaneously measured and illustrated in the bottom part of the movie. Scraping the lumen of the aortic ring with the tip of a glass micro-pipette (approximately at frame 63) caused a rapid Ca^{2+} peak following by a long-lasting decline to the baseline in the ECs adjacent to the injury site, such as the EC enclosed by the orange ROI. As it can be observed on the "Position graph (upper right movie)" when the injury occurs, a drastic position change takes places for the cell surrounded by the tracking ROI. However, despite the clear movement of the EC, the automated data processing procedure enables the ROI to remain positioned over the cell of interest, which leads to a careful measurement of the Ca^{2+} signal. In Figure 1A, a typical CP obtained by the automated tracking system is shown, where the arrow indicates when mechanical injury was performed.

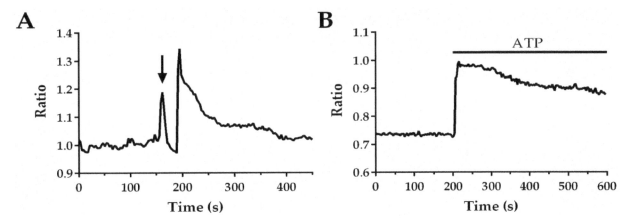

Figure 1. Ca^{2+} profiles from in situ endothelial cells obtained under two different stimuli (**A**) Damaging rat aorta inner wall by using the tip of a glass micro-pipette evokes a long-lasting Ca^{2+} signal in endothelial cells (ECs) located at the injury site. The arrow indicates the time when injury was performed; (**B**) Intracellular Ca^{2+} transient evoked by ATP 300 (µM) in in situ ECs from rat aorta. Calcium profile (CPs) were obtained from one single cell using the automated CP extraction method.

Online Supplementary Video 2 shows a similar procedure used to track an EC in its native microenvironment stimulated by the application of 300 µM of ATP at frame 63. Under this experimental condition, a smallest cell movement was observed. The CP obtained by using the automated CP extraction algorithm is showed in Figure 1B. In the following paragraphs, we will discuss how our Automatic Tracking procedure enables to record the true Ca^{2+} signals induced by the two distinct stimuli without missing any relevant information.

The blocks that will be presented in the Materials and Methods section were merged to conform a functional, easy and user-friendly software application. It is worth pointing out that the proposed blocks are the fundamental components of this proposal to automate the methodology for $[Ca^{2+}]_i$ profiles extraction. To achieve the functionality of the software application, these components were connected with control flows as shown in the high-level flowchart, provided in Flowchart file as Supplementary Material. Programming provides certain degree of freedom, therefore, the implementation of the blocks and their integration into a system may be enhanced as required. The application was built using the Matlab® GUIDE.

2.1. ATP Stimulus

This is a chemical stimulus that has been delivered as described in Materials and Methods. The dimensions of the images are 640 × 480 pixels (px) and they have *. TIFF format. Application of 300 µM ATP to in situ endothelium from rat aorta is able to evoke a biphasic increase in $[Ca^{2+}]_i$, as shown in Figure 2 right (blue line, data obtained with Automatic Tracking approach). However, ECs undergo a little displacement and intensity variations when exposed to bath application of the agonist. For comparison purposes, fifteen cells were selected randomly and their intracellular CPs were obtained using the proposed system and ImageJ program, manually repositioning the ROI every time the cell moves outside of it (ImageJ-Supervised Tracking). The mean square error (MSE) between both obtained data sets is shown in Table 1. Please note that MSE is insignificant compared with the mean of its respective CP extracted with our automated approach ($\overline{CP}$). Although small movements were registered in the cells trajectory, the kinetics of CPs are very similar and the MSE is small, as was expected. This observation demonstrates that this approach, which consists in the simple bath application of the agonist, does not generate significant artifacts and does not induce significant VSMC contraction. It should, however, be pointed out that the magnitude of the initial peak, which is due to Ca^{2+} release from the endoplasmic release, is significantly larger than the amplitude measured without tracking (red line). Cyan line, represent the CP obtained by using ImageJ-Supervised Tracking. In this particular experiment, the single ROI was repositioned at least 3 times along the experiment. As expected, the CP obtained by ImageJ-Supervised Tracking method, overlaps with the CP obtained with our automated approach. This observation reinforces the notion that automatic tracking is recommendable to speed off-line analysis and to gather all the information content of the Ca^{2+} signal.

Table 1. Mean of CPs obtained using the proposed approach ($\overline{CP}$) and the mean square error (MSE) between CPs extracted using ImageJ-Supervised Tracking and our automatic tracking proposal. Adenosine triphosphate (ATP) stimulus experiment.

Cell	$\overline{CP}$	MSE	Cell	$\overline{CP}$	MSE
1	0.9098	1.6242×10^{-4}	9	0.8851	2.9395×10^{-5}
2	1.0111	4.4606×10^{-4}	10	0.8607	5.5402×10^{-5}
3	0.9515	6.0467×10^{-5}	11	0.8728	7.9154×10^{-5}
4	0.9490	4.7925×10^{-5}	12	0.8504	1.1613×10^{-4}
5	0.8713	1.2525×10^{-5}	13	0.8498	1.3611×10^{-5}
6	0.9013	1.7247×10^{-5}	14	0.8648	8.0574×10^{-5}
7	0.8699	2.6627×10^{-5}	15	0.8851	2.2170×10^{-5}
8	0.8481	3.4207×10^{-5}	-	-	-

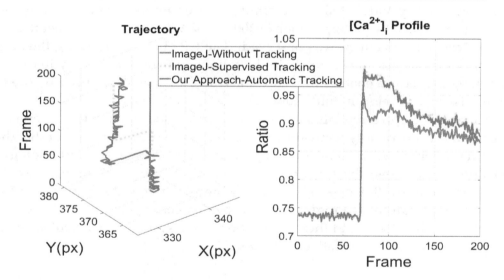

Figure 2. (**Left**), shows the cell trajectory, cell position in X and Y axes respectively, measured in pixels (px) and compared between ImageJ without tracking (red line), ImageJ with manually tracking (cyan line) and trajectory obtained with our automated tracking proposed system (blue line); (**Right**), shows an example of CPs calculated in each condition previously described.

In order to carry on performance evaluation of the system, the density of tracked cells per frame ρ was proposed as figure of merit. The indicator represents an instantaneous measurement of the performance at a given frame i and it is defined in (1).

$$\rho_i = 100 \times \frac{\text{Number of ONLINE cells at frame } i}{\text{Number of binary objects at frame } i} \tag{1}$$

The fluorescence of ECs may decrease or be lost, temporarily or definitively, along a given experiment. A parameter defined as reliability limit (RL) was introduced to provide to the user the possibility to control how many times a cell is allowed to disappear consecutively from the microscope field before it is considered as unreliable to be measured and marked with OFF-LINE status. Health science experts are able to set RL based on their expertise. Usage model of RL is detailed in Section 4.4.

Image sequences of two different ATP experiments with 200 images per excitation wavelength were analyzed. All the detected cells were selected to test the performance of this approach with several reliability limits (RL). The average results are shown in Table 2 and the average number of detected and selected cells was 190.

Table 2. Average performance results with three different reliability limit (RL) values.

RL	ρ_{max} (%)	ρ_{min} (%)	ρ_{mean} (%)
3	104.3956	67.1378	84.8415
5	104.3956	76.4045	89.4309
7	104.3956	79.7753	92.3110

It can be seen that ρmax exceeds the 100%, which is due to the fact that a few cells disappear in some frames thereby causing a mismatch between a given number of cells and a higher number of ROIs than cells. On the other hand, the big amount of noise in some frames produces more binary blobs than ROIs.

2.2. Mechanical Stimulus

In this experiment, a region of the tissue is damaged with the tip of a glass micro-pipette. Detailed experimental procedure is described in Section 4. Materials and Methods. The image sequences are in *.TIFF format with a dimension of 768 × 576 pixels.

Three zones were considered in the image as shown in the Figure 3a and three cells were selected per zone. Zone 2 represents the area onto which the mechanical lesion is imposed, whereas zones 1 and 3 represent the most proximal regions to the damaged site. The fluorescence and position of cells in zones 1 and 3 display a more stable behavior than cells in the zone 2. Some cells of the inner zone (red rectangle on Figure 3a,b) die and then disappear, please see the transition from Figure 3a to Figure 3b; this is reflected as an OFF-LINE status (i.e., blue A001 and A002 ROIs).

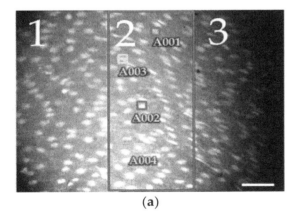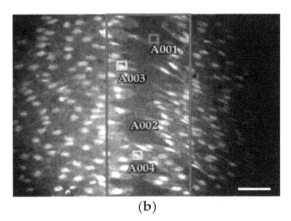

| (a) | (b) |

Figure 3. Images from mechanical stimulus experiment. (**a**) Before and (**b**) after the tissue injury. Suggested zones for the analysis (1, 2 and 3). Blue ROIs, A001 and A002 in (**a**) shown cells in the injured zone that were damaged by the tip of glass micro-pipette and lost Fura-2 fluorescence and then die and disappeared ((**b**) blue ROIs, A001 and A002); these ROIs were marked as OFF-LINE cells. Green ROIs, A003 and A004, enclosed cells adjacent to the injured area, which have partially lost the contact with neighboring ECs, but preserved Fura-2 fluorescence after injury ((**b**) green ROIs, A003 and A004); considered as ON-LINE cells. Bar length is 40 μm.

As the cells moves slightly in the outer zones (1 and 3), it can be seen that the intracellular CPs obtained with ImageJ and with the proposed approach are very similar (Figure 4a). The results obtained from zone 2 show a more notable difference in the structure of the CPs and in the trajectory of the cells (Figure 4b).

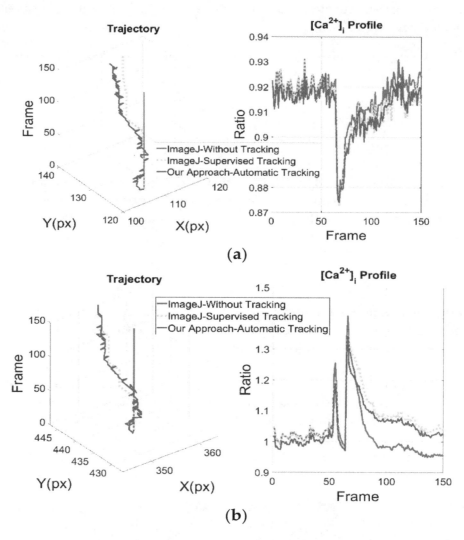

Figure 4. Results per defined zones due to mechanical stimuli: (**a**) Results obtained from zones 1 and 3; (**b**) Results obtained from zone 2. (**Left**), shows trajectories of cells, i.e., calculated position on X and Y axes respectively, measured in pixels (px). (**Right**), shows typical recordings of CP analyzed by: ImageJ without tracking (red line); with manually tracking (cyan line) and automated tracking proposed system (blue line).

Based on the mean squared difference of the results obtained from zones 1 and 3, indicated in Table 3, the proposed scheme delivers consistent results as shown in the previous subsection. While the results of the zone 2 highlights the importance of tracking block since the structure of CPs changes dramatically, it is noteworthy that there are patterns of intracellular CPs associated to different physiological functions. It turns out that a better profile extraction may help a better understanding of cellular behavior [15]. Accordingly, the CP of zone 2 provided by the proposed approach shows a clearly discernible plateau following the initial peak that is missed by ImageJ without tracking. This signal, therefore, reflects the true dynamics of the Ca^{2+} response induced by mechanical stimulation in ECs directly exposed to mechanical injury. The difference in the waveform of the Ca^{2+} signal is rather important, as the Ca^{2+} plateau has long been known to recruit a number of

Ca^{2+}-sensitive decoders involved in cell proliferation and migration, which are the crucial steps for the regeneration process.

As for ATP experiments, two mechanical stimulus experiments with 200 frames per wavelength were used to test the approach. The average results are in Table 4 where the average number of selected cells was 283. As expected, compared to the results of ATP experiments, a lower performance can be which is caused by the more random and more noticeable movement of the cells.

Table 3. Mean of CPs obtained using the proposed approach ($\overline{CP}$) and the mean square error (MSE) between CPs extracted using ImageJ-Supervised Tracking and our automatic tracking proposal. Mechanical stimulus experiment.

Zone	Cell	$\overline{CP}$	MSE
1	1	0.8991	3.5821×10^{-6}
	2	0.8765	4.5569×10^{-6}
	3	0.8937	3.2133×10^{-5}
2	4	1.0551	1.6675×10^{-3}
	5	1.0483	4.2171×10^{-4}
	6	1.0488	9.2146×10^{-4}
3	7	0.9941	3.9861×10^{-6}
	8	0.9676	2.2033×10^{-5}
	9	1.0074	8.7559×10^{-5}

Table 4. Average performance results with different RLs.

RL	ρ_{max} (%)	ρ_{min} (%)	ρ_{mean} (%)
3	100.7117	58.0756	69.9652
5	101.0714	59.7938	73.1079
7	101.0714	61.1684	75.1659

3. Discussion

A multi-target tracking system to automate the intracellular CP extraction from fluorescence image sequences was designed and implemented. The approach was based on logical intersections supported by Kalman estimation which was fed with a constant acceleration model.

A figure of merit was proposed to measure the performance and the system was tested using several image sequences from two different types of experiments, one of them (i.e., ATP experiment) showed the extracted data consistency and the other one (i.e., mechanical lesion) showed the actual need of the cell tracking.

The results allow us to conclude that an important contribution of this work is the automatic relocation of the regions of interest based on statistical measurements which favors massive data extraction and time saving during the analysis. Some appropriate technological improvements of the acquisition systems, for example, faster sample rate, combined with this approach may ensure a meaningful performance improvement, that is, more frames equals more intersections. Finally, the simplicity and efficiency of the proposed approach would allow a real-time application.

In conclusion, this novel algorithm provides an additional tool to accelerate the analysis of the hundreds of endothelial Ca^{2+} signals arising within the intact wall of excised vessels in response to physiological stimuli and to uncover the real dynamics of Ca^{2+} waves at the edge of the lesion site. This approach will be exploited to get more insights into injury-induced intra- and intercellular Ca^{2+} waves, that could in turn lead to the identification of alternative targets to prevent endothelial dysfunction induced by the implantation of medical devices, such as stents and angioplasty, or turbulent blood flow.

4. Materials and Methods

4.1. Dissection of the Aorta

Male Wistar rats aged 2–3 months were provided by the Benemérita Universidad Autónoma de Puebla animal core facilities, where they were kept under conditions of constant environmental temperature and exposed to dark light cycles of 12 h, with water and food *ad libitum*. All the experimental procedures on animals were performed according to protocols approved by the University Animal Care and Use Committee, identification code: BERRSAL71, 18-05-2017. The procedure for obtaining in situ ECs from rat aorta, was developed by Berra-Romani and Cols., 2004 [24] and it is briefly described below. The day of the experiment rats were sacrificed with an overdose of intraperitoneal ketamine-xylazine solution, 0.2 mL per 100 g of weight; subsequently they underwent to an anterior thoracotomy to expose the aortic arch and the heart. The heart was extracted, to cannulate and perfuse the aorta with a physiological salt solution (PSS). Thoracic and abdominal aorta were dissected out and placed in a Petri dish with PSS at room temperature. Using a stereomicroscope (Nikon SMZ-2T), the connective and fatty tissues surrounding the aorta were removed. Subsequently the aorta was cut into ~5 mm wide rings, stored in PSS at room temperature (22–24 °C) and used within 5 h. Using a microdissection scissors, the aortic rings were carefully cut to open and care was taken to avoid any damage to the endothelium and obtain aortic strips with intact endothelium.

4.2. $[Ca^{2+}]_i$ Measurements

The aortic strips with intact endothelium were bathed in PSS with 16 µmol Fura-2/AM for 60 min at room temperature, washed for 30 min to allow intracellular de-esterification of Fura-2/AM and fixed (with the luminal face up) to the bottom of a Petri dish covered by inert silicone (Silgard® 184 Silicone Elastomer, Down Corning, MI, USA) by using four 0.4 mm diameter stainless steel pins. In situ ECs were visualized by an upright epifluorescence Axiolab microscope (Carl Zeiss, Oberkochen, Germany), equipped with a Zeiss 40× Achroplan objective (water-immersion, 2.05 mm working distance, 1.0 numerical aperture). To monitor the $[Ca^{2+}]_i$, ECs were excited alternately at 340 and 380 nm and the emitted light was detected at 510 nm. A neutral density filter (optical density = 1.0) was coupled to the 380 nm filter to approach the intensity of the 340 nm light. A round diaphragm was used to increase the contrast. A filter wheel (Lambda 10, Sutter Instrument, Novato, CA, USA) commanded by a computer positioned alternately along the optical path the two filters that allowed the passage of light respectively at 340 and 380 nm. Custom software, working in the LINUX environment, was used to drive the camera (Extended-ISIS Camera, Photonic Science, Millham, UK) and the filter wheel and to measure and plot in real time (on-line) the fluorescence from ~80–100 drawn manually rectangular "regions of interest" (ROI) enclosing one single cell. $[Ca^{2+}]_i$ was monitored by measuring, for each ROI, the ratio of the mean fluorescence emitted at 510 nm when exciting alternatively at 340 and 380 nm (shortly termed "ratio"). An increase in $[Ca^{2+}]_i$ causes an increase in the ratio. Ratio measurements were performed and plotted on-line every 3 s. The on-line procedure allowed monitoring of $[Ca^{2+}]_i$ at the same time as micro-pipette injury. The fluorescence images obtained at 340 nm and 380 nm were stored on the hard disk and data analysis was carried out later as described in data analysis section. The experiments were performed at room temperature (20–22 °C) to limit time-dependent decreases in the intensity of the fluorescence signal.

4.3. ATP Stimulation of in Situ Endothelial Cells

Medium exchange and administration of ATP (300 µM) was carried out by first removing the bathing medium (2 mL) by a suction pump and then adding the desired solution. The medium could be substituted quickly without producing artifacts in the fluorescence signal because a small meniscus of liquid remained between the tip of the objective and the facing surface of the aorta.

4.4. Mechanical Disruption of in Situ Endothelial Cells

As shown in Reference [15], aortic endothelium was injured under microscopic control by means of a glass micro-pipette with a tip of about 30 μm diameter, driven by an XYZ hydraulic micromanipulator (Narishige Scientific Instrument Lab., Tokyo, Japan). The electrodes were fabricated using a flaming-brown micropipette puller P-1000 (Sutter Instruments, Novato, CA, USA). Images of Fura-2-loaded ECs, together with numbered ROI, were taken before the lesion, to identify the cells facing the injury site. The microelectrode was first positioned almost parallel and very near to the endothelium surface. Then it was moved downward, along the z-axis, until the electrode tip gently touched the endothelium and moved horizontally across the visual field to scrape 1–3 consecutive rows of ECs.

4.4.1. Chemicals

Fura-2/AM was obtained from Molecular Probes (Molecular Probes Europe BV, Leiden, The Netherlands). All other chemicals were purchased from Sigma.

4.4.2. Solution

The composition of the PSS expressed in mmol/L was: NaCl (140), KCl (4.7), MgCl2 (1.2), Glucose (6), CaCl2 (2.5) and HEPES (5). Solution was titrated to pH 7.4 with NaOH.

4.4.3. Data Analysis

Endothelial Ca^{2+} imaging data were processed using an automated data processing procedure. The data processing procedure approach is represented in the block diagram shown in Figure 5. As the first step, the fluorescence images obtained at 340 and 380 nm are separately loaded; thereafter, a calibration stage is inserted to determine the parameters used by the image segmentation.

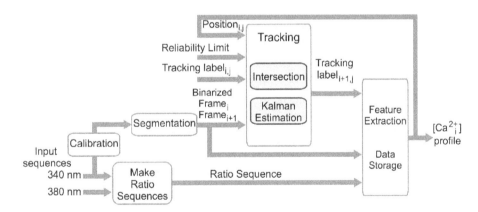

Figure 5. Block diagram of the proposed solution.

The sequence of ratio images is concurrently generated (Make Ratio Sequences). A noise reduction and an adaptive thresholding processes are performed over an image sequence. The resulting binary images are taken as the basis for the tracking scheme based on logical intersections; the tracking scheme is, in turn, enhanced with state estimation by using a Kalman filter. Feature extraction of the ROI and the storage of useful data are performed in the last sub-block.

4.5. Ratio Sequence Block

Fura-2 indicator is excited with λ = 340 nm and λ = 380 nm wavelengths and its fluorescent emissions are both monitored at λ = 510 nm; therefore, two image sequences are generated as experimental data, that is, an image sequence per excitation wavelength.

In this module the pixel to pixel ratio, $\lambda = 340$ nm over $\lambda = 380$ nm excitation wavelength images, is carried out for each frame to generate the Ratio image sequence. At this point, there are three image sequences available: two 8-bit resolution (uint8) *.TIFF and one float type image sequences. The used arrays have $m \times n \times k$ dimensions, where $m = 576$ and $n = 768$ are the images height and width, respectively and k is the number of images per sequence. It is noteworthy to remember that CP are intensity measurements extracted from the ratio images.

As stated above, Fura-2 has a maximum fluorescence when it is excited with 340 nm wavelength, so the contrast of these images is usually better than the contrast of images obtained with 380 nm excitation. Therefore, the 340 nm image sequence was taken as the basis for the calibration, segmentation, tracking and feature extraction blocks. To guarantee the integrity of Ca^{2+} measurements, the 340 nm image sequence is modified only right after the ratio sequence was generated. The 380 nm image sequence was used only to generate the ratiometric image sequence.

4.6. Calibration Block

In this stage, the user manually selects several sample cells, ROIs, from a $\lambda = 340$ nm frame. Then the average of the selected ROIs areas, defined as μ_{Areas}, is used to calculate the window size w to be applied on the next stages, that is, filtering and binarization.

As this is based on area statistics, the more ROIs are selected the better w approximation can be computed. This stage allows the system to be adapted to different cell sizes. Based on the assumption that selected cells are bounded by $w \times w$ squares, the proposed window size is defined as $w = [a \times \sqrt{\mu_{Areas}}]$, where a is scalar.

A set of images was manually segmented by health science experts, aka ground truth segmentation. The same images were also segmented and conditioned automatically with algorithm presented in Section 4.7 and different a values. Then the mean square error between corresponding couples of images was calculated. To simplify the calculation, the binarized images were converted to double data type to use the mmse Matlab built in function. As shown in Figure 6, the minimum mean square error was obtained when $a = 2$.

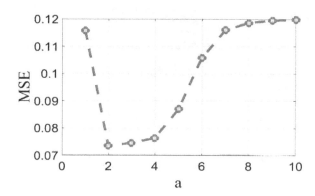

Figure 6. Mean square error (MSE) of binarization and ground truth segmentation for different a value.

4.7. Segmentation

In order to reduce noise, preserve the sharp high frequency detail and enhance the contrast of the images, a median filter in cascade with linear histogram stretching was used before the thresholding process. The median filter square window size was set to $w_{median} = w/8$.

Adaptive binarization based on local statistics was taken as a solution to work with non-uniform illumination images. Since the efficiency of integral image representation has been proven for mean and variance calculation [21], it was used to implement the Niblack technique [22], whose threshold is defined in (2).

$$T(x, y) = m(x, y) + k_T s(x, y)$$ (2)

where m and s stand for local mean and local standard deviation, respectively; the scalar gain was experimentally set to $k_T = 1$.

The last conditioning sub stage was composed of binary image opening with disk of radius 3 as structural element; then, the holes in binary blobs were filled up and finally a dilation with radius 1 disk was applied.

4.8. Tracking Block

The purpose of this block is to follow the selected cells through the image sequences, this task is performed using labels associated to the cells, termed tracking labels. A cell tracking scheme based on intersections was implemented, which proved to be a simple and fast solution, with the drawback of missing cells in subsequent images due to several conditions. In consequence, cell tracking was enhanced using Kalman estimation. The goal of health science experts is to understand the behavior of vascular endothelium, through its Ca^{2+} signals and to discover its regeneration mechanisms under some conditions and stimuli. When endothelium is stimulated, many cells transiently disappear during some frames because of movement artefacts, thereby emitting less fluorescence. In this situation, it is very probable to detect no intersection. Kalman filtering has been successfully used for object tracking in several applications [21–23]. In this work, a Kalman filter is incorporated into the proposed system as a computational tool to keep track of cells in movement even when they momentarily vanish. The position, velocity and acceleration of the cells are estimated in every iteration to be able to predict the position of the current cell in the case it has disappeared.

A counter, per cell, called *reliability counter* is compared with RL to determine the cell status, that is, ONLINE or OFF-LINE. When a cell is selected by the user, it is surrounded by a ROI and the software application sets ONLINE status as an initial condition. Only ONLINE cells are tracked to avoid loss of performance. As of this point, when the *position* of an object is mentioned, it is referring to its *centroid*.

4.8.1. Logical Intersections Approach

As tracking block is based on logical intersections, that is and operations, it works iteratively with two frames of binarized and labeled images, f_{n-1} and f_n. The main goal is to identify the selected cells from f_{n-1} to f_n through the correspondence of the labels, which is termed track_label. A kernel is obtained for each selected cell with the intersection of the current cell with the f_n. When intersection detects only one object, the centroid of that object is obtained in order to get the tracking label in f_n. When more than one or no object is detected, the prediction equations of Kalman filter are used to estimate the position and to find the new label for the current cell in f_n. This process is illustrated in Figure 7.

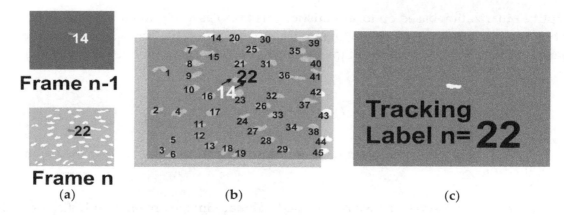

Figure 7. The intersection process to track cells, from frame n-1 to frame n, is illustrated. (**a**) The kernel of a selected cell with track_label = 14 in frame $n - 1$ (blue background and orange highlighted) and the binarized and labeled frame n are the input data, notice that the corresponding label for selected cell will be 22; (**b**) Kernel and binary image are overlapped to show the intersection process, which is a logical and operation between them; (**c**) The outcome is a new kernel, in this case only one object was generated by the intersection; based on its position in frame n the new track_label value is updated to 22. This iterative process is performed per each selected cell (ROI) and each frame.

Once the new tracking label is found, a block termed *store data* is responsible of storing the position and intensity measurement for each cell and resets the reliability counter. If no label is detected with intersections neither with Kalman estimation, it is assumed that the cell disappeared in that frame and the reliability counter value is increased by one. When this counter is equal to the user defined RL, the tracking is considered as unreliable and the status of the current cell is changed to OFFLINE and its new tracking label is no longer searched. The block is initialized with $f_{n-1} = 1$ and $f_n = 2$ and finishes with $f_{n-1} = n - 1$ to $f_n = k$.

4.8.2. Kalman Estimator

Although the cell motion is relatively slow, its randomness entails that the tracking based on intersection may detect in the kernel more than one or none object due to the uncertainty in the target trajectory or acceleration at any given time. Since it is extremely complex to find an exact mathematical model to describe the motion of cells, for tracking purposes, it was assumed that cells have a constant acceleration motion. The deterministic description for two dimensions in state space model is shown in (3).

$$
\begin{bmatrix} x \\ \dot{x} \\ \ddot{x} \\ y \\ \dot{y} \\ \ddot{y} \end{bmatrix}_{n+1} = \begin{bmatrix} 1 & T & \frac{T^2}{2} & 0 & 0 & 0 \\ 0 & 1 & T & 0 & 0 & 0 \\ 0 & 0 & 1 & 0 & 0 & 0 \\ 0 & 0 & 0 & 1 & T & \frac{T^2}{2} \\ 0 & 0 & 0 & 0 & 1 & T \\ 0 & 0 & 0 & 0 & 0 & 1 \end{bmatrix} \begin{bmatrix} x \\ \dot{x} \\ \ddot{x} \\ y \\ \dot{y} \\ \ddot{y} \end{bmatrix}_n + \begin{bmatrix} \frac{T^3 J_x}{3} \\ \frac{T^2 J_x}{2} \\ T J_x \\ \frac{T^3 J_y}{3} \\ \frac{T^2 J_y}{2} \\ T J_y \end{bmatrix}_n
$$

$$
z = \begin{bmatrix} 1 & 0 & 0 & 0 & 0 & 0 \\ 0 & 0 & 0 & 1 & 0 & 0 \end{bmatrix} \begin{bmatrix} x \\ \dot{x} \\ \ddot{x} \\ y \\ \dot{y} \\ \ddot{y} \end{bmatrix}_n
$$

(3)

where J_n is a random change in acceleration, a.k.a. jerk, occurring between time n and $n + 1$. The random jerk J_n has the auto correlation function given by (4) and it is characterized as white noise.

$$J_n \bar{J}_m = \begin{cases} \sigma_J^2 & \text{for } n = m \\ 0 & \text{for } n \neq m \end{cases} \tag{4}$$

With the sampling period of the experimental data, $T = 3$, the observability matrix has full rank, that is, rank = 6, therefore the proposed system model is observable and the Kalman filter can work correctly. Assuming statistical independence between the coordinates x and y, the system covariance matrix is shown in (5) and the observation error covariance matrix in (6).

$$Q = q^T q$$

$$q = \begin{bmatrix} \frac{T^3 \sigma_{Jx}}{3} & \frac{T^2 \sigma_{Jx}}{2} & T\sigma_{Jx} & \frac{T^3 \sigma_{Jy}}{3} & \frac{T^2 \sigma_{Jy}}{2} & T\sigma_{Jy} \end{bmatrix} \tag{5}$$

$$R = \begin{bmatrix} \sigma_x^2 & 0 \\ 0 & \sigma_y^2 \end{bmatrix} \tag{6}$$

4.8.3. Statistical Parameters

The parameters σ_{Jx} and σ_{Jy} belongs to the random nature of the system, for this reason, they were extracted from experimental data. A number $c = 15$ of cells were randomly selected and manually tracked, through $p = 100$ frames, to register their vector positions, x and y. Three numerical derivatives were performed in order to get the jerk vector for each dimension, then the standard deviation of each jerk vector was computed. Finally, σ_{Jx} is the average of those σ_{Jxc}, see (7) and (8). The same process was done for σ_{Jy}.

$$\sigma_{Jx_c}^2 = \frac{1}{p-1} \sum_{i=1}^{p} (Jx_i - \bar{J}_x)^2 \tag{7}$$

$$\sigma_{Jx} = \frac{1}{c} \sum_{n=1}^{c} \sigma_{Jx_n} \tag{8}$$

The state estimation with prediction and correction is performed in every iteration once the current cell has been found with the aim of updating the second and third order derivatives because a single prediction is used only when intersection scheme fails to find the selected cell. For the observation error covariance matrix, the standard deviations were set in such a way that the estimates were very close to measurements and that the influence of Q was incorporated to avoid singularities in the calculus. The parameters are shown in Table 5.

Table 5. Statistical parameters for Kalman filter.

Parameter	σ_{Jx}	σ_{Jy}	σ_x	σ_y
Value	0.0269	0.0993	0.001	0.001

A set of artificial images was generated in order to test the Kalman filter estimation. The image sequence was designed to simulate the trajectory of a cell, the coordinates of the cell were determined mathematically as $x = c$ and $y = A\cos(2\pi f t)$, where c is a constant or a bias, $f = 4$, A was adjusted to produce a 36 pixels peak to peak amplitude and t is a vector with the same number of elements than x. The area of the object in the images was the average cells area obtained from the experimental data.

The estimation was obtained with Kalman filter configured with the statistical parameters described above. In Figure 8 it can be seen the trajectory estimation using the whole scheme predictor–corrector. As some cells disappear during some frames, statistical measurements for Kalman correction are not available along the entire image sequence. Thus, several missed points were used as a focused test of the Kalman prediction, that is, with no evaluation of correction equations.

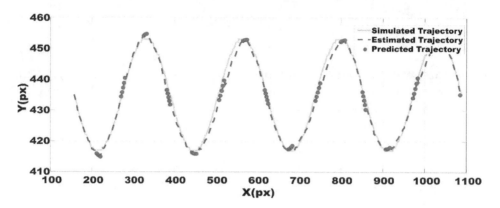

Figure 8. Kalman filter testing. Simulated, estimated and predicted trajectories in solid, dashed and dotted lines, respectively. Position in X and Y axes respectively, measured in pixels (px).

The mean square error between trajectories in both cases is reported in Table 6. By using the centroid as the position of the cell, the reported error is tolerable since it falls inside the average ROI area and it is possible to find again the selected cell. Thus, the Kalman filter has a good performance.

Table 6. Mean square error between trajectories.

Estimator Configuration	MSE$_x$	MSE$_y$
Prediction-Correction	9.6537	0.8958
Only Prediction	10.5347	0.8929

Tracking results with RL = 3 are shown in Figure 9. It can be seen that the value of reliability counter of cell "A007" increased to three, from frame 1 to 4, then its status changed to OFF-LINE and the cell "A007" is not tracked anymore as of frame 5.

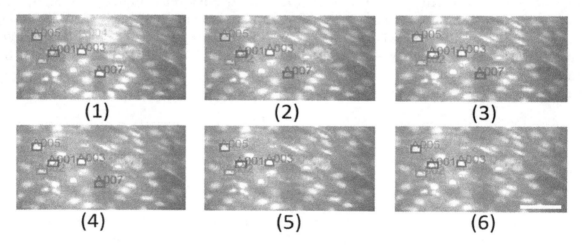

Figure 9. Tracking results through six consecutive frames. Eight cells were selected by the user, only ONLINE cells are labeled, ROI surrounded and tracked. OFFLINE cells are not tracked as shown with cell A007; the cell is visible at frame (**1**), then it vanishes as of frame (**2**) and its reliability counter increases by one between frames (**2**–**4**). At frame (**4**) its reliability counter is equal to RL = 3, then it is tagged as OFFLINE and is not tracked as of frame (**5**). Frame (**6**) shows the tracking of remaining ONLINE cells. Bar length is 40 μm.

4.9. Feature Extraction and Data Storage

This is the last block of the proposed solution. The block receives as input the *new tracking label*. The cell under investigation is identified with that label, the ROI is relocated and the intensity average is computed from the respective ratio frame, that is, the CP is extracted. In this stage, more features can be extracted from the binarized image. Online Supplementary Video 3 shows the analysis complete flow on the software application developed with the approach described throughout this manuscript. This system has, therefore, the potential to help health experts to expand their studies by greatly limiting the movement artefacts due to the manipulation of the extracellular solution or due the delivery of a mechanical stimulus.

Author Contributions: Conceptualization, M.S.-T., J.M.R.-C. and P.G.-G.; Data curation, A.V.-G., F.M. and R.B.-R.; Investigation, M.S.-T., J.M.R.-C., F.M. and R.B.-R.; Methodology, M.S.-T., J.M.R.-C. and R.B.-R.; Software, M.S.-T.; Supervision, J.M.R.-C., P.G.-G. and R.B.-R.; Validation, M.S.-T., A.V.-G., J.M.R.-C. and R.B.-R.; Writing—original draft, M.S.-T.; Writing—review & editing, J.M.R.-C., F.M. and R.B.-R.

Abbreviations

ATP	Adenosine trisphosphate
CP	Calcium profile
Ca^{2+}	Divalent calcium ion
$[Ca^{2+}]_i$	Intracellular Ca^{2+} concentration
ECs	Endothelial cells
EDH	Endothelial-dependent hyperpolarization
NO	Nitric oxide
eNOS	Nitric oxide synthase
PSS	Physiological salt solution
ROI	Region of interest
RL	Reliability limit
VSMC	Vascular smooth muscle cell
MSE	Mean square error

References

1. Pries, A.R.; Secomb, T.W.; Gaehtgens, P. The endothelial surface layer. *Pflug. Arch.* **2000**, *440*, 653–666. [CrossRef] [PubMed]

2. Ayaori, M.; Iwakami, N.; Uto-Kondo, H.; Sato, H.; Sasaki, M.; Komatsu, T.; Iizuka, M.; Takiguchi, S.; Yakushiji, E.; Nakaya, K.; et al. Dipeptidyl peptidase-4 inhibitors attenuate endothelial function as evaluated by flow-mediated vasodilatation in type 2 diabetic patients. *J. Am. Heart Assoc.* **2013**, *2*, 7–10. [CrossRef] [PubMed]

3. Balakumar, P.; Sharma, R.; Kalia, A.N.; Singh, M. Hyperuricemia: Is it a risk factor for vascular endothelial dysfunction and associated cardiovascular disorders? *Curr. Hypertens. Rev.* **2009**, *5*, 1–6. [CrossRef]

4. Dupont, G.; Croisier, H. Spatiotemporal organization of Ca^{2+} dynamics: A modeling-based approach. *HFSP J.* **2010**, *4*, 43–51. [CrossRef] [PubMed]

5. Moccia, F.; Berra-Romani, R.; Tanzi, F. Update on vascular endothelial Ca^{2+} signalling: A tale of ion channels, pumps and transporters. *World J. Biol. Chem.* **2012**, *3*, 127. [CrossRef] [PubMed]

6. Moccia, F.; Tanzi, F.; Munaron, L. Endothelial Remodelling and Intracellular Calcium Machinery. *Curr. Mol. Med.* **2014**, *14*, 457–480. [CrossRef] [PubMed]

7. Berra-Romani, R.; Avelino-Cruz, J.E.; Raqeeb, A.; Della Corte, A.; Cinelli, M.; Montagnani, S.; Guerra, G.; Moccia, F.; Tanzi, F. Ca^{2+}-dependent nitric oxide release in the injured endothelium of excised rat aorta: A

promising mechanism applying in vascular prosthetic devices in aging patients. *BMC Surg.* **2013**, *13*, S40. [CrossRef] [PubMed]

8.   Potenza, D.M.; Guerra, G.; Avanzato, D.; Poletto, V.; Pareek, S.; Guido, D.; Gallanti, A.; Rosti, V.; Munaron, L.; Tanzi, F.; et al. Hydrogen sulphide triggers VEGF-induced intracellular Ca^{2+} signals in human endothelial cells but not in their immature progenitors. *Cell Calcium* **2014**, *56*, 225–234. [CrossRef] [PubMed]

9.   Behringer, E.J. Calcium and electrical signaling in arterial endothelial tubes: New insights into cellular physiology and cardiovascular function. *Microcirculation* **2017**, *24*, e12328. [CrossRef] [PubMed]

10. Bondarenko, A. Sodium-calcium exchanger contributes to membrane hyperpolarization of intact endothelial cells from rat aorta during acetylcholine stimulation. *Br. J. Pharmacol.* **2004**, *143*, 9–18. [CrossRef] [PubMed]

11. Usachev, Y.M.; Marchenko, S.M.; Sage, S.O. Cytosolic calcium concentration in resting and stimulated endothelium of excised intact rat aorta. *J. Physiol.* **1995**, *489*, 309–317. [CrossRef] [PubMed]

12. Berra-Romani, R.; Raqeeb, A.; Avelino-Cruz, J.E.; Moccia, F.; Oldani, A.; Speroni, F.; Taglietti, V.; Tanzi, F. Ca^{2+} signaling in injured in situ endothelium of rat aorta. *Cell Calcium* **2008**, *44*, 298–309. [CrossRef] [PubMed]

13. Berra-Romani, R.; Raqeeb, A.; Guzman-Silva, A.; Torres-Jácome, J.; Tanzi, F.; Moccia, F. Na^+-Ca^{2+} exchanger contributes to Ca^{2+} extrusion in ATP-stimulated endothelium of intact rat aorta. *Biochem. Biophys. Res. Commun.* **2010**, *395*, 126–130. [CrossRef] [PubMed]

14. Huang, T.-Y.; Chu, T.-F.; Chen, H.-I.; Jen, C.J. Heterogeneity of $[Ca^{2+}]_i$ signaling in intact rat aortic endothelium. *FASEB J.* **2000**, *14*, 797–804. [CrossRef] [PubMed]

15. Berra-Romani, R.; Raqeeb, A.; Torres-Jácome, J.; Guzman-Silva, A.; Guerra, G.; Tanzi, F.; Moccia, F. The mechanism of injury-induced intracellular calcium concentration oscillations in the endothelium of excised rat aorta. *J. Vasc. Res.* **2011**, *49*, 65–76. [CrossRef] [PubMed]

16. Bootman, M.D.; Rietdorf, K.; Collins, T.; Walker, S.; Sanderson, M. Ca^{2+}-Sensitive Fluorescent Dyes and Intracellular Ca^{2+} Imaging. *Cold Spring Harb. Protoc.* **2013**, *2013*. [CrossRef] [PubMed]

17. Yeon, S.I.; Kim, J.Y.; Yeon, D.S.; Abramowitz, J.; Birnbaumer, L.; Muallem, S.; Lee, Y.H. Transient receptor potential canonical type 3 channels control the vascular contractility of mouse mesenteric arteries. *PLoS ONE* **2014**, *9*, 17–19. [CrossRef] [PubMed]

18. Laskey, R.E.; Adams, D.J.; Van Breemen, C. Cytosolic $[Ca^{2+}]$ measurements in endothelium of rabbit cardiac valves using imaging fluorescence microscopy. *Am. J. Physiol. Heart Circ. Physiol.* **1994**, *266*, H2130–H2135. [CrossRef] [PubMed]

19. Čehovin, L.; Leonardis, A.; Kristan, M. Visual object tracking performance measures revisited. *IEEE Trans. Image Process.* **2015**, *25*, 1261–1274. [CrossRef]

20. Huang, C.H.; Boyer, E.; Navab, N.; Ilic, S. Human shape and pose tracking using keyframes. In Proceedings of the IEEE Conference on Computer Vision and Pattern Recognition (CVPR), Columbus, OH, USA, 24–27 June 2014; pp. 3446–3453. [CrossRef]

21. Shantaiya, S.; Verma, K.; Mehta, K. Multiple Object Tracking using Kalman Filter and Optical Flow. *Eur. J. Adv. Eng. Technol.* **2015**, *2*, 34–39.

22. Qiang, Y.; Lee, J.Y.; Bartenschlager, R.; Rohr, K.; Feld, I.N. Colocalization analysis and particle tracking on multi-channel fluorescence microscopy images. In Proceedings of the 2017 IEEE 14th International Symposium on Biomedical Imaging (ISBI 2017), Melbourne, VIC, Australia, 18–21 April 2017; pp. 646–649.

23. Ojha, S.; Sakhare, S. Image processing techniques for object tracking in video surveillance—A survey. In Proceedings of the 2015 International Conference on Pervasive Computing (ICPC), Pune, India, 8–10 January 2015; pp. 1–6. [CrossRef]

24. Berra-Romani, R.; Rinaldi, C.; Raqeeb, A.; Castelli, L.; Magistretti, J.; Taglietti, V.; Tanzi, F. The duration and amplitude of the plateau phase of ATP- and ADP-Evoked Ca^{2+} signals are modulated by ectonucleotidases in in situ endothelial cells of rat aorta. *J. Vasc. Res.* **2004**, *41*, 166–173. [CrossRef] [PubMed]

3

T Cell Calcium Signaling Regulation by the Co-Receptor CD5

Claudia M. Tellez Freitas, Deborah K. Johnson and K. Scott Weber *

Department of Microbiology and Molecular Biology, Brigham Young University, Provo, UT 84604, USA; claudiamsmicrobiology@gmail.com (C.M.T.F.); deborahkj@gmail.com (D.K.J.)
* Correspondence: scott_weber@byu.edu

Abstract: Calcium influx is critical for T cell effector function and fate. T cells are activated when T cell receptors (TCRs) engage peptides presented by antigen-presenting cells (APC), causing an increase of intracellular calcium (Ca^{2+}) concentration. Co-receptors stabilize interactions between the TCR and its ligand, the peptide-major histocompatibility complex (pMHC), and enhance Ca^{2+} signaling and T cell activation. Conversely, some co-receptors can dampen Ca^{2+} signaling and inhibit T cell activation. Immune checkpoint therapies block inhibitory co-receptors, such as cytotoxic T-lymphocyte associated antigen 4 (CTLA-4) and programmed death 1 (PD-1), to increase T cell Ca^{2+} signaling and promote T cell survival. Similar to CTLA-4 and PD-1, the co-receptor CD5 has been known to act as a negative regulator of T cell activation and to alter Ca^{2+} signaling and T cell function. Though much is known about the role of CD5 in B cells, recent research has expanded our understanding of CD5 function in T cells. Here we review these recent findings and discuss how our improved understanding of CD5 Ca^{2+} signaling regulation could be useful for basic and clinical research.

Keywords: calcium signaling; T cell receptor (TCR); co-receptors; CD-5; PD-1; CTL-4

---

1. Introduction

T cells are a critical component of the adaptive immune system. T cell responses are influenced by signals that modulate the effects of the T cell receptor (TCR) and peptide-major histocompatibility complex (pMHC) interaction and initiate the transcription of genes involved in cytokine production, proliferation, and differentiation [1–3]. T cell activation requires multiple signals. First, the TCR engages the pMHC leading to tyrosine phosphorylation of CD3 and initiation of the Ca^{2+}/Calcineurin/Nuclear factor of activated T cells (NFAT) or Protein kinase C-theta (PKCθ)/Nuclear factor-κ-light chain enhancer of activated B cells (NF-κB) or Mitogen-activated protein kinase (MAP kinase)/AP-1 pathways [4–6]. Second, cell surface costimulatory molecules, such as co-receptor CD28, amplify TCR-pMHC complex signals and promote stronger intracellular interactions to prevent T cell anergy [7,8]. Finally, cytokines such as interleukin-12 (IL-12), interferon α (INFα), and interleukin-1 (IL-1) promote T cell proliferation, differentiation, and effector functions [6].

Co-receptors such as CD4 and CD8 interact with MHC molecules and additional co-receptors interact with surface ligands present on antigen-presenting cells (APCs) to regulate T cell homeostasis, survival, and effector functions with stimulatory or inhibitory signals [9]. Altering co-receptor levels, balance, or function dramatically affects immune responses and their dysfunction is implicated in autoimmune diseases [10]. Stimulatory co-receptors such as CD28, inducible T cell co-stimulator (ICOS), Tumor necrosis factor receptor superfamily member 9 (TNFRSF9 or 4-1BB), member of the TNR-superfamily receptor (CD134 or OX40), glucocorticoid-induced tumor necrosis factor (TNF) receptor (GITR), CD137, and CD77 promote T cell activation and protective responses [11]. Co-receptor signaling is initiated by the phosphorylation of tyrosine residues located in immunoreceptor tyrosine-based activation motifs (ITAMs) or immunoreceptor tyrosine-based inhibitory motifs (ITIMs) [7,12]. The phosphorylated tyrosines serve as docking sites for spleen tyrosine kinase (Syk) family members such as zeta-chain-associated protein kinase 10 (ZAP-70) and Syk which activate the phospholipase C γ (PLCγ), RAS, and extracellular signal-regulated kinase (ERK) pathways in addition to mobilizing intracellular Ca^{2+} stores [13].

One of the best described T cell co-receptors, CD28, is a stimulatory T cell surface receptor from the Ig superfamily with a single Ig variable-like domain which binds to B7-1 (CD80) and B7-2 (CD86) [2]. Ligand binding phosphorylates CD28 cytoplasmic domain tyrosine motifs such as YMNM and PYAP and initiates binding and activation of phosphatidylinositide 3 kinase (PI3K) which interacts with protein kinase B (Akt) and promotes T cell proliferation and survival [1]. CD28 also activates the NFAT pathway and mobilizes intracellular Ca^{2+} stores through association with growth factor receptor-bound protein 2 (GRB2) and the production of phosphatidylinositol 4,5-bisphosphate (PIP2), the substrate of PLCγ1, respectively [2,14]. Blocking stimulatory co-receptors suppresses T cell effector function. For example, blocking stimulatory CD28 with anti-CD28 antibodies promotes regulatory T cell function and represses activation of auto- and allo-reactive T effector cells after organ transplantation [8,15].

T cells also have inhibitory co-receptors which regulate T cell responses [8]. The best characterized are immunoglobulin (Ig) superfamily members cytotoxic T-lymphocyte-associated protein 4 (CTLA-4) and programmed cell death protein 1 (PD-1) [8,16]. CTLA-4 binds CD80 and CD86 with greater avidity than CD28, and its inhibitory role refines early phase activation signals for proliferation and cytokine production [16–19]. PD-1, another CD28/B7 family member, regulates late phase effector and memory response [20]. Inhibitory co-receptors such as CTLA-4 and PD-1, known as "immune checkpoints", block the interaction between CD28 and its ligands altering downstream secondary T cell activation signals [19]. Therefore, blocking CTLA-4 or PD-1 promotes effector T cell function in immunosuppressive environments [19,21].

There are also a number of co-receptors that have differential modulatory properties. For example, CD5, a lymphocyte glycoprotein expressed on thymocytes and all mature T cells, has contradictory roles at different time points. CD5 expression is set during thymocyte development and decreases the perceived strength of TCR-pMHC signaling in naïve T cells by clustering at the TCR-pMHC complex and reducing TCR downstream signals such as the Ca^{2+} response when its cytoplasmic pseudo-ITAM domain is phosphorylated [22–25]. The CD5 cytoplasmic domain has four tyrosine residues (Y378, Y429, Y411, and Y463), and residues Y429 and Y441 are found in a YSQP-(x8)-YPAL pseudo ITAM motif while other tyrosine residues make up a pseudo-ITIM domain [23]. Phosphorylated tyrosines recruit several effector molecules and may sequester activation kinases away from the TCR complex, effectively reducing activation signaling strength [23]. Recruited proteins include Src homology-2 protein phosphatase-1 (SHP-1), Ras GTPase protein (rasGAP), CBL, casein kinase II (CK2), zeta-chain-associated protein kinase 70 (ZAP70), and PI3K which are involved in regulating both positive and negative TCR-induced responses [26–28]. For example, ZAP-70 phosphorylates other substrates and eventually recruits effector molecules such as PLC gamma and promotes Ca^{2+} signaling and Ras activation which stimulates the ERK pathway and leads to cellular activation [13,29]. Conversely, SHP1 inhibits Ca^{2+} signaling and PKC activation via decreased tyrosine phosphorylation of PLCγ [13,26,30,31]. Further, Y463 serves as a docking site for c-Cb1, a ubiquitin ligase, which is phosphorylated upon CD3–CD5 ligation and leads to increased ubiquitylation and lysosomal/proteasomal degradation of TCR downstream signaling effectors and CD5 itself [32]. Thus, CD5 has a mix of downstream effects that both promote and inhibit T cell activation. Curiously, recent work suggests that in contrast to its initial inhibitory nature, CD5 also co-stimulates resting and mature T cells by augmenting CD3-mediated signaling [25,33–35].

Ca^{2+} is an important second messenger in many cells types, including lymphocytes, and plays a key role in shaping immune responses. In naïve T cells, intracellular Ca^{2+} is maintained at low levels, but when TCR-pMHC complexes are formed, inositol triphosphate (IP3) initiates Ca^{2+} release from intracellular stores of the endoplasmic reticulum (ER) which opens the Ca^{2+} release-activated Ca^{2+} channels (CRAC) and initiates influx of extracellular Ca^{2+} through store-operated Ca^{2+} entry (SOCE) [36–41]. The resulting elevation of intracellular Ca^{2+} levels activates transcription factors involved in T cell proliferation, differentiation, and cytokine production (e.g., nuclear factor of activated cells (NFAT)) [36,37]. Thus, impaired Ca^{2+} mobilization affects T cell development, activation, differentiation, and function [42,43]. Examples of diseases with impaired Ca^{2+} signaling in T cells include systemic lupus erythematosus, type 1 diabetes mellitus, and others [44,45].

In this review, we will focus on CD5 co-receptor signaling and its functional effects on T cell activation. First, we will discuss how the inhibitory co-receptors CTLA-4 and PD-1 modulate T cell function. Then we will compare CTLA-4 and PD-1 function to CD5 function, examine recent findings that expand our understanding of the role of CD5, and assess how these findings apply to T cell Ca^{2+} signaling. Finally, we will consider CD5 Ca^{2+} signaling regulation in T cells and its potential physiological impact on immunometabolism, cell differentiation, homeostasis, and behavior.

2. Roles of Negative Regulatory T Cell Co-Receptors

2.1. Cytotoxic T-Lymphocyte Antigen-4 (CTLA-4)

Cytotoxic T-lymphocyte antigen-4 (CTLA-4, CD152) inhibits early stages of T cell activation by recruiting inhibitory proteins such as SHP-2 and type II serine/threonine phosphatase PP2A that interfere with T cell synapse signaling [21,46–48]. CTLA-4 binds B7, a protein on activated APCs, with higher affinity than the stimulatory co-receptor CD28; the resulting balance between inhibitory and stimulatory signals controls T cell activation or anergy [19,49]. In naïve T cells, CTLA-4 is located in intracellular vesicles which localize at TCR binding sites following antigen recognition and intracellular Ca^{2+} mobilization [19,50]. Like CD28, CTLA-4 aggregates to the central supramolecular activation complex (cSMAC) where it then extrinsically controls activation by decreasing immunological synapse contact time [51–53]. This suppresses proactivation signals by activating ligands (B7-1 and B7-2) and induces the enzyme Inoleamine 2,3-dioxygenase (IDO) which impairs Ca^{2+} mobilization and suppresses T cell activation, ultimately altering IL-2 production and other effector functions in T cells [51,54,55]. CTLA-4 also stimulates production of regulatory cytokines, such as transforming growth factor beta (TGF-β), which inhibit APC presentation and T cell effector function [47,52,53]. Compared to effector T cells (T_{eff}), CTLA-4 is highly expressed in regulatory T cells (T_{reg}) and plays a role in maintaining T_{reg} homeostasis, proliferation, and immune responses [16,56,57]. Total or partial CTLA-4 deficiency inhibits T_{reg}'s ability to control cytokine production and can cause immune dysregulation [58–61]. Thus, CTLA-4 has an important role in the T_{reg} suppressive response [60]. Additionally, CTLA-4 mutations are associated with autoimmune diseases as thoroughly reviewed by Kristiansen et al. [62].

The loss of CTLA-4 results in removal of CTLA-4 competition with CD28 for B7-1 and B7-2 and is implicated in autoimmunity and cancer [15,63]. Because CTLA-4 inhibits TCR signaling, CTLA-4 deficiency leads to T cell overactivation as measured by increased CD3ζ phosphorylation and Ca^{2+} mobilization [64]. Thus, modulating CTLA-4 signaling is an attractive target for immunotherapies that seek to boost or impair early TCR signaling for cancer and autoinflammatory diseases [65,66]. For example, Ipilimunab, an IgG1 antibody-based melanoma treatment, is a T cell potentiator that blocks CTLA-4 to stimulate T cell proliferation and stem malignant disease progression by delaying tumor progression and has been shown to significantly increase life expectancy [19,67,68]. Additionally, Tremelimumab, a noncomplement fixing IgG2 antibody, has been tested alone or in combination with other antibodies such as Durvalumab (a PD-1 inhibitor) and improves antitumor activity in patients with non-small cell lung cancer (NSCLC), melanoma, colon cancer, gastric cancer, and mesothelioma treatment [69–74].

2.2. Programmed Death 1 (PD-1)

Programmed cell death protein-1 (PD-1, CD279) is a 288-amino acid (50–55 KDa) type I transmembrane protein and a member of the B7/CD28 immunoglobulin superfamily expressed on activated T cells, B cells, and myeloid cells [19,75,76]. PD-1 has two known ligands, PD-L1 and PD-L2, which inhibit T cell activation signals [77]. Like CTLA-4, PD-1 also inhibits T cell proliferation and cytokine production (INF-γ, TNF and IL-2) but is expressed at a later phase of T cell activation [19]. PD-1 has an extracellular single immunoglobulin (Ig) superfamily domain and a cytoplasmic domain containing an ITIM and an immunoreceptor tyrosine-based switch motif (ITSM) subunit critical for PD-1 inhibitory function [78]. Upon T cell activation, PD-1 is upregulated and initiates ITIM and ITSM tyrosine interaction with SHP-2 which mediates TCR signaling inhibition by decreasing ERK phosphorylation and intracellular Ca^{2+} mobilization [79,80]. PD-1 can block the activation signaling pathways PI3K-Akt and Ras-Mek-ERK, which inhibit or regulate T cell activation [79,81]. Thus, engagement of PD-1 by its ligand affects intracellular Ca^{2+} mobilization, IL-2 and TNF-α production, supporting PD-1's inhibitory role in TCR strength-mediated signals [82].

PD-1 signaling also affects regulatory T cell (T_{reg}) homeostasis, expansion, and function [83]. T_{reg} activation and proliferation are impacted by PD-1 expression which enhances their development and function while inhibiting T effector cells [75,84]. PD-1, PD-L, and T_{regs} help terminate immune responses [85]. Thus, PD-1 deficiency results not only in increased T cell activation, but in the breakdown of tolerance and the development of autoimmunity in diseases such as multiple sclerosis and systemic lupus erythematosus [85–89]. PD-1 and its ligands protect tissues from autoimmune attacks by regulating T cell activation and inducing and maintaining peripheral tolerance [90,91]. Studies done in PD-1-deficient mice observed the development of lupus-like glomerulonephritis and arthritis, cardiomyopathy, autoimmune hydronephrosis, and Type I diabetes, among other ailments [92–94]. PD-1 protects against autoimmunity and promotes T_{reg} function. [85]. Enhancing T_{reg} response with a PD-L1 agonist shows therapeutic potential for asthma and other autoimmune disorders [85,95]. Because PD-1 specifically modulates lymphocyte function, effective FDA-approved monoclonal antibodies targeting PD-1 are clinically available (i.e., Pembrolizumab and Nivolumab) to treat advanced malignancies [20]. Not only does blocking PD-1 decrease immunotolerance of tumor cells, it also increases cytotoxic T lymphocyte antitumor activity [20].

3. CD5: A Contradictory Co-Receptor

3.1. Overview of CD5 Signaling and Ca^{2+} Mobilization in T Cells

CD5, known as Ly-1 antigen in mice or as Leu-1 in humans, is a type I transmembrane glycoprotein (67 kDa) expressed on the surface of thymocytes, mature T cells, and a subset of B cells (B-1a) [96,97]. Although CD5 was discovered over 30 years ago, it was only in the last decade that CD5 gained attention as a key T cell activation regulator [98,99]. CD5 expression is set in the thymus during positive selection and correlates with how tightly the thymocyte TCR binds to self-peptide-MHC (self-pMHC); greater TCR affinity for self-peptide leads to increased CD5 expression in double positive (DP) thymocytes [100]. In other words, DP thymocytes that receive strong activation signals through their TCR express more CD5 than those DP thymocytes that receive weak TCR signals [100]. CD5 knockout mice ($CD5^{-/-}$) have a defective negative and positive selection process, and therefore their thymocytes are hyper-responsive to TCR stimulation with increased Ca^{2+} mobilization, proliferation, and cytokine production [23,98]. On the other hand, because of the increased TCR avidity for self-pMHC, mature T cells with high CD5 expression ($CD5^{hi}$) (peripheral or postpositive selection T cells) respond to foreign peptide with increased survival and activation compared to mature T cells with low CD5 expression ($CD5^{lo}$) [34,101]. Therefore, CD5 is a negative regulator of TCR signaling in the thymus and modulates mature T cell response in the periphery [23,34,100,102].

While CTLA-4 and PD-1 belong to the immunoglobulin (Ig) family, CD5 belongs to group B of the scavenger receptor cysteine-rich (SRCR) superfamily and contains three extracellular SRCR domains [30,96,103]. The cytoplasmic tail of CD5 contains several tyrosine residues which mediate the negative regulatory activity independent of extracellular engagement [100,104,105]. As CD5 physically associates with TCRζ/CD3 complex upon TCR and pMHC interaction, the tyrosine residues in both

TCRζ and CD5 are phosphorylated by tyrosine kinases associated with the complex [30,106–110]. This interaction is so intrinsic to T cell signaling that CD5 expression levels are proportional to the degree of TCRζ phosphorylation, IL-2 production capacity, and ERK phosphorylation which are critical for CD3-mediated signaling [33,111]. It is unknown whether posttranslational modifications, such as conserved domain 1 and domain 2 glycosylations, impact CD5 signaling [112,113]. CD5 is present in membrane lipids rafts of mature T cells where, upon activation, it helps augment TCR signaling, increases Ca^{2+} mobilization, and upregulates ZAP-70/LAT (linker for activation of T cells) activation [114–116]. This suggests that CD5 is not only a negative regulator in thymocytes, but also appears to positively influence T cell immune response to foreign antigens [117,118]. See Figure 1.

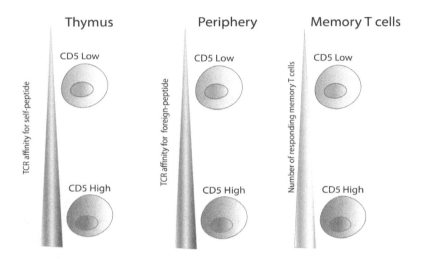

Figure 1. Effects of CD5 on different stages of T cell development. CD5 expression on thymocytes is directly proportional to the signaling intensity of the TCR:self-pMHC interaction. In the periphery, T cells with higher CD5 levels (CD5hi) are better responders to foreign-peptide. Long-lived memory cells populations are enriched for CD5hi T cells [34,102,119].

CD5 has three known ligands: CD72, a glycoprotein expressed by B cells, CD5 ligand or CD5L, an activation antigen expressed on splenocytes, and CD5 itself [120–122]. Crosslinking CD5L to CD5 increases intracellular Ca^{2+} concentrations [30,120,121,123,124]. Early studies with anti-CD5 monoclonal antibodies also demonstrated enhanced Ca^{2+} mobilization and proliferation, suggesting that CD5 co-stimulates and increases the T cell activation signal [125,126]. Following TCR:pMHC interaction, CD5 cytoplasmic ITAM and ITIM like-domains are phosphorylated by p56lck and bound by Src homology 2 (SH2) domain-containing protein tyrosine phosphatase (SHP-1) [108,127,128]. However, while SHP-1 affects Ca^{2+} mobilization and is a purported down-regulator of thymocyte activation, recent findings suggest that SHP-1 is not necessary for CD5 signaling as T cells deficient in SHP-1 have normal CD5 expression and continue to signal normally [26,129]. Thus, while CD5 is not a SHP-1 substrate and SHP-1 is likely unnecessary for CD5 signaling, CD5 signaling results in increased Ca^{2+} mobilization. It has yet to be resolved how CD5 can act as an inhibiting co-receptor in the thymus and as an activating co-receptor in the periphery.

3.2. CD5 as a Ca^{2+} Signaling Modulator

As previously mentioned, CD5 expression levels are set in the thymus during T cell development and are maintained on peripheral lymphocytes [117]. CD5 expression in T cells plays an important role during development and primes naïve T cells for responsiveness in the periphery [35,111,130]. CD5hi T cells have the highest affinity for self-peptides and respond with increased cytokine production and proliferation to infection [101,131,132].

Our laboratory works with two TCR transgenic mouse lines with different levels of CD5 expression: LLO56 (CD5hi) and LLO118 (CD5lo) [111,117,130]. While LLO56 (CD5hi) and LLO118

(CD5^{lo}) have similar affinity for the same immunodominant epitope (listeriolysin O amino acids 190–205 or LLO$_{190-205}$) from *Listeria monocytogenes*, on day 7 of primary response, LLO118 (CD5^{lo}) has approximately three times the number of responding cells compared to LLO56 (CD5^{hi}), and conversely, on day 4 during secondary infection, LLO56 (CD5^{hi}) has approximately fifteen times more cells than LLO118 (CD5^{lo}) [130]. This difference is not due to differential proliferative capacity, rather LLO56 (CD5^{hi}) has higher levels of apoptosis during the primary response [130]. Thus, LLO56 CD5^{hi} and LLO118 CD5^{lo}'s capacity to respond to infection appears to be regulated by their CD5 expression levels [117]. LLO56 (CD5^{hi}) thymocytes have greater affinity for self-peptide, which primes them to be highly apoptotic [130].

Recently we reported that in response to foreign peptide, LLO56 (CD5^{hi}) naïve T cells have higher intracellular Ca^{2+} mobilization than LLO118 (CD5^{lo}), which correlates with increased rate of apoptosis of LLO56 (CD5^{hi}), as Ca^{2+} overloaded mitochondria release cytochrome c which activates caspase and nuclease enzymes, thus initiating the apoptotic pathways [35,133,134]. LLO56 (CD5^{hi}) naïve T cell increased Ca^{2+} mobilization also provides additional support to the idea that CD5^{hi} T cells have an enhanced response to foreign peptide [35,134]. This supports previous research that found that upon T cell activation, increased CD5 expression is correlated with greater basal TCRζ phosphorylation, increased ERK phosphorylation, and more IL-2 production [101,111].

Thus, unlike CTLA-4 and PD-1 which are expressed only on activated T cells in the periphery during early and late phases of immune response, respectively, CD5 is set during T cell development, and influences T cells both during thymic development and during postthymic immune responses [19, 101,111] (see Figure 2). CD5 not only has an important inhibitory role in the thymus, but also appears to positively influence the T cell population response; for example, more CD5^{hi} T cells populate the memory T cell repertoire because CD5^{hi} naïve T cells have a stronger primary response [34,135]. CD5 finetunes the sensitivity of TCR signaling to pMHC, altering intracellular Ca^{2+} mobilization and NFAT transcription, key players in T cell effector function [19,64,126]. As Ca^{2+} signaling plays a key role in T cell activation and function, controlling Ca^{2+} mobilization in T cells through CD5 expression could influence diverse areas of clinical research including metabolism, cancer treatments, and even cognitive behavior.

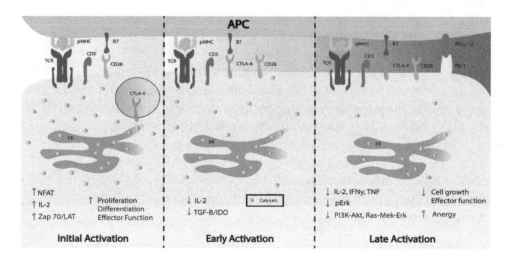

Figure 2. Inhibiting co-receptors modulate T cell activation by increasing (green arrows) or decreasing activity (red arrows). CD5 is present in naïve T cells and localizes to the TCR:pMHC complex during activation. Initial activation cascades signal for the release of CTLA-4 from vesicles to the cell surface while the transcription factor NFAT transcribes PD-1. CTLA-4 provides inhibitory signals during early activation while PD-1 is expressed later and inhibits later stages of T cell activation. The initial Ca^{2+} mobilization is decreased by CTLA-4 and PD-1 downstream signals. A more detailed illustration of the calcium signaling pathway (i.e., IP3, STIM 1/2, CRAC channel, calmodulin, etc.) is outlined in Figure 3.

4. Physiological Impact of CD5 Expression in T Cells

4.1. Metabolism

Naive T cells are in a quiescent state and rely on oxidative phosphorylation (OXPHOS) to generate ATP for survival [136,137]. Upon TCR-pMHC interaction, T cells undergo metabolic reprograming to meet energetic demands by switching from OXPHOS to glycolysis [138]. Glycolysis is a rapid source of ATP and regulates posttranscriptional production of INF-γ, a critical effector cytokine [139]. Following the immune response, most effector T cells undergo apoptosis while a subset become quiescent memory T cells. Memory T cells have lower energetic requirements and rely on OXPHOS and Fatty Acid Oxidation (FAO) to enhance mitochondrial capacity for maintenance and survival [140].

Ca^{2+} signaling is a key second messenger in T cell activation and Ca^{2+} ions also modulate T cell metabolism through CRAC channel activity and NFAT activation [3,141]. During TCR-pMHC binding Ca^{2+} is released from the endoplasmic reticulum (ER) where it is absorbed by the mitochondria and initiates an influx of extracellular Ca^{2+} [3]. First, the rise of cytoplasmic Ca^{2+} activates stromal interaction molecule 1 (STIM1) located on the ER membrane to interact with the CRAC channel located on the cell membrane [142]. The release of the ER store and resulting extracellular Ca^{2+} influx increases the intracellular Ca^{2+} concentration and promotes AMPK (adenosine monophosphates activated protein kinase) expression and CaMKK (calmodulin-dependent protein kinase kinase) activity [3,142,143]. AMPK senses cellular energy levels through the ratio of AMP to ATP and generates ATP by inhibiting ATP-dependent pathways and stimulating catabolic pathways [144]. This indirectly controls T cell fate as AMPK indirectly inhibits mTOR (mammalian target of rapamycin complex) [145]. Because mTOR coordinates the metabolic cues that control T cell homeostasis, it plays a critical role in T cell fate [146]. T cells that are TSC1 (Tuberous sclerosis complex 1)-deficient show metabolic alterations through increased glucose uptake and glycolytic flux [147].

The rise of cytoplasmic Ca^{2+} also encourages mitochondria to uptake cytoplasmic Ca^{2+} through the mitochondrial Ca^{2+} uniporter (MCU) [148]. This MCU uptake increases Ca^{2+} influx by depleting Ca^{2+} near the ER which further activates the CRAC channels and promotes STIM1 oligomerization [3,149–151]. Ca^{2+} uptake in the mitochondria also enhances the function of the tricarboxylic acid cycle (TAC), which generates more ATP through OXPHOS [152,153]. OXPHOS is maintained by a glycolysis product, phosphoenolpyruvate (PEP), which sustains TCR-mediated Ca^{2+}-NFAT signaling by inhibiting the sarcoendoplasmic reticulum (SR) calcium transport ATPase (SERCA) pump, thus promoting T cell effector function [154,155]. Downregulation of calmodulin kinase, CaMKK2, which controls NFAT signaling, decreases glycolytic flux, glucose uptake, and lactate and citrate metabolic processes [156]. Ca^{2+} may also orchestrate the metabolic reprogramming of naïve T cells by promoting glycolysis and OXPHOS through the SOCE/calcineurin pathway which controls the expression of glucose transporters GLUT1/GLUT3 and transcriptional co-regulator proteins important for the expression of electron transport chain complexes required for mitochondria respiration [141].

Co-receptor stimulation plays a pivotal role in T cell metabolism and function. A decrease in T cell Ca^{2+} signaling represses glycolysis and affects T cell effector function [152]. PD-1 and CTLA-4 depress Ca^{2+} signaling and glycolysis while promoting FAO and antibodies against CTLA-4 and PD-1 increase Ca^{2+} mobilization and glycolysis during T cell activation [157,158]. Like CTLA-4 and PD-1, CD5 modulatory function has the potential to influence T cell metabolism. Analysis of gene families modulated by CD5 in B cells found that CD5 upregulates metabolic-related genes including VEFG, Wnt signaling pathways genes, MAPK cascade genes, I-kB/NF-kB cascade genes, TGF β signaling genes, and adipogenesis process genes [159]. Therefore, proliferation differences correlated with CD5 expression in T cells may be caused by improved metabolic function as CD5[lo] T cells seem to be more quiescent than CD5[hi] T cells [160]. Although not much is known about how CD5 alters metabolic function in T cells, signaling strength differences of CD5[hi] and CD5[lo] T cell populations correlate with intracellular Ca^{2+} mobilization during activation and influence their immune response [35,111,130]. This implies that different metabolic processes may be initiated which would influence proliferation, memory cell generation, and cytokine production. Figure 3 summarizes how Ca^{2+} may be mobilized in $CD5^{hi}$ and $CD5^{lo}$ naïve T cells and the role Ca^{2+} may play on metabolism.

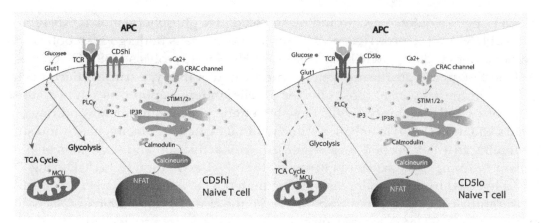

Figure 3. CD5 expression levels in naïve T cells may influence T cell metabolism and function. Differential levels of CD5 result in differences in Ca^{2+} mobilization in naïve T cells. $CD5^{hi}$ naïve T cells have higher Ca^{2+} influx than $CD5^{lo}$ naïve T cells upon TCR:pMHC interaction [35]. Ca^{2+} signaling plays a significant role in T cell activation and influences metabolism and T cell function. Differential Ca^{2+} mobilization and expression of calcineurin and NFAT affect glycolysis and mitochondrial respiration (hypothetical levels of metabolic activation are shown with dashed (low) or solid (high) arrows), suggesting CD5 expression may affect metabolic reprograming during T cell activation [141].

4.2. Neuroimmunology

The field of neuroimmunology examines the interplay between the immune system and the central nervous system (CNS) [161]. The adaptive immune system does influence the CNS as cognition is impaired by the absence of mature T cells [162]. In wild type mice, there is an increase in the number of T cells present in the meninges during the learning process, in stark contrast to mice with T helper 2 cytokine deficiencies (such as IL-4 and IL-13) who have decreased T cell recruitment and impaired learning [163]. Furthermore, regulation of T cell activation and cytokine production critically assists neuronal function and behavior, suggesting that manipulation of T cells could be a potential therapeutic target in treating neuroimmunological diseases [164,165].

T cells go through several microenvironments before reaching the CNS [166]. Many of the signal interactions present in these microenvironments affect T cell function and involve changes in intracellular Ca^{2+} levels [166,167]. In experimental autoimmune encephalitis (EAE), a model for human multiple sclerosis, autoreactive T cells have Ca^{2+} fluctuations throughout their journey to the CNS [166]. Prior to reaching the CNS, T cells interact with splenic stroma cells that do not display the cognate auto-antigen and this interaction produces short-lived low Ca^{2+} mobilization spikes [166]. Following entrance into the CNS, T cells encounter autoantigen-presenting cells and have sustained Ca^{2+} mobilization which results in NFAT translocation and T cell activation [166,168]. EAE mice display reduced social interaction and cognition demonstrating that autoimmune response impairs neuronal function and organismal behavior [169].

Inhibitory T cell co-receptors are implicated in CNS dysregulation and disease. Varicella zoster virus (VZV) infection is characterized by lifelong persistence in neurons. VZV increases the expression of CTLA-4 and PD-1 in infected T cells which reduces IL-2 production and increases T cell anergy [170,171]. PD-1-deficient mice ($Pdcd1^{-/-}$) have increased T cell activation, leading to greater intracellular Ca^{2+} mobilization, and as previously discussed, increased glycolysis [86]. PD-1 deficiency causes elevated concentration of aromatic amino acids in the serum, specifically tryptophan and tyrosine, which decreases their availability in the brain where they are important for the synthesis of neurotransmitters such as dopamine and serotonin; consequently, there is an increase in anxiety-like behavior and fear in Pdcd1$^{-/-}$ mice [86]. Therefore, increased T cell activation caused by PD-1 deficiency can affect brain function and thus, affects cognitive behavior [86].

4.3. Cancer

T cells are critical components of the immune response to cancer. Helper T cells directly activate killer T cells to eradicate tumors and are essential in generating a strong antitumor response alone or in concert with killer T cells by promoting killer T cell activation, infiltration, persistence, and memory formation [172–177]. Tumor-specific T cells may not mount a robust response towards cancerous cells because the tumor microenvironment has numerous immunosuppressive factors; cancerous cells also downregulate cell surface co-stimulatory and MHC proteins which suppresses T cell activation [178–182]. Potent antitumor immune checkpoint blockade therapies using CTLA-4 and PD-1 monoclonal antibodies augment T cell response by suppressing the co-receptors' inhibitory signals, thereby promoting increased Ca^{2+} mobilization, glycolysis, and activation [183,184]. CTLA-4 monoclonal antibodies such as ipilimumab (Yervoy) and tremelimumab block B7-interaction and have been used to treat melanoma [47,185,186]. The monoclonal antibody pembrolizumab is highly selective for PD-1 and prevents PD-1 from engaging PD-L1 and PD-L2, thus enhancing T cell immune response [19,187,188]. Further research will address whether combining anti-CTLA-4 and anti-PD-1 antibodies will improve cancer treatments [19].

As previously mentioned, Ca^{2+} is critical for T cell activation and immune response. Manipulating Ca^{2+} signaling to enhance T cell-directed immune response against cancer is an intriguing notion, yet the means to target the Ca^{2+} response of specific cells without tampering with the metabolic processes of other cells remains elusive [189]. Antitumor activity of tumor-infiltrating lymphocytes (TIL) is inversely related to CD5 expression [99]. CD5 levels in naïve T cells are constantly tuned in the periphery by interactions with self pMHC complexes to maintain homeostasis; therefore, CD5 expression on TILs can be downregulated in response to low affinity for cancer antigens [190–192]. Thus, the majority of TILs are CD5^{lo} which increase their reactivity while CD5^{hi} TILs do not elicit a Ca^{2+} response and become anergic and are unable to eliminate malignant cells [99,192]. While downregulation of CD5 on TILs enhances antitumor T cell activity, CD5^{lo} T cells are also more likely to experience activation-induced cell death (AICD) as CD5 protects T cells from overstimulation [23]. To maximize TIL effectiveness, the inhibitory effects of CD5 could be blocked by neutralizing monoclonal antibodies or soluble CD5-Fc molecules combined with soluble FAS-Fc molecules to reduce the inherent AICD [23,193,194]. Soluble human CD5 (shCD5) may have a similar effect but avoids targeting issues by blocking CD5-mediated interaction via a "decoy receptor" effect. Mice constitutively expressing shCD5 had reduced melanoma and thyoma tumor cell growth and increased numbers of CD4$^+$ and CD8$^+$ T cells [195]. Wild type mice treated with an injection of recombinant shCD5 also had reduced tumor growth [195]. Finally, CD5-deficient mice engrafted with B16-F10 melanoma cells had slower tumor growth compared to wild type C57BL/6 mice [196]. This evidence suggests that CD5, along with PD-1 and CTLA-4, may be a potential target to specifically modulate T cell Ca^{2+} mobilization in an immunosuppressive tumor setting.

4.4. Microbiome

The gut microbiome, including the bacteria and their products, forms a dynamic beneficial symbiosis with the immune system influencing host genes and cellular response. The gut microbiome shapes and directs immune responses while the immune system dictates the bacterial composition of the gut microbiome [197]. As the gut is the major symbiotic system intersecting the immune system and microbiota, understanding their connection has implications for immune system development and function as the gut microbiome is involved in protecting against pathogens, influencing states of inflammation, and even affecting cancer patient outcomes [198,199].

The gut microbiome primes immune responses [200]. Alteration in the microbial composition can induce changes in T cell function in infectious disease, autoimmunity, and cancer [201]. For example, mice treated with antibiotics which restrict or reduce the microbial environment exhibit impaired immune response because their T cells have altered TCR signaling and compromised intracellular Ca^{2+} mobilization in infectious disease and cystic fibrosis models [202–204]. In contrast, administering oral antibiotics to mice with EAE increases the frequency of $CD5^+$ B cell subpopulations in distal lymphoid sites and confers disease protection [205]. In cancer, the microbiome also influences patient response to immune checkpoint inhibitors such as CTLA-4 and PD-1 [206,207]. Mice and melanoma patients immunized or populated with *Bacteriodes fragilis* respond better to treatment with Ipilimumab, a monoclonal antibody against CTLA-4 [198]. Similarly, tumor-specific immunity improved when anti-PD-1/PD-L1 monoclonal antibodies where used in the presence of *Bifidobacterium* [208].

Though little is known about how CD5 influences T cell interaction with the microbiome, some tantalizing details are available. As specific bacterium promotes cancer regression during CTLA-4 and PD-1 checkpoint blockades, a CD5 blockade in conjunction with bacterial selection may also improve immune response. Such studies would lead to novel immunotherapeutic treatments for cancer and autoimmune diseases.

5. Conclusions

CD5, widely known as an inhibitory co-receptor in the thymus, appears to modulate the signaling intensity of peripheral T cells by increasing Ca^{2+} signaling activity and efficacy of $CD5^{hi}$ T cells. CD5 expression levels in the periphery correlates with intracellular Ca^{2+} mobilization, suggesting that CD5 promotes peripheral T cell activation and immune response. As such, CD5 may be a novel checkpoint therapy to regulate T cell activation and metabolism through altering Ca^{2+} mobilization, and could be used to affect neurological behavior, alter microbiome interactions, and treat cancer and autoinflammatory diseases. While this paper focuses on the role of co-receptor CD5 effects on calcium signaling and activation of T cells, CD5 itself may be regulated through posttranslational modifications, such as N-glycosylation, which may affect Ca^{2+} mobilization, T cell metabolism, activation, and function. In the future it would be interesting to determine the role of other posttranslational modifications (e.g., N-glycosylation, S-glutathionylation, lipidation) in CD5 signaling.

Author Contributions: C.M.T.F. is the first author and wrote the manuscript, D.K.J. contributed additional material and editing help, K.S.W. helped with the plan for the manuscript and editing and is the corresponding author.

Acknowledgments: We thank Kiara Vaden Whitley, Jeralyn Jones Fransen, Tyler Cox and Josie Tueller for their critical reviews of this manuscript.

Abbreviations

CTLA-4	Cytotoxic T-lymphocyte antigen 4
CD	Cluster of differenciation
PD-1	Programmed cell death protein 1
AMP	Adenosine monophosphate
ATP	Adenosine triphosphate
CaMKK	Calmodulin-dependent protein kinase kinase
AMPK	AMP-activated protein kinase
SOCE	Store-operated calcium channels
CRAC	Calcium$^+$-release-activated channel
STIM	Stromal interaction molecule
SERCA	Sarcoendoplasmic reticulum calcium transport ATPase
ER	Endoplasmic reticulum
NFAT	Nuclear factor of activated T cells
INF-γ	Interferon gamma

Abbreviations

TNF	Tumor necrosis factor
IL-2	Interleukin 2
GLUT1	Glucose transporter 1
GLUT3	Glucose transporter 3
TIL	Tumor infiltrating lymphocytes
ERK	Extracellular signal-regulated kinases

References

1. Chen, L.; Flies, D.B. Molecular mechanisms of T cell co-stimulation and co-inhibition. *Nat. Rev. Immunol.* **2013**, *13*, 227–242. [CrossRef] [PubMed]
2. Beyersdorf, N.; Kerkau, T.; Hünig, T. CD28 co-stimulation in T cell homeostasis: A recent perspective. *Immunotargets Ther.* **2015**, *4*, 111–122. [PubMed]
3. Fracchia, K.M.; Pai, C.Y.; Walsh, C.M. Modulation of T cell metabolism and function through calcium signaling. *Front. Immunol.* **2013**, *4*, 324. [CrossRef] [PubMed]
4. Cunningham, A.J.; Lafferty, K.J. Letter: Cellular proliferation can be an unreliable index of immune competence. *J. Immunol.* **1974**, *112*, 436–437. [CrossRef] [PubMed]
5. Nakayama, T.; Yamashita, M. The TCR-mediated signaling pathways that control the direction of helper T cell differentiation. *Semin. Immunol.* **2010**, *22*, 303–309. [CrossRef] [PubMed]
6. Goral, S. The three-signal hypothesis of lymphocyte activation/targets for immunosuppression. *Dial. Transplant.* **2011**, *40*, 14–16. [CrossRef]
7. Pennock, N.D.; White, J.T.; Cross, E.W.; Cheney, E.E.; Tamburini, B.A.; Kedl, R.M. T cell responses: Naïve to memory and everything in between. *Adv. Physiol. Educ.* **2013**, *37*, 273–283. [CrossRef] [PubMed]
8. Sharpe, A.H.; Abbas, A.K. T cell costimulation—Biology, therapeutic potential, and challenges. *N. Engl. J. Med.* **2006**, *355*, 973–975. [CrossRef] [PubMed]
9. Artyomov, M.N.; Lis, M.; Devadas, S.; Davis, M.M.; Chakraborty, A.K. CD4 and CD8 binding to MHC molecules primarily acts to enhance LCK delivery. *Proc. Natl. Acad. Sci. USA* **2010**, *107*, 16916–16921. [CrossRef] [PubMed]
10. Ravetch, J.V.; Lanier, L.L. Immune inhibitory receptors. *Science* **2000**, *290*, 84–89. [CrossRef] [PubMed]
11. Mellman, I.; Coukos, G.; Dranoff, G. Cancer immunotherapy comes of age. *Nature* **2011**, *480*, 480. [CrossRef] [PubMed]
12. Fuertes Marraco, S.A.; Neubert, N.J.; Verdeil, G.; Speiser, D.E. Inhibitory receptors beyond T cell exhaustion. *Front. Immunol.* **2015**, *6*, 310. [CrossRef] [PubMed]
13. Barrow, A.D.; Trowsdale, J. You say ITAM and I say ITIM, let's call the whole thing off: The ambiguity of immunoreceptor signalling. *Eur. J. Immunol.* **2006**, *36*, 1646–1653. [CrossRef] [PubMed]
14. Esensten, J.H.; Helou, Y.A.; Chopra, G.; Weiss, A.; Bluestone, J.A. CD28 costimulation: From mechanism to therapy. *Immunity* **2016**, *44*, 973–988. [CrossRef] [PubMed]
15. Dilek, N.; Poirier, N.; Hulin, P.; Coulon, F.; Mary, C.; Ville, S.; Vie, H.; Clémenceau, B.; Blancho, G.; Vanhove, B. Targeting CD28, CTLA-4 and PD-L1 costimulation differentially controls immune synapses and function of human regulatory and conventional t cells. *PLoS ONE* **2013**, *8*, e83139. [CrossRef] [PubMed]
16. Chambers, C.A.; Sullivan, T.J.; Allison, J.P. Lymphoproliferation in CTLA-4-deficient mice is mediated by costimulation-dependent activation of CD4+ T cells. *Immunity* **1997**, *7*, 885–895. [CrossRef]
17. Lindsten, T.; Lee, K.P.; Harris, E.S.; Petryniak, B.; Craighead, N.; Reynolds, P.J.; Lombard, D.B.; Freeman, G.J.; Nadler, L.M.; Gray, G.S.; et al. Characterization of CTLA-4 structure and expression on human T cells. *J. Immunol.* **1993**, *151*, 3489–3499. [PubMed]
18. Boise, L.H.; Minn, A.J.; Noel, P.J.; June, C.H.; Accavitti, M.A.; Lindsten, T.; Thompson, C.B. CD28 costimulation can promote T cell survival by enhancing the expression of Bcl-XL. *Immunity* **1995**, *3*, 87–98. [CrossRef]
19. Buchbinder, E.I.; Desai, A. CTLA-4 and PD-1 pathways: Similarities, differences, and implications of their inhibition. *Am. J. Clin. Oncol.* **2016**, *39*, 98–106. [CrossRef] [PubMed]
20. Iwai, Y.; Hamanishi, J.; Chamoto, K.; Honjo, T. Cancer immunotherapies targeting the PD-1 signaling pathway. *J. Biomed. Sci.* **2017**, *24*, 26. [CrossRef] [PubMed]

21.   Chambers, C.A.; Kuhns, M.S.; Egen, J.G.; Allison, J.P. CTLA-4-mediated inhibition in regulation of T cell responses: Mechanisms and manipulation in tumor immunotherapy. *Annu. Rev. Immunol.* **2001**, *19*, 565–594. [CrossRef] [PubMed]

22.   Brossard, C.; Semichon, M.; Trautmann, A.; Bismuth, G. CD5 inhibits signaling at the immunological synapse without impairing its formation. *J. Immunol.* **2003**, *170*, 4623–4629. [CrossRef] [PubMed]

23.   Tabbekh, M.; Mokrani-Hammani, M.B.; Bismuth, G.; Mami-Chouaib, F. T cell modulatory properties of CD5 and its role in antitumor immune responses. *Oncoimmunology* **2013**, *2*, e22841. [CrossRef] [PubMed]

24.   Mahoney, K.M.; Freeman, G.J.; McDermott, D.F. The next immune-checkpoint inhibitors: PD-1/PD-L1 blockade in melanoma. *Clin. Ther.* **2015**, *37*, 764–782. [CrossRef] [PubMed]

25.   De Wit, J.; Souwer, Y.; van Beelen, A.J.; de Groot, R.; Muller, F.J.; Klaasse Bos, H.; Jorritsma, T.; Kapsenberg, M.L.; de Jong, E.C.; van Ham, S.M. CD5 costimulation induces stable Th17 development by promoting IL-23R expression and sustained STAT3 activation. *Blood* **2011**, *118*, 6107–6114. [CrossRef] [PubMed]

26.   Perez-Villar, J.J.; Whitney, G.S.; Bowen, M.A.; Hewgill, D.H.; Aruffo, A.A.; Kanner, S.B. CD5 negatively regulates the T cell antigen receptor signal transduction pathway: Involvement of SH2-containing phosphotyrosine phosphatase SHP-1. *Mol. Cell. Biol.* **1999**, *19*, 2903–2912. [CrossRef] [PubMed]

27.   Gary-Gouy, H.; Harriague, J.; Dalloul, A.; Donnadieu, E.; Bismuth, G. CD5-negative regulation of B cell receptor signaling pathways originates from tyrosine residue Y429 outside an immunoreceptor tyrosine-based inhibitory motif. *J. Immunol.* **2002**, *168*, 232–239. [CrossRef] [PubMed]

28.   Dennehy, K.M.; Broszeit, R.; Garnett, D.; Durrheim, G.A.; Spruyt, L.L.; Beyers, A.D. Thymocyte activation induces the association of phosphatidylinositol 3-kinase and pp120 with CD5. *Eur. J. Immunol.* **1997**, *27*, 679–686. [CrossRef] [PubMed]

29.   Samelson, L.E. Signal transduction mediated by the T cell antigen receptor: The role of adapter proteins. *Annu. Rev. Immunol.* **2002**, *20*, 371–394. [CrossRef] [PubMed]

30.   Burgess, K.E.; Yamamoto, M.; Prasad, K.V.S.; Rudd, C.E. CD5 acts as a tyrosine kinase substrate within a receptor complex comprising T cell receptor ζ-chain CD3 and protein-tyrosine kinases P56LCK and P59FYN. *Proc. Natl. Acad. Sci. USA* **1992**, *89*, 9311–9315. [CrossRef] [PubMed]

31.   Consuegra-Fernandez, M.; Aranda, F.; Simoes, I.; Orta, M.; Sarukhan, A.; Lozano, F. CD5 as a Target for Immune-Based Therapies. *Crit. Rev. Immunol.* **2015**, *35*, 85–115. [CrossRef] [PubMed]

32.   Roa, N.S.; Ordonez-Rueda, D.; Chavez-Rios, J.R.; Raman, C.; Garcia-Zepeda, E.A.; Lozano, F.; Soldevila, G. The carboxy-terminal region of CD5 is required for c-CBL mediated TCR signaling downmodulation in thymocytes. *Biochem. Biophys. Res. Commun.* **2013**, *432*, 52–59. [CrossRef] [PubMed]

33.   Berney, S.M.; Schaan, T.; Wolf, R.E.; Kimpel, D.L.; van der Heyde, H.; Atkinson, T.P. CD5 (OKT1) augments CD3-mediated intracellular signaling events in human T lymphocytes. *Inflammation* **2001**, *25*, 215–221. [CrossRef] [PubMed]

34.   Azzam, H.S.; DeJarnette, J.B.; Huang, K.; Emmons, R.; Park, C.-S.; Sommers, C.L.; El-Khoury, D.; Shores, E.W.; Love, P.E. Fine tuning of TCR signaling by CD5. *J. Immunol.* **2001**, *166*, 5464–5472. [CrossRef] [PubMed]

35.   Freitas, C.M.T.; Hamblin, G.J.; Raymond, C.M.; Weber, K.S. Naive helper T cells with high CD5 expression have increased calcium signaling. *PLoS ONE* **2017**, *12*, e0178799. [CrossRef] [PubMed]

36.   Feske, S. Calcium signalling in lymphocyte activation and disease. *Nat. Rev. Immunol.* **2007**, *7*, 690–702. [CrossRef] [PubMed]

37.   Joseph, N.; Reicher, B.; Barda-Saad, M. The calcium feedback loop and T cell activation: How cytoskeleton networks control intracellular calcium flux. *Biochim. Biophys. Acta Biomembr.* **2014**, *1838*, 557–568. [CrossRef] [PubMed]

38.   Vig, M.; Kinet, J.-P. Calcium signaling in immune cells. *Nat. Immunol.* **2009**, *10*, 21–27. [CrossRef] [PubMed]

39.   Wolf, I.M.A.; Guse, A.H. Ca^{2+} microdomains in T-lymphocytes. *Front. Oncol.* **2017**, *7*, 73. [CrossRef] [PubMed]

40.   Hogan, P.G.; Lewis, R.S.; Rao, A. Molecular basis of calcium signaling in lymphocytes: STIM and ORAI. *Annu. Rev. Immunol.* **2010**, *28*, 491–533. [CrossRef] [PubMed]

41.   Oh-hora, M.; Rao, A. Calcium signaling in lymphocytes. *Curr. Opin. Immunol.* **2008**, *20*, 250–258. [CrossRef] [PubMed]

42.   Janeway, C.A., Jr. The co-receptor function of CD4. *Semin. Immunol.* **1991**, *3*, 153–160. [PubMed]

43. Moran, A.E.; Hogquist, K.A. T cell receptor affinity in thymic development. *Immunology* **2012**, *135*, 261–267. [CrossRef] [PubMed]

44. Kyttaris, V.C.; Zhang, Z.; Kampagianni, O.; Tsokos, G.C. Calcium signaling in systemic lupus erythematosus T cells: A treatment target. *Arthritis Rheum.* **2011**, *63*, 2058–2066. [CrossRef] [PubMed]

45. Demkow, U.; Winklewski, P.; Ciepiela, O.; Popko, K.; Lipinska, A.; Kucharska, A.; Michalska, B.; Wasik, M. Modulatory effect of insulin on T cell receptor mediated calcium signaling is blunted in long lasting type 1 diabetes mellitus. *Pharmacol. Rep.* **2012**, *64*, 150–156. [CrossRef]

46. Parry, R.V.; Chemnitz, J.M.; Frauwirth, K.A.; Lanfranco, A.R.; Braunstein, I.; Kobayashi, S.V.; Linsley, P.S.; Thompson, C.B.; Riley, J.L. CTLA-4 and PD-1 receptors inhibit T cell activation by distinct mechanisms. *Mol. Cell. Biol.* **2005**, *25*, 9543–9553. [CrossRef] [PubMed]

47. Grosso, J.F.; Jure-Kunkel, M.N. CTLA-4 blockade in tumor models: An overview of preclinical and translational research. *Cancer Immun.* **2013**, *13*, 5. [PubMed]

48. Rudd, C.E.; Taylor, A.; Schneider, H. CD28 and CTLA-4 coreceptor expression and signal transduction. *Immunol. Rev.* **2009**, *229*, 12–26. [CrossRef] [PubMed]

49. Jago, C.B.; Yates, J.; Olsen Saraiva CÂMara, N.; Lechler, R.I.; Lombardi, G. Differential expression of CTLA-4 among T cell subsets. *Clin. Exp. Immunol.* **2004**, *136*, 463–471. [CrossRef] [PubMed]

50. Linsley, P.S.; Bradshaw, J.; Greene, J.; Peach, R.; Bennett, K.L.; Mittler, R.S. Intracellular trafficking of CTLA-4 and focal localization towards sites of TCR engagement. *Immunity* **1996**, *4*, 535–543. [CrossRef]

51. Schneider, H.; Smith, X.; Liu, H.; Bismuth, G.; Rudd, C.E. CTLA-4 disrupts ZAP70 microcluster formation with reduced T cell/APC dwell times and calcium mobilization. *Eur. J. Immunol.* **2008**, *38*, 40–47. [CrossRef] [PubMed]

52. Grohmann, U.; Orabona, C.; Fallarino, F.; Vacca, C.; Calcinaro, F.; Falorni, A.; Candeloro, P.; Belladonna, M.L.; Bianchi, R.; Fioretti, M.C.; et al. CTLA-4-Ig regulates tryptophan catabolism in vivo. *Nat. Immunol.* **2002**, *3*, 1097–1101. [CrossRef] [PubMed]

53. Chen, W.; Jin, W.; Wahl, S.M. Engagement of cytotoxic T lymphocyte-associated antigen 4 (CTLA-4) induces transforming growth factor beta (TGF-β) production by murine CD4$^+$ T cells. *J. Exp. Med.* **1998**, *188*, 1849–1857. [CrossRef] [PubMed]

54. Hryniewicz, A.; Boasso, A.; Edghill-Smith, Y.; Vaccari, M.; Fuchs, D.; Venzon, D.; Nacsa, J.; Betts, M.R.; Tsai, W.-P.; Heraud, J.-M.; et al. CTLA-4 blockade decreases TGF-β, IDO, and viral RNA expression in tissues of SIVmac251-infected macaques. *Blood* **2006**, *108*, 3834–3842. [CrossRef] [PubMed]

55. Iken, K.; Liu, K.; Liu, H.; Bizargity, P.; Wang, L.; Hancock, W.W.; Visner, G.A. Indoleamine 2,3-dioxygenase and metabolites protect murine lung allografts and impair the calcium mobilization of T cells. *Am. J. Respir. Cell Mol. Biol.* **2012**, *47*, 405–416. [CrossRef] [PubMed]

56. Walker, L.S.K.; Sansom, D.M. Confusing signals: Recent progress in CTLA-4 biology. *Trends Immunol.* **2015**, *36*, 63–70. [CrossRef] [PubMed]

57. Cederbom, L.; Hall, H.; Ivars, F. CD4$^+$CD25$^+$ regulatory T cells down-regulate co-stimulatory molecules on antigen-presenting cells. *Eur. J. Immunol.* **2000**, *30*, 1538–1543. [CrossRef]

58. Burnett, D.L.; Parish, I.A.; Masle-Farquhar, E.; Brink, R.; Goodnow, C.C. Murine LRBA deficiency causes CTLA-4 deficiency in Tregs without progression to immune dysregulation. *Immunol. Cell Biol.* **2017**, *95*, 775–778. [CrossRef] [PubMed]

59. Verma, N.; Burns, S.O.; Walker, L.S.K.; Sansom, D.M. Immune deficiency and autoimmunity in patients with CTLA-4 (CD152) mutations. *Clin. Exp. Immunol.* **2017**, *190*, 1–7. [CrossRef] [PubMed]

60. Wing, K.; Onishi, Y.; Prieto-Martin, P.; Yamaguchi, T.; Miyara, M.; Fehervari, Z.; Nomura, T.; Sakaguchi, S. CTLA-4 control over Foxp3$^+$ regulatory T cell function. *Science* **2008**, *322*, 271–275. [CrossRef] [PubMed]

61. Sojka, D.K.; Hughson, A.; Fowell, D.J. CTLA-4 is Required by CD4$^+$CD25$^+$ treg to control CD4$^+$ T cell lymphopenia-induced proliferation. *Eur. J. Immunol.* **2009**, *39*, 1544–1551. [CrossRef] [PubMed]

62. Kristiansen, O.P.; Larsen, Z.M.; Pociot, F. CTLA-4 in autoimmune diseases–a general susceptibility gene to autoimmunity? *Genes Immun.* **2000**, *1*, 170–184. [CrossRef] [PubMed]

63. Chikuma, S. CTLA-4, an essential immune-checkpoint for T cell activation. *Curr. Top. Microbiol. Immunol.* **2017**, *410*, 99–126. [PubMed]

64. Tai, X.; Van Laethem, F.; Pobezinsky, L.; Guinter, T.; Sharrow, S.O.; Adams, A.; Granger, L.; Kruhlak, M.; Lindsten, T.; Thompson, C.B.; et al. Basis of CTLA-4 function in regulatory and conventional CD4$^+$ T cells. *Blood* **2012**, *119*, 5155–5163. [CrossRef] [PubMed]

65. Lo, B.; Abdel-Motal, U.M. Lessons from CTLA-4 deficiency and checkpoint inhibition. *Curr. Opin. Immunol.* **2017**, *49*, 14–19. [CrossRef] [PubMed]

66. Avogadri, F.; Yuan, J.; Yang, A.; Schaer, D.; Wolchok, J.D. Modulation of CTLA-4 and GITR for cancer immunotherapy. *Curr. Top. Microbiol. Immunol.* **2011**, *344*, 211–244. [PubMed]

67. Royal, R.E.; Levy, C.; Turner, K.; Mathur, A.; Hughes, M.; Kammula, U.S.; Sherry, R.M.; Topalian, S.L.; Yang, J.C.; Lowy, I.; et al. Phase 2 trial of single agent Ipilimumab (anti-CTLA-4) for locally advanced or metastatic pancreatic adenocarcinoma. *J. Immunother.* **2010**, *33*, 828–833. [CrossRef] [PubMed]

68. Le, D.T.; Lutz, E.; Uram, J.N.; Sugar, E.A.; Onners, B.; Solt, S.; Zheng, L.; Diaz, L.A., Jr.; Donehower, R.C.; Jaffee, E.M.; et al. Evaluation of ipilimumab in combination with allogeneic pancreatic tumor cells transfected with a GM-CSF gene in previously treated pancreatic cancer. *J. Immunother.* **2013**, *36*, 382–389. [CrossRef] [PubMed]

69. Chung, K.Y.; Gore, I.; Fong, L.; Venook, A.; Beck, S.B.; Dorazio, P.; Criscitiello, P.J.; Healey, D.I.; Huang, B.; Gomez-Navarro, J.; et al. Phase II study of the anti-cytotoxic T-lymphocyte-associated antigen 4 monoclonal antibody, tremelimumab, in patients with refractory metastatic colorectal cancer. *J. Clin. Oncol.* **2010**, *28*, 3485–3490. [CrossRef] [PubMed]

70. Ribas, A.; Camacho, L.H.; Lopez-Berestein, G.; Pavlov, D.; Bulanhagui, C.A.; Millham, R.; Comin-Anduix, B.; Reuben, J.M.; Seja, E.; Parker, C.A.; et al. Antitumor activity in melanoma and anti-self responses in a phase I trial with the anti-cytotoxic T lymphocyte-associated antigen 4 monoclonal antibody CP-675,206. *J. Clin. Oncol.* **2005**, *23*, 8968–8977. [CrossRef] [PubMed]

71. Calabro, L.; Morra, A.; Fonsatti, E.; Cutaia, O.; Fazio, C.; Annesi, D.; Lenoci, M.; Amato, G.; Danielli, R.; Altomonte, M.; et al. Efficacy and safety of an intensified schedule of tremelimumab for chemotherapy-resistant malignant mesothelioma: An open-label, single-arm, phase 2 study. *Lancet Respir. Med.* **2015**, *3*, 301–309. [CrossRef]

72. Comin-Anduix, B.; Escuin-Ordinas, H.; Ibarrondo, F.J. Tremelimumab: Research and clinical development. *OncoTargets Ther.* **2016**, *9*, 1767–1776.

73. Ribas, A.; Comin-Anduix, B.; Chmielowski, B.; Jalil, J.; de la Rocha, P.; McCannel, T.A.; Ochoa, M.T.; Seja, E.; Villanueva, A.; Oseguera, D.K.; et al. Dendritic cell vaccination combined with CTLA4 blockade in patients with metastatic melanoma. *Clin. Cancer Res.* **2009**, *15*, 6267–6276. [CrossRef] [PubMed]

74. Antonia, S.; Goldberg, S.B.; Balmanoukian, A.; Chaft, J.E.; Sanborn, R.E.; Gupta, A.; Narwal, R.; Steele, K.; Gu, Y.; Karakunnel, J.J.; et al. Safety and antitumour activity of durvalumab plus tremelimumab in non-small cell lung cancer: A multicentre, phase 1b study. *Lancet Oncol.* **2016**, *17*, 299–308. [CrossRef]

75. Dong, Y.; Sun, Q.; Zhang, X. PD-1 and its ligands are important immune checkpoints in cancer. *Oncotarget* **2017**, *8*, 2171–2186. [CrossRef] [PubMed]

76. Shi, L.; Chen, S.; Yang, L.; Li, Y. The role of PD-1 and PD-L1 in T cell immune suppression in patients with hematological malignancies. *J. Hematol. Oncol.* **2013**, *6*, 74. [CrossRef] [PubMed]

77. Keir, M.E.; Butte, M.J.; Freeman, G.J.; Sharpe, A.H. PD-1 and its ligands in tolerance and immunity. *Annu. Rev. Immunol.* **2008**, *26*, 677–704. [CrossRef] [PubMed]

78. Okazaki, T.; Wang, J. PD-1/PD-L pathway and autoimmunity. *Autoimmunity* **2005**, *38*, 353–357. [CrossRef] [PubMed]

79. Boussiotis, V.A. Molecular and biochemical aspects of the PD-1 checkpoint pathway. *N. Engl. J. Med.* **2016**, *375*, 1767–1778. [CrossRef] [PubMed]

80. Wang, S.-F.; Fouquet, S.; Chapon, M.; Salmon, H.; Regnier, F.; Labroquère, K.; Badoual, C.; Damotte, D.; Validire, P.; Maubec, E.; et al. Early T cell signalling is reversibly altered in PD-1+ T lymphocytes infiltrating human tumors. *PLoS ONE* **2011**, *6*, e17621. [CrossRef] [PubMed]

81. Gorentla, B.K.; Zhong, X.-P. T cell receptor signal transduction in T lymphocytes. *J. Clin. Cell. Immunol.* **2012**, *2012*, 005.

82. Wei, F.; Zhong, S.; Ma, Z.; Kong, H.; Medvec, A.; Ahmed, R.; Freeman, G.J.; Krogsgaard, M.; Riley, J.L. Strength of PD-1 signaling differentially affects T cell effector functions. *Proc. Natl. Acad. Sci. USA* **2013**, *110*, E2480–E2489. [CrossRef] [PubMed]

83. Cochain, C.; Chaudhari, S.M.; Koch, M.; Wiendl, H.; Eckstein, H.-H.; Zernecke, A. Programmed cell death-1 deficiency exacerbates T cell activation and atherogenesis despite expansion of regulatory T cells in atherosclerosis-prone mice. *PLoS ONE* **2014**, *9*, e93280. [CrossRef] [PubMed]

84. Asano, T.; Kishi, Y.; Meguri, Y.; Yoshioka, T.; Iwamoto, M.; Maeda, Y.; Yagita, H.; Tanimoto, M.; Koreth, J.; Ritz, J.; et al. PD-1 signaling has a critical role in maintaining regulatory T cell homeostasis; implication for treg depletion therapy by PD-1 blockade. *Blood* **2015**, *126*, 848.

85. Francisco, L.M.; Sage, P.T.; Sharpe, A.H. The PD-1 pathway in tolerance and autoimmunity. *Immunol. Rev.* **2010**, *236*, 219–242. [CrossRef] [PubMed]

86. Miyajima, M.; Zhang, B.; Sugiura, Y.; Sonomura, K.; Guerrini, M.M.; Tsutsui, Y.; Maruya, M.; Vogelzang, A.; Chamoto, K.; Honda, K.; et al. Metabolic shift induced by systemic activation of T cells in PD-1-deficient mice perturbs brain monoamines and emotional behavior. *Nat. Immunol.* **2017**, *18*, 1342–1352. [CrossRef] [PubMed]

87. Riella, L.V.; Paterson, A.M.; Sharpe, A.H.; Chandraker, A. Role of the PD-1 pathway in the immune response. *Am. J. Transplant.* **2012**, *12*, 2575–2587. [CrossRef] [PubMed]

88. Kroner, A.; Mehling, M.; Hemmer, B.; Rieckmann, P.; Toyka, K.V.; Maurer, M.; Wiendl, H. A PD-1 polymorphism is associated with disease progression in multiple sclerosis. *Ann. Neurol* **2005**, *58*, 50–57. [CrossRef] [PubMed]

89. Pawlak-Adamska, E.; Nowak, O.; Karabon, L.; Pokryszko-Dragan, A.; Partyka, A.; Tomkiewicz, A.; Ptaszkowski, J.; Frydecka, I.; Podemski, R.; Dybko, J.; et al. *PD-1* gene polymorphic variation is linked with first symptom of disease and severity of relapsing-remitting form of MS. *J. Neuroimmunol.* **2017**, *305*, 115–127. [CrossRef] [PubMed]

90. Dai, S.; Jia, R.; Zhang, X.; Fang, Q.; Huang, L. The PD-1/PD-Ls pathway and autoimmune diseases. *Cell. Immunol.* **2014**, *290*, 72–79. [CrossRef] [PubMed]

91. Gianchecchi, E.; Delfino, D.V.; Fierabracci, A. Recent insights into the role of the PD-1/PD-L1 pathway in immunological tolerance and autoimmunity. *Autoimmun. Rev.* **2013**, *12*, 1091–1100. [CrossRef] [PubMed]

92. Wang, J.; Yoshida, T.; Nakaki, F.; Hiai, H.; Okazaki, T.; Honjo, T. Establishment of NOD-Pdcd1$^{-/-}$ mice as an efficient animal model of type I diabetes. *Proc. Natl. Acad. Sci. USA* **2005**, *102*, 11823–11828. [CrossRef] [PubMed]

93. Okazaki, T.; Otaka, Y.; Wang, J.; Hiai, H.; Takai, T.; Ravetch, J.V.; Honjo, T. Hydronephrosis associated with antiurothelial and antinuclear autoantibodies in BALB/c-Fcgr2b$^{-/-}$Pdcd1$^{-/-}$ mice. *J. Exp. Med.* **2005**, *202*, 1643–1648. [CrossRef] [PubMed]

94. Nishimura, H.; Okazaki, T.; Tanaka, Y.; Nakatani, K.; Hara, M.; Matsumori, A.; Sasayama, S.; Mizoguchi, A.; Hiai, H.; Minato, N.; et al. Autoimmune dilated cardiomyopathy in PD-1 receptor-deficient mice. *Science* **2001**, *291*, 319–322. [CrossRef] [PubMed]

95. Xiao, Y.; Yu, S.; Zhu, B.; Bedoret, D.; Bu, X.; Francisco, L.M.; Hua, P.; Duke-Cohan, J.S.; Umetsu, D.T.; Sharpe, A.H.; et al. RGMb is a novel binding partner for PD-L2 and its engagement with PD-L2 promotes respiratory tolerance. *J. Exp. Med.* **2014**, *211*, 943–959. [CrossRef] [PubMed]

96. Masuda, K.; Kishimoto, T. CD5: A new partner for IL-6. *Immunity* **2016**, *44*, 720–722. [CrossRef] [PubMed]

97. Huang, H.J.; Jones, N.H.; Strominger, J.L.; Herzenberg, L.A. Molecular cloning of Ly-1, a membrane glycoprotein of mouse T lymphocytes and a subset of B cells: Molecular homology to its human counterpart Leu-1/T1 (CD5). *Proc. Natl. Acad. Sci. USA* **1987**, *84*, 204–208. [CrossRef] [PubMed]

98. Tarakhovsky, A.; Kanner, S.B.; Hombach, J.; Ledbetter, J.A.; Muller, W.; Killeen, N.; Rajewsky, K. A role for CD5 in TCR-mediated signal transduction and thymocyte selection. *Science* **1995**, *269*, 535–537. [CrossRef] [PubMed]

99. Dalloul, A. CD5: A safeguard against autoimmunity and a shield for cancer cells. *Autoimmun. Rev.* **2009**, *8*, 349–353. [CrossRef] [PubMed]

100. Bhandoola, A.; Bosselut, R.; Yu, Q.; Cowan, M.L.; Feigenbaum, L.; Love, P.E.; Singer, A. CD5-mediated inhibition of TCR signaling during intrathymic selection and development does not require the CD5 extracellular domain. *Eur. J. Immunol.* **2002**, *32*, 1811–1817. [CrossRef]

101. Mandl, J.N.; Monteiro, J.P.; Vrisekoop, N.; Germain, R.N. T cell positive selection uses self-ligand binding strength to optimize repertoire recognition of foreign antigens. *Immunity* **2013**, *38*, 263–274. [CrossRef] [PubMed]

102. Henderson, J.G.; Opejin, A.; Jones, A.; Gross, C.; Hawiger, D. CD5 instructs extrathymic regulatory T cell development in response to self and tolerizing antigens. *Immunity* **2015**, *42*, 471–483. [CrossRef] [PubMed]

103. Gringhuis, S.I.; de Leij, L.F.; Wayman, G.A.; Tokumitsu, H.; Vellenga, E. The Ca^{2+}/calmodulin-dependent kinase type IV is involved in the CD5-mediated signaling pathway in human T lymphocytes. *J. Biol. Chem.* **1997**, *272*, 31809–31820. [CrossRef] [PubMed]

104. Hassan, N.J.; Simmonds, S.J.; Clarkson, N.G.; Hanrahan, S.; Puklavec, M.J.; Bomb, M.; Barclay, A.N.; Brown, M.H. CD6 regulates T cell responses through activation-dependent recruitment of the positive regulator SLP-76. *Mol. Cell. Biol.* **2006**, *26*, 6727–6738. [CrossRef] [PubMed]

105. Pena-Rossi, C.; Zuckerman, L.A.; Strong, J.; Kwan, J.; Ferris, W.; Chan, S.; Tarakhovsky, A.; Beyers, A.D.; Killeen, N. Negative regulation of CD4 lineage development and responses by CD5. *J. Immunol.* **1999**, *163*, 6494–6501. [PubMed]

106. Davies, A.A.; Ley, S.C.; Crumpton, M.J. CD5 is phosphorylated on tyrosine after stimulation of the T cell antigen receptor complex. *Proc. Natl. Acad. Sci. USA* **1992**, *89*, 6368–6372. [CrossRef] [PubMed]

107. Samelson, L.E.; Phillips, A.F.; Luong, E.T.; Klausner, R.D. Association of the fyn protein-tyrosine kinase with the T cell antigen receptor. *Proc. Natl. Acad. Sci. USA* **1990**, *87*, 4358–4362. [CrossRef] [PubMed]

108. Raab, M.; Yamamoto, M.; Rudd, C.E. The T cell antigen CD5 acts as a receptor and substrate for the protein-tyrosine kinase p56lck. *Mol.Cell. Biol.* **1994**, *14*, 2862–2870. [CrossRef] [PubMed]

109. Beyers, A.D.; Spruyt, L.L.; Williams, A.F. Molecular associations between the T-lymphocyte antigen receptor complex and the surface antigens CD2, CD4, or CD8 and CD5. *Proc. Natl. Acad. Sci. USA* **1992**, *89*, 2945–2949. [CrossRef] [PubMed]

110. Spertini, F.; Stohl, W.; Ramesh, N.; Moody, C.; Geha, R.S. Induction of human T cell proliferation by a monoclonal antibody to CD5. *J. Immunol.* **1991**, *146*, 47–52. [PubMed]

111. Persaud, S.P.; Parker, C.R.; Lo, W.-L.; Weber, K.S.; Allen, P.M. Intrinsic CD4+ T cell sensitivity and response to pathogen are set and sustained by avidity for thymic and peripheral self-pMHC. *Nat. Immunol.* **2014**, *15*, 266–274. [CrossRef] [PubMed]

112. Calvo, J.; Padilla, O.; Places, L.; Vigorito, E.; Vila, J.M.; Vilella, R.; Mila, J.; Vives, J.; Bowen, M.A.; Lozano, F. Relevance of individual CD5 extracellular domains on antibody recognition, glycosylation and co-mitogenic signalling. *Tissue Antigen.* **1999**, *54*, 16–26. [CrossRef]

113. McAlister, M.S.; Brown, M.H.; Willis, A.C.; Rudd, P.M.; Harvey, D.J.; Aplin, R.; Shotton, D.M.; Dwek, R.A.; Barclay, A.N.; Driscoll, P.C. Structural analysis of the CD5 antigen—Expression, disulphide bond analysis and physical characterisation of CD5 scavenger receptor superfamily domain 1. *Eur J. Biochem.* **1998**, *257*, 131–141. [CrossRef] [PubMed]

114. Cho, J.-H.; Kim, H.-O.; Surh, C.D.; Sprent, J. T cell receptor-dependent regulation of lipid rafts controls naive CD8+ T cell homeostasis. *Immunity* **2010**, *32*, 214–226. [CrossRef] [PubMed]

115. Yashiro-Ohtani, Y.; Zhou, X.-Y.; Toyo-oka, K.; Tai, X.-G.; Park, C.-S.; Hamaoka, T.; Abe, R.; Miyake, K.; Fujiwara, H. Non-CD28 costimulatory molecules present in T cell rafts induce T cell costimulation by enhancing the association of TCR with rafts. *J. Immunol.* **2000**, *164*, 1251–1259. [CrossRef] [PubMed]

116. König, R. Chapter 315—Signal Transduction in T Lymphocytes A2—Bradshaw, Ralph A. In *Handbook of Cell Signaling*, 2nd ed.; Dennis, E.A., Ed.; Academic Press: San Diego, CA, USA, 2010; pp. 2679–2688.

117. Milam, A.V.; Allen, P.M. Functional heterogeneity in CD4+ T cell responses against a bacterial pathogen. *Front. Immunol* **2015**, *6*, 621. [CrossRef] [PubMed]

118. Lozano, F.; Simarro, M.; Calvo, J.; Vila, J.M.; Padilla, O.; Bowen, M.A.; Campbell, K.S. CD5 signal transduction: Positive or negative modulation of antigen receptor signaling. *Crit. Rev. Immunol.* **2000**, *20*, 347–358. [CrossRef] [PubMed]

119. Hogquist, K.A.; Jameson, S.C. The self-obsession of T cells: How TCR signaling thresholds affect fate decisions in the thymus and effector function in the periphery. *Nat. Immunol.* **2014**, *15*, 815–823. [CrossRef] [PubMed]

120. Van de Velde, H.; von Hoegen, I.; Luo, W.; Parnes, J.R.; Thielemans, K. The B-cell surface protein CD72/Lyb-2 is the ligand for CD5. *Nature* **1991**, *351*, 662–665. [CrossRef] [PubMed]

121. Biancone, L.; Bowen, M.A.; Lim, A.; Aruffo, A.; Andres, G.; Stamenkovic, I. Identification of a novel inducible cell-surface ligand of CD5 on activated lymphocytes. *J. Exp. Med.* **1996**, *184*, 811–819. [CrossRef] [PubMed]

122. Brown, M.H.; Lacey, E. A ligand for CD5 is CD5. *J. Immunol.* **2010**, *185*, 6068–6074. [CrossRef] [PubMed]

123. Luo, W.; Van de Velde, H.; von Hoegen, I.; Parnes, J.R.; Thielemans, K. Ly-1 (CD5), a membrane glycoprotein of mouse T lymphocytes and a subset of B cells, is a natural ligand of the B cell surface protein Lyb-2 (CD72). *J. Immunol.* **1992**, *148*, 1630–1634. [PubMed]

124. Vandenberghe, P.; Verwilghen, J.; Van Vaeck, F.; Ceuppens, J.L. Ligation of the CD5 or CD28 molecules on resting human T cells induces expression of the early activation antigen CD69 by a calcium- and tyrosine kinase-dependent mechanism. *Immunology* **1993**, *78*, 210–217. [PubMed]

125. Ceuppens, J.L.; Baroja, M.L. Monoclonal antibodies to the CD5 antigen can provide the necessary second signal for activation of isolated resting T cells by solid-phase-bound OKT3. *J. Immunol.* **1986**, *137*, 1816–1821. [PubMed]

126. June, C.H.; Rabinovitch, P.S.; Ledbetter, J.A. CD5 antibodies increase intracellular ionized calcium concentration in T cells. *J. Immunol.* **1987**, *138*, 2782–2792. [PubMed]

127. Reth, M. Antigen receptor tail clue. *Nature* **1989**, *338*, 383–384. [CrossRef] [PubMed]

128. Unkeless, J.C.; Jin, J. Inhibitory receptors, ITIM sequences and phosphatases. *Curr. Opin. Immunol* **1997**, *9*, 338–343. [CrossRef]

129. Dong, B.; Somani, A.K.; Love, P.E.; Zheng, X.; Chen, X.; Zhang, J. CD5-mediated inhibition of TCR signaling proceeds normally in the absence of SHP-1. *Int. J. Mol. Med.* **2016**, *38*, 45–56. [CrossRef] [PubMed]

130. Weber, K.S.; Li, Q.J.; Persaud, S.P.; Campbell, J.D.; Davis, M.M.; Allen, P.M. Distinct CD4$^+$ helper T cells involved in primary and secondary responses to infection. *Proc. Natl. Acad. Sci. USA* **2012**, *109*, 9511–9516. [CrossRef] [PubMed]

131. Fulton, R.B.; Hamilton, S.E.; Xing, Y.; Best, J.A.; Goldrath, A.W.; Hogquist, K.A.; Jameson, S.C. The TCR's sensitivity to self peptide–MHC dictates the ability of naive CD8$^+$ T cells to respond to foreign antigens. *Nat. Immunol.* **2014**, *16*, 107–117. [CrossRef] [PubMed]

132. Palin, A.C.; Love, P.E. CD5 helps aspiring regulatory T cells ward off unwelcome cytokine advances. *Immunity* **2015**, *42*, 395–396. [CrossRef] [PubMed]

133. Mattson, M.P.; Chan, S.L. Calcium orchestrates apoptosis. *Nat. Cell Biol.* **2003**, *5*, 1041–1043. [CrossRef] [PubMed]

134. Orrenius, S.; Nicotera, P. The calcium ion and cell death. *J. Neural Transm. Suppl.* **1994**, *43*, 1–11. [PubMed]

135. Zhao, C.; Davies, J.D. A peripheral CD4$^+$ T cell precursor for naive, memory, and regulatory T cells. *J. Exp. Med.* **2010**, *207*, 2883–2894. [CrossRef] [PubMed]

136. Wahl, D.R.; Byersdorfer, C.A.; Ferrara, J.L.M.; Opipari, A.W.; Glick, G.D. Distinct metabolic programs in activated T cells: Opportunities for selective immunomodulation. *Immunol. Rev.* **2012**, *249*, 104–115. [CrossRef] [PubMed]

137. Pearce, E.L.; Pearce, E.J. Metabolic pathways in immune cell activation and quiescence. *Immunity* **2013**, *38*, 633–643. [CrossRef] [PubMed]

138. Van der Windt, G.J.; Pearce, E.L. Metabolic switching and fuel choice during T cell differentiation and memory development. *Immunol. Rev.* **2012**, *249*, 27–42. [CrossRef] [PubMed]

139. Chang, C.-H.; Curtis, J.D.; Maggi, L.B., Jr.; Faubert, B.; Villarino, A.V.; O'Sullivan, D.; Huang, S.C.-C.; van der Windt, G.J.W.; Blagih, J.; Qiu, J.; et al. Posttranscriptional control of T cell effector function by aerobic glycolysis. *Cell* **2013**, *153*, 1239–1251. [CrossRef] [PubMed]

140. Almeida, L.; Lochner, M.; Berod, L.; Sparwasser, T. Metabolic pathways in T cell activation and lineage differentiation. *Semin. Immunol.* **2016**, *28*, 514–524. [CrossRef] [PubMed]

141. Vaeth, M.; Maus, M.; Klein-Hessling, S.; Freinkman, E.; Yang, J.; Eckstein, M.; Cameron, S.; Turvey, S.E.; Serfling, E.; Berberich-Siebelt, F.; et al. Store-operated Ca^{2+} entry controls clonal expansion of T cells through metabolic reprogramming. *Immunity* **2017**, *47*, 664–679. [CrossRef] [PubMed]

142. Feske, S.; Skolnik, E.Y.; Prakriya, M. Ion channels and transporters in lymphocyte function and immunity. *Nat. Rev. Immunol.* **2012**, *12*, 532–547. [CrossRef] [PubMed]

143. Tamás, P.; Hawley, S.A.; Clarke, R.G.; Mustard, K.J.; Green, K.; Hardie, D.G.; Cantrell, D.A. Regulation of the energy sensor AMP-activated protein kinase by antigen receptor and Ca^{2+} in T lymphocytes. *J. Exp. Med.* **2006**, *203*, 1665–1670. [CrossRef] [PubMed]

144. Ma, E.H.; Poffenberger, M.C.; Wong, A.H.; Jones, R.G. The role of AMPK in T cell metabolism and function. *Curr. Opin. Immunol.* **2017**, *46*, 45–52. [CrossRef] [PubMed]

145. Huang, J.; Manning, B.D. The TSC1–TSC2 complex: A molecular switchboard controlling cell growth. *Biochem. J.* **2008**, *412*, 179–190. [CrossRef] [PubMed]

146. Chi, H. Regulation and function of mTOR signalling in T cell fate decision. *Nat. Rev. Immunol.* **2012**, *12*, 325–338. [CrossRef] [PubMed]

147. MacIver, N.J.; Blagih, J.; Saucillo, D.C.; Tonelli, L.; Griss, T.; Rathmell, J.C.; Jones, R.G. The liver kinase B1 is a central regulator of T cell development, activation, and metabolism. *J. Immunol.* **2011**, *187*, 4187–4198. [CrossRef] [PubMed]

148. Kirichok, Y.; Krapivinsky, G.; Clapham, D.E. The mitochondrial calcium uniporter is a highly selective ion channel. *Nature* **2004**, *427*, 360–364. [CrossRef] [PubMed]

149. Gilabert, J.A.; Bakowski, D.; Parekh, A.B. Energized mitochondria increase the dynamic range over which inositol 1,4,5-trisphosphate activates store-operated calcium influx. *EMBO J.* **2001**, *20*, 2672–2679. [CrossRef] [PubMed]

150. Gilabert, J.A.; Parekh, A.B. Respiring mitochondria determine the pattern of activation and inactivation of the store-operated Ca^{2+} current I (CRAC). *EMBO J.* **2000**, *19*, 6401–6407. [CrossRef] [PubMed]

151. Singaravelu, K.; Nelson, C.; Bakowski, D.; de Brito, O.M.; Ng, S.W.; di Capite, J.; Powell, T.; Scorrano, L.; Parekh, A.B. Mitofusin 2 regulates STIM1 migration from the Ca^{2+} store to the plasma membrane in cells with depolarized mitochondria. *J. Biol. Chem.* **2011**, *286*, 12189–12201. [CrossRef] [PubMed]

152. Dimeloe, S.; Burgener, A.V.; Grahlert, J.; Hess, C. T cell metabolism governing activation, proliferation and differentiation; a modular view. *Immunology* **2017**, *150*, 35–44. [CrossRef] [PubMed]

153. Jouaville, L.S.; Pinton, P.; Bastianutto, C.; Rutter, G.A.; Rizzuto, R. Regulation of mitochondrial ATP synthesis by calcium: Evidence for a long-term metabolic priming. *Proc. Natl. Acad. Sci. USA* **1999**, *96*, 13807–13812. [CrossRef] [PubMed]

154. Ho, P.C.; Bihuniak, J.D.; Macintyre, A.N.; Staron, M.; Liu, X.; Amezquita, R.; Tsui, Y.C.; Cui, G.; Micevic, G.; Perales, J.C.; et al. Phosphoenolpyruvate is a metabolic checkpoint of anti-tumor T cell responses. *Cell* **2015**, *162*, 1217–1228. [CrossRef] [PubMed]

155. Rumi-Masante, J.; Rusinga, F.I.; Lester, T.E.; Dunlap, T.B.; Williams, T.D.; Dunker, A.K.; Weis, D.D.; Creamer, T.P. Structural basis for activation of calcineurin by calmodulin. *J. Mol. Biol.* **2012**, *415*, 307–317. [CrossRef] [PubMed]

156. Racioppi, L.; Means, A.R. Calcium/calmodulin-dependent protein kinase kinase 2: Roles in signaling and pathophysiology. *J. Biol. Chem.* **2012**, *287*, 31658–31665. [CrossRef] [PubMed]

157. Chang, C.-H.; Qiu, J.; O'Sullivan, D.; Buck, M.D.; Noguchi, T.; Curtis, J.D.; Chen, Q.; Gindin, M.; Gubin, M.M.; van der Windt, G.J.W.; et al. Metabolic competition in the tumor microenvironment is a driver of cancer progression. *Cell* **2015**, *162*, 1229–1241. [CrossRef] [PubMed]

158. Patsoukis, N.; Bardhan, K.; Chatterjee, P.; Sari, D.; Liu, B.; Bell, L.N.; Karoly, E.D.; Freeman, G.J.; Petkova, V.; Seth, P.; et al. PD-1 alters T cell metabolic reprogramming by inhibiting glycolysis and promoting lipolysis and fatty acid oxidation. *Nat. Commun.* **2015**, *6*, 6692. [CrossRef] [PubMed]

159. Gary-Gouy, H.; Sainz-Perez, A.; Marteau, J.-B.; Marfaing-Koka, A.; Delic, J.; Merle-Beral, H.; Galanaud, P.; Dalloul, A. Natural phosphorylation of CD5 in chronic lymphocytic leukemia B cells and analysis of CD5-regulated genes in a B cell line suggest a role for CD5 in malignant phenotype. *J. Immunol.* **2007**, *179*, 4335–4344. [CrossRef] [PubMed]

160. Palmer, M.J.; Mahajan, V.S.; Chen, J.; Irvine, D.J.; Lauffenburger, D.A. Signaling thresholds govern heterogeneity in IL-7-receptor-mediated responses of naive $CD8^+$ T cells. *Immunol Cell. Biol.* **2011**, *89*, 581–594. [CrossRef] [PubMed]

161. Kipnis, J.; Gadani, S.; Derecki, N.C. Pro-cognitive properties of T cells. *Nat. Rev. Immunol.* **2012**, *12*, 663–669. [CrossRef] [PubMed]

162. Kipnis, J.; Cohen, H.; Cardon, M.; Ziv, Y.; Schwartz, M. T cell deficiency leads to cognitive dysfunction: Implications for therapeutic vaccination for schizophrenia and other psychiatric conditions. *Proc. Natl. Acad. Sci. USA* **2004**, *101*, 8180–8185. [CrossRef] [PubMed]

163. Brombacher, T.M.; Nono, J.K.; De Gouveia, K.S.; Makena, N.; Darby, M.; Womersley, J.; Tamgue, O.; Brombacher, F. IL-13–mediated regulation of learning and memory. *J. Immunol.* **2017**, *198*, 2681–2688. [CrossRef] [PubMed]

164. Oliveira-dos-Santos, A.J.; Matsumoto, G.; Snow, B.E.; Bai, D.; Houston, F.P.; Whishaw, I.Q.; Mariathasan, S.; Sasaki, T.; Wakeham, A.; Ohashi, P.S.; et al. Regulation of T cell activation, anxiety, and male aggression by RGS2. *Proc. Natl. Acad. Sci. USA* **2000**, *97*, 12272–12277. [CrossRef] [PubMed]

165. Filiano, A.J.; Gadani, S.P.; Kipnis, J. How and why do T cells and their derived cytokines affect the injured and healthy brain? *Nat. Rev. Neurosci.* **2017**, *18*, 375. [CrossRef] [PubMed]

166. Kyratsous, N.I.; Bauer, I.J.; Zhang, G.; Pesic, M.; Bartholomäus, I.; Mues, M.; Fang, P.; Wörner, M.; Everts, S.; Ellwart, J.W.; et al. Visualizing context-dependent calcium signaling in encephalitogenic T cells in vivo by two-photon microscopy. *Proc. Natl. Acad. Sci. USA* **2017**, *114*, E6381–E6389. [CrossRef] [PubMed]

167. Smedler, E.; Uhlén, P. Frequency decoding of calcium oscillations. *Biochim. Biophys. Acta Gen. Subj.* **2014**, *1840*, 964–969. [CrossRef] [PubMed]

168. Pesic, M.; Bartholomaus, I.; Kyratsous, N.I.; Heissmeyer, V.; Wekerle, H.; Kawakami, N. 2-photon imaging of phagocyte-mediated T cell activation in the CNS. *J. Clin. Investig.* **2013**, *123*, 1192–1201. [CrossRef] [PubMed]

169. de Bruin, N.M.W.J.; Schmitz, K.; Schiffmann, S.; Tafferner, N.; Schmidt, M.; Jordan, H.; Häußler, A.; Tegeder, I.; Geisslinger, G.; Parnham, M.J. Multiple rodent models and behavioral measures reveal unexpected responses to FTY720 and DMF in experimental autoimmune encephalomyelitis. *Behav. Brain Res.* **2016**, *300*, 160–174. [CrossRef] [PubMed]

170. Schub, D.; Janssen, E.; Leyking, S.; Sester, U.; Assmann, G.; Hennes, P.; Smola, S.; Vogt, T.; Rohrer, T.; Sester, M.; et al. Altered phenotype and functionality of varicella zoster virus–specific cellular immunity in individuals with active infection. *J. Infect. Dis.* **2015**, *211*, 600–612. [CrossRef] [PubMed]

171. Schub, D.; Fousse, M.; Faßbender, K.; Gärtner, B.C.; Sester, U.; Sester, M.; Schmidt, T. CTLA-4-expression on VZV-specific T cells in CSF and blood is specifically increased in patients with VZV related central nervous system infections. *Eur. J. Immunol.* **2018**, *48*, 151–160. [CrossRef] [PubMed]

172. Koebel, C.M.; Vermi, W.; Swann, J.B.; Zerafa, N.; Rodig, S.J.; Old, L.J.; Smyth, M.J.; Schreiber, R.D. Adaptive immunity maintains occult cancer in an equilibrium state. *Nature* **2007**, *450*, 903–907. [CrossRef] [PubMed]

173. Mattes, J.; Hulett, M.; Xie, W.; Hogan, S.; Rothenberg, M.E.; Foster, P.; Parish, C. Immunotherapy of cytotoxic T cell-resistant tumors by T helper 2 cells: An eotaxin and STAT6-dependent process. *J. Exp. Med.* **2003**, *197*, 387–393. [CrossRef] [PubMed]

174. Hung, K.; Hayashi, R.; Lafond-Walker, A.; Lowenstein, C.; Pardoll, D.; Levitsky, H. The central role of CD4+ T cells in the antitumor immune response. *J. Exp. Med.* **1998**, *188*, 2357–2368. [CrossRef] [PubMed]

175. Scholler, J.; Brady, T.L.; Binder-Scholl, G.; Hwang, W.T.; Plesa, G.; Hege, K.M.; Vogel, A.N.; Kalos, M.; Riley, J.L.; Deeks, S.G.; et al. Decade-long safety and function of retroviral-modified chimeric antigen receptor T cells. *Sci. Transl. Med.* **2012**, *4*, 132ra153. [CrossRef] [PubMed]

176. Ho, Y.C.; Shan, L.; Hosmane, N.N.; Wang, J.; Laskey, S.B.; Rosenbloom, D.I.S.; Lai, J.; Blankson, J.N.; Siliciano, J.D.; Siliciano, R.F. Replication-competent noninduced proviruses in the latent reservoir increase barrier to HIV-1 cure. *Cell* **2013**, *155*, 540–551. [CrossRef] [PubMed]

177. Huetter, G.; Nowak, D.; Mossner, M.; Ganepola, S.; Muessig, A.; Allers, K.; Schneider, T.; Hofmann, J.; Kuecherer, C.; Blau, O.; et al. Long-Term Control of HIV by CCR5 Δ32/Δ32 Stem-Cell Transplantaion. *N. Engl. J. Med.* **2009**, *360*, 692–698. [CrossRef] [PubMed]

178. Ahmadzadeh, M.; Johnson, L.A.; Heemskerk, B.; Wunderlich, J.R.; Dudley, M.E.; White, D.E.; Rosenberg, S.A. Tumor antigen-specific CD8 T cells infiltrating the tumor express high levels of PD-1 and are functionally impaired. *Blood* **2009**, *114*, 1537–1544. [CrossRef] [PubMed]

179. Baitsch, L.; Baumgaertner, P.; Devevre, E.; Raghav, S.K.; Legat, A.; Barba, L.; Wieckowski, S.; Bouzourene, H.; Deplancke, B.; Romero, P.; et al. Exhaustion of tumor-specific CD8+ T cells in metastases from melanoma patients. *J. Clin. Investig.* **2011**, *121*, 2350–2360. [CrossRef] [PubMed]

180. Staveley-O'Carroll, K.; Sotomayor, E.; Montgomery, J.; Borrello, I.; Hwang, L.; Fein, S.; Pardoll, D.; Levitsky, H. Induction of antigen-specific T cell anergy: An early event in the course of tumor progression. *Proc. Natl. Acad. Sci. USA* **1998**, *95*, 1178–1183. [CrossRef] [PubMed]

181. Rosenberg, S.A.; Yang, J.C.; Sherry, R.M.; Kammula, U.S.; Hughes, M.S.; Phan, G.Q.; Citrin, D.E.; Restifo, N.P.; Robbins, P.F.; Wunderlich, J.R.; et al. Durable complete responses in heavily pretreated patients with metastatic melanoma using T cell transfer immunotherapy. *Clin. Cancer Res.* **2011**, *17*, 4550–4557. [CrossRef] [PubMed]

182. Dudley, M.E.; Wunderlich, J.R.; Robbins, P.F.; Yang, J.C.; Hwu, P.; Schwartzentruber, D.J.; Topalian, S.L.; Sherry, R.; Restifo, N.P.; Hubicki, A.M.; et al. Cancer regression and autoimmunity in patients after clonal repopulation with antitumor lymphocytes. *Science* **2002**, *298*, 850–854. [CrossRef] [PubMed]

183. Postow, M.A.; Callahan, M.K.; Wolchok, J.D. Immune checkpoint blockade in cancer therapy. *J. Clin. Oncol.* **2015**, *33*, 1974–1982. [CrossRef] [PubMed]

184. Wei, S.C.; Levine, J.H.; Cogdill, A.P.; Zhao, Y.; Anang, N.-A.A.S.; Andrews, M.C.; Sharma, P.; Wang, J.; Wargo, J.A.; Pe'er, D.; et al. Distinct cellular mechanisms underlie anti-CTLA-4 and anti-PD-1 checkpoint blockade. *Cell* **2017**, *170*, 1120–1133.e17. [CrossRef] [PubMed]

185. Barbee, M.S.; Ogunniyi, A.; Horvat, T.Z.; Dang, T.O. Current status and future directions of the immune checkpoint inhibitors ipilimumab, pembrolizumab, and nivolumab in oncology. *Ann. Pharmacother.* **2015**, *49*, 907–937. [CrossRef] [PubMed]

186. Sangro, B.; Gomez-Martin, C.; de la Mata, M.; Inarrairaegui, M.; Garralda, E.; Barrera, P.; Riezu-Boj, J.I.; Larrea, E.; Alfaro, C.; Sarobe, P.; et al. A clinical trial of CTLA-4 blockade with tremelimumab in patients with hepatocellular carcinoma and chronic hepatitis C. *J. Hepatol.* **2013**, *59*, 81–88. [CrossRef] [PubMed]

187. Reck, M.; Rodríguez-Abreu, D.; Robinson, A.G.; Hui, R.; Csőszi, T.; Fülöp, A.; Gottfried, M.; Peled, N.; Tafreshi, A.; Cuffe, S.; et al. Pembrolizumab versus chemotherapy for PD-L1–positive non–small-cell lung cancer. *N. Engl. J. Med.* **2016**, *375*, 1823–1833. [CrossRef] [PubMed]

188. Hersey, P.; Gowrishankar, K. Pembrolizumab joins the anti-PD-1 armamentarium in the treatment of melanoma. *Future Oncol.* **2015**, *11*, 133–140. [CrossRef] [PubMed]

189. Rooke, R. Can calcium signaling be harnessed for cancer immunotherapy? *Biochim. Biophys. Acta Mol. Cell Res.* **2014**, *1843*, 2334–2340. [CrossRef] [PubMed]

190. Ernst, B.; Lee, D.S.; Chang, J.M.; Sprent, J.; Surh, C.D. The peptide ligands mediating positive selection in the thymus control T cell survival and homeostatic proliferation in the periphery. *Immunity* **1999**, *11*, 173–181. [CrossRef]

191. Smith, K.; Seddon, B.; Purbhoo, M.A.; Zamoyska, R.; Fisher, A.G.; Merkenschlager, M. Sensory adaptation in naive peripheral CD4 T cells. *J. Exp. Med.* **2001**, *194*, 1253–1261. [CrossRef] [PubMed]

192. Dorothée, G.; Vergnon, I.; El Hage, F.; Chansac, B.L.M.; Ferrand, V.; Lécluse, Y.; Opolon, P.; Chouaib, S.; Bismuth, G.; Mami-Chouaib, F. In situ sensory adaptation of tumor-infiltrating T lymphocytes to peptide-MHC levels elicits strong antitumor reactivity. *J. Immunol.* **2005**, *174*, 6888–6897. [CrossRef] [PubMed]

193. Friedlein, G.; El Hage, F.; Vergnon, I.; Richon, C.; Saulnier, P.; Lecluse, Y.; Caignard, A.; Boumsell, L.; Bismuth, G.; Chouaib, S.; et al. Human CD5 protects circulating tumor antigen-specific CTL from tumor-mediated activation-induced cell death. *J. Immunol.* **2007**, *178*, 6821–6827. [CrossRef] [PubMed]

194. Axtell, R.C.; Webb, M.S.; Barnum, S.R.; Raman, C. Cutting edge: Critical role for CD5 in experimental autoimmune encephalomyelitis: Inhibition of engagement reverses disease in mice. *J. Immunol.* **2004**, *173*, 2928–2932. [CrossRef] [PubMed]

195. Simoes, I.T.; Aranda, F.; Carreras, E.; Andres, M.V.; Casado-Llombart, S.; Martinez, V.G.; Lozano, F. Immunomodulatory effects of soluble CD5 on experimental tumor models. *Oncotarget* **2017**, *8*, 108156–108169. [CrossRef] [PubMed]

196. Tabbekh, M.; Franciszkiewicz, K.; Haouas, H.; Lecluse, Y.; Benihoud, K.; Raman, C.; Mami-Chouaib, F. Rescue of tumor-infiltrating lymphocytes from activation-induced cell death enhances the antitumor CTL response in CD5-deficient mice. *J. Immunol.* **2011**, *187*, 102–109. [CrossRef] [PubMed]

197. Round, J.L.; Mazmanian, S.K. The gut microbiome shapes intestinal immune responses during health and disease. *Nat. Rev. Immunol.* **2009**, *9*, 313–323. [CrossRef] [PubMed]

198. Vétizou, M.; Pitt, J.M.; Daillère, R.; Lepage, P.; Waldschmitt, N.; Flament, C.; Rusakiewicz, S.; Routy, B.; Roberti, M.P.; Duong, C.P.M.; et al. Anticancer immunotherapy by CTLA-4 blockade relies on the gut microbiota. *Science* **2015**, *350*, 1079–1084. [CrossRef] [PubMed]

199. Botticelli, A.; Zizzari, I.; Mazzuca, F.; Ascierto, P.A.; Putignani, L.; Marchetti, L.; Napoletano, C.; Nuti, M.; Marchetti, P. Cross-talk between microbiota and immune fitness to steer and control response to anti PD-1/PDL-1 treatment. *Oncotarget* **2017**, *8*, 8890–8899. [CrossRef] [PubMed]

200. Kosiewicz, M.M.; Dryden, G.W.; Chhabra, A.; Alard, P. Relationship between gut microbiota and development of T cell associated disease. *FEBS Lett.* **2014**, *588*, 4195–4206. [CrossRef] [PubMed]

201. Lathrop, S.K.; Bloom, S.M.; Rao, S.M.; Nutsch, K.; Lio, C.-W.; Santacruz, N.; Peterson, D.A.; Stappenbeck, T.S.; Hsieh, C.-S. Peripheral education of the immune system by colonic commensal microbiota. *Nature* **2011**, *478*, 250. [CrossRef] [PubMed]

202. Gonzalez-Perez, G.; Lamousé-Smith, E.S.N. Gastrointestinal microbiome dysbiosis in infant mice alters peripheral CD8$^+$ T cell receptor signaling. *Front. Immunol.* **2017**, *8*, 265. [CrossRef] [PubMed]

203. Huang, T.; Wei, B.; Velazquez, P.; Borneman, J.; Braun, J. Commensal microbiota alter the abundance and TCR responsiveness of splenic naïve CD4$^+$ T lymphocytes. *Clin. Immunol.* **2005**, *117*, 221–230. [CrossRef] [PubMed]

204. Bazett, M.; Bergeron, M.-E.; Haston, C.K. Streptomycin treatment alters the intestinal microbiome, pulmonary T cell profile and airway hyperresponsiveness in a cystic fibrosis mouse model. *Sci. Rep.* **2016**, *6*, 19189. [CrossRef] [PubMed]

205. Ochoa-Repáraz, J.; Mielcarz, D.W.; Haque-Begum, S.; Kasper, L.H. Induction of a regulatory B cell population in experimental allergic encephalomyelitis by alteration of the gut commensal microflora. *Gut Microbes* **2010**, *1*, 103–108. [CrossRef] [PubMed]

206. Allison, J.P.; Krummel, M.F. The Yin and Yang of T cell costimulation. *Science* **1995**, *270*, 932–933. [CrossRef] [PubMed]

207. Allison, J.P. Checkpoints. *Cell* **2015**, *162*, 1202–1205. [CrossRef] [PubMed]

208. Sivan, A.; Corrales, L.; Hubert, N.; Williams, J.B.; Aquino-Michaels, K.; Earley, Z.M.; Benyamin, F.W.; Lei, Y.M.; Jabri, B.; Alegre, M.-L.; et al. Commensal Bifidobacterium promotes antitumor immunity and facilitates anti–PD-L1 efficacy. *Science* **2015**, *350*, 1084–1089. [CrossRef] [PubMed]

4

The Role of Endothelial Ca²⁺ Signaling in Neurovascular Coupling: A View from the Lumen

Germano Guerra [1], Angela Lucariello [2], Angelica Perna [1], Laura Botta [3], Antonio De Luca [2] and Francesco Moccia [3,*]

[1]   Department of Medicine and Health Sciences "Vincenzo Tiberio", University of Molise, via F. De Santis,
      86100 Campobasso, Italy; germano.guerra@unimol.it (G.G.); angelicaperna@gmail.com (A.P.)
[2]   Department of Mental Health and Preventive Medicine, Section of Human Anatomy,
      University of Campania "L. Vanvitelli", 81100 Naples, Italy; angela.lucariello@gmail.com (A.L.);
      antonio.deluca@unicampania.it (A.D.L.)
[3]   Laboratory of General Physiology, Department of Biology and Biotechnology "L. Spallanzani",
      University of Pavia, via Forlanini 6, 27100 Pavia, Italy; laura.botta@unipv.it
*   Correspondence: francesco.moccia@unipv.it

Abstract: Background: Neurovascular coupling (NVC) is the mechanism whereby an increase in neuronal activity (NA) leads to local elevation in cerebral blood flow (CBF) to match the metabolic requirements of firing neurons. Following synaptic activity, an increase in neuronal and/or astrocyte Ca²⁺ concentration leads to the synthesis of multiple vasoactive messengers. Curiously, the role of endothelial Ca²⁺ signaling in NVC has been rather neglected, although endothelial cells are known to control the vascular tone in a Ca²⁺-dependent manner throughout peripheral vasculature. Methods: We analyzed the literature in search of the most recent updates on the potential role of endothelial Ca²⁺ signaling in NVC. Results: We found that several neurotransmitters (i.e., glutamate and acetylcholine) and neuromodulators (e.g., ATP) can induce dilation of cerebral vessels by inducing an increase in endothelial Ca²⁺ concentration. This, in turn, results in nitric oxide or prostaglandin E2 release or activate intermediate and small-conductance Ca²⁺-activated K⁺ channels, which are responsible for endothelial-dependent hyperpolarization (EDH). In addition, brain endothelial cells express multiple transient receptor potential (TRP) channels (i.e., TRPC3, TRPV3, TRPV4, TRPA1), which induce vasodilation by activating EDH. Conclusions: It is possible to conclude that endothelial Ca²⁺ signaling is an emerging pathway in the control of NVC.

Keywords: neurovascular coupling; neuronal activity; brain endothelial cells; Ca²⁺ signaling; glutamate; acetylcholine; ATP; nitric oxide; endothelial-dependent hyperpolarization; TRP channels

1. Introduction

The brain comprises only 2% of the total body mass, yet it accounts for 20% of the overall energy metabolism [1]. As the brain has a limited capacity to store energy and lacks survival mechanisms that render other organs, such as heart and liver, more tolerant to short periods of anoxia or ischemia, the continuous supply of oxygen (O_2) and nutrients and removal of catabolic waste are critical to maintain neuronal integrity and overall brain function. Accordingly, the brain receives up to 20% of cardiac output and consumes ≈20% and ≈25% of total body's O_2 and glucose [2–4]. Brain functions cease within seconds after the interruption of cerebral blood flow (CBF), while irreversible neuronal injury occurs within minutes [2,3]. Cerebral autoregulation is the mechanism whereby CBF remains relatively stable in spite of physiological fluctuations in arterial pressure, at least within a certain range. Thus, cerebral arteries constrict in response to an increase in arterial pressure and relax upon a decrease in blood pressure [4,5]. A subtler mechanism, known as functional hyperemia or neurovascular coupling (NVC), intervenes to locally increase the rate of CBF to active brain areas, thereby ensuring adequate matching between the enhanced metabolic needs of neural cells and blood supply [2,4,6]. Through NVC, a local elevation in neuronal activity (NA) causes a significant vasodilation of neighboring microvessels, which increases CBF and generates the blood oxygenation level-dependent (BOLD) signals that are used to monitor brain function through functional magnetic resonance imaging [7,8]. NVC is finely orchestrated by an intercellular signaling network comprised of neurons, astrocytes and vascular cells (endothelial cells, smooth muscle cells and pericytes), which altogether form the neurovascular unit (NVU) [4,6,9]. Synaptically-activated neurons may signal to adjacent vessels either directly or through the interposition of glial cells. Recent evidence has shown that astrocytes modulate NVC at arteriole levels, whereas they mediate neuronal-to-vascular communication at capillaries [2,4,10–14]. NA controls cerebrovascular tone through a number of Ca^{2+}-dependent vasoactive mediators, which regulate the contractile state of either vascular smooth muscle cells (VSMCs) (arteries and arterioles) or pericytes (capillaries). These include nitric oxide (NO) and the arachidonic acid (AA) derivatives, prostaglandin E2 (PGE2), epoxyeicosatrienoic acids (EETs) and (20-HETE) [2,4,9]. While the role played by astrocytes and mural cells, i.e., VSMCs and pericytes, in CBF regulation has been extensively investigated, the contribution of microvascular endothelial cells to NVC has been largely underestimated [7]. This is quite surprising as the endothelium regulates the vascular tone in both systemic and pulmonary circulation by lining the innermost layer of all blood vessels [15–17]. Endothelial cells decode a multitude of chemical (e.g., transmitters and autacoids) and physical (e.g., shear stress pulsatile stretch) signals by generating spatio-temporally-patterned intracellular Ca^{2+} signals, which control VSMC contractility by inducing the synthesis and release of the vasorelaxing factors, NO, prostacyclin (or PGI2), carbon monoxide and hydrogen sulfide, as well as the vasoconstricting prostaglandin H2 and thromboxane A2 [15,18–21]. Moreover, an increase in sub-membranal Ca^{2+} concentration stimulates intermediate- and small-conductance Ca^{2+}-activated K^+ channels (IK_{Ca} and SK_{Ca}, respectively), thereby causing an abrupt hyperpolarization in endothelial membrane potential, which is electronically transmitted to adjacent VSMCs and causes vessel dilation [16,22,23]. Herein, we aim at surveying the role of endothelial Ca^{2+} signaling in NVC by examining the physiological stimuli, e.g., neurotransmitters, neuromodulators and dietary molecules, that modulate CBF through an increase in intracellular Ca^{2+} concentration ($[Ca^{2+}]_i$) in brain microvascular endothelial cells.

2. Neurovascular Coupling

2.1. Cerebral Circulation and the Neurovascular Unit

The concept of NVU was first introduced during the first Stroke Progress Review Group meeting of the National Institute of Neurological Disorders and Stroke of the NIH (July 2001) to highlight the relevance to CBF regulation of the intimate association between neurons and cerebral vessels [4]. The NVU represents a functional unit where neurons, interneurons and astrocytes are in close proximity and are functionally coupled to smooth muscle cells, pericytes, endothelial cells and extracellular matrix (Figure 1). Within the NVU, neurons, glial and vascular cells establish a mutual influence on each other to ensure a highly efficient system to match blood and nutrient supply to the local needs of brain cells [2,4,6]. The interaction between neurons, glial cells and blood vessels may, however, vary along the cerebrovascular tree. Large cerebral arteries arise from Circle of Willis at the base of the brain and give rise to a heavily-interconnected network of pial arteries and arterioles, which run along the cortical surface. Cerebral vessels are lined by a single layer of endothelial cells (*tunica intima*), which is separated from an intermediate layer of smooth muscle cells (*tunica media*) by the internal elastic lamina. An outermost layer mainly comprised of collagen fibers and fibroblasts and enriched in perivascular nerves, known *as tunica adventitia*, is separated from the brain by the Virchow–Robin space, which is an extension of the sub-arachnoid space [4,6]. Penetrating arterioles branch off smaller pial arteries and dive into the brain tissue, thereby giving rise to parenchymal arterioles. The muscular component of the vascular wall is reduced to a single layer of VSMCs in intracerebral arterioles: of note, endothelial cells may project through the internal elastic lamina by establishing direct connections, known as myoendothelial projections, with the adjoining smooth muscle cells. Myoendothelial projections are enriched with gap junctions and allow the bidirectional transfer of information between endothelial cells and VSMCs [24]. While penetrating arterioles are still separated from the substance of the brain by the Virchow–Robin space, perivascular astrocytic processes (i.e., end feet) enter in contact and fuse together with the basal lamina of parenchymal arterioles, so that the perivascular space is obliterated [4,6]. Relevant to local CBF regulation, pyramidal cells and γ-aminobutyric acid (GABA) interneurons provide extensive innervation to parenchymal arterioles, often with the interposition of glial cells [4,10,25–28]. In addition, sub-cortical neurons from locus coeruleus, raphe nucleus and basal forebrain may also send projection fibers, containing respectively acetylcholine, norepinephrine and 5-hydroxytryptamine (5-HT), to intracortical microvessels and surrounding astrocytes [29]. Finally, parenchymal arterioles supply the cerebral circulation by giving rise to a dense network of intercommunicating capillary vessels, which are composed only of specialized endothelial cells and lack VSMCs. However, ≈30% of the brain capillary surface is covered by spatially-isolated contractile cells, i.e., pericytes, while the remaining endothelial capillary tubes are almost entirely wrapped by astrocytic end feet [30,31]. Astrocytes and pericytes located outside of brain capillaries may be innervated by local neurons, as described for astrocytes and VSMCs situated in the upper districts of the vascular tree [2]. The average capillary density of human brain amounts to ≈400 capillaries mm^{-2}, which is enough to ensure that each neuron is endowed with its own capillary and to reduce on average to less than 15 μM the diffusion distance for O_2, nutrients and catabolic waste [32,33]. Therefore, brain microvascular endothelial cells are ideally positioned to sense neurotransmitters released by axonal terminals during local synaptic activity and by sub-cortical projections provided that they express their specific membrane receptors. The following increase in $[Ca^{2+}]_i$ could, in turn, drive the synthesis of Ca^{2+}-dependent vasoactive mediators. At the same time, brain microvascular endothelial cells have the potential to regulate local CBF independently of NA by detecting changes in blood flow variations and in the concentration of blood-borne agonists through distinct patterns of intracellular Ca^{2+} elevation.

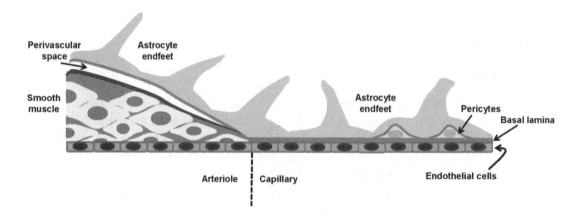

Figure 1. Cellular composition of the neurovascular unit. The vascular wall presents a different structure in arterioles and capillaries, which control the local supply of cerebral blood. Smooth muscle cells form one or more continuous layers around arterioles and change in their contractile state determine vessel diameter and regulate blood perfusion. Capillary diameter is regulated by contractile pericytes, which extend longitudinally and circumferentially along the capillary wall. Astrocyte end feet envelope arterioles and capillaries and are able to release vasoactive mediators, which regulate the contractile state of smooth muscle cells (arterioles) and pericytes (capillary) in response to neuronal activity.

2.2. Cellular and Biochemical Pathways of Neurovascular Coupling: The Role of Neurons and Astrocytes in Arterioles and Capillaries

The mechanisms whereby synaptic activity controls the microvascular tone vary depending on the brain structure and may differ along the vascular tree (Figure 2) [2,4,9]. An increase in intracellular Ca^{2+} concentration ($[Ca^{2+}]_i$) within the dendritic tree is the crucial signal that triggers the synthesis and release of vasoactive messengers in response to excitatory synaptic inputs [2,4,9,34]. For instance, glutamate stimulates post-synaptic N-methyl-D-aspartate (NMDA) and a-amino-3-hydroxy-5-methyl-4-isoxazol epropionic acid (AMPA) receptors to induce extracellular Ca^{2+} entry and recruit the Ca^{2+}/calmodulin (Ca^{2+}/CaM)-dependent neuronal nitric oxide (NO) synthase (nNOS) in hippocampal and cerebellar GABA interneurons [10,25,28,33,35]. The following NO release elicits arteriole vasodilation by inducing VSMC hyperpolarization and relaxation through a soluble guanylate cyclase/protein kinase G (PKG)-dependent mechanism (see below) [13,26,35–37]. Conversely, NMDA receptor (NMDAR)-mediated Ca^{2+} entry in synaptically-activated pyramidal neurons of the somatosensory cortex engages cyclooxygenase 2, which catalyzes the synthesis of the powerful vasodilator, prostaglandin E2 (PGE2), which acts through EP2 and EP4 receptors on VSMCs [27,28]. Intriguingly, NO plays a permissive role in PGE2-dependent vasodilation by maintaining the hemodynamic response during sustained NA [25,38]. Long-lasting synaptic activity could lead to an increase in $[Ca^{2+}]_i$ within perisynaptic astrocytic processes (Figure 2), which lags behind the onset of CBF, but is able to activate phospholipase A2 (PLA2) and cleave AA from the plasma membrane. AA, in turn, diffuses to adjacent VSMCs and is converted into the vasoconstricting messenger, 20-hydroxyeicosatetraenoic (20-HETE), by cytochrome P450 4A (CYP4A) [33,38]. However, neuronal-derived NO inhibits CYP4A, thereby preventing 20-HETE formation and maintaining PGE2-dependent vasodilation [9,33,38]. The mechanism(s) whereby NA induces astrocytic Ca^{2+} signals is still unclear. Glutamate has been predicted to increase astrocyte $[Ca^{2+}]_i$ by binding to metabotropic glutamate receptors (mGluRs) 1 and 5 (mGluR1 and mGluR5, respectively), which stimulate phospholipase Cβ (PLCβ) through Gqα monomer and induce inositol-1,4,5-trisphosphate ($InsP_3$)-dependent Ca^{2+} release from the endoplasmic reticulum (ER) [9,39]. However, mGluR1 and mGluR5 are lacking in adult astrocytes, and the genetic deletion of type 2 $InsP_3$ receptor ($InsP_3R2$), which represents the primary $InsP_3R$ isoform in glial cells, does not prevent NVC [2,30,39]. Nevertheless, there is indisputable evidence that astrocytes require mGluRs to drive the hemodynamic response to sensory stimulation also in adult mice [24,40,41]. Furthermore, alternative mechanisms may drive astrocyte Ca^{2+} signaling, including NMDARs, purinergic receptors and multiple transient receptor potential (TRP) channels [14,39]. We refer the reader to a number of exhaustive and recent reviews about the controversial role of astrocyte $[Ca^{2+}]_i$ in NVC [14,30,39,42]. Intriguingly, astrocytes are able to sense transmural pressure across the vascular wall through vanilloid TRP 4 (TRPV4) channels [43,44], which are located on their perivascular end feet. It has, therefore, been proposed that the initial hemodynamic response to NA activates these mechanosensitive Ca^{2+}-permeable channels, thereby causing an increase in astrocyte $[Ca^{2+}]_i$ and recruiting the Ca^{2+}-dependent PLA2 [14]. In addition, synaptically-released ATP could mobilize ER Ca^{2+} by activating P2Y2 and P2Y4 receptors, which are located on astrocytic end feet wrapped around cerebral vessels [45].

Although it has long been thought that CBF regulation occurs at the arteriole level [6], recent studies have convincingly shown that most of the hydraulic resistance that must be decreased in order to increase cortical perfusion is located in the capillary bed, which are wrapped by contractile pericytes [30,46,47]. This model makes physiological sense as, on average, firing neurons are remarkably closer to capillaries than to arterioles (8–23 μm away versus 70–160 μm) [47]. Therefore, capillaries are located in a much more suitable position to rapidly detect NA and initiate the hemodynamic response that ultimately generates BOLD signals. In addition, the regulation of CBF at the capillary level could selectively increase CBF only in active areas, thereby finely matching the local tissue O_2 supply to local cerebral demand [30,46,47]. The cerebellar cortex is composed of three layers: molecular layer, Purkinje cells and granular layer, respectively, from outermost to innermost [48]. Recent studies demonstrated that, at the molecular layer, synaptic activity caused an increase in astrocytic end feet Ca^{2+} concentration by inducing Ca^{2+} entry through P2X1 channels. This spatially-restricted Ca^{2+} signal, in turn, recruited phospholipase D2 (PLD2) and diacylglycerol lipase to synthesize AA, which was then metabolized by cyclooxygenase 1 into PGE2. Finally, PGE2 evoked capillary dilation by binding to EP4 receptors, which were presumably located on pericytes [13]. Conversely, in the granular layer, synaptic activity induced robust NO release by promoting NMDARs-mediated Ca^{2+} entry in granule cells without astrocyte involvement [11].

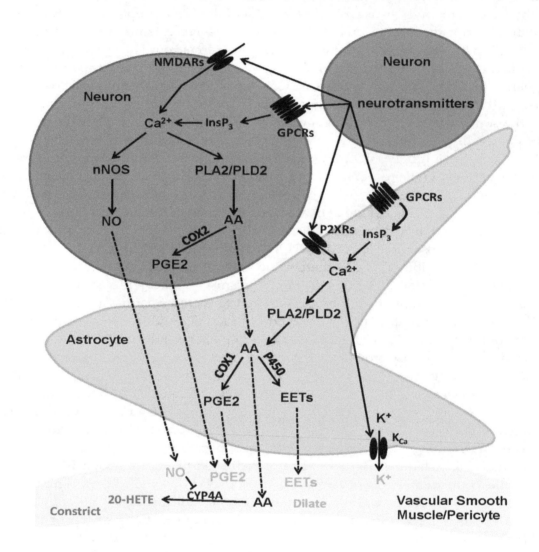

Figure 2. The mechanisms by which neurons and astrocytes stimulate arteriole and capillary dilation in response to synaptic activity. Synaptic activity increases intracellular Ca^{2+} levels within the postsynaptic neuron by stimulating metabotropic G-protein-coupled receptors (GPCRs), ionotropic receptors (e.g., NMDARs) or L-type voltage-operated Ca^{2+} channels (VOCs). This increase in $[Ca^{2+}]_i$ leads to the synthesis of NO and PGE2, which may relax both smooth muscle cells (arterioles) and pericytes (capillaries). Synaptically-released neurotransmitters may also increase $[Ca^{2+}]_i$ in perisynaptic astrocytes, thereby triggering NO release and PGE2/EET production. AA, which may be synthesized by PLA2 in both neurons and astrocytes, may be converted in the vasoconstricting factor, 20-HETE, in perivascular cells. Abbreviations: 20-HETE: 20-hydroxyeicosatetraenoic acid; AA: arachidonic acid; COX1: cyclooxygenase 1; COX2: cyclooxygenase 2; EETs: epoxyeicosatrienoic acids; GPCRs; G-protein-coupled receptors; InsP3: inositol-1,4,5-trisphosphate; K_{Ca}: Ca^{2+}-activated intermediate and small conductance K^+ channels; NO: nitric oxide; nNOS: neuronal NO synthase; P450: cytochrome P450; PGE2: prostaglandin E2; PLA2: phospholipase A2; PLD2: phospholipase D2; VOCs: L-type voltage-operated Ca^{2+} channels.

2.3. Is There a Role for Endothelial Ca^{2+} Signaling in Neurovascular Coupling?

Vascular endothelial cells control vascular tone by releasing a myriad of vasoactive mediators in response to an increase in $[Ca^{2+}]_i$ evoked by either chemical or mechanical inputs [16,18–20,49]. For instance, endothelial Ca^{2+} signals trigger the selective increase in blood flow to skeletal, respiratory and cardiac muscles induced by physical exercise/training [15]. Curiously, it is still unclear whether and how endothelial Ca^{2+} signals play any role in translating NA into vasoactive signals in the brain [7,24]. Luminal perfusion of neurotransmitters and neuromodulators, such as acetylcholine, ATP, ADP and bradykinin, results in the dilation of cerebral arterioles by inducing the activation of specific Gq-protein-coupled receptors (GPCRs), which are located on the endothelial membrane [29,50–53]. GPCRs stimulate PLCβ to synthesize the ER Ca^{2+}-releasing messenger, InsP₃, thereby triggering the Ca^{2+}-mediated signaling cascade that leads to the synthesis and release of most endothelial-derived vasoactive messengers [16,18–20,49,54]. Nevertheless, only scarce information is available about the role of endothelial Ca^{2+} signaling in the hemodynamic response to NA. As described elsewhere [50], synaptically-activated neurons and/or astrocytes could directly stimulate brain microvessels by releasing mediators that traverse the VSMC or pericyte layers and bind to receptors located on the abluminal side of endothelium. Interestingly, brain microvascular endothelial cells may also express NMDARs [55,56] and mGluR1 [57–60], although their potential contribution to NVC has been barely appreciated [4]. Accordingly, recent studies demonstrated that endothelial NMDARs participate in glutamate-induced vasodilation of intraparenchymal arterioles by mediating Ca^{2+} entry and subsequent eNOS activation [55,56]. Finally, brain endothelium could actively mediate the retrograde propagation of the vasodilation signal from the capillaries feeding the sites of NA to the upstream pial vessels (arteries and arterioles) that supply the activated area [6,7,61]. Retrograde vasodilation into the proximal arterial supply is required to achieve an optimally-localized increase in blood flow to active neurons and to avoid a "flow steal" from interconnected vascular networks [4,62]. Retrograde propagation in peripheral vessels is accomplished by endothelial Ca^{2+} signals that impinge on two distinct components to drive the conducted vasomotor response: (1) the Ca^{2+}-dependent rapid activation of EDH, which is restricted to stimulated endothelial cells, but is rapidly transmitted to more remote sites along the vessel and (2) a slower intercellular Ca^{2+} wave that spreads vasodilation via endothelial release of NO and PGE2 [7,63,64]. In the following section, we will discuss the evidence supporting the contribution of endothelial Ca^{2+} signals in NVC and the possibility that interendothelial Ca^{2+} waves are involved in the control of cerebrovessel diameters and microhemodynamics during intense synaptic activity.

3. The Role of Endothelial Ca^{2+} Signaling in Neurovascular Coupling

3.1. Endothelial Ca^{2+} Signaling in Brief

It has long been known that an increase in endothelial $[Ca^{2+}]_i$ delivers the crucial signal to induce the synthesis of multiple vasoactive mediators [16–20]. For instance, intracellular Ca^{2+} signals recruit the Ca^{2+}/CaM-dependent endothelial NOS (eNOS) and the Ca^{2+}-dependent phospholipase A2 (PLA₂) to generate NO and prostaglandin I2 (PGI₂ or prostacyclin), respectively [16,19]. The Ca^{2+} response to extracellular autacoids typically consists of an initial Ca^{2+} peak, which is due to InsP₃-dependent ER Ca^{2+} release, followed by sustained Ca^{2+} entry through store-operated Ca^{2+} channels (Figure 3) [65–67]. Store-operated Ca^{2+} entry (SOCE) is a major Ca^{2+} entry pathway in endothelial cells, being activated by any stimulus leading to the depletion of the ER Ca^{2+} pool [68–70]. The dynamic interplay between InsP₃-dependent Ca^{2+} release, which could be amplified by adjoining ryanodine receptors (RyRs) through the process of Ca^{2+}-induced Ca^{2+} release (CICR) (Figure 3), and SOCE results in biphasic Ca^{2+} signals [71,72] or repetitive $[Ca^{2+}]_i$ oscillations in brain vascular endothelium [73–75].

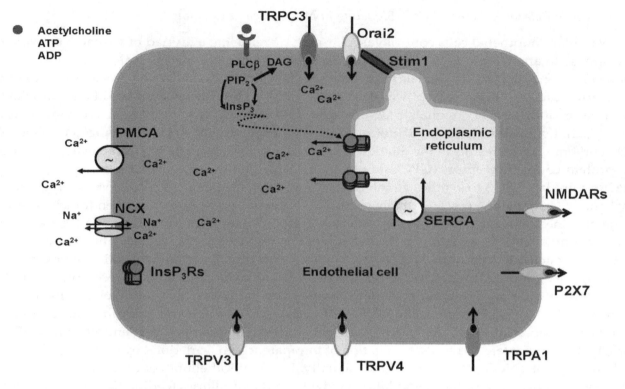

Figure 3. The Ca^{2+} signaling toolkit in brain microvascular endothelial cells. There is scarce information available regarding the molecular components of the Ca^{2+} signaling toolkit in brain microvascular endothelial cells. A recent investigation, however, provided a thorough characterization of the Ca^{2+} machinery in bEND5 cells [76], which represent an established mouse brain microvascular endothelial cell line. Further information was obtained by the analysis of endothelial Ca^{2+} signals in rodent parenchymal arterioles and in the human hCMEC/D3 cell line. Extracellular autacoids bind to specific G-protein-coupled receptors, such as M-AchRs and P2Y1 receptors, thereby activating $PLC\beta$, which in turn cleaves PIP_2 into $InsP_3$ and DAG. $InsP_3$ triggers ER-dependent Ca^{2+} releasing by gating $InsP_3Rs$, while DAG could activate TRPC3. The $InsP_3$-dependent drop in ER Ca^{2+} levels induces SOCE, which is mediated by the interaction between Stim1 and Orai2 in bEND5 cells. Moreover, extracellular Ca^{2+} entry may occur through TRPV3, TRPV4 and TRPA1, which are coupled to either eNOS or EDH [77,78]. Finally, brain microvascular endothelial cells may express Ca^{2+}-permeable ionotropic receptors, such as NMDARs and P2X7 receptors. The elevation in $[Ca^{2+}]_i$ decays to the baseline via the concerted interaction between SERCA and PMCA pumps, as well as through NCX [72,79,80]. Abbreviations: $InsP_3$, inositol-1,4,5-trisphosphate; DAG, diacylglycerol; $InsP_3Rs$, $InsP_3$ receptors; NCX, Na^+–Ca^{2+} exchanger; PMCA, plasma membrane Ca^{2+} ATPase; PIP_2, phosphatidylinositol-4,5-bisphosphate; $PLC\beta$, phospholipase $C\beta$; SERCA, sarco-endoplasmic reticulum Ca^{2+}-ATPase. The thicker line connecting PIP_2 to $InsP_3$ indicates a high amount of second messenger produced upon $PLC\beta$ activation.

3.1.1. The Endothelial Ca^{2+} Toolkit in Brain Microvascular Endothelial Cells: Endogenous Ca^{2+} Release

Scarce information is available regarding the composition of the Ca^{2+} toolkit in brain endothelial cells [74]. For instance, human brain microvascular endothelial cells expressed $InsP_3R1$ and displayed ER-dependent Ca^{2+} release in response to an increase in $InsP_3$ levels [81]. Conversely, only $InsP_3R2$ was expressed in capillary endothelium of rat hippocampus, while RyRs could not be detected in the same study [82]. Nevertheless, RyRs sustained $InsP_3$-dependent Ca^{2+} release in rat brain microvascular endothelial cells in vitro [71], which suggests that the brain endothelial Ca^{2+} toolkit could undergo substantial remodeling in cell cultures [19]. A recent investigation carried out a thorough investigation of the Ca^{2+} toolkit in bEND5 cells [76], a widely-employed mouse brain microvascular endothelial cell line [83]. This analysis revealed that bEND5 cells expressed both

InsP$_3$R1 and InsP$_3$R2, while they lacked InsP$_3$R2 and RyRs [76]. ER-mobilized Ca^{2+} may lead to an increase in mitochondrial Ca^{2+} concentration due to the physical interaction between InsP$_3$Rs and the Ca^{2+}-permeable voltage-dependent anion channels (VDACs), which are embedded in the outer mitochondrial membrane. InsP$_3$Rs-released Ca^{2+} is transferred by VDAC1 into the intermembrane space, from which it is routed towards the mitochondrial matrix though the mitochondrial Ca^{2+} uniporter (MCU) [84,85]. This ER-to-mitochondria Ca^{2+} transfer boosts cellular bioenergetics by stimulating intramitochondrial Ca^{2+}-dependent dehydrogenases, such as oxoglutarate dehydrogenase, NAD-isocitrate dehydrogenase and pyruvate dehydrogenase [86–88]. Subsequently, endothelial mitochondria may contribute to silently (i.e., without a global increase in [Ca^{2+}]$_i$) refill the ER in a sarco-endoplasmic reticulum Ca^{2+}-ATPase (SERCA)-mediated fashion by releasing Ca^{2+} though the mitochondrial Na$^+$/Ca^{2+} exchanger [89,90]. Ca^{2+} entry through the MCU is driven by the negative (i.e., -180 mV) membrane potential ($\Delta\Psi$) that exists across the inner mitochondrial membrane [86]. Mitochondrial content in cerebrovascular endothelium is significantly higher as compared to other vascular districts [91], and InsP$_3$Rs-mediated mitochondrial Ca^{2+} signals arise in rat [92,93] and human [94] brain capillary endothelial cells. Moreover, mitochondrial depolarization hampers the ER-to-mitochondrial Ca^{2+} shuttle, thereby causing a remarkable increase in [Ca^{2+}]$_i$ in rat brain microvascular endothelial cells [95]. An additional mode of Ca^{2+}-mediated cross-talk in endothelial cells may be established between ER and the acidic vesicles of the endolysosomal (EL) Ca^{2+} store [96,97]. The Ca^{2+}-releasing messenger, nicotinic acid adenine dinucleotide phosphate (NAADP), mobilizes EL Ca^{2+} by gating two-pore channels 1 (TPC1) or TPC2 in several types of endothelial cells [54,98–100]. Lysosomal Ca^2 release may be amplified by adjacent ER-embedded InsP$_3$Rs through CICR [97]. The role of lysosomal Ca^{2+} signaling in brain microvascular endothelial cells remains, however, elusive.

3.1.2. The Endothelial Ca^{2+} Toolkit in Brain Microvascular Endothelial Cells: Endogenous Ca^{2+} Release

Likewise, the molecular structure of SOCE in brain vascular endothelium remains to be fully elucidated. Typically, endothelial SOCE is comprised of Stromal interaction molecule 1 and 2 (Stim1 and Stim2, respectively), which sense the drop in ER Ca^{2+} concentration ([Ca^{2+}]$_{ER}$) and Orai1-2 channels, which provide the Ca^{2+}-permeable channel-forming subunits on the plasma membrane [65, 76,101–104]. More specifically, Stim2 is activated by small fluctuations in [Ca^{2+}]$_{ER}$, drives resting Ca^{2+} entry and maintains basal Ca^{2+} levels in endothelial cells [103], while Stim1 is engaged by massive ER Ca^{2+} depletion and sustains agonists-induced extracellular Ca^{2+} entry [65,76,101,102, 104]. However, exceptions to this widespread model may exist. For instance, Stim2 expression is rather modest, and Orai2 represents the only Orai isoform endowed to mouse brain microvascular cells [76]. Of note, Orai2 constitutes the prominent pore-forming subunit of store-operated channels also in mouse neurons [105], which is consistent with the notion that endothelial cells are sensitive to both environmental cues and epigenetic modifications [106]. As widely illustrated in [106,107], the endothelial phenotype is unique in its plasticity and depends on both site-specific signal inputs delivered by the surrounding milieu (which may be diluted in cell culture) and by site-specific epigenetic modifications (i.e., DNA methylation, histone methylation and histone acetylation), which persist under in vitro culture conditions. Similar to other endothelial cells types [108,109], SOCE is constitutively activated to refill the InsP$_3$-dependent ER Ca^{2+} pool and maintains basal Ca^{2+} levels in mouse brain microvascular endothelial cells [76]. AA may induce extracellular Ca^{2+} entry in vascular endothelial cells by prompting Orai1 to interact with Orai3 independently on Stim1 [110]. Nevertheless, Orai3 is not expressed in bEND5 cells, and therefore, this mechanism is unlikely to work in NVC [76]. The endothelial SOCE machinery could involve additional components, such as members of the canonical TRP (TRPC) sub-family of non-selective cation channels, of which seven isoforms exist (TRPC1-7) [68,111]. For instance, Stim1 could also recruit TRPC1 and TRPC4 to assemble into a ternary complex [67,112], whose Na$^+$/Ca^{2+} permeability is determined by Orai1 [113,114]. TRPC1 was expressed in bEND5 cells, while TRPC4 was absent [76]. However, TRPC1 requires Orai1 to be recruited into the SOCE complex upon ER Ca^{2+} depletion [112,115]. It is, therefore, unlikely that it contributes to SOCE in mouse brain microvascular endothelial cells, which lack Orai1 [76], while it may assemble with polycystic TRP2 (TRPP2) to form a stretch-sensitive Ca^{2+}-permeable channel [116]. In addition to SOCE, PLCβ-dependent signaling may lead to the activation of diacylglycerol (DAG)-sensitive Ca^{2+} channels, such as TRPC3 [117,118] and TRPC6 [119,120]. The TRP superfamily of non-selective cation channels consists of 28 members that are classified into six sub-families based on their amino acid sequence homology and structural homology [121,122]. These subfamilies are designated as canonical (TRPC1-7), vanilloid (TRPV1-6), melastatin (TRPM1-8), ankyrin (TRPA1), mucolipin (TRPML1-3) and polycystin (TRPP; TRPP2, TRPP3 and TRPP5) [78,121]. Endothelial cells from different vascular beds may dispose of distinct complements of TRP channels to respond to a myriad of different chemical, mechanical and thermal stimuli [19,78,123,124]. Vascular tone in the brain is specifically regulated by a restricted number of TRP channels, including TRPC3, TRPV3, TRPV4 and TRPA1 (Figure 3), which may stimulate NO release and/or activate IK_{Ca} and SK_{Ca} channels to trigger endothelial-dependent hyperpolarization (EDH) [19,77,78,124]. Endothelial TRPV4 has a potentially relevant role in NVC as it can be activated by AA and its cytochrome P450 epoxygenases-metabolites, i.e., epoxyeicosatrienoic acids (EETs), which evoke vasodilation in intraparenchymal arterioles [9,125]. Finally, human [126], mouse [55,56,127] and rat [128] brain endothelial cells express functional NMDARs (Figure 3) [55], which may mediate glutamate-induced extracellular Ca^{2+} entry. Likewise, P2X7 receptors were recently found in both hCMEC/D3 cells (Figure 3) [129], an immortalized human brain endothelial cell line, and in rat brain endothelial cells in situ [130]. Therefore, the multifaceted endothelial Ca^{2+} toolkit, being located at the interface between neuronal projections and flowing blood, is in an ideal position to trigger and/or modulate NVC by sensing synaptically-released neurotransmitters and blood-borne autacoids.

3.2. Endothelial NMDA Receptors Trigger Glutamate-Induced Nitric Oxide-Mediated Vasodilation

NO represents the major mediator whereby endothelial cells control the vascular tone in large conduit arteries (up to 100%), while its contribution to agonists and/or flow-induced vasodilation decreases (up to 20–50%) as the vascular tree branches into the network of arterioles and capillaries that locally control blood flow [131,132]. NO release is mainly sustained by SOCE rather than by ER-dependent Ca^{2+} release (Figure 4) [133–135], and an oscillatory increase in $[Ca^{2+}]_i$ is the typical waveform that leads to extracellular autacoids-induced NO liberation from vascular endothelial cells [136,137]. In addition to SOCE, TRPC3 [138] and TRPV4 [139] may also evoke extracellular Ca^{2+} entry-mediated eNOS activation, NO release and endothelium-dependent vasodilation. Once released, NO diffuses to adjoining VSMCs to stimulate soluble guanylate cyclase and induce cyclic guanosine-3′,5′-monophosphate (cGMP) production (Figure 4). Then, cGMP activates protein kinase (PKG), which phosphorylates multiple targets to prevent the Ca^{2+}-dependent recruitment of myosin light chain kinase and induce VSMC relaxation. For instance, PKG-dependent phosphorylation inhibits the increase in VSMC $[Ca^{2+}]_i$ induced by L-type voltage-operated Ca^{2+} channels and $InsP_3Rs$; in addition, PKG phosphorylates SERCA, thereby boosting cytosolic Ca^{2+} sequestration into the ER lumen [16,140]. In addition, PKG-dependent phosphorylation stimulates large-conductance Ca^{2+}-activated K^+ channels (BK_{Ca}), thereby inducing VSMC hyperpolarization and vessel dilation [16,140]. Finally, NO accelerates SERCA-mediated Ca^{2+} reuptake through S-nitrosylation of its cysteine thiols [140]. It is generally assumed that neuronal-derived NO induces vasodilation in the hippocampus and cerebellum [10,25,26,28,33,35–37] and plays a permissive role in PGE2-induced vasorelaxation [9,33,38]. However, alternative sources, e.g., eNOS, have been implicated in glutamate-induced NVC [141,142]. Moreover, glutamate evoked endothelium-dependent vasodilation in the presence of D-serine, a NMDAR co-agonist, in isolated middle cerebral arteries and in brain slice parenchymal arterioles [55,143,144]. NMDARs are heterotetramers comprising seven distinct subunits (GluN1, GluN2A–D and GluN3A and B); GluN1 is strictly required for the assembly of a functional channel and for correct trafficking of the other subunits [144–146].

NMDAR activation requires binding of glutamate to GluN1 and of a co-agonist, D-serine or glycine, to GluN2 [145]. Two GluN1 subunits associate with GluN2A and GluN2B to form neuronal NMDARs, which are therefore sensitive to extracellular Mg^{2+} block [145]. As a consequence, NMDAR activation requires simultaneous binding of synaptically-released glutamate and release of Mg^{2+} inhibition by AMPARs-dependent membrane depolarization [145]. However, in non-neuronal cells, GluN1 subunits assemble with GluN2C and GluN2D, which confer a lower sensitivity to Mg^{2+}, while incorporation of GluN3 further decreases the inhibitory effect of extracellular Mg^{2+} and limits Ca^{2+} permeability [144,145]. NMDAR subunits (i.e., GluN1 and GluN2A-D) have been detected in brain endothelial cells in vitro [126–128,143,147,148] and in cerebral cortex in situ [127] (Figure 3). Intriguingly, NMDARs are more abundant on the basolateral endothelial membrane, which place them in the most suitable position to mediate direct neuronal-to-vascular communication [127]. Recently, Anderson's group demonstrated that glutamate and NMDA induced vasodilation in isolated middle cerebral arteries and in brain slices penetrating arterioles only in the presence of the NMDAR co-agonist, D-serine [55]. This feature could explain why previous investigations, which omitted D-serine or glycine from the bathing solution, failed to observe NMDA-induced dilation of cerebral vessels [149]. Subsequently, the same group showed that astrocytic Ca^{2+} signals induced D-serine release, which was in turn able to activate endothelial NMDARs, thereby activating eNOS in a Ca^{2+}-dependent manner [56,127]. Interestingly, NO evoked dilation of cortical penetrating arterioles by suppressing 20-hydroxyeicosatetraenoic acid (20-HETE) synthesis and boosting PGE2-induced vasorelaxation [56,127]. Future work will have to assess whether eNOS is also activated in response to somatosensory stimulation in vivo and to elucidate the physiological transmitter that induces the Ca^{2+} response in astrocytes. However, these findings clearly show that endothelial NMDARs may control CBF by engaging eNOS at the arteriole level.

3.3. Intracellular Ca^{2+} Signals Drive Acetylcholine-Induced Nitric Oxide Release from Brain Microvascular Endothelial Cells

The cortex receives a widespread acetylcholine innervation mainly arising from the basal forebrain nucleus [28,150]. Basal forebrain acetylcholine neurons broadly projects on intraparenchymal arterioles and capillaries and are, therefore, optimally suited to directly control CBF [28,29,150]. Accordingly, electrical stimulation of basal forebrain neurons results in dilation of intracortical arterioles and increases local CBF [53,151,152]. Acetylcholine-induced cerebral vasodilation is mediated by Gq-coupled muscarinic M5 receptors (M5-AchRs) [53,153], which activate eNOS and induce NO release from cerebrovascular endothelium, in mouse and pig [53,150]. Acetylcholine evokes NO release by triggering repetitive $[Ca^{2+}]_i$ oscillations in several types of endothelial cells [137,154,155]. A recent study focused on bEND5 cells to unravel how acetylcholine induces Ca^{2+}-dependent NO release from cerebrovascular endothelium [76,156,157]. Acetylcholine triggered intracellular oscillations in $[Ca^{2+}]_i$, which were driven by rhythmical InsP3-dependent ER Ca^{2+} release and maintained by SOCE (Figure 3) [76]. The Ca^{2+} response to acetylcholine was mediated by M3-AchRs, which represent another PLCβ-coupled M-AchR isoform. Conversely, nicotine did not cause any detectable increase in $[Ca^{2+}]_i$ [76], as also observed in other endothelial cell types [158]. Acetylcholine-induced Ca^{2+} oscillations led to robust NO release, with eNOS requiring both intracellular Ca^{2+} release and extracellular Ca^{2+} entry to be fully activated [76]. These data were partially confirmed in hCMEC/D3 cells, which displayed a biphasic increase in $[Ca^{2+}]_i$ in response to acetylcholine [76]; of note, human brain microvascular endothelial cells express M5-AchRs [52]. Our preliminary data suggest that the distinct waveforms of acetylcholine-induced Ca^{2+} signals in human vs. mouse brain endothelial cells reflect crucial differences in their Ca^{2+} toolkit. Accordingly, hCMEC/D3 only express InsP3R3, while they lack InsP3R1 and InsP3R2 [159]). InsP3R2, which shows the sharpest dependence on ambient Ca^{2+} and is the most sensitive InsP3R isoform to InsP3, has long been known as the main oscillatory Ca^{2+} unit [160,161]. Conversely, InsP3R3, which is not inhibited by surrounding Ca^{2+}, tends to suppress intracellular Ca^{2+} oscillations [160,161]. Therefore, endothelial Ca^{2+} signaling can be truly regarded as a crucial determinant for acetylcholine-induced NVC.

3.4. Endothelial Ca^{2+} Signals Could Mediate ATP-Induced Vasodilation

Following synaptic activity, neurons and astrocytes release ATP, which serves as a modulator of cellular excitability, synaptic strength and plasticity [162]. ATP is rapidly (200 ms) hydrolyzed by extracellular ectonucleotidases into ADP and adenosine [162], all these mediators being able to physiologically increase CBF [52,163]. Luminal application of ATP and ADP has long been known to induce endothelium-dependent vasodilation in isolated cerebral vessels. For instance, ATP and ADP promoted vasodilation in rat middle cerebral arteries by stimulating NO release, although ATP could also act through cytochrome P450-metabolites [51]. Of note, extraluminal administration of ATP evoked a biphasic vasomotor response in rat penetrating arterioles, consisting of an initial transient vasoconstriction followed by local vasodilation, which was subsequently propagated to upstream locations ($\approx$500 μm) along the vascular wall [164,165]. ATP-induced local vasoconstriction was mediated by ionotropic P2X receptors on VSMCs, while endothelial P2Y1 receptors triggered local vasodilation [164]. ATP induced vasodilation by stimulating NO release and stimulating EET production to activate EDH (see below); EETs were also responsible for the upstream propagation of the vasomotor response [164]. These findings were supported by a recent study, showing that the hemodynamic response to whisker stimulation in the mouse somatosensory cortex required P2Y1 receptor-dependent eNOS activation. Previous administration of fluoroacetate, a glial-specific metabolic toxin, prevented eNOS-dependent functional hyperemia, thereby suggesting that ATP was mainly released by perivascular astrocytes [166]. P2Y1 receptors are GPCRs, which bind to both ATP and ADP and control the vascular tone by inducing intracellular Ca^{2+} signals in endothelial cells through distinct signaling pathways [167–169]. For instance, P2Y1 receptors induced $InsP_3$-dependent ER Ca^{2+} release in rat cardiac microvascular endothelial cells [167], whereas they stimulated cyclic nucleotide-gated channels in the H5V endothelial cell line, in primary cultures of bovine aortic endothelial cells and in mouse aorta endothelial cells in situ [169]. Moreover, ATP induced local and conducted vasodilation in hamster cheek pouch arterioles by triggering an increase in endothelial $[Ca^{2+}]_i$ both locally and $\approx$1200 μm upstream along the same vessel (see also below) [170]. Therefore, it is conceivable that endothelial Ca^{2+} signals mediate P2Y1 receptor-dependent NO release and functional hyperemia in the somatosensory cortex.

3.5. TRP Channels Trigger Endothelial-Dependent Hyperpolarization in Cerebrovascular Endothelial Cells

EDH provides the largest contribution to endothelial vasorelaxing mechanisms in resistance-sized arteries and arterioles, as shown in coronary, renal and mesenteric circulation [22,23,131,171]. EDH is initiated by an increase in endothelial $[Ca^{2+}]_i$, which stimulates IK_{Ca} ($KC_{a3.1}$) and SK_{Ca} ($KC_{a2.3}$) channels to hyperpolarize the endothelial cell membrane (Figure 4). Hyperpolarizing current spreads from vascular endothelium to overlying smooth muscle cells to trigger VSMC relaxation and vessel dilation by inhibiting voltage-dependent Ca^{2+} entry (Figure 4) [22,23,77]. This vasorelaxing mechanism requires a tightly-regulated disposition of the Ca^{2+} sources and the Ca^{2+}-sensitive decoders, i.e., SK_{Ca} and IK_{Ca} channels, which effect membrane hyperpolarization. To achieve such a precise spatial arrangement, endothelial cells extend cellular protrusions through the internal elastic lamina, which establish a heterocellular coupling with adjacent smooth muscle cells through connexin-based myo-endothelial gap-junctions (MEGJs) [22,23,77]. The endothelial ER also protrudes into these myo-endothelial microdomain sites and forms spatially-discrete $InsP_3$-sensitive Ca^{2+} pools, which are juxtaposed with IK_{Ca} channels, which are located on the endothelial protrusions traversing the holes in the vascular wall [23,172]. Conversely, SK_{Ca} channels are distributed throughout the endothelial cell membrane, although they are enriched at MEGJs and may, therefore, sense ER-released Ca^{2+} [173]. $InsP_3Rs$ are constitutively activated to produce repetitive, spatially-restricted Ca^{2+} release events, termed Ca^{2+} pulsars, whose frequency can be increased by extracellular vasoactive agonists, such as acetylcholine. Spontaneous $InsP_3$-driven Ca^{2+} pulsars are selectively coupled to IK_{Ca} and SK_{Ca} channels, thereby hyperpolarizing the endothelial membrane potential at myo-endothelial projections and inducing dilation in adjoining VSMCs [172]. K^+ signaling in the myo-endothelial space could then be boosted by the stimulation of endothelial inward rectifying K^+ (K_{ir}) channels or of Na^+/K^+ ATPase in VSMCs. In addition to $InsP_3$-induced ER-dependent Ca^{2+} release, IK_{Ca} and SK_{Ca} channels can be activated by extracellular Ca^{2+} entry through TRPV4 channels, which is also largely expressed at MEGJs [174]. EDH does not only evoke local vasodilation at the site of endothelial stimulation; the hyperpolarizing current spreads along vascular endothelium to upstream arteries (up to $\approx$2 mm) and drives the retrograde vasodilation, which is ultimately responsible for the drop in vascular resistance that increases blood supply to active regions [175–177]. Preliminary data revealed that the same clustered architecture of TRPV4 and IK_{Ca} channels is maintained at MEGJs of rodent cerebral arteries [24,78,178–180], which suggests that discrete $InsP_3Rs$-mediated Ca^{2+} pulsars arise also in parenchymal vessels [24]. Moreover, cerebral MEGJs are enriched with TRPA1 channels, which may also support EDH in cortical circulation [178,181]. EDH mediates ATP-, UTP- and SLIGR (a selective agonist of protease-activated receptor 2)-induced vasodilation in rat middle cerebral arteries [164,180,182–185] and acetylcholine-induced vasodilation in mouse posterior cerebral arteries [186].

3.5.1. TRPV4

TRPV4 is emerging as a major regulator of CBF [24,39,163] and represents one of the most important Ca^{2+}-entry pathways in vascular endothelial cells [111,187–191]. TRPV4 channels are polymodal non-selective cation channels that mediate Ca^{2+} entry in response to distinct chemical, thermal and mechanical stimuli [54,125,187,191–194], thereby integrating the diverse surrounding cues acting on the vascular wall. For instance, TRPV4 may be activated by EETs, which are produced via AA epoxygenation by cytochrome P450 epoxygenase enzymes. Briefly, an increase in endothelial $[Ca^{2+}]_i$ induced by either mechanical or chemical stimuli may stimulate the Ca^{2+}-dependent phospholipase A2 (PLA2), which cleaves AA from membrane phospholipids [77,125]. AA is, turn, is converted into EETs, such as 5,6-EET, 8,9-EET, 11,12-EET and 14,15-EET, by cytochrome P450 2C9 or cytochrome P450 2J2 [125]. In particular, 5,6-EET, 8,9-EET and 11,12-EET were shown to stimulate TRPV4-mediated non-selective cation channels and intracellular Ca^{2+} signals in endothelial cells from several vascular beds [191,195–198]. Luminal application of UTP, which is a selective agonist of P2Y2 receptors, in rat middle cerebral arteries induced TRPV4-mediated Ca^{2+} entry across the luminal and abluminal face of the endothelial monolayer. TRPV4-mediated Ca^{2+} entry, in turn, activated PLA2, whose activation was polarized to the abluminal side, to activate EDH and induce vasodilation [199,200]. Moreover, TRPV4-mediated Ca^{2+} entry was necessary to activate IK_{Ca} and SK_{Ca} channels and induce the vasodilatory response to acetylcholine in mouse posterior cerebral arteries [186]. TRPV4 is, therefore, the most likely candidate to trigger ATP-induced retrograde vasodilation in rat parenchymal arterioles, which is initiated by EETs and involves IK_{Ca}, but not SK_{Ca}, channels [164]. Intriguingly, EETs mediate vasodilation in rodent cortical arterioles [27,42,201,202]. It has been proposed that NA-induced elevation in astrocyte $[Ca^{2+}]_i$ engages PLA2 to liberate AA, thereby resulting in EET synthesis, either within astrocytes or adjacent smooth muscle cells [33,42]. Future work is necessary to assess whether: (1) astrocyte-derived EETs also activate endothelial TRPV4 within the NVU; and/or (2) neural activity directly stimulates brain endothelial cells to produce EET synthesis and activate TRPV4. A recent investigation suggested that EDH was responsible for propagating the hemodynamic response to somatosensory stimulation from the capillary bed to upstream penetrating arterioles and pial arteries [7,203]. However, IK_{Ca} and IK_{Ca} currents could not be recorded in mouse brain capillary endothelial cells [204]. These findings strongly suggest that: (1) EDH is restricted to pial arteries and arterioles, but it cannot be activated in the capillary bed; and (2) an alternative mechanism spreads the vasomotor signal from brain capillaries to upstream vessels (see below) [204].

ENDOTHELIAL CELL

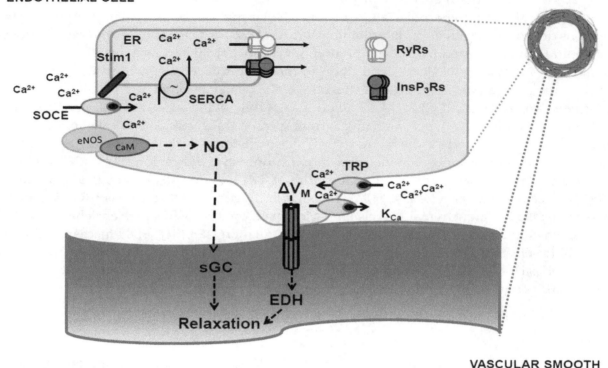

VASCULAR SMOOTH
MUSCLE CELL

Figure 4. Ca^{2+}-regulated endothelium-dependent vasodilation. Agonists-induced InsP$_3$-dependent ER Ca^{2+} depletion in vascular endothelial cells leads to store-operated Ca^{2+} entry (SOCE). SOCE is tightly coupled to eNOS, thereby triggering robust NO release. NO, in turn, diffuses towards adjacent vascular smooth muscle cells (VSMCs) at myo-endothelial projections and activates soluble guanylyl cyclase (sGC) to induce vasorelaxation. Extracellular agonists may also activate TRP channels (e.g., TRPC3, TRPV3, TRPV4 and TRPA1), which is preferentially coupled to Ca^{2+}-activated K^+ channels (K_{Ca}), such as IK_{Ca} and SK_{Ca}. Endothelial hyperpolarization spreads through myo-endothelial gap junctions to adjoining VSMCs to induce vasorelaxation according to a mechanism known endothelial-dependent hyperpolarization (EDH).

3.5.2. TRPA1, TRPC3 and TRPV3

In addition to TRPV4, TRPA1 channels are also enriched at MEGJs in rat cerebral arteries and may induce EDH in isolated vessels [78]. Accordingly, allyl isothiocyanate (AITC) (mustard oil), a selective TRPA1 agonist, induced local Ca^{2+} entry events, IK_{Ca} channel activation and vasodilation in rat cerebral arteries [205,206]. TRPA1 was physiologically activated by superoxide anions generated by the reactive oxygen species-generating enzyme NADPH oxidase isoform 2 (NOX2), which colocalizes with TRPA1 in cerebral endothelium, but not in other vascular districts [178]. Furthermore, EDH in cerebral arteries may be elicited by Ca^{2+} influx through TRPC3 and TRPV3 channels [78]. TRPC3 is gated by DAG to conduct extracellular Ca^{2+} into vascular endothelial cells upon PLC activation [117,118,138,207]. A recent investigation revealed that TRPC3 contributed to ATP-induced vasodilation in mouse middle cerebral arteries and posterior cerebral arteries by delivering the Ca^{2+} necessary for IK_{Ca} and SK_{Ca} channel activation. IK_{Ca} channels mediated the initial phase of ATP-induced hyperpolarization, whereas SK_{Ca} channels sustained the delayed phase of endothelial hyperpolarization [208]. IK_{Ca} and SK_{Ca} channels in rat brain vessels may also be activated by TRPV3. Accordingly, carvacrol, a monoterpenoid phenol compound highly concentrated in the essential oil of oregano, caused vasodilation by selectively activating TRPV3 in rat isolated posterior cerebral and superior cerebellar arteries. TRPV3-mediated extracellular Ca^{2+} entry, in turn, engaged IK_{Ca} and SK_{Ca} channels to trigger EDH. Unlike TRPV4 and TRPA1, TRPV3 was not specifically localized to MEGJs, but was distributed throughout the endothelial membrane [209]. In addition to carvacrol, TRPV3, as well as TRPA1 are sensitive to several dietary agonists. For instance, TRPV3 may also be activated by eugenol (clove oil) and thymol (found in thyme) [210], whereas allicin (garlic) and cinnamaldehyde induce TRPA1-mediated Ca^{2+} entry [211]. These observations, therefore, suggest that dietary manipulation of TRP channels could be exploited to rescue CBF in cerebrovascular pathologies [6,212].

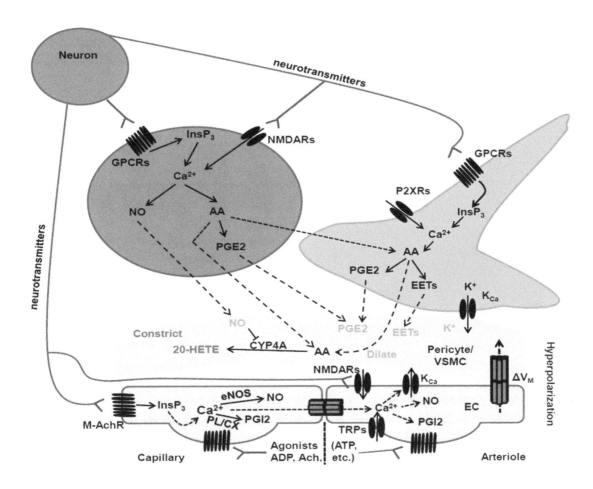

Figure 5. The putative role of endothelial Ca^{2+} signaling in neurovascular coupling. Synaptic activity leads to arteriole and capillary vasodilation by inducing an increase in $[Ca^{2+}]_i$ in postsynaptic neurons and perisynaptic astrocytes, as shown in Figure 2. Recent evidence indicated that synaptically-released glutamate may activate endothelial NMDARs, thereby eliciting the Ca^{2+}-dependent activation of eNOS, in intraparenchymal arterioles. Moreover, acetylcholine may induce Ca^{2+}-dependent NO release from brain endothelial cells by initiating the concerted interplay between InsP3Rs and SOCE [76]. This local vasodilation may be spread to more remote sites (≈ 500 μm) through the initiation of an interendothelial Ca^{2+} wave, which ignites NO release and PGI2 production as long as it travels along the endothelial monolayer. Moreover, this propagating Ca^{2+} sweep could induce vasodilation by also stimulating K_{Ca} channels and evoking EDH in arterioles. Finally, blood-borne autacoids and dietary agonists could induce vasodilation by, respectively, binding to their specific GPCRs and stimulating multiple TRP channels (TRPC3, TRPV3, TRPV4, TRPA1) to initiate EDH. Abbreviations: 20-HETE: 20-hydroxyeicosatetraenoic acid; AA: arachidonic acid; CX: cyclooxygenases 1 and 2; EETs: epoxyeicosatrienoic acids; GPCRs; G-protein-coupled receptors; InsP3: inositol-1,4,5-trisphosphate; K_{Ca}: Ca^{2+}-activated intermediate and small conductance K^+ channels; NO: nitric oxide; nNOS: neuronal NO synthase; P450: cytochrome P450; PGE2: prostaglandin E2; PL: phospholipase A2; PLD2: phospholipase D2; TRPs: TRP channels; VSMC: vascular smooth muscle cell.

4. How to Integrate Endothelial Ca^{2+} Signals in the Current Models of Neurovascular Coupling and Propagated Vasodilation

The role of endothelial Ca^{2+} signaling in NVC has largely been neglected as most authors focused on other cellular components of the NVU, such as neurons, interneurons, astrocytes, smooth muscle cells and, more recently, pericytes [2,4,6,9,30]. However, the literature discussed in the previous paragraphs suggests that the endothelial Ca^{2+} toolkit could play a crucial role in detecting and translating NA into a local vasoactive signal that is then propagated to upstream vessels [7,24]. In order to interpret the precise function of the ion channel network that controls CBF, it is mandatory to understand when and where the hemodynamic response starts. As mentioned earlier, synaptic activity-evoked NVC could be initiated at the capillary level and conducted upstream [12,47,203,213–217]. This hypothesis makes physiological sense as capillaries are closer to active neurons compared to arterioles and could represent the earliest vascular component to detect NA. In addition, endothelial signaling underpins long-range and almost unattenuated propagation (up to 2 mm) of vasomotor responses along peripheral vessels [22,62,63,176,177,218]. Endothelial engagement during NVC could, therefore, fulfill the same function observed in peripheral circulation, i.e., initiating (or contributing to initiate) vasodilation in proximity of the most metabolically active areas and conducting the vasomotor signal to upstream feeding arteries and arterioles to ensure an adequate increase in local blood supply (Figure 5) [62,177].

We believe that there is wide evidence to conclude that acetylcholine, which is liberated by cholinergic afferents emanated from the basal forebrain neurons during alert wakefulness, arousal, learning and attentional effort [28,150], increases CBF by inducing an increase in endothelial [Ca^{2+}]$_I$, which results in robust NO release [53,76,150,151,153,157]. This model does not rule out the possibility that the sub-population of acetylcholine and NO-synthesizing basal forebrain neurons that send projections onto cortical microvessels could directly control CBF though NO liberation [150,219]; in addition, acetylcholine and NO-synthesizing fibers may also contact NO cortical interneurons, which could contribute to NO-dependent vasodilation [150,220]. Similar to acetylcholine, glutamate could induce NMDARs-mediated NO release from brain endothelium [55,56,127]. Endothelial-derived NO could represent the alternative, i.e., non-neuronal, source of NO supporting vasodilation in cerebellar [141] and cortical [142] parenchymal arterioles. Future work will have to assess whether capillary brain endothelial cells express functional NMDARs and generate NO in response to glutamate stimulation. Intriguingly, the role of glial Ca^{2+} signals [13], which lead to the indispensable release of D-serine in arterioles, and of NO in glutamate-induced capillary dilation have been recently reported in cerebellum [11]. Alternatively, glutamate-induced endothelial-dependent vasodilation could contribute to pre-dilate upstream pial arteries and parenchymal arterioles prior to the retrograde propagation of the vasomotor signal from the capillary bed. Retrograde propagation of the initial vasodilation to upstream cortical vessels still represents a matter of hot debate [4,7,221]. Hillman's group demonstrated that light-dye disruption of endothelial lining dampens propagation of stimulus (electrical hind paw stimulation)-induced vasodilation in pial arteries in vivo, thereby blunting the increase in CBF [203]. In Section 2.3, we anticipated that endothelial Ca^{2+} signals mediate retrograde propagation of the vasomotor response by engaging two distinct mechanisms: (1) IK$_{Ca}$ and SK$_{Ca}$ channels, which effect the fast component of conducted vasodilation to upstream feeding vessels (i.e., pial arteries and arterioles); and (2) an intercellular Ca^{2+} wave that mediates the slower component of conducted vasodilation by inducing the release of NO and PGI2 from vascular endothelial cells [7,63,64]. EDH cannot, however, be initiated by brain capillary endothelial cells, which lack functional IK$_{Ca}$ and SK$_{Ca}$ channels [204]. Nelson's group recently revealed that a modest increase in the extracellular K$^+$ concentration ($\approx$ 10 mM), which is likely to truly reflect NA, activated brain capillary cell endothelial inward rectifier K$^+$ (K$_{IR}$2.1) channels to generate a hyperpolarizing signal that rapidly propagated to upstream arterioles (monitored up to $\approx$500 μm) to cause vasodilation and increase CBF in vivo [204]. Therefore, it appears that K$_{IR}$2.1, rather than IK$_{Ca}$ and SK$_{Ca}$, channels mediate the fast component of conducted vasodilation in cerebral circulation. Of note, the hyperpolarization conduction velocity was in the same range, $\approx$2 mm/s, as the propagation speed of retrograde pial artery dilation [204,213]. The secondary slow component of conducted vasodilation is sustained by interendothelial Ca^{2+} waves, as discussed elsewhere [4,7,62]. Earlier work carried out by monitoring endothelial Ca^{2+} with a Ca^{2+}-sensitive dye, i.e., Fura-2, showed that local delivery of acetylcholine triggered a Ca^{2+} wave travelling bidirectionally along the endothelium of hamster feed arteries for no less than 1 mm [222]. This intercellular Ca^{2+} wave, in turn, sustained the slow component of conducted vasodilation by promoting NO and PGI2 release [223]. The use of a genetic Ca^{2+} indicator, GCaMP2, confirmed these data in cremaster muscle arterioles in vivo. Accordingly, acetylcholine induced a local increase in endothelial $[Ca^{2+}]_i$ that activated K_{Ca} channels to induce rapidly-conducting vasodilation at distances >1 mm and travelled along the endothelium as an intercellular Ca^{2+} wave for 300–400 μm. As shown in previous studies [222,223], the intercellular Ca^{2+} wave preceded a secondary vasodilation that was mediated by NO and PGE2 [63]. These findings led to the suggestion that the initial hyperpolarization conducts vasodilation far away (> several millimeters) from the local site of stimulation (including daughter and parent branches), whereas the slower Ca^{2+} wave encompasses only the vessel segments closer to the active region, thereby finely tuning the magnitude and duration of the vasodilatory response [62,63]. A recent study showed that removal of extracellular Ca^{2+} may induce intercellular Ca^{2+} waves in immortalized (RBE4) and primary (from bovine origin) brain microvascular endothelial cells [224]. Although this finding is not sufficient to confirm that interendothelial Ca^{2+} waves may propagate NA-induced local vasodilation to upstream arteries and arterioles, it confirms that brain endothelial cells are able to generate this type of propagating Ca^{2+} signal, which is likely to impinge on $InsP_3Rs$ [225–227]. Future work will have to assess whether: (1) an intercellular Ca^{2+} wave contributes to propagating the hemodynamic response from the capillary bed to feeding vessels; and (2) if so, whether EDH is stimulated by the incoming Ca^{2+} wave in parenchymal arterioles to sustain the vasomotor response.

5. Conclusions

Neurovascular coupling is the crucial process to adjust local CBF to the metabolic requirements of active neurons and maintain brain function. Although functional magnetic resonance imaging is routinely employed to monitor the changes in neuronal spiking activity, BOLD signals actually reflect the increases in CBF induced by NA. Therefore, understanding the cellular and molecular bases of NVC is mandatory to interpret the complex relationship between neuronal firing, metabolism and blood flow in both physiological and pathological conditions [6–8]. Intracellular Ca^{2+} signaling has long been known to drive NVC by coupling synaptic activity with the production of vasoactive messengers. Emerging evidence, however, suggests that the endothelial Ca^{2+} toolkit could also be recruited by neurotransmitters (i.e., glutamate and acetylcholine) and neuromodulators (e.g., ATP) to induce NO and PGE2 release or activate EDH. Future work will benefit from the availability of transgenic mice selectively expressing a genetic Ca^{2+} indicator, e.g., GCaMP2, in vascular endothelial cells [63,174]. The combination of endothelial GCaMP2-expressing mice with the novel advances in imaging techniques (i.e., two- or three-photon microscopy) will permit assessing whether synaptic activity increases endothelial $[Ca^{2+}]_i$ in vivo. It will also be important to decipher the role played by IK_{Ca} and SK_{Ca} channels in the hemodynamic response to NA. Earlier work showed that EDH is involved in local and retrograde vasodilation in parenchymal arterioles and middle cerebral arteries, but is absent in capillaries, where functional hyperemia is likely to initiate in response to sensory stimulation. The local vasodilation induced by NA increases the strength of laminar shear stress acting on vascular endothelium, a mechanism that leads to TRPV4-dependent NO release and EDH in peripheral circulation [191,228]. There is, however, conflicting evidence regarding flow-induced vasodilation in pial arteries and parenchymal arteries in the brain [229–231]. Nevertheless, EDH could be exploited by dietary manipulation to treat the vascular dysfunctions associated with aging and neurodegenerative disorders, which cause cognitive impairment by halting CBF [2,3,6,212]. For instance, NVC is severely impaired in Alzheimer's disease due to a defect in NO release [217,232,233]. A recent study revealed that the intracellular Ca^{2+} toolkit is severely compromised in rat brain microvascular endothelial cells exposed to amyloid-beta (Aβ) peptide [93], whose accumulation in brain parenchyma and in the cerebrovasculature represents a major pathogenic factor in AD [4]. Future work will have to assess whether dietary and/or pharmacological manipulation is able to rescue or halt functional hyperemia in these subjects by activating the TRP channels (i.e., TRPC3, TRPV3, TRPV4 and TRPA1), which are coupled to EDH in the brain. An emerging area of research for NVC is represented by the vasodilating role of mitochondrial Ca^{2+} in brain microvascular endothelial cells. A recent series of studies revealed that mitochondrial depolarization, induced by BMS-291095 (BMS), an opener of mitochondrial ATP-dependent K^+ (K_{ATP}) channels, prevented Ca^{2+} accumulation within the mitochondrial matrix, thereby resulting in a large increase in $[Ca^{2+}]_i$ in rat cerebral arteries endothelial cells. This mitochondrial-derived Ca^{2+} signal, in turn, induced eNOS activation, NO release and endothelium-dependent vasodilation [95]. Intriguingly, subsequent work demonstrated that mitochondrial-dependent NO release and vasodilation were impaired in cerebral arteries of obese Zucker rats [234]. The physiological stimulus responsible for mitochondrial-induced NO release in cerebral endothelium is yet to be elucidated, but it could be implicated in many other cerebrovascular disorders. Finally, astrocyte-released AA is the precursor of multiple vasoactive messengers, such as PGE2, EETs and 20-HETE, which are all involved in NVC, although their contributions depend on the vascular segment and/or the brain area. However, AA was shown to induce Ca^{2+}-dependent NO release by directly activating TRPV4 in several types of endothelial cells [54,235]. Some evidence suggested that AA was able to induce Ca^{2+} signals in human brain microvascular endothelial cells [94]. Given the key role of this lipid mediator within the NVU, we predict that future work will unveil that AA-dependent endothelial Ca^{2+} signals may also contribute to NVC.

Abbreviations

20-HETE	20-hydroxyeicosatetraenoic acid
AA	arachidonic acid
AMPA	a-amino-3-hydroxy-5-methyl-4-isoxazol epropionic acid
BK_{Ca}	large-conductance Ca^{2+}-activated K^+ channels
BOLD	blood oxygenation level dependent
CaM	Calmodulin
CBF	cerebral blood flow
cGMP	cyclic guanosine-3′,5′-monophosphate
CICR	Ca^{2+}-induced Ca^{2+} release
CYP4A	cytochrome P450 4A
DAG	Diacylglycerol
EDH	endothelial-dependent hyperpolarization
EETs	epoxyeicosatrienoic acids
EL	Endolysosomal
eNOS	neuronal NO synthase
ER	endoplasmic reticulum
GPCRs	G-protein-coupled receptors
IK_{Ca}	intermediate-conductance Ca^{2+}-activated K^+ channels
$InsP_3$	inositol-1,4,5-trisphosphate
$InsP_3Rs$	$InsP_3$ receptors
K_{Ca}	Ca^{2+}-activated K^+ channels
K_{ir}	inward rectifying K^+ channels
M-AchRs	muscarinic acetylcholine receptors
MCU	mitochondrial Ca^{2+} uniporter

MEGJs	myo-endothelial gap-junctions
mGluRs	metabotropic glutamate receptors
NA	neuronal activity
NAADP	nicotinic acid adenine dinucleotide phosphate
NMDA	N-methyl-D-aspartate
NMDARs	NMDA receptors
NO	nitric oxide
nNOS	neuronal NO synthase
NOX2	NADPH oxidase isoform 2
NVC	neurovascular coupling
NVU	neurovascular unit
O_2	Oxygen
P450	cytochrome P450
PGE2	prostaglandin E2
PIP_2	phosphatidylinositol-4,5-bisphosphate
PLA2	phospholipase A2
PLD2	phospholipase D2
RyRs	ryanodine receptors
SERCA	sarco-endoplasmic reticulum Ca^{2+}-ATPase
SK_{Ca}	small-conductance Ca^{2+}-activated K^+ channels
SOCE	store-operated Ca^{2+} entry
Stim	stromal interaction molecule
TPCs	two-pore channels
TRPs	TRP channels
TRPA1	ankyrin TRP 1
TRPC	canonical TRP
TRPV4	vanilloid TRP 4
VDACs	voltage-dependent anion channels
VSMC	vascular smooth muscle cells

References

1. Attwell, D.; Laughlin, S.B. An energy budget for signaling in the grey matter of the brain. *J. Cereb. Blood Flow Metab.* **2001**, *21*, 1133–1145. [CrossRef] [PubMed]
2. Kisler, K.; Nelson, A.R.; Montagne, A.; Zlokovic, B.V. Cerebral blood flow regulation and neurovascular dysfunction in Alzheimer disease. *Nat. Rev. Neurosci.* **2017**, *18*, 419–434. [CrossRef] [PubMed]
3. Zlokovic, B.V. Neurovascular pathways to neurodegeneration in Alzheimer's disease and other disorders. *Nat. Rev. Neurosci.* **2011**, *12*, 723–738. [CrossRef] [PubMed]
4. Iadecola, C. The Neurovascular Unit Coming of Age: A Journey through Neurovascular Coupling in Health and Disease. *Neuron* **2017**, *96*, 17–42. [CrossRef] [PubMed]
5. Koller, A.; Toth, P. Contribution of flow-dependent vasomotor mechanisms to the autoregulation of cerebral blood flow. *J. Vasc. Res.* **2012**, *49*, 375–389. [CrossRef] [PubMed]
6. Iadecola, C. Neurovascular regulation in the normal brain and in Alzheimer's disease. *Nat. Rev. Neurosci.* **2004**, *5*, 347–360. [CrossRef] [PubMed]
7. Hillman, E.M. Coupling mechanism and significance of the BOLD signal: A status report. *Annu. Rev. Neurosci.* **2014**, *37*, 161–181. [CrossRef] [PubMed]
8. Attwell, D.; Iadecola, C. The neural basis of functional brain imaging signals. *Trends Neurosci.* **2002**, *25*, 621–625. [CrossRef]
9. Attwell, D.; Buchan, A.M.; Charpak, S.; Lauritzen, M.; Macvicar, B.A.; Newman, E.A. Glial and neuronal control of brain blood flow. *Nature* **2010**, *468*, 232–243. [CrossRef] [PubMed]
10. Rancillac, A.; Rossier, J.; Guille, M.; Tong, X.K.; Geoffroy, H.; Amatore, C.; Arbault, S.; Hamel, E.; Cauli, B. Glutamatergic Control of Microvascular Tone by Distinct GABA Neurons in the Cerebellum. *J. Neurosci.* **2006**, *26*, 6997–7006. [CrossRef] [PubMed]

11. Mapelli, L.; Gagliano, G.; Soda, T.; Laforenza, U.; Moccia, F.; D'Angelo, E.U. Granular Layer Neurons Control Cerebellar Neurovascular Coupling Through an NMDA Receptor/NO-Dependent System. *J. Neurosci.* **2017**, *37*, 1340–1351. [CrossRef] [PubMed]

12. Hall, C.N.; Reynell, C.; Gesslein, B.; Hamilton, N.B.; Mishra, A.; Sutherland, B.A.; O'Farrell, F.M.; Buchan, A.M.; Lauritzen, M.; Attwell, D. Capillary pericytes regulate cerebral blood flow in health and disease. *Nature* **2014**, *508*, 55–60. [CrossRef] [PubMed]

13. Mishra, A.; Reynolds, J.P.; Chen, Y.; Gourine, A.V.; Rusakov, D.A.; Attwell, D. Astrocytes mediate neurovascular signaling to capillary pericytes but not to arterioles. *Nat. Neurosci.* **2016**, *19*, 1619–1627. [CrossRef] [PubMed]

14. Rosenegger, D.G.; Gordon, G.R. A slow or modulatory role of astrocytes in neurovascular coupling. *Microcirculation* **2015**, *22*, 197–203. [CrossRef] [PubMed]

15. Laughlin, M.H.; Davis, M.J.; Secher, N.H.; van Lieshout, J.J.; Arce-Esquivel, A.A.; Simmons, G.H.; Bender, S.B.; Padilla, J.; Bache, R.J.; Merkus, D.; et al. Peripheral circulation. *Compr. Physiol.* **2012**, *2*, 321–447. [PubMed]

16. Khaddaj Mallat, R.; Mathew John, C.; Kendrick, D.J.; Braun, A.P. The vascular endothelium: A regulator of arterial tone and interface for the immune system. *Crit. Rev. Clin. Lab. Sci.* **2017**, *54*, 458–470. [CrossRef] [PubMed]

17. Mancardi, D.; Pla, A.F.; Moccia, F.; Tanzi, F.; Munaron, L. Old and new gasotransmitters in the cardiovascular system: Focus on the role of nitric oxide and hydrogen sulfide in endothelial cells and cardiomyocytes. *Curr. Pharm. Biotechnol.* **2011**, *12*, 1406–1415. [CrossRef] [PubMed]

18. Vanhoutte, P.M.; Tang, E.H. Endothelium-dependent contractions: When a good guy turns bad! *J. Physiol.* **2008**, *586*, 5295–5304. [CrossRef] [PubMed]

19. Moccia, F.; Berra-Romani, R.; Tanzi, F. Update on vascular endothelial Ca^{2+} signalling: A tale of ion channels, pumps and transporters. *World J. Biol. Chem.* **2012**, *3*, 127–158. [CrossRef] [PubMed]

20. Moccia, F.; Tanzi, F.; Munaron, L. Endothelial remodelling and intracellular calcium machinery. *Curr. Mol. Med.* **2014**, *14*, 457–480. [CrossRef] [PubMed]

21. Altaany, Z.; Moccia, F.; Munaron, L.; Mancardi, D.; Wang, R. Hydrogen sulfide and endothelial dysfunction: Relationship with nitric oxide. *Curr. Med. Chem.* **2014**, *21*, 3646–3661. [CrossRef] [PubMed]

22. Behringer, E.J. Calcium and electrical signaling in arterial endothelial tubes: New insights into cellular physiology and cardiovascular function. *Microcirculation* **2017**, *24*. [CrossRef] [PubMed]

23. Garland, C.J.; Dora, K.A. EDH: Endothelium-dependent hyperpolarization and microvascular signalling. *Acta Physiologica* **2017**, *219*, 152–161. [CrossRef] [PubMed]

24. Longden, T.A.; Hill-Eubanks, D.C.; Nelson, M.T. Ion channel networks in the control of cerebral blood flow. *J. Cereb. Blood Flow Metab.* **2016**, *36*, 492–512. [CrossRef] [PubMed]

25. Cauli, B.; Hamel, E. Revisiting the role of neurons in neurovascular coupling. *Front. Neuroenerg.* **2010**, *2*, 9. [CrossRef] [PubMed]

26. Cauli, B.; Tong, X.K.; Rancillac, A.; Serluca, N.; Lambolez, B.; Rossier, J.; Hamel, E. Cortical GABA interneurons in neurovascular coupling: Relays for subcortical vasoactive pathways. *J. Neurosci.* **2004**, *24*, 8940–8949. [CrossRef] [PubMed]

27. Lecrux, C.; Toussay, X.; Kocharyan, A.; Fernandes, P.; Neupane, S.; Levesque, M.; Plaisier, F.; Shmuel, A.; Cauli, B.; Hamel, E. Pyramidal neurons are "neurogenic hubs" in the neurovascular coupling response to whisker stimulation. *J. Neurosci.* **2011**, *31*, 9836–9847. [CrossRef] [PubMed]

28. Lecrux, C.; Hamel, E. Neuronal networks and mediators of cortical neurovascular coupling responses in normal and altered brain states. *Philos. Trans. R. Soc. Lond. B Biol. Sci.* **2016**, *371*. pii: 20150350. [CrossRef] [PubMed]

29. Hamel, E. Perivascular nerves and the regulation of cerebrovascular tone. *J. Appl. Physiol. (1985)* **2006**, *100*, 1059–1064. [CrossRef] [PubMed]

30. Nortley, R.; Attwell, D. Control of brain energy supply by astrocytes. *Curr. Opin. Neurobiol.* **2017**, *47*, 80–85. [CrossRef] [PubMed]

31. Mathiisen, T.M.; Lehre, K.P.; Danbolt, N.C.; Ottersen, O.P. The perivascular astroglial sheath provides a complete covering of the brain microvessels: An electron microscopic 3D reconstruction. *Glia* **2010**, *58*, 1094–1103. [CrossRef] [PubMed]

32. Zlokovic, B.V. Neurovascular mechanisms of Alzheimer's neurodegeneration. *Trends Neurosci.* **2005**, *28*, 202–208. [CrossRef] [PubMed]

33. Dalkara, T.; Alarcon-Martinez, L. Cerebral microvascular pericytes and neurogliovascular signaling in health and disease. *Brain Res.* **2015**, *1623*, 3–17. [CrossRef] [PubMed]

34. Lauritzen, M. Reading vascular changes in brain imaging: Is dendritic calcium the key? *Nat. Rev. Neurosci.* **2005**, *6*, 77–85. [CrossRef] [PubMed]

35. Lourenco, C.F.; Ledo, A.; Barbosa, R.M.; Laranjinha, J. Neurovascular-neuroenergetic coupling axis in the brain: Master regulation by nitric oxide and consequences in aging and neurodegeneration. *Free Radic. Biol. Med.* **2017**, *108*, 668–682. [CrossRef] [PubMed]

36. Yang, G.; Chen, G.; Ebner, T.J.; Iadecola, C. Nitric oxide is the predominant mediator of cerebellar hyperemia during somatosensory activation in rats. *Am. J. Physiol.* **1999**, *277 (Pt 2)*, R1760–R1770. [CrossRef] [PubMed]

37. Lovick, T.A.; Brown, L.A.; Key, B.J. Neurovascular relationships in hippocampal slices: Physiological and anatomical studies of mechanisms underlying flow-metabolism coupling in intraparenchymal microvessels. *Neuroscience* **1999**, *92*, 47–60. [CrossRef]

38. Duchemin, S.; Boily, M.; Sadekova, N.; Girouard, H. The complex contribution of NOS interneurons in the physiology of cerebrovascular regulation. *Front. Neural Circ.* **2012**, *6*, 51. [CrossRef] [PubMed]

39. Filosa, J.A.; Morrison, H.W.; Iddings, J.A.; Du, W.; Kim, K.J. Beyond neurovascular coupling, role of astrocytes in the regulation of vascular tone. *Neuroscience* **2016**, *323*, 96–109. [CrossRef] [PubMed]

40. Wang, X.; Lou, N.; Xu, Q.; Tian, G.F.; Peng, W.G.; Han, X.; Kang, J.; Takano, T.; Nedergaard, M. Astrocytic Ca^{2+} signaling evoked by sensory stimulation in vivo. *Nat. Neurosci.* **2006**, *9*, 816–823. [CrossRef] [PubMed]

41. Takano, T.; Tian, G.F.; Peng, W.; Lou, N.; Libionka, W.; Han, X.; Nedergaard, M. Astrocyte-mediated control of cerebral blood flow. *Nat. Neurosci.* **2006**, *9*, 260–267. [CrossRef] [PubMed]

42. Mishra, A. Binaural blood flow control by astrocytes: Listening to synapses and the vasculature. *J. Physiol.* **2017**, *595*, 1885–1902. [CrossRef] [PubMed]

43. Kim, K.J.; Iddings, J.A.; Stern, J.E.; Blanco, V.M.; Croom, D.; Kirov, S.A.; Filosa, J.A. Astrocyte contributions to flow/pressure-evoked parenchymal arteriole vasoconstriction. *J. Neurosci.* **2015**, *35*, 8245–8257. [CrossRef] [PubMed]

44. Kim, K.J.; Ramiro Diaz, J.; Iddings, J.A.; Filosa, J.A. Vasculo-Neuronal Coupling: Retrograde Vascular Communication to Brain Neurons. *J. Neurosci.* **2016**, *36*, 12624–12639. [CrossRef] [PubMed]

45. Simard, M.; Arcuino, G.; Takano, T.; Liu, Q.S.; Nedergaard, M. Signaling at the gliovascular interface. *J. Neurosci.* **2003**, *23*, 9254–9262. [PubMed]

46. Attwell, D.; Mishra, A.; Hall, C.N.; O'Farrell, F.M.; Dalkara, T. What is a pericyte? *J. Cereb. Blood Flow Metab.* **2016**, *36*, 451–455. [CrossRef] [PubMed]

47. Hamilton, N.B.; Attwell, D.; Hall, C.N. Pericyte-mediated regulation of capillary diameter: A component of neurovascular coupling in health and disease. *Front. Neuroenerg.* **2010**, *2*. pii: 5. [CrossRef] [PubMed]

48. D'Angelo, E. The organization of plasticity in the cerebellar cortex: From synapses to control. *Prog. Brain Res.* **2014**, *210*, 31–58. [PubMed]

49. Berra-Romani, R.; Avelino-Cruz, J.E.; Raqeeb, A.; Della Corte, A.; Cinelli, M.; Montagnani, S.; Guerra, G.; Moccia, F.; Tanzi, F. Ca(2)(+)-dependent nitric oxide release in the injured endothelium of excised rat aorta: A promising mechanism applying in vascular prosthetic devices in aging patients. *BMC Surg.* **2013**, *13* (Suppl. 2), S40. [CrossRef] [PubMed]

50. Andresen, J.; Shafi, N.I.; Bryan, R.M., Jr. Endothelial influences on cerebrovascular tone. *J. Appl. Physiol. (1985)* **2006**, *100*, 318–327. [CrossRef] [PubMed]

51. You, J.; Johnson, T.D.; Childres, W.F.; Bryan, R.M., Jr. Endothelial-mediated dilations of rat middle cerebral arteries by ATP and ADP. *Am. J. Physiol.* **1997**, *273 (Pt 2)*, H1472–H1477. [CrossRef] [PubMed]

52. Elhusseiny, A.; Cohen, Z.; Olivier, A.; Stanimirovic, D.B.; Hamel, E. Functional acetylcholine muscarinic receptor subtypes in human brain microcirculation: Identification and cellular localization. *J. Cereb. Blood Flow Metab.* **1999**, *19*, 794–802. [CrossRef] [PubMed]

53. Elhusseiny, A.; Hamel, E. Muscarinic–but not nicotinic–acetylcholine receptors mediate a nitric oxide-dependent dilation in brain cortical arterioles: A possible role for the M5 receptor subtype. *J. Cereb. Blood Flow Metab.* **2000**, *20*, 298–305. [CrossRef] [PubMed]

54. Zuccolo, E.; Dragoni, S.; Poletto, V.; Catarsi, P.; Guido, D.; Rappa, A.; Reforgiato, M.; Lodola, F.; Lim, D.; Rosti, V.; et al. Arachidonic acid-evoked Ca^{2+} signals promote nitric oxide release and proliferation in human endothelial colony forming cells. *Vascul. Pharmacol.* **2016**, *87*, 159–171. [CrossRef] [PubMed]

55. LeMaistre, J.L.; Sanders, S.A.; Stobart, M.J.; Lu, L.; Knox, J.D.; Anderson, H.D.; Anderson, C.M. Coactivation of NMDA receptors by glutamate and D-serine induces dilation of isolated middle cerebral arteries. *J. Cereb. Blood Flow Metab.* **2012**, *32*, 537–547. [CrossRef] [PubMed]

56. Stobart, J.L.; Lu, L.; Anderson, H.D.; Mori, H.; Anderson, C.M. Astrocyte-induced cortical vasodilation is mediated by D-serine and endothelial nitric oxide synthase. *Proc. Natl. Acad. Sci. USA* **2013**, *110*, 3149–3154. [CrossRef] [PubMed]

57. Beard, R.S., Jr.; Reynolds, J.J.; Bearden, S.E. Metabotropic glutamate receptor 5 mediates phosphorylation of vascular endothelial cadherin and nuclear localization of beta-catenin in response to homocysteine. *Vascul. Pharmacol.* **2012**, *56*, 159–167. [CrossRef] [PubMed]

58. Lv, J.M.; Guo, X.M.; Chen, B.; Lei, Q.; Pan, Y.J.; Yang, Q. The Noncompetitive AMPAR Antagonist Perampanel Abrogates Brain Endothelial Cell Permeability in Response to Ischemia: Involvement of Claudin-5. *Cell. Mol. Neurobiol.* **2016**, *36*, 745–753. [CrossRef] [PubMed]

59. Collard, C.D.; Park, K.A.; Montalto, M.C.; Alapati, S.; Buras, J.A.; Stahl, G.L.; Colgan, S.P. Neutrophil-derived glutamate regulates vascular endothelial barrier function. *J. Biol. Chem.* **2002**, *277*, 14801–14811. [CrossRef] [PubMed]

60. Gillard, S.E.; Tzaferis, J.; Tsui, H.C.; Kingston, A.E. Expression of metabotropic glutamate receptors in rat meningeal and brain microvasculature and choroid plexus. *J. Comp. Neurol.* **2003**, *461*, 317–332. [CrossRef] [PubMed]

61. Itoh, Y.; Suzuki, N. Control of brain capillary blood flow. *J. Cereb. Blood Flow Metab.* **2012**, *32*, 1167–1176. [CrossRef] [PubMed]

62. Segal, S.S. Integration and Modulation of Intercellular Signaling Underlying Blood Flow Control. *J. Vasc. Res.* **2015**, *52*, 136–157. [CrossRef] [PubMed]

63. Tallini, Y.N.; Brekke, J.F.; Shui, B.; Doran, R.; Hwang, S.M.; Nakai, J.; Salama, G.; Segal, S.S.; Kotlikoff, M.I. Propagated endothelial Ca^{2+} waves and arteriolar dilation in vivo: Measurements in Cx40BAC GCaMP2 transgenic mice. *Circ. Res.* **2007**, *101*, 1300–1309. [CrossRef] [PubMed]

64. Wolfle, S.E.; Chaston, D.J.; Goto, K.; Sandow, S.L.; Edwards, F.R.; Hill, C.E. Non-linear relationship between hyperpolarisation and relaxation enables long distance propagation of vasodilatation. *J. Physiol.* **2011**, *589 (Pt 10)*, 2607–2623. [CrossRef] [PubMed]

65. Abdullaev, I.F.; Bisaillon, J.M.; Potier, M.; Gonzalez, J.C.; Motiani, R.K.; Trebak, M. STIM1 and ORAI1 mediate CRAC currents and store-operated calcium entry important for endothelial cell proliferation. *Circ. Res.* **2008**, *103*, 1289–1299. [CrossRef] [PubMed]

66. Zuccolo, E.; Di Buduo, C.; Lodola, F.; Orecchioni, S.; Scarpellino, G.; Kheder, D.A.; Poletto, V.; Guerra, G.; Bertolini, F.; Balduini, A.; et al. Stromal Cell-Derived Factor-1alpha Promotes Endothelial Colony-Forming Cell Migration Through the Ca(2+)-Dependent Activation of the Extracellular Signal-Regulated Kinase 1/2 and Phosphoinositide 3-Kinase/AKT Pathways. *Stem Cells Dev.* **2018**, *27*, 23–34. [CrossRef] [PubMed]

67. Freichel, M.; Suh, S.H.; Pfeifer, A.; Schweig, U.; Trost, C.; Weissgerber, P.; Biel, M.; Philipp, S.; Freise, D.; Droogmans, G.; et al. Lack of an endothelial store-operated Ca^{2+} current impairs agonist-dependent vasorelaxation in TRP4-/- mice. *Nat. Cell Biol.* **2001**, *3*, 121–127. [CrossRef] [PubMed]

68. Blatter, L.A. Tissue Specificity: SOCE: Implications for Ca^{2+} Handling in Endothelial Cells. *Adv. Exp. Med. Biol.* **2017**, *993*, 343–361. [PubMed]

69. Moccia, F.; Lodola, F.; Dragoni, S.; Bonetti, E.; Bottino, C.; Guerra, G.; Laforenza, U.; Rosti, V.; Tanzi, F. Ca^{2+} signalling in endothelial progenitor cells: A novel means to improve cell-based therapy and impair tumour vascularisation. *Curr. Vasc. Pharmacol.* **2014**, *12*, 87–105. [CrossRef] [PubMed]

70. Moccia, F.; Dragoni, S.; Lodola, F.; Bonetti, E.; Bottino, C.; Guerra, G.; Laforenza, U.; Rosti, V.; Tanzi, F. Store-dependent Ca(2+) entry in endothelial progenitor cells as a perspective tool to enhance cell-based therapy and adverse tumour vascularization. *Curr. Med. Chem.* **2012**, *19*, 5802–5018. [CrossRef] [PubMed]

71. Brailoiu, E.; Shipsky, M.M.; Yan, G.; Abood, M.E.; Brailoiu, G.C. Mechanisms of modulation of brain microvascular endothelial cells function by thrombin. *Brain Res.* **2017**, *1657*, 167–175. [CrossRef] [PubMed]

72. Domotor, E.; Abbott, N.J.; Adam-Vizi, V. Na^+-Ca^{2+} exchange and its implications for calcium homeostasis in primary cultured rat brain microvascular endothelial cells. *J. Physiol.* **1999**, *515 (Pt 1)*, 147–155. [CrossRef] [PubMed]

73. De Bock, M.; Culot, M.; Wang, N.; Bol, M.; Decrock, E.; De Vuyst, E.; da Costa, A.; Dauwe, I.; Vinken, M.; Simon, A.M.; et al. Connexin channels provide a target to manipulate brain endothelial calcium dynamics and blood-brain barrier permeability. *J. Cereb. Blood Flow Metab.* **2011**, *31*, 1942–1957. [CrossRef] [PubMed]

74. De Bock, M.; Wang, N.; Decrock, E.; Bol, M.; Gadicherla, A.K.; Culot, M.; Cecchelli, R.; Bultynck, G.; Leybaert, L. Endothelial calcium dynamics, connexin channels and blood-brain barrier function. *Prog. Neurobiol.* **2013**, *108*, 1–20. [CrossRef] [PubMed]

75. Scharbrodt, W.; Abdallah, Y.; Kasseckert, S.A.; Gligorievski, D.; Piper, H.M.; Boker, D.K.; Deinsberger, W.; Oertel, M.F. Cytosolic Ca^{2+} oscillations in human cerebrovascular endothelial cells after subarachnoid hemorrhage. *J. Cereb. Blood Flow Metab.* **2009**, *29*, 57–65. [CrossRef] [PubMed]

76. Zuccolo, E.; Lim, D.; Kheder, D.A.; Perna, A.; Catarsi, P.; Botta, L.; Rosti, V.; Riboni, L.; Sancini, G.; Tanzi, F.; et al. Acetylcholine induces intracellular Ca$_{2+}$ oscillations and nitric oxide release in mouse brain endothelial cells. *Cell Calcium* **2017**, *66*, 33–47. [CrossRef] [PubMed]

77. Earley, S. Endothelium-dependent cerebral artery dilation mediated by transient receptor potential and Ca^{2+}-activated K$^+$ channels. *J. Cardiovasc. Pharmacol.* **2011**, *57*, 148–153. [CrossRef] [PubMed]

78. Earley, S.; Brayden, J.E. Transient receptor potential channels in the vasculature. *Physiol. Rev.* **2015**, *95*, 645–690. [CrossRef] [PubMed]

79. Manoonkitiwongsa, P.S.; Whitter, E.F.; Wareesangtip, W.; McMillan, P.J.; Nava, P.B.; Schultz, R.L. Calcium-dependent ATPase unlike ecto-ATPase is located primarily on the luminal surface of brain endothelial cells. *Histochem. J.* **2000**, *32*, 313–324. [CrossRef] [PubMed]

80. Moccia, F.; Berra-Romani, R.; Baruffi, S.; Spaggiari, S.; Signorelli, S.; Castelli, L.; Magistretti, J.; Taglietti, V.; Tanzi, F. Ca^{2+} uptake by the endoplasmic reticulum Ca^{2+}-ATPase in rat microvascular endothelial cells. *Biochem. J.* **2002**, *364 (Pt 1)*, 235–244. [CrossRef] [PubMed]

81. Haorah, J.; Knipe, B.; Gorantla, S.; Zheng, J.; Persidsky, Y. Alcohol-induced blood-brain barrier dysfunction is mediated via inositol 1,4,5-triphosphate receptor (IP3R)-gated intracellular calcium release. *J. Neurochem.* **2007**, *100*, 324–336. [CrossRef] [PubMed]

82. Hertle, D.N.; Yeckel, M.F. Distribution of inositol-1,4,5-trisphosphate receptor isotypes and ryanodine receptor isotypes during maturation of the rat hippocampus. *Neuroscience* **2007**, *150*, 625–638. [CrossRef] [PubMed]

83. Czupalla, C.J.; Liebner, S.; Devraj, K. In vitro models of the blood-brain barrier. *Methods Mol. Biol.* **2014**, *1135*, 415–437. [PubMed]

84. Pedriali, G.; Rimessi, A.; Sbano, L.; Giorgi, C.; Wieckowski, M.R.; Previati, M.; Pinton, P. Regulation of Endoplasmic Reticulum-Mitochondria Ca^{2+} Transfer and Its Importance for Anti-Cancer Therapies. *Front. Oncol.* **2017**, *7*, 180. [CrossRef] [PubMed]

85. Dong, Z.; Shanmughapriya, S.; Tomar, D.; Siddiqui, N.; Lynch, S.; Nemani, N.; Breves, S.L.; Zhang, X.; Tripathi, A.; Palaniappan, P.; et al. Mitochondrial Ca^{2+} Uniporter Is a Mitochondrial Luminal Redox Sensor that Augments MCU Channel Activity. *Mol. Cell* **2017**, *65*, 1014–1028.e7. [CrossRef] [PubMed]

86. De Stefani, D.; Rizzuto, R.; Pozzan, T. Enjoy the Trip: Calcium in Mitochondria Back and Forth. *Annu. Rev. Biochem.* **2016**, *85*, 161–192. [CrossRef] [PubMed]

87. Moccia, F. Remodelling of the Ca^{2+} Toolkit in Tumor Endothelium as a Crucial Responsible for the Resistance to Anticancer Therapies. *Curr. Signal Transduct. Ther.* **2017**, *12*. [CrossRef]

88. Kluge, M.A.; Fetterman, J.L.; Vita, J.A. Mitochondria and endothelial function. *Circ. Res.* **2013**, *112*, 1171–1188. [CrossRef] [PubMed]

89. Malli, R.; Frieden, M.; Hunkova, M.; Trenker, M.; Graier, W.F. Ca^{2+} refilling of the endoplasmic reticulum is largely preserved albeit reduced Ca^{2+} entry in endothelial cells. *Cell Calcium* **2007**, *41*, 63–76. [CrossRef] [PubMed]

90. Malli, R.; Frieden, M.; Osibow, K.; Zoratti, C.; Mayer, M.; Demaurex, N.; Graier, W.F. Sustained Ca^{2+} transfer across mitochondria is Essential for mitochondrial Ca^{2+} buffering, sore-operated Ca^{2+} entry, and Ca^{2+} store refilling. *J. Biol. Chem.* **2003**, *278*, 44769–44779. [CrossRef] [PubMed]

91. Busija, D.W.; Rutkai, I.; Dutta, S.; Katakam, P.V. Role of Mitochondria in Cerebral Vascular Function: Energy Production, Cellular Protection, and Regulation of Vascular Tone. *Compr. Physiol.* **2016**, *6*, 1529–1548. [PubMed]

92. Gerencser, A.A.; Adam-Vizi, V. Selective, high-resolution fluorescence imaging of mitochondrial Ca^{2+} concentration. *Cell Calcium* **2001**, *30*, 311–321. [CrossRef] [PubMed]

93. Fonseca, A.C.; Moreira, P.I.; Oliveira, C.R.; Cardoso, S.M.; Pinton, P.; Pereira, C.F. Amyloid-beta disrupts calcium and redox homeostasis in brain endothelial cells. *Mol. Neurobiol.* **2015**, *51*, 610–622. [CrossRef] [PubMed]

94. Evans, J.; Ko, Y.; Mata, W.; Saquib, M.; Eldridge, J.; Cohen-Gadol, A.; Leaver, H.A.; Wang, S.; Rizzo, M.T. Arachidonic acid induces brain endothelial cell apoptosis via p38-MAPK and intracellular calcium signaling. *Microvasc. Res.* **2015**, *98*, 145–158. [CrossRef] [PubMed]

95. Katakam, P.V.; Wappler, E.A.; Katz, P.S.; Rutkai, I.; Institoris, A.; Domoki, F.; Gaspar, T.; Grovenburg, S.M.; Snipes, J.A.; Busija, D.W. Depolarization of mitochondria in endothelial cells promotes cerebral artery vasodilation by activation of nitric oxide synthase. *Arterioscler. Thromb. Vasc. Biol.* **2013**, *33*, 752–759. [CrossRef] [PubMed]

96. Ronco, V.; Potenza, D.M.; Denti, F.; Vullo, S.; Gagliano, G.; Tognolina, M.; Guerra, G.; Pinton, P.; Genazzani, A.A.; Mapelli, L.; et al. A novel Ca(2)(+)-mediated cross-talk between endoplasmic reticulum and acidic organelles: Implications for NAADP-dependent Ca(2)(+) signalling. *Cell Calcium* **2015**, *57*, 89–100. [CrossRef] [PubMed]

97. Zuccolo, E.; Lim, D.; Poletto, V.; Guerra, G.; Tanzi, F.; Rosti, V.; Moccia, F. Acidic Ca_{2+} stores interact with the endoplasmic reticulum to shape intracellular Ca^{2+} signals in human endothelial progenitor cells. *Vascul. Pharmacol.* **2015**, *75*, 70–71. [CrossRef]

98. Di Nezza, F.; Zuccolo, E.; Poletto, V.; Rosti, V.; De Luca, A.; Moccia, F.; Guerra, G.; Ambrosone, L. Liposomes as a Putative Tool to Investigate NAADP Signaling in Vasculogenesis. *J. Cell. Biochem.* **2017**, *118*, 3722–3729. [CrossRef] [PubMed]

99. Favia, A.; Desideri, M.; Gambara, G.; D'Alessio, A.; Ruas, M.; Esposito, B.; Del Bufalo, D.; Parrington, J.; Ziparo, E.; Palombi, F.; et al. VEGF-induced neoangiogenesis is mediated by NAADP and two-pore channel-2-dependent Ca^{2+} signaling. *Proc. Natl. Acad. Sci. USA* **2014**, *111*, E4706–E4715. [CrossRef] [PubMed]

100. Favia, A.; Pafumi, I.; Desideri, M.; Padula, F.; Montesano, C.; Passeri, D.; Nicoletti, C.; Orlandi, A.; Del Bufalo, D.; Sergi, M.; et al. NAADP-Dependent Ca(2+) Signaling Controls Melanoma Progression, Metastatic Dissemination and Neoangiogenesis. *Sci. Rep.* **2016**, *6*, 18925. [CrossRef] [PubMed]

101. Lodola, F.; Laforenza, U.; Bonetti, E.; Lim, D.; Dragoni, S.; Bottino, C.; Ong, H.L.; Guerra, G.; Ganini, C.; Massa, M.; et al. Store-operated Ca^{2+} entry is remodelled and controls in vitro angiogenesis in endothelial progenitor cells isolated from tumoral patients. *PLoS ONE* **2012**, *7*, e42541. [CrossRef] [PubMed]

102. Li, J.; Cubbon, R.M.; Wilson, L.A.; Amer, M.S.; McKeown, L.; Hou, B.; Majeed, Y.; Tumova, S.; Seymour, V.A.L.; Taylor, H.; et al. ORAI1 and CRAC channel dependence of VEGF-activated Ca^{2+} entry and endothelial tube formation. *Circ. Res.* **2011**, *108*, 1190–1198. [CrossRef] [PubMed]

103. Brandman, O.; Liou, J.; Park, W.S.; Meyer, T. STIM2 is a feedback regulator that stabilizes basal cytosolic and endoplasmic reticulum Ca^{2+} levels. *Cell* **2007**, *131*, 1327–1339. [CrossRef] [PubMed]

104. Antigny, F.; Jousset, H.; Konig, S.; Frieden, M. Thapsigargin activates Ca(2)+ entry both by store-dependent, STIM1/ORAI1-mediated, and store-independent, TRPC3/PLC/PKC-mediated pathways in human endothelial cells. *Cell Calcium* **2011**, *49*, 115–127. [CrossRef] [PubMed]

105. Moccia, F.; Zuccolo, E.; Soda, T.; Tanzi, F.; Guerra, G.; Mapelli, L.; Lodola, F.; D'Angelo, E. Stim and Orai proteins in neuronal Ca(2)+ signaling and excitability. *Front. Cell. Neurosci.* **2015**, *9*, 153. [CrossRef] [PubMed]

106. Aird, W.C. Endothelial cell heterogeneity. *Cold Spring Harb. Perspect. Med.* **2012**, *2*, a006429. [CrossRef] [PubMed]

107. Regan, E.R.; Aird, W.C. Dynamical systems approach to endothelial heterogeneity. *Circ. Res.* **2012**, *111*, 110–130. [CrossRef] [PubMed]

108. Moccia, F.; Berra-Romani, R.; Baruffi, S.; Spaggiari, S.; Adams, D.J.; Taglietti, V.; Tanzi, F. Basal nonselective cation permeability in rat cardiac microvascular endothelial cells. *Microvasc. Res.* **2002**, *64*, 187–197. [CrossRef] [PubMed]

109. Zuccolo, E.; Bottino, C.; Diofano, F.; Poletto, V.; Codazzi, A.C.; Mannarino, S.; Campanelli, R.; Fois, G.; Marseglia, G.L.; Guerra, G.; et al. Constitutive Store-Operated Ca(2+) Entry Leads to Enhanced Nitric Oxide Production and Proliferation in Infantile Hemangioma-Derived Endothelial Colony-Forming Cells. *Stem Cells Dev.* **2016**, *25*, 301–319. [CrossRef] [PubMed]

110. Li, J.; Bruns, A.F.; Hou, B.; Rode, B.; Webster, P.J.; Bailey, M.A.; Appleby, H.L.; Moss, N.K.; Ritchie, J.E.; Yuldasheva, N.Y.; et al. ORAI3 Surface Accumulation and Calcium Entry Evoked by Vascular Endothelial Growth Factor. *Arterioscler. Thromb. Vasc. Biol.* **2015**, *35*, 1987–1994. [CrossRef] [PubMed]

111. Moccia, F. Endothelial Ca(2+) Signaling and the Resistance to Anticancer Treatments: Partners in Crime. *Int. J. Mol. Sci.* **2018**, *19*. pii: E217. [CrossRef] [PubMed]

112. Sundivakkam, P.C.; Freichel, M.; Singh, V.; Yuan, J.P.; Vogel, S.M.; Flockerzi, V.; Malik, A.B.; Tiruppathi, C. The Ca(2+) sensor stromal interaction molecule 1 (STIM1) is necessary and sufficient for the store-operated Ca(2+) entry function of transient receptor potential canonical (TRPC) 1 and 4 channels in endothelial cells. *Mol. Pharmacol.* **2012**, *81*, 510–526. [CrossRef] [PubMed]

113. Cioffi, D.L.; Wu, S.; Chen, H.; Alexeyev, M.; St Croix, C.M.; Pitt, B.R.; Uhlig, S.; Stevens, T. ORAI1 determines calcium selectivity of an endogenous TRPC heterotetramer channel. *Circ. Res.* **2012**, *110*, 1435–1444. [CrossRef] [PubMed]

114. Xu, N.; Cioffi, D.L.; Alexeyev, M.; Rich, T.C.; Stevens, T. Sodium entry through endothelial store-operated calcium entry channels: Regulation by ORAI1. *Am. J. Physiol. Cell Physiol.* **2015**, *308*, C277–C288. [CrossRef] [PubMed]

115. Cioffi, D.L.; Barry, C.; Stevens, T. Store-operated calcium entry channels in pulmonary endothelium: The emerging story of TRPCS and Orai1. *Adv. Exp. Med. Biol.* **2010**, *661*, 137–154. [PubMed]

116. Berrout, J.; Jin, M.; O'Neil, R.G. Critical role of TRPP2 and TRPC1 channels in stretch-induced injury of blood-brain barrier endothelial cells. *Brain Res.* **2012**, *1436*, 1–12. [CrossRef] [PubMed]

117. Dragoni, S.; Laforenza, U.; Bonetti, E.; Lodola, F.; Bottino, C.; Guerra, G.; Borghesi, A.; Stronati, M.; Rosti, V.; Tanzi, F.; et al. Canonical transient receptor potential 3 channel triggers vascular endothelial growth factor-induced intracellular Ca^{2+} oscillations in endothelial progenitor cells isolated from umbilical cord blood. *Stem Cells Dev.* **2013**, *22*, 2561–2580. [CrossRef] [PubMed]

118. Moccia, F.; Lucariello, A.; Guerra, G. TRPC3-mediated Ca(2+) signals as a promising strategy to boost therapeutic angiogenesis in failing hearts: The role of autologous endothelial colony forming cells. *J. Cell. Physiol.* **2017**. [CrossRef]

119. Hamdollah Zadeh, M.A.; Glass, C.A.; Magnussen, A.; Hancox, J.C.; Bates, D.O. VEGF-mediated elevated intracellular calcium and angiogenesis in human microvascular endothelial cells in vitro are inhibited by dominant negative TRPC6. *Microcirculation* **2008**, *15*, 605–614. [CrossRef] [PubMed]

120. Chaudhuri, P.; Rosenbaum, M.A.; Birnbaumer, L.; Graham, L.M. Integration of TRPC6 and NADPH oxidase activation in lysophosphatidylcholine-induced TRPC5 externalization. *Am. J. Physiol. Cell Physiol.* **2017**, *313*, C541–C555. [CrossRef] [PubMed]

121. Gees, M.; Colsoul, B.; Nilius, B. The role of transient receptor potential cation channels in Ca^{2+} signaling. *Cold Spring Harb. Perspect. Biol.* **2010**, *2*, a003962. [CrossRef] [PubMed]

122. Parenti, A.; De Logu, F.; Geppetti, P.; Benemei, S. What is the evidence for the role of TRP channels in inflammatory and immune cells? *Br. J. Pharmacol.* **2016**, *173*, 953–969. [CrossRef] [PubMed]

123. Di, A.; Malik, A.B. TRP channels and the control of vascular function. *Curr. Opin. Pharmacol.* **2010**, *10*, 127–132. [CrossRef] [PubMed]

124. Wong, C.O.; Yao, X. TRP channels in vascular endothelial cells. *Adv. Exp. Med. Biol.* **2011**, *704*, 759–780. [PubMed]

125. Ellinsworth, D.C.; Earley, S.; Murphy, T.V.; Sandow, S.L. Endothelial control of vasodilation: Integration of myoendothelial microdomain signalling and modulation by epoxyeicosatrienoic acids. *Pflugers Arch.* **2014**, *466*, 389–405. [CrossRef] [PubMed]

126. Sharp, C.D.; Hines, I.; Houghton, J.; Warren, A.; Jackson, T.H. t.; Jawahar, A.; Nanda, A.; Elrod, J.W.; Long, A.; Chi, A.; et al. Glutamate causes a loss in human cerebral endothelial barrier integrity through activation of NMDA receptor. *Am. J. Physiol. Heart Circ. Physiol.* **2003**, *285*, H2592–H2598. [CrossRef] [PubMed]

127. Lu, L.; Hogan-Cann, A.D.; Globa, A.K.; Lu, P.; Nagy, J.I.; Bamji, S.X.; Anderson, C.M. Astrocytes drive cortical vasodilatory signaling by activating endothelial NMDA receptors. *J. Cereb. Blood Flow Metab.* **2017**. [CrossRef] [PubMed]

128. Krizbai, I.A.; Deli, M.A.; Pestenacz, A.; Siklos, L.; Szabo, C.A.; Andras, I.; Joo, F. Expression of glutamate receptors on cultured cerebral endothelial cells. *J. Neurosci. Res.* **1998**, *54*, 814–819. [CrossRef]

129. Yang, F.; Zhao, K.; Zhang, X.; Zhang, J.; Xu, B. ATP Induces Disruption of Tight Junction Proteins via IL-1 Beta-Dependent MMP-9 Activation of Human Blood-Brain Barrier In Vitro. *Neural Plast.* **2016**, *2016*, 8928530. [CrossRef] [PubMed]

130. Zhao, H.; Zhang, X.; Dai, Z.; Feng, Y.; Li, Q.; Zhang, J.H.; Liu, X.; Chen, Y.; Feng, H. P2X7 Receptor Suppression Preserves Blood-Brain Barrier through Inhibiting RhoA Activation after Experimental Intracerebral Hemorrhage in Rats. *Sci. Rep.* **2016**, *6*, 23286. [CrossRef] [PubMed]

131. De Wit, C.; Wolfle, S.E. EDHF and gap junctions: Important regulators of vascular tone within the microcirculation. *Curr. Pharm. Biotechnol.* **2007**, *8*, 11–25. [CrossRef] [PubMed]

132. Shu, X.; Keller, T.C. t.; Begandt, D.; Butcher, J.T.; Biwer, L.; Keller, A.S.; Columbus, L.; Isakson, B.E. Endothelial nitric oxide synthase in the microcirculation. *Cell. Mol. Life Sci.* **2015**, *72*, 4561–4575. [CrossRef] [PubMed]

133. Dedkova, E.N.; Blatter, L.A. Nitric oxide inhibits capacitative Ca_{2+} entry and enhances endoplasmic reticulum Ca^{2+} uptake in bovine vascular endothelial cells. *J. Physiol.* **2002**, *539 (Pt 1)*, 77–91. [CrossRef] [PubMed]

134. Parekh, A.B. Functional consequences of activating store-operated CRAC channels. *Cell Calcium* **2007**, *42*, 111–121. [CrossRef] [PubMed]

135. Lin, S.; Fagan, K.A.; Li, K.X.; Shaul, P.W.; Cooper, D.M.; Rodman, D.M. Sustained endothelial nitric-oxide synthase activation requires capacitative Ca^{2+} entry. *J. Biol. Chem.* **2000**, *275*, 17979–17985. [CrossRef] [PubMed]

136. Kimura, C.; Oike, M.; Ohnaka, K.; Nose, Y.; Ito, Y. Constitutive nitric oxide production in bovine aortic and brain microvascular endothelial cells: A comparative study. *J. Physiol.* **2004**, *554 (Pt 3)*, 721–730. [CrossRef] [PubMed]

137. Boittin, F.X.; Alonso, F.; Le Gal, L.; Allagnat, F.; Beny, J.L.; Haefliger, J.A. Connexins and M3 muscarinic receptors contribute to heterogeneous Ca(2+) signaling in mouse aortic endothelium. *Cell. Physiol. Biochem.* **2013**, *31*, 166–178. [CrossRef] [PubMed]

138. Yeon, S.I.; Kim, J.Y.; Yeon, D.S.; Abramowitz, J.; Birnbaumer, L.; Muallem, S.; Lee, Y.H. Transient receptor potential canonical type 3 channels control the vascular contractility of mouse mesenteric arteries. *PLoS ONE* **2014**, *9*, e110413. [CrossRef] [PubMed]

139. Zhang, D.X.; Mendoza, S.A.; Bubolz, A.H.; Mizuno, A.; Ge, Z.D.; Li, R.; Warltier, D.C.; Suzuki, M.; Gutterman, D.D. Transient receptor potential vanilloid type 4-deficient mice exhibit impaired endothelium-dependent relaxation induced by acetylcholine in vitro and in vivo. *Hypertension* **2009**, *53*, 532–538. [CrossRef] [PubMed]

140. Zhao, Y.; Vanhoutte, P.M.; Leung, S.W. Vascular nitric oxide: Beyond eNOS. *J. Pharmacol. Sci.* **2015**, *129*, 83–94. [CrossRef] [PubMed]

141. Iadecola, C.; Zhang, F.; Xu, X. Role of nitric oxide synthase-containing vascular nerves in cerebrovasodilation elicited from cerebellum. *Am. J. Physiol.* **1993**, *264 (Pt 2)*, R738–R346. [CrossRef] [PubMed]

142. De Labra, C.; Rivadulla, C.; Espinosa, N.; Dasilva, M.; Cao, R.; Cudeiro, J. Different sources of nitric oxide mediate neurovascular coupling in the lateral geniculate nucleus of the cat. *Front. Syst. Neurosci.* **2009**, *3*, 9. [CrossRef] [PubMed]

143. Fiumana, E.; Parfenova, H.; Jaggar, J.H.; Leffler, C.W. Carbon monoxide mediates vasodilator effects of glutamate in isolated pressurized cerebral arterioles of newborn pigs. *Am. J. Physiol. Heart Circ. Physiol.* **2003**, *284*, H1073–H1079. [CrossRef] [PubMed]

144. Hogan-Cann, A.D.; Anderson, C.M. Physiological Roles of Non-Neuronal NMDA Receptors. *Trends Pharmacol. Sci.* **2016**, *37*, 750–767. [CrossRef] [PubMed]

145. Paoletti, P.; Bellone, C.; Zhou, Q. NMDA receptor subunit diversity: Impact on receptor properties, synaptic plasticity and disease. *Nat. Rev. Neurosci.* **2013**, *14*, 383–400. [CrossRef] [PubMed]

146. Fukaya, M.; Kato, A.; Lovett, C.; Tonegawa, S.; Watanabe, M. Retention of NMDA receptor NR2 subunits in the lumen of endoplasmic reticulum in targeted NR1 knockout mice. *Proc. Natl. Acad. Sci. USA* **2003**, *100*, 4855–4860. [CrossRef] [PubMed]

147. Neuhaus, W.; Freidl, M.; Szkokan, P.; Berger, M.; Wirth, M.; Winkler, J.; Gabor, F.; Pifl, C.; Noe, C.R. Effects of NMDA receptor modulators on a blood-brain barrier in vitro model. *Brain Res.* **2011**, *1394*, 49–61. [CrossRef] [PubMed]

148. Kamat, P.K.; Kalani, A.; Tyagi, S.C.; Tyagi, N. Hydrogen Sulfide Epigenetically Attenuates Homocysteine-Induced Mitochondrial Toxicity Mediated Through NMDA Receptor in Mouse Brain Endothelial (bEnd3) Cells. *J. Cell. Physiol.* **2015**, *230*, 378–394. [CrossRef] [PubMed]

149. Busija, D.W.; Bari, F.; Domoki, F.; Louis, T. Mechanisms involved in the cerebrovascular dilator effects of N-methyl-D-aspartate in cerebral cortex. *Brain Res. Rev.* **2007**, *56*, 89–100. [CrossRef] [PubMed]

150. Hamel, E. Cholinergic modulation of the cortical microvascular bed. *Prog. Brain Res.* **2004**, *145*, 171–178. [PubMed]

151. Zhang, F.; Xu, S.; Iadecola, C. Role of nitric oxide and acetylcholine in neocortical hyperemia elicited by basal forebrain stimulation: Evidence for an involvement of endothelial nitric oxide. *Neuroscience* **1995**, *69*, 1195–1204. [CrossRef]

152. Iadecola, C.; Zhang, F. Permissive and obligatory roles of NO in cerebrovascular responses to hypercapnia and acetylcholine. *Am. J. Physiol.* **1996**, *271 (Pt 2)*, R990–R1001. [CrossRef] [PubMed]

153. Yamada, M.; Lamping, K.G.; Duttaroy, A.; Zhang, W.; Cui, Y.; Bymaster, F.P.; McKinzie, D.L.; Felder, C.C.; Deng, C.X.; Faraci, F.M.; et al. Cholinergic dilation of cerebral blood vessels is abolished in M(5) muscarinic acetylcholine receptor knockout mice. *Proc. Natl. Acad. Sci. USA* **2001**, *98*, 14096–14101. [CrossRef] [PubMed]

154. Sandow, S.L.; Senadheera, S.; Grayson, T.H.; Welsh, D.G.; Murphy, T.V. Calcium and endothelium-mediated vasodilator signaling. *Adv. Exp. Med. Biol.* **2012**, *740*, 811–831. [PubMed]

155. Kasai, Y.; Yamazawa, T.; Sakurai, T.; Taketani, Y.; Iino, M. Endothelium-dependent frequency modulation of Ca^{2+} signalling in individual vascular smooth muscle cells of the rat. *J. Physiol.* **1997**, *504 (Pt 2)*, 349–357. [CrossRef] [PubMed]

156. Weikert, S.; Freyer, D.; Weih, M.; Isaev, N.; Busch, C.; Schultze, J.; Megow, D.; Dirnagl, U. Rapid Ca^{2+}-dependent NO-production from central nervous system cells in culture measured by NO-nitrite/ozone chemoluminescence. *Brain Res.* **1997**, *748*, 1–11. [CrossRef]

157. Radu, B.M.; Osculati, A.M.M.; Suku, E.; Banciu, A.; Tsenov, G.; Merigo, F.; Di Chio, M.; Banciu, D.D.; Tognoli, C.; Kacer, P.; et al. All muscarinic acetylcholine receptors (M1–M5) are expressed in murine brain microvascular endothelium. *Sci. Rep.* **2017**, *7*, 5083. [CrossRef] [PubMed]

158. Moccia, F.; Frost, C.; Berra-Romani, R.; Tanzi, F.; Adams, D.J. Expression and function of neuronal nicotinic ACh receptors in rat microvascular endothelial cells. *Am. J. Physiol. Heart Circ. Physiol.* **2004**, *286*, H486–H491. [CrossRef] [PubMed]

159. Laforenza, U.; Pellavio, G.; Moccia, F. Muscarinic M5 receptors trigger acetylcholine-induced Ca^{2+} signals and nitric oxide release in human brain microvascular endothelial cells. 2017; manuscript in preparation.

160. Uhlen, P.; Fritz, N. Biochemistry of calcium oscillations. *Biochem. Biophys. Res. Commun.* **2010**, *396*, 28–32. [CrossRef] [PubMed]

161. Mikoshiba, K. IP3 receptor/Ca^{2+} channel: From discovery to new signaling concepts. *J. Neurochem.* **2007**, *102*, 1426–1046. [CrossRef] [PubMed]

162. Guzman, S.J.; Gerevich, Z. P2Y Receptors in Synaptic Transmission and Plasticity: Therapeutic Potential in Cognitive Dysfunction. *Neural Plast.* **2016**, *2016*, 1207393. [CrossRef] [PubMed]

163. Munoz, M.F.; Puebla, M.; Figueroa, X.F. Control of the neurovascular coupling by nitric oxide-dependent regulation of astrocytic Ca(2+) signaling. *Front. Cell. Neurosci.* **2015**, *9*, 59. [CrossRef] [PubMed]

164. Dietrich, H.H.; Horiuchi, T.; Xiang, C.; Hongo, K.; Falck, J.R.; Dacey, R.G., Jr. Mechanism of ATP-induced local and conducted vasomotor responses in isolated rat cerebral penetrating arterioles. *J. Vasc. Res.* **2009**, *46*, 253–264. [CrossRef] [PubMed]

165. Dietrich, H.H.; Kajita, Y.; Dacey, R.G., Jr. Local and conducted vasomotor responses in isolated rat cerebral arterioles. *Am. J. Physiol.* **1996**, *71*, H1109–H1116. [CrossRef] [PubMed]

166. Toth, P.; Tarantini, S.; Davila, A.; Valcarcel-Ares, M.N.; Tucsek, Z.; Varamini, B.; Ballabh, P.; Sonntag, W.E.; Baur, J.A.; Csiszar, A.; et al. Purinergic glio-endothelial coupling during neuronal activity: Role of P2Y1 receptors and eNOS in functional hyperemia in the mouse somatosensory cortex. *Am. J. Physiol. Heart Circ. Physiol.* **2015**, *309*, H1837–H1845. [CrossRef] [PubMed]

167. Moccia, F.; Baruffi, S.; Spaggiari, S.; Coltrini, D.; Berra-Romani, R.; Signorelli, S.; Castelli, L.; Taglietti, V.; Tanzi, F. P2y1 and P2y2 receptor-operated Ca^{2+} signals in primary cultures of cardiac microvascular endothelial cells. *Microvasc. Res.* **2001**, *61*, 240–252. [CrossRef] [PubMed]

168. Marrelli, S.P. Mechanisms of endothelial P2Y(1)- and P2Y(2)-mediated vasodilatation involve differential $[Ca^{2+}]i$ responses. *Am. J. Physiol. Heart Circ. Physiol.* **2001**, *281*, H1759–H1766. [CrossRef] [PubMed]

169. Kwan, H.Y.; Cheng, K.T.; Ma, Y.; Huang, Y.; Tang, N.L.; Yu, S.; Yao, X. CNGA2 contributes to ATP-induced noncapacitative Ca^{2+} influx in vascular endothelial cells. *J. Vasc. Res.* **2010**, *47*, 148–156. [CrossRef] [PubMed]

170. Duza, T.; Sarelius, I.H. Conducted dilations initiated by purines in arterioles are endothelium dependent and require endothelial Ca^{2+}. *Am. J. Physiol. Heart Circ. Physiol.* **2003**, *285*, H26–H37. [CrossRef] [PubMed]

171. Kohler, R.; Hoyer, J. The endothelium-derived hyperpolarizing factor: Insights from genetic animal models. *Kidney Int.* **2007**, *72*, 145–150. [CrossRef] [PubMed]

172. Ledoux, J.; Taylor, M.S.; Bonev, A.D.; Hannah, R.M.; Solodushko, V.; Shui, B.; Tallini, Y.; Kotlikoff, M.I.; Nelson, M.T. Functional architecture of inositol 1,4,5-trisphosphate signaling in restricted spaces of myoendothelial projections. *Proc. Natl. Acad. Sci. USA* **2008**, *105*, 9627–9632. [CrossRef] [PubMed]

173. Dora, K.A.; Gallagher, N.T.; McNeish, A.; Garland, C.J. Modulation of endothelial cell KCa3.1 channels during endothelium-derived hyperpolarizing factor signaling in mesenteric resistance arteries. *Circ. Res.* **2008**, *102*, 1247–1255. [CrossRef] [PubMed]

174. Sonkusare, S.K.; Bonev, A.D.; Ledoux, J.; Liedtke, W.; Kotlikoff, M.I.; Heppner, T.J.; Hill-Eubanks, D.C.; Nelson, M.T. Elementary Ca^{2+} signals through endothelial TRPV4 channels regulate vascular function. *Science* **2012**, *336*, 597–601. [CrossRef] [PubMed]

175. Behringer, E.J.; Segal, S.S. Spreading the signal for vasodilatation: Implications for skeletal muscle blood flow control and the effects of ageing. *J. Physiol.* **2012**, *590*, 6277–6284. [CrossRef] [PubMed]

176. Behringer, E.J.; Segal, S.S. Tuning electrical conduction along endothelial tubes of resistance arteries through Ca(2+)-activated K(+) channels. *Circ. Res.* **2012**, *110*, 1311–1321. [CrossRef] [PubMed]

177. Kapela, A.; Behringer, E.J.; Segal, S.S.; Tsoukias, N.M. Biophysical properties of microvascular endothelium: Requirements for initiating and conducting electrical signals. *Microcirculation* **2017**, *25*. [CrossRef] [PubMed]

178. Sullivan, M.N.; Gonzales, A.L.; Pires, P.W.; Bruhl, A.; Leo, M.D.; Li, W.; Oulidi, A.; Boop, F.A.; Feng, Y.; Jaggar, J.H.; et al. Localized TRPA1 channel Ca^{2+} signals stimulated by reactive oxygen species promote cerebral artery dilation. *Science Signal.* **2015**, *8*, ra2. [CrossRef] [PubMed]

179. Hannah, R.M.; Dunn, K.M.; Bonev, A.D.; Nelson, M.T. Endothelial SK(Ca) and IK(Ca) channels regulate brain parenchymal arteriolar diameter and cortical cerebral blood flow. *J. Cereb. Blood Flow Metab.* **2011**, *31*, 1175–1186. [CrossRef] [PubMed]

180. Sokoya, E.M.; Burns, A.R.; Setiawan, C.T.; Coleman, H.A.; Parkington, H.C.; Tare, M. Evidence for the involvement of myoendothelial gap junctions in EDHF-mediated relaxation in the rat middle cerebral artery. *Am. J. Physiol. Heart Circ. Physiol.* **2006**, *291*, H385–H393. [CrossRef] [PubMed]

181. Bagher, P.; Beleznai, T.; Kansui, Y.; Mitchell, R.; Garland, C.J.; Dora, K.A. Low intravascular pressure activates endothelial cell TRPV4 channels, local Ca^{2+} events, and IKCa channels, reducing arteriolar tone. *Proc. Natl. Acad. Sci. USA* **2012**, *109*, 18174–18179. [CrossRef] [PubMed]

182. Golding, E.M.; Marrelli, S.P.; You, J.; Bryan, R.M., Jr. Endothelium-derived hyperpolarizing factor in the brain: A new regulator of cerebral blood flow? *Stroke* **2002**, *33*, 661–663. [PubMed]

183. McNeish, A.J.; Dora, K.A.; Garland, C.J. Possible role for K^+ in endothelium-derived hyperpolarizing factor-linked dilatation in rat middle cerebral artery. *Stroke* **2005**, *36*, 1526–1532. [CrossRef] [PubMed]

184. Marrelli, S.P.; Eckmann, M.S.; Hunte, M.S. Role of endothelial intermediate conductance KCa channels in cerebral EDHF-mediated dilations. *Am. J. Physiol. Heart Circ. Physiol.* **2003**, *285*, H1590–H1599. [CrossRef] [PubMed]

185. McNeish, A.J.; Sandow, S.L.; Neylon, C.B.; Chen, M.X.; Dora, K.A.; Garland, C.J. Evidence for involvement of both IKCa and SKCa channels in hyperpolarizing responses of the rat middle cerebral artery. *Stroke* **2006**, *37*, 1277–1282. [CrossRef] [PubMed]

186. Zhang, L.; Papadopoulos, P.; Hamel, E. Endothelial TRPV4 channels mediate dilation of cerebral arteries: Impairment and recovery in cerebrovascular pathologies related to Alzheimer's disease. *Br. J. Pharmacol.* **2013**, *170*, 661–670. [CrossRef] [PubMed]

187. Dragoni, S.; Guerra, G.; Fiorio Pla, A.; Bertoni, G.; Rappa, A.; Poletto, V.; Bottino, C.; Aronica, A.; Lodola, F.; Cinelli, M.P.; et al. A functional Transient Receptor Potential Vanilloid 4 (TRPV4) channel is expressed in human endothelial progenitor cells. *J. Cell. Physiol.* **2015**, *230*, 95–104. [CrossRef] [PubMed]

188. Thodeti, C.K.; Matthews, B.; Ravi, A.; Mammoto, A.; Ghosh, K.; Bracha, A.L.; Ingber, D.E. TRPV4 channels mediate cyclic strain-induced endothelial cell reorientation through integrin-to-integrin signaling. *Circ. Res.* **2009**, *104*, 1123–1130. [CrossRef] [PubMed]

189. He, D.; Pan, Q.; Chen, Z.; Sun, C.; Zhang, P.; Mao, A.; Zhu, Y.; Li, H.; Lu, C.; Xie, M.; et al. Treatment of hypertension by increasing impaired endothelial TRPV4-KCa2.3 interaction. *EMBO Mol. Med.* **2017**, *9*, 1491–1503. [CrossRef] [PubMed]

190. Ho, W.S.; Zheng, X.; Zhang, D.X. Role of endothelial TRPV4 channels in vascular actions of the endocannabinoid, 2-arachidonoylglycerol. *Br. J. Pharmacol.* **2015**, *172*, 5251–5264. [CrossRef] [PubMed]

191. Mendoza, S.A.; Fang, J.; Gutterman, D.D.; Wilcox, D.A.; Bubolz, A.H.; Li, R.; Suzuki, M.; Zhang, D.X. TRPV4-mediated endothelial Ca^{2+} influx and vasodilation in response to shear stress. *Am. J. Physiol. Heart Circ. Physiol.* **2010**, *298*, H466–H476. [CrossRef] [PubMed]

192. Zheng, X.; Zinkevich, N.S.; Gebremedhin, D.; Gauthier, K.M.; Nishijima, Y.; Fang, J.; Wilcox, D.A.; Campbell, W.B.; Gutterman, D.D.; Zhang, D.X. Arachidonic acid-induced dilation in human coronary arterioles: Convergence of signaling mechanisms on endothelial TRPV4-mediated Ca^{2+} entry. *J. Am. Heart Assoc.* **2013**, *2*, e000080. [CrossRef] [PubMed]

193. Watanabe, H.; Vriens, J.; Suh, S.H.; Benham, C.D.; Droogmans, G.; Nilius, B. Heat-evoked activation of TRPV4 channels in a HEK293 cell expression system and in native mouse aorta endothelial cells. *J. Biol. Chem.* **2002**, *277*, 47044–47051. [CrossRef] [PubMed]

194. White, J.P.; Cibelli, M.; Urban, L.; Nilius, B.; McGeown, J.G.; Nagy, I. TRPV4: Molecular Conductor of a Diverse Orchestra. *Physiol. Rev.* **2016**, *96*, 911–973. [CrossRef] [PubMed]

195. Watanabe, H.; Vriens, J.; Prenen, J.; Droogmans, G.; Voets, T.; Nilius, B. Anandamide and arachidonic acid use epoxyeicosatrienoic acids to activate TRPV4 channels. *Nature* **2003**, *424*, 434–438. [CrossRef] [PubMed]

196. Graier, W.F.; Simecek, S.; Sturek, M. Cytochrome P450 mono-oxygenase-regulated signalling of Ca^{2+} entry in human and bovine endothelial cells. *J. Physiol.* **1995**, *482 (Pt 2)*, 259–274. [CrossRef] [PubMed]

197. Vriens, J.; Owsianik, G.; Fisslthaler, B.; Suzuki, M.; Janssens, A.; Voets, T.; Morisseau, C.; Hammock, B.D.; Fleming, I.; Busse, R.; et al. Modulation of the Ca^2 permeable cation channel TRPV4 by cytochrome P450 epoxygenases in vascular endothelium. *Circ. Res.* **2005**, *97*, 908–915. [CrossRef] [PubMed]

198. Earley, S.; Pauyo, T.; Drapp, R.; Tavares, M.J.; Liedtke, W.; Brayden, J.E. TRPV4-dependent dilation of peripheral resistance arteries influences arterial pressure. *Am. J. Physiol. Heart Circ. Physiol.* **2009**, *297*, H1096–H1102. [CrossRef] [PubMed]

199. Marrelli, S.P.; O'Neil R, G.; Brown, R.C.; Bryan, R.M., Jr. PLA2 and TRPV4 channels regulate endothelial calcium in cerebral arteries. *Am. J. Physiol. Heart Circ. Physiol.* **2007**, *292*, H1390–H1397. [CrossRef] [PubMed]

200. You, J.; Marrelli, S.P.; Bryan, R.M., Jr. Role of cytoplasmic phospholipase A2 in endothelium-derived hyperpolarizing factor dilations of rat middle cerebral arteries. *J. Cereb. Blood Flow Metab.* **2002**, *22*, 1239–1247. [CrossRef] [PubMed]

201. Liu, X.; Li, C.; Gebremedhin, D.; Hwang, S.H.; Hammock, B.D.; Falck, J.R.; Roman, R.J.; Harder, D.R.; Koehler, R.C. Epoxyeicosatrienoic acid-dependent cerebral vasodilation evoked by metabotropic glutamate receptor activation in vivo. *Am. J. Physiol. Heart Circ. Physiol.* **2011**, *301*, H373–H381. [CrossRef] [PubMed]

202. Peng, X.; Zhang, C.; Alkayed, N.J.; Harder, D.R.; Koehler, R.C. Dependency of cortical functional hyperemia to forepaw stimulation on epoxygenase and nitric oxide synthase activities in rats. *J. Cereb. Blood Flow Metab.* **2004**, *24*, 509–517. [CrossRef] [PubMed]

203. Chen, B.R.; Kozberg, M.G.; Bouchard, M.B.; Shaik, M.A.; Hillman, E.M. A critical role for the vascular endothelium in functional neurovascular coupling in the brain. *J. Am. Heart Assoc.* **2014**, *3*, e000787. [CrossRef] [PubMed]

204. Longden, T.A.; Dabertrand, F.; Koide, M.; Gonzales, A.L.; Tykocki, N.R.; Brayden, J.E.; Hill-Eubanks, D.; Nelson, M.T. Capillary K(+)-sensing initiates retrograde hyperpolarization to increase local cerebral blood flow. *Nat. Neurosci.* **2017**, *20*, 717–726. [CrossRef] [PubMed]

205. Earley, S.; Gonzales, A.L.; Crnich, R. Endothelium-dependent cerebral artery dilation mediated by TRPA1 and Ca^{2+}-Activated K$^+$ channels. *Circ. Res.* **2009**, *104*, 987–994. [CrossRef] [PubMed]

206. Qian, X.; Francis, M.; Solodushko, V.; Earley, S.; Taylor, M.S. Recruitment of dynamic endothelial Ca^{2+} signals by the TRPA1 channel activator AITC in rat cerebral arteries. *Microcirculation* **2013**, *20*, 138–148. [CrossRef] [PubMed]

207. Senadheera, S.; Kim, Y.; Grayson, T.H.; Toemoe, S.; Kochukov, M.Y.; Abramowitz, J.; Housley, G.D.; Bertrand, R.L.; Chadha, P.S.; Bertrand, P.P.; et al. Transient receptor potential canonical type 3 channels facilitate endothelium-derived hyperpolarization-mediated resistance artery vasodilator activity. *Cardiovasc. Res.* **2012**, *95*, 439–447. [CrossRef] [PubMed]

208. Kochukov, M.Y.; Balasubramanian, A.; Abramowitz, J.; Birnbaumer, L.; Marrelli, S.P. Activation of endothelial transient receptor potential C3 channel is required for small conductance calcium-activated potassium

channel activation and sustained endothelial hyperpolarization and vasodilation of cerebral artery. *J. Am. Heart Assoc.* **2014**, *3*, e000913. [CrossRef] [PubMed]

209. Pires, P.W.; Sullivan, M.N.; Pritchard, H.A.; Robinson, J.J.; Earley, S. Unitary TRPV3 channel Ca^{2+} influx events elicit endothelium-dependent dilation of cerebral parenchymal arterioles. *Am. J. Physiol. Heart Circ. Physiol.* **2015**, *309*, H2031–H2041. [CrossRef] [PubMed]

210. Yang, P.; Zhu, M.X. Trpv3. In *Handbook of Experimental Pharmacology*; Springer: Berlin, Germany, 2014; Volume 222, pp. 273–291.

211. Zygmunt, P.M.; Hogestatt, E.D. Trpa1. In *Handbook of Experimental Pharmacology*; Springer: Berlin, Germany, 2014; Volume 222, pp. 583–630.

212. Iadecola, C. The pathobiology of vascular dementia. *Neuron* **2013**, *80*, 844–866. [CrossRef] [PubMed]

213. Chen, B.R.; Bouchard, M.B.; McCaslin, A.F.; Burgess, S.A.; Hillman, E.M. High-speed vascular dynamics of the hemodynamic response. *NeuroImage* **2011**, *54*, 1021–1030. [CrossRef] [PubMed]

214. Hillman, E.M.; Devor, A.; Bouchard, M.B.; Dunn, A.K.; Krauss, G.W.; Skoch, J.; Bacskai, B.J.; Dale, A.M.; Boas, D.A. Depth-resolved optical imaging and microscopy of vascular compartment dynamics during somatosensory stimulation. *NeuroImage* **2007**, *35*, 89–104. [CrossRef] [PubMed]

215. Stefanovic, B.; Hutchinson, E.; Yakovleva, V.; Schram, V.; Russell, J.T.; Belluscio, L.; Koretsky, A.P.; Silva, A.C. Functional reactivity of cerebral capillaries. *J. Cereb. Blood Flow Metab.* **2008**, *28*, 961–972. [CrossRef] [PubMed]

216. O'Herron, P.; Chhatbar, P.Y.; Levy, M.; Shen, Z.; Schramm, A.E.; Lu, Z.; Kara, P. Neural correlates of single-vessel haemodynamic responses in vivo. *Nature* **2016**, *534*, 378–382. [CrossRef] [PubMed]

217. Kisler, K.; Nelson, A.R.; Rege, S.V.; Ramanathan, A.; Wang, Y.; Ahuja, A.; Lazic, D.; Tsai, P.S.; Zhao, Z.; Zhou, Y.; et al. Pericyte degeneration leads to neurovascular uncoupling and limits oxygen supply to brain. *Nat. Neurosci.* **2017**, *20*, 406–416. [CrossRef] [PubMed]

218. Behringer, E.J.; Socha, M.J.; Polo-Parada, L.; Segal, S.S. Electrical conduction along endothelial cell tubes from mouse feed arteries: Confounding actions of glycyrrhetinic acid derivatives. *Br. J. Pharmacol.* **2012**, *166*, 774–787. [CrossRef] [PubMed]

219. Vaucher, E.; Hamel, E. Cholinergic basal forebrain neurons project to cortical microvessels in the rat: Electron microscopic study with anterogradely transported Phaseolus vulgaris leucoagglutinin and choline acetyltransferase immunocytochemistry. *J. Neurosci.* **1995**, *15*, 7427–7441. [PubMed]

220. Vaucher, E.; Linville, D.; Hamel, E. Cholinergic basal forebrain projections to nitric oxide synthase-containing neurons in the rat cerebral cortex. *Neuroscience* **1997**, *79*, 827–836. [CrossRef]

221. Jensen, L.J.; Holstein-Rathlou, N.H. The vascular conducted response in cerebral blood flow regulation. *J. Cereb. Blood Flow Metab.* **2013**, *33*, 649–656. [CrossRef] [PubMed]

222. Uhrenholt, T.R.; Domeier, T.L.; Segal, S.S. Propagation of calcium waves along endothelium of hamster feed arteries. *Am. J. Physiol. Heart Circ. Physiol.* **2007**, *292*, H1634–H1640. [CrossRef] [PubMed]

223. Domeier, T.L.; Segal, S.S. Electromechanical and pharmacomechanical signalling pathways for conducted vasodilatation along endothelium of hamster feed arteries. *J. Physiol.* **2007**, *579*, 175–186. [CrossRef] [PubMed]

224. De Bock, M.; Culot, M.; Wang, N.; da Costa, A.; Decrock, E.; Bol, M.; Bultynck, G.; Cecchelli, R.; Leybaert, L. Low extracellular Ca^{2+} conditions induce an increase in brain endothelial permeability that involves intercellular Ca^{2+} waves. *Brain Res.* **2012**, *1487*, 78–87. [CrossRef] [PubMed]

225. Berra-Romani, R.; Raqeeb, A.; Torres-Jácome, J.; Guzman-Silva, A.; Guerra, G.; Tanzi, F.; Moccia, F. The mechanism of injury-induced intracellular calcium concentration oscillations in the endothelium of excised rat aorta. *J. Vasc. Res.* **2012**, *49*, 65–76. [CrossRef] [PubMed]

226. Francis, M.; Waldrup, J.R.; Qian, X.; Solodushko, V.; Meriwether, J.; Taylor, M.S. Functional Tuning of Intrinsic Endothelial Ca^{2+} Dynamics in Swine Coronary Arteries. *Circ. Res.* **2016**, *118*, 1078–1090. [CrossRef] [PubMed]

227. Wilson, C.; Saunter, C.D.; Girkin, J.M.; McCarron, J.G. Clusters of specialized detector cells provide sensitive and high fidelity receptor signaling in the intact endothelium. *FASEB J.* **2016**, *30*, 2000–2013. [CrossRef] [PubMed]

228. Kohler, R.; Hoyer, J. Role of TRPV4 in the Mechanotransduction of Shear Stress in Endothelial Cells. In *TRP Ion Channel Function in Sensory Transduction and Cellular Signaling Cascades*; Liedtke, W.B., Heller, S., Eds.; CRC Press: Boca Raton, FL, USA, 2007.

229. Bryan, R.M., Jr.; Marrelli, S.P.; Steenberg, M.L.; Schildmeyer, L.A.; Johnson, T.D. Effects of luminal shear stress on cerebral arteries and arterioles. *Am. J. Physiol. Heart Circ. Physiol.* **2001**, *280*, H2011–H2022. [CrossRef] [PubMed]

230. Bryan, R.M., Jr.; Steenberg, M.L.; Marrelli, S.P. Role of endothelium in shear stress-induced constrictions in rat middle cerebral artery. *Stroke* **2001**, *32*, 1394–1400. [CrossRef] [PubMed]

231. Ngai, A.C.; Winn, H.R. Modulation of cerebral arteriolar diameter by intraluminal flow and pressure. *Circ. Res.* **1995**, *77*, 832–840. [CrossRef] [PubMed]

232. Lourenco, C.F.; Ledo, A.; Barbosa, R.M.; Laranjinha, J. Neurovascular uncoupling in the triple transgenic model of Alzheimer's disease: Impaired cerebral blood flow response to neuronal-derived nitric oxide signaling. *Exp. Neurol.* **2017**, *291*, 36–43. [CrossRef] [PubMed]

233. Lourenco, C.F.; Ledo, A.; Dias, C.; Barbosa, R.M.; Laranjinha, J. Neurovascular and neurometabolic derailment in aging and Alzheimer's disease. *Front. Aging Neurosci.* **2015**, *7*, 103. [CrossRef] [PubMed]

234. Katakam, P.V.; Domoki, F.; Snipes, J.A.; Busija, A.R.; Jarajapu, Y.P.; Busija, D.W. Impaired mitochondria-dependent vasodilation in cerebral arteries of Zucker obese rats with insulin resistance. *Am. J. Physiol. Regul. Integr. Comp. Physiol.* **2009**, *296*, R289–R298. [CrossRef] [PubMed]

235. Mottola, A.; Antoniotti, S.; Lovisolo, D.; Munaron, L. Regulation of noncapacitative calcium entry by arachidonic acid and nitric oxide in endothelial cells. *FASEB J.* **2005**, *19*, 2075–2077. [CrossRef] [PubMed]

5

HDAC Inhibition Improves the Sarcoendoplasmic Reticulum Ca^{2+}-ATPase Activity in Cardiac Myocytes

Viviana Meraviglia [1,†], Leonardo Bocchi [2,†], Roberta Sacchetto [3], Maria Cristina Florio [1],
Benedetta M. Motta [1], Corrado Corti [1], Christian X. Weichenberger [1], Monia Savi [2], Yuri D'Elia [1],
Marcelo D. Rosato-Siri [1], Silvia Suffredini [1], Chiara Piubelli [1], Giulio Pompilio [4,5],
Peter P. Pramstaller [1], Francisco S. Domingues [1], Donatella Stilli [2,*,‡] and Alessandra Rossini [1,*,‡]

[1] Institute for Biomedicine, Eurac Research, 39100 Bolzano, Italy (affiliated institute of the University of
     Lübeck, 23562 Lübeck, Germany); viviana.meraviglia@eurac.edu (V.M.); cristiflorio@gmail.com (M.C.F.);
     benedetta.motta@eurac.edu (B.M.M.); corrado.corti@eurac.edu (C.C.);
     christian.weichenberger@eurac.edu (C.X.W.); yuri.delia@eurac.edu (Y.D.);
     Marcelo.RosatoSiri@eurac.edu (M.D.R.-S.); silvia.suffredini@alice.it (S.S.);
     chiara.piubelli@sacrocuore.it (C.P.); peter.pramstaller@eurac.edu (P.P.P.);
     francisco.domingues@eurac.edu (F.S.D.)
[2] Department of Chemistry, Life Sciences and Environmental Sustainability, University of Parma,
     43124 Parma, Italy; leonardo.bocchi@unipr.it (L.B.); monia.savi@unipr.it (M.S.)
[3] Department of Comparative Biomedicine and Food Science, University of Padova, 35020 Legnaro (Padova),
     Italy; roberta.sacchetto@unipd.it
[4] Vascular Biology and Regenerative Medicine Unit, Centro Cardiologico Monzino, IRCCS, 20138 Milano,
     Italy; giulio.pompilio@ccfm.it
[5] Dipartimento di Scienze Cliniche e di Comunità, Università degli Studi di Milano, 20122 Milano, Italy
*   Correspondence: donatella.stilli@unipr.it (D.S.); alessandra.rossini@eurac.edu (A.R.);
†   These authors contributed equally to this work.
‡   These authors contributed equally to this work.

Abstract: SERCA2a is the Ca^{2+} ATPase playing the major contribution in cardiomyocyte (CM) calcium removal. Its activity can be regulated by both modulatory proteins and several post-translational modifications. The aim of the present work was to investigate whether the function of SERCA2 can be modulated by treating CMs with the histone deacetylase (HDAC) inhibitor suberanilohydroxamic acid (SAHA). The incubation with SAHA (2.5 μM, 90 min) of CMs isolated from rat adult hearts resulted in an increase of SERCA2 acetylation level and improved ATPase activity. This was associated with a significant improvement of calcium transient recovery time and cell contractility. Previous reports have identified K464 as an acetylation site in human SERCA2. Mutants were generated where K464 was substituted with glutamine (Q) or arginine (R), mimicking constitutive acetylation or deacetylation, respectively. The K464Q mutation ameliorated ATPase activity and calcium transient recovery time, thus indicating that constitutive K464 acetylation has a positive impact on human SERCA2a (hSERCA2a) function. In conclusion, SAHA induced deacetylation inhibition had a positive impact on CM calcium handling, that, at least in part, was due to improved SERCA2 activity. This observation can provide the basis for the development of novel pharmacological approaches to ameliorate SERCA2 efficiency.

Keywords: SERCA2; acetylation; HDAC inhibition; ATPase activity; calcium transients; cardiomyocyte mechanics

1. Introduction

The Sarco(Endo)plasmic Reticulum (SR) Ca^{2+} ATPase (SERCA) is a Ca^{2+} pump that uses the energy of ATP hydrolysis to translocate two Ca^{2+} ions from the cytosol to the SR lumen. It is composed of a single polypeptide weighing approximately 110 kDa and organized into four different domains: one transmembrane domain composed of ten alpha-helices, one nucleotide (ATP) binding site, one phosphorylation/catalytic domain that drives ATP hydrolysis, and one actuator domain involved in the gating mechanism that regulates calcium binding and release [1,2]. In vertebrates, three different SERCA genes have been identified so far, showing high degree of conservation among species: *ATP2A1*, *ATP2A2* and *ATP2A3*, encoding respectively for the SERCA1, SERCA2 and SERCA3 proteins. Thanks to alternative splicing mechanisms, they can give rise to different protein isoforms, whose expression is regulated during development and in a tissue specific manner [3]. SERCA2a is the most expressed isoform in cardiac muscle [3] and is the responsible for the level of SR Ca^{2+} load and the majority of the cytosolic calcium removal after contraction (70% in human, up to 90% in mouse and rats), thus representing a key player for cardiomyocyte (CM) relaxation [4]. Alterations in SERCA2a are recognized among the major factors contributing to ventricular dysfunction in several heart diseases, like diabetic cardiomyopathy [5,6] and heart failure [7,8].

Owing its essential role in cardiac physiology, the mechanisms regulating SERCA2a have been extensively studied [9]. Evidence has been provided that the pump activity can be directly modulated by several post-translational modifications [10]. Specifically, sumoylation [11] and glutathionylation [12] increase the activity, while glycosylation [13] and nitration [14] seem to decrease SERCA2a function. A recent study by Foster and coworkers [15] identified three acetylation sites within the structure of SERCA2a. However, the potential impact of SERCA2a acetylation on intracellular Ca^{2+} cycling has never been evaluated.

The enzymes responsible for protein acetylation and deacetylation are histone acetyltransferases (HATs) and histone deacetylases (HDACs), respectively. HATs and HDACs act not only on histones, but also on other nuclear and cytoplasmic proteins, such as transcription factors and structural proteins [16].

The acetylation balance carried out by HATs and HDACs plays a role in the physiology of adult CMs. For instance, it has been demonstrated that HDAC inhibition prevented ventricular arrhythmias and restored conduction defects of the heart in a model of dystrophic mice [17].

The specific aim of the present work was to investigate the impact of deacetylation inhibition on SERCA2 activity. Specifically, adult ventricular CMs isolated from normal rat hearts were exposed to the HDAC general inhibitor suberanilohydroxamic acid (SAHA, or Vorinostat). Additionally, three stable lines of HEK cells were produced to express the human SERCA2a either wild type or with mutations mimicking constitutive acetylation/deacetylation on a selected lysine residue.

In this manuscript, we demonstrate for the first time that HDAC inhibition promotes improved efficiency in cytosolic Ca^{2+} removal, along with increased SERCA2 acetylation and ATPase activity.

2. Results

2.1. Effect of SAHA Treatment on SERCA2a Acetylation and Function in Control Rat Cardiomyocytes

Cardiomyocytes (CMs) isolated from normal adult rat hearts were used as model to investigate whether SAHA-promoted HDAC inhibition can impact SERCA2a acetylation and function. Following enzymatic dissociation, CMs were either untreated (CTR) or incubated with 2.5 μM SAHA (CTR+SAHA) for 90 min. As expected [18], SAHA administration induced higher tubulin acetylation level (Figure 1a).

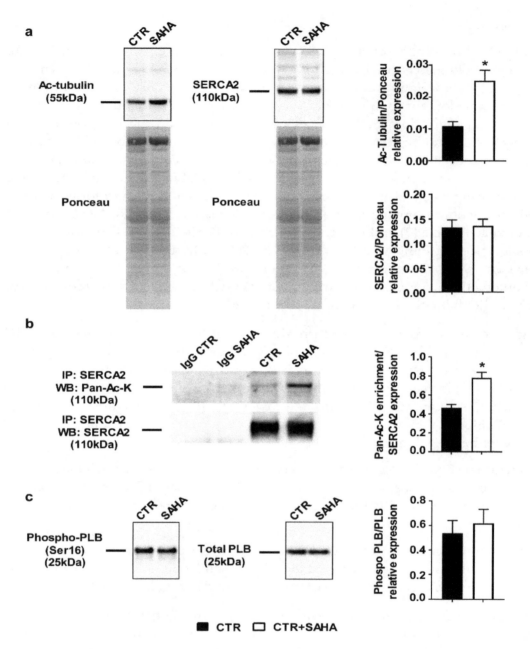

Figure 1. Effect of SAHA treatment on cardiomyocytes isolated from adult rat hearts. (**a**) Western blot panels (**on the left**) and densitometric analysis (**on the right**) showing SERCA2 and Ac-Tubulin protein expression after SAHA treatment ($n = 7$). Mann–Whitney U-test: * $p < 0.005$ vs. CTR; (**b**) Immunoprecipitation experiments and densitometric analysis (**on the right**) indicating higher acetylation level of SERCA2 after SAHA treatment ($n = 5$). Mann–Whitney U-test: * $p < 0.005$ vs. CTR; (**c**) Western blot analysis showing the expression of phosphorylated phospholamban (Phospho-PLB, Ser16) compared to total phospholamban (PLB) in adult rat CMs after SAHA treatment ($n = 7$).

Co-immunoprecipitation experiments revealed that SAHA increased SERCA2 acetylation level (Figure 1b and Figure S1), without inducing significant changes in SERCA2 protein expression (Figure 1a). The ratio of phosphorylated phospholamban (PLB)/total phospholamban (Figure 1c) remained unchanged.

To investigate whether increased acetylation could also affect SERCA2 functional properties, ATPase activity was measured on microsomes [19] isolated from both CTR and CTR+SAHA CMs. HDAC inhibition resulted in an increase of ATPase activity when microsomes were exposed to 10 µM calcium concentration (corresponding to pCa5) [20,21] (Figure 2a).

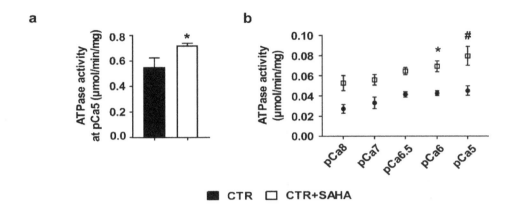

Figure 2. Effect of SAHA treatment on SERCA2 ATPase activity evaluated in cardiomyocytes isolated from adult rat hearts and HL-1 cells. (**a**) ATPase activity assay performed on microsomes isolated from adult rat CMs either untreated or treated with SAHA at pCa5. Experiments were performed on 3 independent CM sets per group and repeated twice. Unpaired Student's t-test: * $p < 0.005$ vs. CTR; (**b**) ATPase activity assay performed on microsomes isolated from HL-1 cells at different pCa. Each point represents the mean ± SEM of at least 4 independent experiments; Two-way ANOVA followed by Sidak's multiple comparison: * $p < 0.05$ vs. SAHA pCa6; # $p < 0.05$ vs. SAHA pCa5. All data are presented as mean ± SEM.

In order to confirm our result on SERCA2 functional properties, we decided to perform an additional set of experiments measuring ATPase activity on microsomes isolated from HL-1 cells, derived from the AT-1 mouse atrial cardiomyocyte tumor lineage. These cells partially maintain an adult cardiac phenotype and are able to contract [22]. The calcium-dependence of ATPase activity on HL-1 cells, either untreated (CTR) or treated for 90 min with 2.5 μM SAHA, was analyzed at different calcium concentration (from pCa8 to pCa5). ATPase activity of microsomes extracted from HL-1 cells was increased after SAHA treatment in comparison with CTR and the difference reached statistical significance when microsomes were exposed to 1 and 10 μM calcium concentration (corresponding to pCa6 and pCa5, respectively; Figure 2b).

2.2. Effect of SAHA Treatment on Calcium Transients and Cell Mechanics in CMs Isolated from Adult Rat Hearts

We then investigated whether SAHA treatment affected CM functional parameters that directly depend on SERCA2 activity, namely calcium transients and cell contractility. The amplitude and the time to peak (TTP) of the calcium transient were comparable in CTR and CTR+SAHA groups, while the rate of cytosolic calcium clearing was significantly higher in SAHA-treated cardiomyocytes (Figure 3a,c). Specifically, SAHA induced a 21% decrease in the time constant tau, as well as a significant reduction in the time to 10%, 50% and 90% of fluorescence signal decay (BL10, BL50, BL90; Figure 3c). Consistent with this finding, SAHA also affected CM mechanics during the re-lengthening phase, as documented by the significant increase in the maximal rate of re-lengthening (+dl/dt$_{max}$, approximately 16%) associated with a decrease in the time to 10%, 50% and 90% of re-lengthening (Figure 3b,d). Conversely, the average diastolic sarcomere length, the fraction of shortening and the maximal rate of shortening were comparable in CTR and CTR+SAHA cardiomyocytes (Figure 3d). Of note, SAHA exposure, at the conditions used in the present manuscript (90 min, 2.5 μM), did not affect cardiomyocytes diastolic nor systolic calcium concentration assessed by Fura-2 dye (Figure S2).

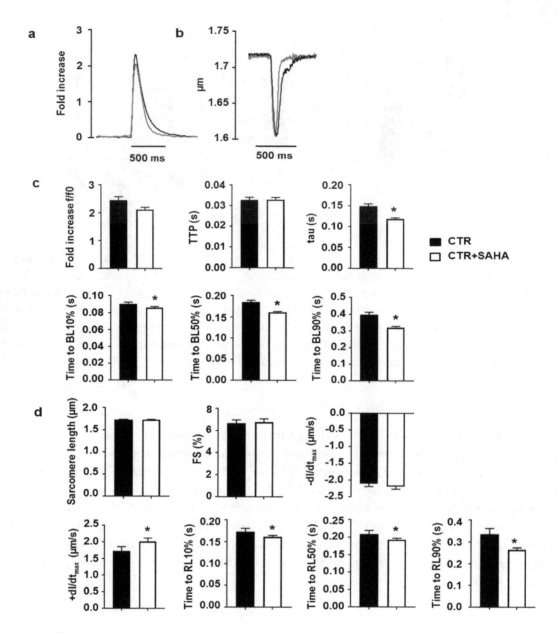

Figure 3. Effect of SAHA treatment on calcium transients and cell mechanics in CMs isolated from adult rat hearts. Representative examples of calcium transients (**a**) normalized traces: fold increase and sarcomere shortening (**b**) recorded from CTR (black line) and CTR+SAHA (red line) ventricular myocytes. (**c**) Calcium transient parameters: amplitude of calcium transient (expressed as calcium peak fluorescence normalized to baseline fluorescence, f/f0: fold increase), time to peak of the calcium transient (TTP), time constant of the rate of intracellular Ca^{2+} clearing (tau), and recovery phase of Ca^{2+} transients indicated as time to 10% (BL10), 50% (BL50) and 90% (BL90) of fluorescence signal decay. (**d**) Cell mechanics: diastolic sarcomere length and fraction of shortening (FS), maximal rates of shortening ($-dl/dt_{max}$) and re-lengthening ($+dl/dt_{max}$), time of re-lengthening at 10% (RL10), 50% (RL50) and 90% (RL90). All data are presented as mean $\pm$ SEM and analysed using General Linear Model (GLM) ANOVA for repeated measurements ($n = 45$ untreated CMs, $n = 70$ CMs treated with SAHA): * $p < 0.005$ vs. CTR.

2.3. Nε-Lysine Acetylation Sites of Human SERCA2a

To support the hypothesis that direct acetylation can boost SERCA2a function, we searched in PHOSIDA and PhosphoSitePlus for reported Nε-lysine acetylation sites in human SERCA2. PHOSIDA Post Translation Modification Database reports acetylation at K464, while PhosphoSitePlus reports acetylation at K31, K33, K218, K464, K476 and K514. All these residues are mostly conserved in mammals (Figure S3). Out of these candidates, site K464 was selected as the target for further analysis given two previous independent reports identifying K464 as an acetylation site in human SERCA2 [23,24]. Homology modeling results indicate that K464 is solvent accessible at the surface of the nucleotide binding-domain (Figure 4a and Figure S4).

To assess the impact of Nε-lysine acetylation on hSERCA2a function, two different mutants were generated where lysine in position 464 was changed into glutamine (Q) or arginine (R) to mimic constitutive acetylation or constitutive deacetylation, respectively [15]. The expression of either the hSERCA2a wild type form (WT) and the mutant forms (K464Q/R) was induced in HEK human cells (Figure 4b), expressing lower levels of the SERCA2 isoform compared to other commonly used human cell lines, such as HeLa. SERCA2 expression was significantly increased in all stable transfected HEK cells compared to non-transfected cells (NT; Figure 4b). However, although the observed difference did not reach statistical significance, the expression of SERCA2 appeared lower in both mutants compared to the HEK transfected with WT (Figure 4b).

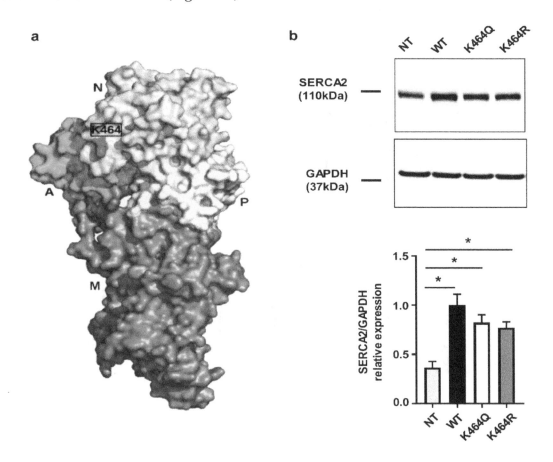

Figure 4. Analysis of human SERCA2a mutants transfected in HEK cells: bioinformatics prediction and protein expression. (**a**) Molecular surface view of predicted hSERCA2a. Domains are visualized in different colors: Actuator (A) in light blue, transmembrane (M) in dark blue, phosphorylation (P) in yellow and nucleotide-binding (N) in beige. K464 is highlighted in red; (**b**) Western blot analysis of SERCA2 expression in stable transfected HEK cells. Densitometry is reported in the bar graph ($n = 8$). Ordinary one-way ANOVA followed by Bonferroni's multiple comparison: $p < 0.005$ vs. NT.

2.4. Effects of K464 Mutation on hSERCA2a ATPase Activity and HEK Calcium Transients

The effects of mutagenesis were evaluated on ATPase activity measured in the microsomal fraction isolated from transfected HEK cells. In order to take into consideration the possible impact of different level of protein expression on hSERCA2a function, the values of ATPase activity recorded for each sample were normalized to the correspondent SERCA2 expression evaluated by Western blot on cells obtained from the same sample (Figure 4b). Taking into account the variable amount of SERCA2 expression in each cell line and for each experiment, we observed that SERCA2 function was significantly ameliorated in mutants where K464 were mutated into Q compared with WT (Figure 5a). Conversely, the K464R mutant did not show any differences in comparison with the WT (Figure 5a). Thus, constitutive acetylation of hSERCA2a K464 improved hSERCA2a ATPase activity, while K464 mutation into R (constitutive deacetylation) did not have a significant effect (Figure 5a). Importantly, untransfected HEK cells responded with a transient increase of intracellular calcium concentration to caffeine puff (1 s, 10 mM, Figure 5b). Therefore, we further evaluated the effects of SERCA2a mutagenesis on calcium transients induced by caffeine in HEK cells loaded with the calcium sensitive dye Fluo-4. Again, in order to consider the effects of different SERCA2 protein levels in HEK mutants, calcium transients were corrected on the mean SERCA2 expression detected by Western Blot in the correspondent cell line. The constitutive acetylation of SERCA2 K646Q promoted an acceleration of Ca^{2+} transient recovery phase at BL10%, BL50% compared to WT and K464R (Figure 5c), thus indicating that constitutive acetylation of K464 has a positive impact on hSERCA2a function.

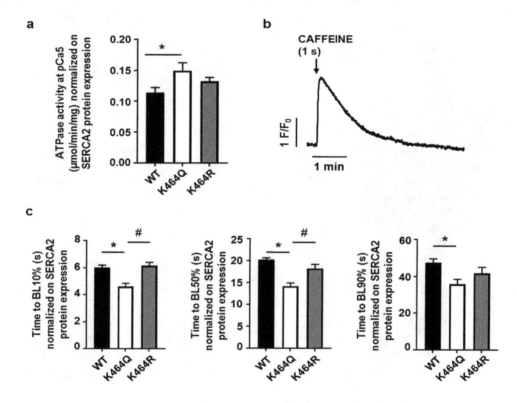

Figure 5. ATPase activity on microsomes and calcium transients evaluated in HEK cells transfected with human SERCA2a mutants. (**a**) ATPase activity assay performed on microsomes isolated from stable transfected HEK cells at pCa5 normalized on SERCA2 protein expression. Experiments were performed on 4 independent HEK sets per group and repeated twice. Ordinary one-way ANOVA followed by Bonferroni's multiple comparison: $p < 0.005$ WT vs. K464Q. All data are presented as mean $\pm$ SEM; (**b**) Representative trace of the rise in intracellular calcium concentration evoked by a caffeine pulse in untransfected HEK cells; (**c**) Calcium transients in SERCA2a K464 mutants normalized on SERCA2 protein expression. K464 is mutated either into Q or R, mimicking constitutive acetylation and deacetylation, respectively ($n = 99$ WT, $n = 69$ K464Q and $n = 36$ K464R). Time to 10% (BL10) and time to 90% (BL90) were analysed by Kruskal-Wallis followed by Dunn's multiple comparisons test: * $p < 0.005$ WT vs. K464Q, # $p < 0.005$ K464Q vs. K464R. Time to 50% (BL50) was analysed by ordinary one-way ANOVA followed by Bonferroni's multiple comparisons test: * $p < 0.005$ WT vs. K464Q, # $p < 0.005$ K464Q vs. K464R. All data are presented as mean $\pm$ SEM.

3. Discussion

SERCA2 activity has been recently suggested to be affected by acetylation/deacetylation mechanisms [11,15], although no experimental evidence has so far been provided. In the present work, we show that HDAC inhibition promotes SERCA2 acetylation. This is associated with a significant increase of microsomal ATPase activity (which is recognized to be mainly due to SERCA2 function [25]) and a significant amelioration of calcium clearing parameters measured in CMs isolated from normal adult rats. Our results indicate for the first time a relation between HDACs, SERCA2 acetylation and its function. In addition, lysine 464 has been mutated in the human SERCA2a (hSERCA2a) coding sequence transfected in a HEK stable expression system, in order to provide evidence for a causal association between SERCA2 deacetylation and the ATPase activity of the pump.

In our CM experimental model, SERCA2 acetylation was promoted by SAHA (Vorinostat), a general HDAC inhibitor, acting on class I and II HDACs [26]. SAHA use was approved by the US FDA in October 2006 for the treatment of refractory cutaneous T-cell lymphoma [27]. Although several in vitro studies [28,29] have reported that 2.5 μM SAHA is able to inhibit the growth of cancer cells, it has also been reported that 2 μM SAHA is able to reduce CM death after simulated ischaemia/reperfusion injury [30]. In our work, the decision to treat adult CMs with 2.5 μM SAHA for 90 min was taken on the basis of current literature showing that SAHA in the range of 1 to 5 μM increases histone acetylation in an in vitro model within one hour, reaching the plateau effect at 3 h [31]. Therefore, we assumed that 90 min are sufficient to achieve the acetylation increase of all the other HDAC potential targets, including cytoplasmic substrates and to preserve cell viability. We demonstrated higher levels of SERCA2 acetylation following SAHA treatment in adult CMs from CTR rats by immunoprecipitation and Western blot for Acetylated-Lysine. The SERCA2 ATPase activity has been then explored using isolated microsomes from normal adult rat CMs, where an increase has been shown after SAHA treatment. Although it is a common praxis to perform measures at pCa5 [21,32–34], we decided to test the calcium-dependence of ATPase activity in order to confirm our results on SERCA2 ATPase activity also using microsomes isolated from HL-1 cells, as additional in vitro model. Specifically, HL-1 cells, derived from mouse atrial cardiomyocyte tumor lineage [22], represent a more abundant source of cardiomyocytes compared to freshly isolated rat adult CMs.

The treatment with SAHA ameliorated intracellular Ca^{2+} dynamics and, accordingly, contractile performance of adult rat CMs. These findings support the hypothesis that the inhibition of SERCA2 deacetylation induced by SAHA plays a crucial role in promoting a more efficient SR calcium re-uptake, which in turn results in a higher rate of re-lengthening and shorter relaxation times [35].

In view of these results, one should also consider that Gupta and colleagues recently observed an enhanced calcium sensitivity of myofilaments following HDAC inhibition. This aspect can play a key role in explaining the improved contractile efficiency of SAHA treated CMs [36]. On the other hand, an increase of myofilament calcium sensitivity would be expected to delay SR Ca^{2+} uptake and consequently prolong twitch duration, thus having an opposite effect compared to our observation. Nevertheless, it is important to notice that Gupta and coworkers have demonstrated that acetylation increases myofilament Ca^{2+} sensitivity in a model of skinned cardiac papillary muscle fibers [36]. It is known that the muscle skinning process leads to the loss of all the other sensitizing and desensitizing intracellular components that synergistically act in determining the effective calcium sensitivity of myofilaments [37]. Therefore it is conceivable that the discrepancy between our and Gupta results depends, at least in part, on the use of a whole cell model instead of a reductionist model consisting only of free calcium ions and myofilaments.

It is now well recognized that Ca^{2+} handling not only governs contractile events, but can also directly or indirectly influence ion channel function in cardiomyocytes, resulting in electrophysiological effects. An important component of Na^+ channel regulation is due to Ca^{2+}, calmodulin (CaM) and CaM-dependent protein kinase II (CaMKII) pathway that affects channel function [38]. L-type Ca^{2+} (ICaL) and sodium(Na^+)-Ca^{2+} exchanger (NCX) currents are Ca^{2+}-sensitive, as well as the slowly activating delayed rectifier current (IKs), which plays an important role in regulating action potential duration [39,40]. In this work, we did not perform electrophysiological measurements on isolated ventricular myocytes that could help us in evaluating the impact on cardiac ion channels and pumps of SERCA2a activity amelioration induced by SAHA deacetylation inhibition. However, several findings reported in the present study can be useful to speculate on this topic. We observed that the amplitude of the calcium transient, which is dependent on the amplitude of Ca^{2+} entry, was comparable in control and SAHA-treated cardiomyocytes, suggesting that SAHA does not induce substantial changes in the calcium entry, through L-type Ca^{2+} channels and reverse activity of NCX. In addition, ryanodine receptors should not be affected by SAHA treatment, as indirectly suggested by the calcium transient rate of rise, measured as time to peak fluorescence, showing similar values in treated and untreated cells. Also, it should be considered that diastolic and systolic cytosolic Ca^{2+} levels were unchanged in SAHA treated cells as compared to control cardiomyocytes. This observation, besides confirming that calcium homeostasis is maintained even in the presence of a faster SERCA2 calcium re-uptake, should imply no changes in both the intracellular Na^+ concentration that is tightly coupled to calcium concentration regulation, via electrogenic Na^+/Ca^{2+} exchange, and Ca^{2+}-sensitive IKs current.

Interestingly, it has been recently shown that SAHA administration in vivo in a mouse model of muscular dystrophy can reduce the appearance of cardiac arrhythmias by reverting the remodeling of connexins and sodium channels [17]. Although the mechanisms of action are still unclear, those findings, together with our results, support the positive effect of SAHA treatment on cardiac function.

In order to explore possible acetylation sites influencing human SERCA2 activity, 3D models of hSERCA2a were produced by bioinformatic prediction methods, identifying the residue 464 as the most promising candidate. Residue K464 is located at the surface of the N domain in hSERCA2a. Although homology modeling of the different conformational states does not reveal an obvious functional impact for the K464 acetylation, the site is expected to locate on the side of a large cleft in state E2P [1,41] at the interface between domains N, P and M, near a putative phospholamban binding site (Figure S4). The potential effect of the K464 acetylation in protein electrostatics, flexibility and phospholamban binding affinity should be further investigated. Indeed, this lysine, conserved in all mammals, has already been reported as acetylated in the guinea pig heart, although no evidence of its functional role has been provided [15].

We produced three mutant HEK lines expressing WT hSERCA2a, or hSERCA2a where K464 was mutated into Q or R, mimicking constitutive acetylated and deacetylated state of the residue, respectively [15]. Of note, HEK cells have already been used by other groups as a valuable model to overexpress mutant SERCA and to study the consequent effects on the protein pumping activity [11]. HEK cells overexpressing mutated SERCA have also been recently used as a tool to investigate intracellular calcium dynamics [42]. Importantly, despite the debate on the presence of functional RyRs in HEK cells [43–45], we found that the application of a pulse of caffeine caused the abrupt increase of intracellular calcium concentration. Therefore, we decided to expose HEK cells transfected with WT hSERCA2a or mutated hSERCA2a to caffeine in order to increase the intracellular calcium concentration and then compare the time of recovery in WT compared to mutants. It is important to notice that, although the recovery phase in HEK cells was considerably longer than that observed in adult cardiomyocytes, SERCA has been described as the major player in the recovery of calcium transients long up to 4 s [46]. We cannot exclude a possible involvement of other Ca^{2+} removal systems (such as mitochondria and Peripheral Membrane Calcium ATPase) as well as of Ca^{2+} buffering in restoring basal conditions. However, given that in our experimental conditions the hSERCA2a pump was overexpressed, we can reasonably assume that the recovery phase in (transfected) HEK cells at least partially depends on the activity of transfected hSERCA2a. Intriguingly, K464Q mutants exhibited a higher ATPase activity and an acceleration of Ca^{2+} transient recovery phase compared to WT, supporting the hypothesis that the acetylation is a positive regulator of SERCA2 function. However, it is important to notice that SAHA could inhibit the acetylation of different lysine residues at the same time. Thus, mutating only one lysine at a time might only partially mimic the global effect obtained after HDAC inhibition on CMs.

Interestingly, Hajjar and coworkers demonstrated that sumoylation of lysine residue 585 enhances SERCA2a stability and activity [11]. It has been recently shown that HDAC inhibition can stimulate protein sumoylation possibly through acetylation of the SUMO machinery [47]. Thus, our findings in isolated SAHA-treated cardiomyocytes sustain the idea that HDAC-inhibition could support other approaches, such as SUMO-1 overexpression [11], in the treatment of heart disease associated with SERCA2 impairment.

The main limitation of the present study is the fact that acetylation is only one of the factors contributing to SERCA2 function in health and disease. Additionally, we cannot exclude a possible action of SAHA on other proteins, such as SERCA2 interacting proteins or transporters involved in calcium dynamics, such as the sodium-calcium exchanger [48]. However, SERCA2 is recognized as the main player in calcium re-uptake during the relaxation/recovery phase in CMs, removing approximately 70–90% of calcium ions from the cytosol in species including humans, mice and rats [4]. This observation let us hypothesize that SERCA2 is the key factor in SR reuptake whose function is modulated by SAHA. The fact that the time to peak (TTP) of calcium transients recorded in isolated rat cardiomyocytes remained unchanged following SAHA administration seems also to indicate that RyRs are not affected by the treatment, at least under our experimental conditions. Further, our results indicate that phosphorylated and total PLB remain unchanged after treatment, thus suggesting that PKA activity might not be affected by SAHA. Nevertheless, deeper investigation are required in order to explain whether acetylation can directly or indirectly modulate other mechanisms of SERCA2 regulation, like CAMKII mediated phosphorylation. Importantly, it has been recently hypothesized that SERCA2 can be potentially regulated by microRNAs [49]. Concomitantly, experimental evidence has been provided that knock-out mice for miR-22 expression exhibit prolonged calcium transients and a decrease of SERCA transcript induced by indirect interaction [50]. More recently, it has been shown that miR-275 can directly target SERCA in mosquito gut [51], thus providing new basis for further studies on SERCA regulation by microRNA in vertebrates and humans. In this context, it should be reported that SAHA is well recognized as a microRNA modulator [52–54]. Further studies are required to understand whether HDAC inhibition can exert its action on SERCA and calcium dynamics by acting on microRNA target. Additionally, other potential mechanism of action should be taken into consideration for future studies, such as the role of the TGF-beta pathway, already shown to be involved in cardiac calcium signaling [55,56] and to be modulated by HDAC inhibitors in organs other than heart [57–59].

However, we would like to underline the fact that the aim of the present paper was to evaluate whether HDAC inhibition can modulate cardiac cell function by acting on SERCA2. SERCA2 is reasonably one of the multiple targets of SAHA action that can result in the enhancement or in the inhibition of many proteins, the net effect on EC coupling depending on the combination of the single contributions and on the model used.

Another question remaining unanswered is which enzymes are responsible for SERCA2a acetylation/deacetylation. It is known that SAHA at low concentration is a potent inhibitor of the class I HDAC isoforms 1 and 8 (IC50 < 20 nM), while it is active on other class I and class II HDACs at higher concentrations [60]. It is also well recognized that class II HDAC have both nuclear and cytoplasmic localization [61]. However, although traditionally considered as nuclear enzymes, class I HDACs have been recently demonstrated to be present in the endoplasmic reticulum [62]. Therefore, it is conceivable that both class of enzymes might act on SERCA2, contributing to the control of its acetylation status. Further studies are required to elucidate this point.

In conclusion, the results here demonstrate for the first time that SERCA2 acetylation and activity are increased following HDAC inhibition promoted by SAHA. Importantly, our findings suggest that the modulation of acetylation can be a novel strategy to treat diseases where calcium handling is affected, such as diabetic cardiomyopathy [6,63], dilated cardiomyopathy [64], ischemia/reperfusion injury [65], cardiac hypertrophy [66] and heart failure [67].

4. Materials and Methods

4.1. Rat Population

This study was carried out in strict accordance with the recommendations in the Guide for the Care and Use of Laboratory Animals (National Institute of Health, Bethesda, MD, USA, revised 1996). The investigation was approved by the Veterinary Animal Care and Use Committee of the University of Parma-Italy (Prot. No. 59/12; 9 May 2012) and conforms to the National Ethical Guidelines of the Italian Ministry of Health. All efforts were made to minimize animal suffering.

The study population consisted of 5 male Wistar rats (*Rattus norvegicus*) aged 12–14 weeks, weighing 379 ± 8 g, individually housed in a temperature-controlled room at 22–24 °C, with the light on between 7.00 AM and 7.00 PM.

4.2. Rat Cardiomyocyte Isolation

Individual left ventricular (LV) cardiomyocytes (CMs) were enzymatically isolated by collagenase perfusion from the adult rat hearts in accordance with a procedure previously described [68]. Briefly, rats were anesthetised with ketamine chloride (Imalgene, Merial, Milan, Italy; 40 mg/kg ip) plus medetomidine hydrochloride (Domitor, Pfizer Italia S.r.l., Latina, Italy; 0.15 mg/kg ip). The heart was then removed and rapidly perfused at 37 °C by means of an aortic cannula with the following sequence of solutions: (1) a calcium-free solution for 5 min to remove the blood. Calcium-free solution contains the following (in mM): 126 NaCl, 22 dextrose, 5 $MgCl_2$, 4.4 KCl, 20 taurine, 5 creatine, 5 sodium pyruvate, 1 NaH_2PO_4, and 24 HEPES (pH 7.4, adjusted with NaOH), and the solution was gassed with 100% O_2; (2) a low-calcium solution (0.1 mM) plus 1 mg/mL type 2 collagenase (Worthington Biochemical, NJ, USA) and 0.1 mg/mL type XIV protease (Sigma-Aldrich, Milan, Italy) for about 20 min; (3) an enzyme-free, low-calcium solution for 5 min. The left ventricle was then minced and shaken for 10 min. The cells were filtered through a nylon mesh, re-suspended in low-calcium solutions (0.1 mM) for 30 min, and then used for recording cell mechanics and calcium transients. Following enzymatic dissociation, CMs were either untreated or incubated with 2.5 μM SAHA for 90 min. A fraction of cells from each experimental group was washed three times with low-calcium solution and centrifuged ($42 \times$ g for 5 min). After removing the supernatant, the pellet was stored at -80 °C for subsequent microsome isolation (see the Section 4.6). The remaining cardiomyocytes were used for recording cell mechanics and calcium transients, as described in the Section 4.8.

4.3. HL-1 Cardiomyocyte Culture

HL-1 cells, derived from the AT-1 mouse atrial cardiomyocyte tumor lineage [22], were kindly donated by Prof. W.C. Claycomb. Cells were grown as previously described [22] in Claycomb Medium (Sigma-Aldrich S.r.l., Milan, Italy) supplemented with 100 μM norepinephrine (Sigma-Aldrich), 10% FBS (Sigma-Aldrich), 1% *v/v* penicillin streptomycin (Invitrogen by Thermo Fisher Scientific, Monza, Italy) and 2 mM L-glutamine (Invitrogen). Cells were cultured at 37 °C in a humid atmosphere of 5% CO_2 and 95% air. Cells were plated onto gelatin/fibronectin-coated flasks at a density of 10,000 cells/cm^2. When cells reached 90% of confluency, culture medium was replaced and supplemented for 90 min with 2.5 μM SAHA.

4.4. Western Blot Analysis

Cells were washed in PBS and harvested in protein extraction buffer (10 mM Tris-HCl pH 7.4, 150 mM NaCl, 1% Igepal CA630, 1% sodium deoxycholate, 0.1% SDS (Sodium Dodecyl Sulphate) and 1% Glycerol supplemented with protease/phosphatase inhibitor mix (Roche Diagnostics S.p.A., Milan, Italy). Proteins were quantified by BCA protein kit (Thermo Fisher Scientific, Monza, Italy) following manufacturer's instructions. Then, 30 μg of protein extracts were separated by SDS-PAGE on precast gradient (4–12%) gels (Invitrogen) using MOPS running buffer (Invitrogen) and transferred onto nitrocellulose membranes (Bio-Rad, Segrate, Milan, Italy) in Transfer buffer (Life Technologies by

Thermo Fisher Scientific) supplemented with 10% (v/v) methanol (Sigma-Aldrich, Milan, Italy). After blocking in 5% BSA in PBS containing 0.05% Tween 20 (1 h at room temperature), membranes were incubated overnight at 4 °C with primary antibodies: SERCA2 (1:1000, SantaCruz, Dallas, TX, USA, sc-376235), Acetylated-tubulin (1:1000, Sigma-Aldrich # T7451), GAPDH (glyceraldehyde-3-phosphate dehydrogenase, 1:2000, SantaCruz, Dallas, TX, USA, sc-32233), Pan-Ac-K (Acetylated-Lysine, 1:1000, Cell Signaling, Danvers, MA, USA, #9441), Phospho-PLB (Ser16) (Phospho-Phospholamban (Ser16), 1:5000, Merck Millipore, Billerica, MA, USA, # 07-052) and PLB (Phospholamban, 1:8000, Abcam, Hong Kong, China, #ab2865). Blots were washed three times in PBS-Tween buffer and then incubated with appropriate horseradish peroxidase conjugated secondary antibody (SantaCruz) for 1 h at room temperature. Detection was performed by enhanced chemiluminescence system (Supersignal West Dura Extended Duration Substrate, Thermo Scientific). Results were quantified by Image Lab software 5.2.1 (BioRad, Segrate-Milan, Italy).

4.5. Immunoprecipitation

Co-immunoprecipitation experiments were performed using protein-Agarose accordingly to the manufacturer's protocol (Roche Diagnostics S.p.A., Milan, Italy). In particular, 1 mg of protein extracts was co-immunoprecipitated with 10 µg of anti-SERCA2 antibody (SantaCruz, Dallas, TX, USA). Negative controls were performed with the same amount of protein extract (derived from rat cardiomyocytes) and were immunoprecipitated with the corresponding purified IgG antisera (SantaCruz) in the absence of primary antibody. Immunoprecipitated samples were resolved by SDS–PAGE, transferred onto nitrocellulose membrane (Bio-Rad, Segrate-Milan; Italy) and western blots were performed as described in the Section 4.4.

4.6. Isolation of the Microsomal Fraction

Microsome isolation from rat cardiomyocytes and HL-1 cells was performed according to Maruyama and MacLennan (1988) with minor modifications [69]. Specifically, cells were washed twice with PBS and then homogenized with 30 strokes in a glass Dounce homogenizer in 10 mM Tris-HCl, pH 7.5, 0.5 mM MgCl$_2$. The homogenate was diluted with an equal volume of a solution of 10 mM Tris-HCl, pH 7.5, 0.5 M sucrose, 300 mM KCl. The suspension was centrifuged at 10,000× g for 30 min at 4 °C to pellet nuclei and mitochondria. The pellet was discarded and the supernatant was further centrifuged at 100,000× g for 150 min at 4 °C to sediment the microsomal fraction. The pellet was re-suspended in a solution containing 0.25 M sucrose and 10 mM MOPS. All solutions were enriched in protease inhibitors (Roche). The microsome concentration was measured using the BCA protein kit (Thermo Fisher Scientific) following the manufacturer's instructions.

4.7. ATP/NADH Coupled Assay for Calcium ATPase Activity

The ATPase activity of the microsomal fraction isolated from HL-1 cells (25 µg/mL), was measured by spectrophotometric determination of NADH oxidation coupled to an ATP regenerating system, as previously described [70]. We used a Beckman DU 640 spectrophotometer adjusted at a wavelength of 340 nm. The assay was performed at 37 °C in a final volume of 1 mL of a buffer containing 20 mM histidine pH 7.2, 5 mM MgCl$_2$, 0.5 mM EGTA, 100 mM KCl, 2 mM ATP, 0.5 mM phosphoenolpyruvate, 0.15 mM NADH, 1.4 units of pyruvate kinase/lactic dehydrogenase, in the presence of 2 µg/mL A23187 Ca^{2+}-ionophore. ATPase activity was obtained subtracting the basal activity (measured in the presence of EGTA without calcium added) from the maximal activity and was expressed as micromoles/min/mg of protein. For investigating the calcium-dependence of ATPase activity, the concentration of free calcium was varied from pCa8 to pCa5 using EGTA buffered solutions.

NADH coupled ATPase assay protocol was then adapted for use on a 96-well microplate reader, according to the method described by Kiianitsa and coworkers [71]. We used the microplate protocol to analyze the ATPase activity of the microsomal fraction isolated from adult rat ventricular cardiomyocytes (5 µg/well) and from HEK293 cells (20 µg/well) expressing either Wild Type or mutated human SERCA2a (see Section 4.10). The absorbance change at 340 nm was monitored using the EnVision (Perkin Elmer, Waltham, MA, USA) plate reader. The experiments were performed at 37 °C in a final volume of 200 μL at pCa5 [21] on the same buffer described above. Technical duplicates were performed for each experiment.

4.8. Rat Cardiomyocyte Mechanics and Calcium Transients

CM mechanical properties were assessed by using the IonOptix fluorescence and contractility systems (IonOptix, Milton, MA, USA). CMs were placed in a chamber mounted on the stage of an inverted microscope (Nikon-Eclipse TE2000-U, Nikon Instruments, Florence, Italy) and superfused (1 mL/min at 37 °C) with a Tyrode solution containing (in mM): 140 NaCl, 5.4 KCl, 1 $MgCl_2$, 5 HEPES, 5.5 glucose, and 1 $CaCl_2$ (pH 7.4, adjusted with NaOH). Only rod-shaped myocytes with clear edges and average sarcomere length ≥ 1.7 μm were selected for the analysis. All the selected myocytes did not show spontaneous contractions. The cells were field stimulated at a frequency of 0.5 Hz by constant current pulses (2 ms in duration, and twice diastolic threshold in intensity) delivered by platinum electrodes placed on opposite sides of the chamber, connected to a MyoPacer Field Stimulator (IonOptix). The stimulated myocyte was displayed on a computer monitor using an IonOptix MyoCam camera. Load-free contraction of myocytes was measured with the IonOptix system, which captures sarcomere length dynamics via a Fast Fourier Transform algorithm. Sampling rate was fixed at 1 KHz. A total of 115 control isolated ventricular myocytes (of which 45 were used as untreated control and 70 were exposed to SAHA) were analyzed to compute the following parameters: mean diastolic sarcomere length and fraction of shortening (FS), maximal rates of shortening and re-lengthening ($\pm dl/dt_{max}$), and time at 10%, 50% and 90% of re-lengthening (RL10, RL50 and RL90, respectively). Steady-state contraction of myocytes was achieved before data recording by means of a 10 s conditioning stimulation.

Calcium transients were measured simultaneously with cell motion. Ca^{2+} transients were detected by epifluorescence after loading the myocytes with Fluo-3 AM (10 μmol/L; Invitrogen, Carlsbad, CA, USA) for 30 min. Excitation length was 480 nm, with emission collected at 535 nm using a 40× oil objective lens (NA: 1.3). Fluo-3 signals were expressed as normalized fluorescence (f/f0: fold increase). The time course of the fluorescence signal decay was described by a single exponential equation, and the time constant (tau) was used as a measure of the rate of intracellular calcium clearing [72]. The time to peak of the calcium transients (TTP) and the times to 10%, 50% and 90% of fluorescence signal decay were also measured (time to BaseLine fluorescence BL10, BL50 and BL90, respectively).

4.9. Bioinformatics Analysis

The post-translation modification databases PHOSIDA [73] and PhosphoSitePlus [74] were used for the identification of acetylation sites investigated in human SERCA2. The location of the lysine acetylation was investigated in human SERCA2a structural models derived by homology modelling based on rabbit homologue templates with 83% sequence identity to hSERCA2a, and using MODELLER version 9.11 [75].

4.10. Mutagenesis

SERCA2a mutants were generated by PCR-site directed mutagenesis (for a list of primers see Table S1) and verified by complete sequencing (Eurofins Genomics, Ebersberg, Germany). The SERCA2a wild-type coding sequence (NM_001681.3; SER-WT) as well as the SERCA2a coding sequence with Lys 464 residue mutated into glutamine (SER-K464Q) or arginine (SER-K464R) were subsequently cloned into the pCDNA3 vector carrying Geneticin resistance (Life Technologies) for subsequent cell transfection. Transfections were performed on HEK-MSR1 cells (Human Embryonic Kidney 293 cells) that stably express human Macrophage Scavenger Receptor 1, facilitating cell adhesion in blasticidin selection (5 μg/mL, Invitrogen) [76]. Lipofectamine LTX (Invitrogen) was used for transfection following manufacturer's instructions. Stable cell lines were selected 24 h after transfections in culture medium composed by: MEM (Invitrogen) supplemented with 10% FBS (Sigma-Aldrich), 1% v/v glutamine (Invitrogen), 1% v/v minimum non-essential amino acids (Invitrogen) and 450 µg/mL Geneticin (Invitrogen).

4.10.1. SERCA2 Mutant Calcium Transients

HEK-MSR1 stably transfected with K464 onto SERCA2 sequence mutated in Q or R were plated on the bottom of glass dishes (25 mm diameter) and grown for 2 days in MEM medium supplemented with FBS 10%. Cytosolic Ca^{2+} dynamics were evaluated in Fluo-4 AM loaded cells. Briefly, cells were loaded with 2.5 µM of Fluo-4 AM (Molecular Probes, Invitrogen) for 30 min at 37 °C and then washed with Tyrode's solution containing in mM: 10 D-(+)-glucose, 5.0 Hepes, 140 NaCl, 5.4 KCl, 1.2 $MgCl_2$, 1.8 $CaCl_2$, pH adjusted to 7.3 with NaOH. Dishes were mounted in a perfusion chamber and placed on the stage of an epifluorescence microscope (Nikon) equipped with a 75 W Xenon lamp and connected to a CCD camera (Cool SnapTM EZ Photometrics, Tucson, AZ, USA). Cells were excited at 488 nm wavelength, fluorescence emission was measured at 520 nm and recorded using the MetaFluor Software (Molecular Devices, Sunnyvale, CA, USA); imaging was scanned repeatedly at 20 Hz. This low temporal acquisition is justified by the longer calcium transient in HEK cells lasting on average 1 min and 20 s. Calcium transients were evoked by a 1 s caffeine-pulse (10 mM). Temperature was maintained at 36.5 ± 0.5 °C throughout the experiment.

The background fluorescence signal was fitted to a linear regression and then subtracted. Each trace was subsequently resampled and denoised with an Rbf interpolator (f = 20, e = 1, s = 1). The baseline was calculated per-trace using the median value of the first 250 ms before the excitation and then subtracted.

Traces which didn't reach a peak value of at least 2 times the baseline level, or dropped under 90% of the absolute value of the baseline were discarded. We then calculated the time to decay at the first intersection of 10%, 50% and 90% of the curves, thus defining time to BL10, BL50 and BL90. Only traces that successfully recovered to BL90 within the recording timeline (90 s) were considered.

4.10.2. Statistical Analysis

All data are reported as means ± standard error (SEM). Normality of the data was evaluated by D'Agostino & Pearson normality test. For two groups, significance was assessed by two-tailed unpaired t-test or non-parametric Mann–Whitney U-test when appropriate. For more than two groups either two-way ANOVA, parametric one-way ANOVA or Kruskal-Wallis non parametric test were used followed by appropriate post-hoc individual comparisons (GraphPad Prism software 6.03). General Linear Model (GLM) ANOVA for repeated measurements was used to compare cell contractility and calcium transient data (IBM-SPSS 24.0, SPSS Inc., Chicago, IL, USA). The details on the specific test used are reported in the figure legend of each specific experiment. A p-value < 0.05 was considered statistically significant.

Acknowledgments: The authors would like to thank William C. Claycomb for providing the HL-1 cell line. This work was supported by the Department of Innovation, Research and Universities of the Autonomous Province of Bolzano-South Tyrol (Italy) and local research funding from University of Parma (Italy) (FIL2014 and FIL2016). The founding sponsors had no role in the design of the study, in the collection, analyses, or interpretation of data, in the writing of the manuscript, and in the decision to publish the results.

Author Contributions: Viviana Meraviglia, Leonardo Bocchi, Donatella Stilli and Alessandra Rossini conceived and designed the experiments; Viviana Meraviglia, Leonardo Bocchi, Roberta Sacchetto, Maria Cristina Florio, Benedetta M. Motta, Corrado Corti, Monia Savi, Marcelo D. Rosato-Siri, Silvia Suffredini, Chiara Piubelli performed the experiments; Viviana Meraviglia, Leonardo Bocchi, Roberta Sacchetto, Marcelo D. Rosato-Siri, Yuri D'Elia analyzed the data; Christian X. Weichenberger and Francisco S. Domingues performed the bioinformatics analysis; Viviana Meraviglia, Francisco S. Domingues, Donatella Stilli and Alessandra Rossini wrote the paper; Giulio Pompilio and Peter P. Pramstaller critically and substantively revised the paper; Donatella Stilli and Alessandra Rossini supervised and managed the project.

References

1. Toyoshima, C. Structural aspects of ion pumping by Ca^{2+}-ATPase of sarcoplasmic reticulum. *Arch. Biochem. Biophys.* **2008**, *476*, 3–11. [CrossRef] [PubMed]
2. Asahi, M.; Sugita, Y.; Kurzydlowski, K.; De Leon, S.; Tada, M.; Toyoshima, C.; MacLennan, D.H. Sarcolipin regulates sarco(endo)plasmic reticulum Ca^{2+}-ATPase (SERCA) by binding to transmembrane helices alone or in association with phospholamban. *Proc. Natl. Acad. Sci. USA* **2003**, *100*, 5040–5045. [CrossRef] [PubMed]
3. Periasamy, M.; Kalyanasundaram, A. SERCA pump isoforms: Their role in calcium transport and disease. *Muscle Nerve* **2007**, *35*, 430–442. [CrossRef] [PubMed]
4. Bers, D.M. Cardiac excitation-contraction coupling. *Nature* **2002**, *415*, 198–205. [CrossRef] [PubMed]
5. Sulaiman, M.; Matta, M.J.; Sunderesan, N.R.; Gupta, M.P.; Periasamy, M.; Gupta, M. Resveratrol, an activator of SIRT1, upregulates sarcoplasmic calcium ATPase and improves cardiac function in diabetic cardiomyopathy. *Am. J. Physiol. Heart Circ. Physiol.* **2010**, *298*, H833–H843. [CrossRef] [PubMed]
6. Savi, M.; Bocchi, L.; Mena, P.; Dall'Asta, M.; Crozier, A.; Brighenti, F.; Stilli, D.; Del Rio, D. In vivo administration of urolithin A and B prevents the occurrence of cardiac dysfunction in streptozotocin-induced diabetic rats. *Cardiovasc. Diabetol.* **2017**, *16*, 80. [CrossRef] [PubMed]
7. Del Monte, F.; Harding, S.E.; Schmidt, U.; Matsui, T.; Kang, Z.B.; Dec, G.W.; Gwathmey, J.K.; Rosenzweig, A.; Hajjar, R.J. Restoration of contractile function in isolated cardiomyocytes from failing human hearts by gene transfer of SERCA2a. *Circulation* **1999**, *100*, 2308–2311. [CrossRef]
8. Park, W.J.; Oh, J.G. SERCA2a: A prime target for modulation of cardiac contractility during heart failure. *BMB Rep.* **2013**, *46*, 237–243. [CrossRef] [PubMed]
9. Periasamy, M.; Bhupathy, P.; Babu, G.J. Regulation of sarcoplasmic reticulum Ca^{2+} ATPase pump expression and its relevance to cardiac muscle physiology and pathology. *Cardiovasc. Res.* **2008**, *77*, 265–273. [CrossRef] [PubMed]
10. Stammers, A.N.; Susser, S.E.; Hamm, N.C.; Hlynsky, M.W.; Kimber, D.E.; Kehler, D.S.; Duhamel, T.A. The regulation of sarco(endo)plasmic reticulum calcium-ATPases (SERCA). *Can. J. Physiol. Pharmacol.* **2015**, *93*, 843–854. [CrossRef] [PubMed]
11. Kho, C.; Lee, A.; Jeong, D.; Oh, J.G.; Chaanine, A.H.; Kizana, E.; Park, W.J.; Hajjar, R.J. SUMO1-dependent modulation of SERCA2a in heart failure. *Nature* **2011**, *477*, 601–605. [CrossRef] [PubMed]
12. Adachi, T.; Weisbrod, R.M.; Pimentel, D.R.; Ying, J.; Sharov, V.S.; Schoneich, C.; Cohen, R.A. S-Glutathiolation by peroxynitrite activates SERCA during arterial relaxation by nitric oxide. *Nat. Med.* **2004**, *10*, 1200–1207. [CrossRef] [PubMed]
13. Bidasee, K.R.; Zhang, Y.; Shao, C.H.; Wang, M.; Patel, K.P.; Dincer, U.D.; Besch, H.R., Jr. Diabetes increases formation of advanced glycation end products on Sarco(endo)plasmic reticulum Ca^{2+}-ATPase. *Diabetes* **2004**, *53*, 463–473. [CrossRef] [PubMed]
14. Knyushko, T.V.; Sharov, V.S.; Williams, T.D.; Schoneich, C.; Bigelow, D.J. 3-Nitrotyrosine modification of SERCA2a in the aging heart: A distinct signature of the cellular redox environment. *Biochemistry* **2005**, *44*, 13071–13081. [CrossRef] [PubMed]
15. Foster, D.B.; Liu, T.; Rucker, J.; O'Meally, R.N.; Devine, L.R.; Cole, R.N.; O'Rourke, B. The cardiac acetyl-lysine proteome. *PLoS ONE* **2013**, *8*, e67513. [CrossRef] [PubMed]
16. Minucci, S.; Pelicci, P.G. Histone deacetylase inhibitors and the promise of epigenetic (and more) treatments for cancer. *Nat. Rev. Cancer* **2006**, *6*, 38–51. [CrossRef] [PubMed]
17. Colussi, C.; Berni, R.; Rosati, J.; Straino, S.; Vitale, S.; Spallotta, F.; Baruffi, S.; Bocchi, L.; Delucchi, F.; Rossi, S.; et al. The histone deacetylase inhibitor suberoylanilide hydroxamic acid reduces cardiac arrhythmias in dystrophic mice. *Cardiovasc. Res.* **2010**, *87*, 73–82. [CrossRef] [PubMed]
18. McLendon, P.M.; Ferguson, B.S.; Osinska, H.; Bhuiyan, M.S.; James, J.; McKinsey, T.A.; Robbins, J. Tubulin hyperacetylation is adaptive in cardiac proteotoxicity by promoting autophagy. *Proc. Natl. Acad. Sci. USA* **2014**, *111*, E5178–E5186. [CrossRef] [PubMed]

19. Bianchini, E.; Testoni, S.; Gentile, A.; Cali, T.; Ottolini, D.; Villa, A.; Brini, M.; Betto, R.; Mascarello, F.; Nissen, P.; et al. Inhibition of ubiquitin proteasome system rescues the defective sarco(endo)plasmic reticulum Ca^{2+}-ATPase (SERCA1) protein causing Chianina cattle pseudomyotonia. *J. Biol. Chem.* **2014**, *289*, 33073–33082. [CrossRef] [PubMed]

20. Lytton, J.; Westlin, M.; Burk, S.E.; Shull, G.E.; MacLennan, D.H. Functional comparisons between isoforms of the sarcoplasmic or endoplasmic reticulum family of calcium pumps. *J. Biol. Chem.* **1992**, *267*, 14483–14489. [PubMed]

21. Kho, C.; Lee, A.; Jeong, D.; Oh, J.G.; Gorski, P.A.; Fish, K.; Sanchez, R.; DeVita, R.J.; Christensen, G.; Dahl, R.; et al. Small-molecule activation of SERCA2a SUMOylation for the treatment of heart failure. *Nat. Commun.* **2015**, *6*, 7229. [CrossRef] [PubMed]

22. Claycomb, W.C.; Lanson, N.A., Jr.; Stallworth, B.S.; Egeland, D.B.; Delcarpio, J.B.; Bahinski, A.; Izzo, N.J., Jr. HL-1 cells: A cardiac muscle cell line that contracts and retains phenotypic characteristics of the adult cardiomyocyte. *Proc. Natl. Acad. Sci. USA* **1998**, *95*, 2979–2984. [CrossRef] [PubMed]

23. Mertins, P.; Qiao, J.W.; Patel, J.; Udeshi, N.D.; Clauser, K.R.; Mani, D.R.; Burgess, M.W.; Gillette, M.A.; Jaffe, J.D.; Carr, S.A. Integrated proteomic analysis of post-translational modifications by serial enrichment. *Nat. Methods* **2013**, *10*, 634–637. [CrossRef] [PubMed]

24. Choudhary, C.; Kumar, C.; Gnad, F.; Nielsen, M.L.; Rehman, M.; Walther, T.C.; Olsen, J.V.; Mann, M. Lysine acetylation targets protein complexes and co-regulates major cellular functions. *Science* **2009**, *325*, 834–840. [CrossRef] [PubMed]

25. Lytton, J.; Westlin, M.; Hanley, M.R. Thapsigargin inhibits the sarcoplasmic or endoplasmic reticulum Ca-ATPase family of calcium pumps. *J. Biol. Chem.* **1991**, *266*, 17067–17071. [PubMed]

26. Munshi, A.; Tanaka, T.; Hobbs, M.L.; Tucker, S.L.; Richon, V.M.; Meyn, R.E. Vorinostat, a histone deacetylase inhibitor, enhances the response of human tumor cells to ionizing radiation through prolongation of gamma-H2AX foci. *Mol. Cancer Ther.* **2006**, *5*, 1967–1974. [CrossRef] [PubMed]

27. Witt, O.; Milde, T.; Deubzer, H.E.; Oehme, I.; Witt, R.; Kulozik, A.; Eisenmenger, A.; Abel, U.; Karapanagiotou-Schenkel, I. Phase I/II intra-patient dose escalation study of vorinostat in children with relapsed solid tumor, lymphoma or leukemia. *Klinische Pädiatrie* **2012**, *224*, 398–403. [CrossRef] [PubMed]

28. Butler, L.M.; Zhou, X.; Xu, W.S.; Scher, H.I.; Rifkind, R.A.; Marks, P.A.; Richon, V.M. The histone deacetylase inhibitor SAHA arrests cancer cell growth, up-regulates thioredoxin-binding protein-2, and down-regulates thioredoxin. *Proc. Natl. Acad. Sci. USA* **2002**, *99*, 11700–11705. [CrossRef] [PubMed]

29. Richon, V.M.; Sandhoff, T.W.; Rifkind, R.A.; Marks, P.A. Histone deacetylase inhibitor selectively induces p21WAF1 expression and gene-associated histone acetylation. *Proc. Natl. Acad. Sci. USA* **2000**, *97*, 10014–10019. [CrossRef] [PubMed]

30. Xie, M.; Kong, Y.; Tan, W.; May, H.; Battiprolu, P.K.; Pedrozo, Z.; Wang, Z.V.; Morales, C.; Luo, X.; Cho, G.; et al. Histone deacetylase inhibition blunts ischemia/reperfusion injury by inducing cardiomyocyte autophagy. *Circulation* **2014**, *129*, 1139–1151. [CrossRef] [PubMed]

31. Tiffon, C.; Adams, J.; van der Fits, L.; Wen, S.; Townsend, P.; Ganesan, A.; Hodges, E.; Vermeer, M.; Packham, G. The histone deacetylase inhibitors vorinostat and romidepsin downmodulate IL-10 expression in cutaneous T-cell lymphoma cells. *Br. J. Pharmacol.* **2011**, *162*, 1590–1602. [CrossRef] [PubMed]

32. Simonides, W.S.; van Hardeveld, C. An assay for sarcoplasmic reticulum Ca^{2+}-ATPase activity in muscle homogenates. *Anal. Biochem.* **1990**, *191*, 321–331. [CrossRef]

33. Kennedy, D.; Omran, E.; Periyasamy, S.M.; Nadoor, J.; Priyadarshi, A.; Willey, J.C.; Malhotra, D.; Xie, Z.; Shapiro, J.I. Effect of chronic renal failure on cardiac contractile function, calcium cycling, and gene expression of proteins important for calcium homeostasis in the rat. *J. Am. Soc. Nephrol.* **2003**, *14*, 90–97. [CrossRef] [PubMed]

34. Kennedy, D.J.; Vetteth, S.; Xie, M.; Periyasamy, S.M.; Xie, Z.; Han, C.; Basrur, V.; Mutgi, K.; Fedorov, V.; Malhotra, D.; et al. Ouabain decreases sarco(endo)plasmic reticulum calcium ATPase activity in rat hearts by a process involving protein oxidation. *Am. J. Physiol. Heart Circ. Physiol.* **2006**, *291*, H3003–H3011. [CrossRef] [PubMed]

35. Frank, K.F.; Bolck, B.; Erdmann, E.; Schwinger, R.H. Sarcoplasmic reticulum Ca^{2+}-ATPase modulates cardiac contraction and relaxation. *Cardiovasc. Res.* **2003**, *57*, 20–27. [CrossRef]

36. Gupta, M.P.; Samant, S.A.; Smith, S.H.; Shroff, S.G. HDAC4 and PCAF bind to cardiac sarcomeres and play a role in regulating myofilament contractile activity. *J. Biol. Chem.* **2008**, *283*, 10135–10146. [CrossRef] [PubMed]

37. Chung, J.H.; Biesiadecki, B.J.; Ziolo, M.T.; Davis, J.P.; Janssen, P.M. Myofilament Calcium Sensitivity: Role in Regulation of In vivo Cardiac Contraction and Relaxation. *Front. Physiol.* **2016**, *7*, 562. [CrossRef] [PubMed]

38. Chen-Izu, Y.; Shaw, R.M.; Pitt, G.S.; Yarov-Yarovoy, V.; Sack, J.T.; Abriel, H.; Aldrich, R.W.; Belardinelli, L.; Cannell, M.B.; Catterall, W.A.; et al. Na⁺ channel function, regulation, structure, trafficking and sequestration. *J. Physiol.* **2015**, *593*, 1347–1360. [CrossRef] [PubMed]

39. Kennedy, M.; Bers, D.M.; Chiamvimonvat, N.; Sato, D. Dynamical effects of calcium-sensitive potassium currents on voltage and calcium alternans. *J. Physiol.* **2017**, *595*, 2285–2297. [CrossRef] [PubMed]

40. Bers, D.M.; Chen-Izu, Y. Sodium and calcium regulation in cardiac myocytes: From molecules to heart failure and arrhythmia. *J. Physiol.* **2015**, *593*, 1327–1329. [CrossRef] [PubMed]

41. Olesen, C.; Picard, M.; Winther, A.M.; Gyrup, C.; Morth, J.P.; Oxvig, C.; Moller, J.V.; Nissen, P. The structural basis of calcium transport by the calcium pump. *Nature* **2007**, *450*, 1036–1042. [CrossRef] [PubMed]

42. Ying, J.; Tong, X.; Pimentel, D.R.; Weisbrod, R.M.; Trucillo, M.P.; Adachi, T.; Cohen, R.A. Cysteine-674 of the sarco/endoplasmic reticulum calcium ATPase is required for the inhibition of cell migration by nitric oxide. *Arterioscler. Thromb. Vasc. Biol.* **2007**, *27*, 783–790. [CrossRef] [PubMed]

43. Tong, J.; Du, G.G.; Chen, S.R.; MacLennan, D.H. HEK-293 cells possess a carbachol- and thapsigargin-sensitive intracellular Ca²⁺ store that is responsive to stop-flow medium changes and insensitive to caffeine and ryanodine. *Biochem. J.* **1999**, *343 Pt 1*, 39–44. [CrossRef] [PubMed]

44. Querfurth, H.W.; Haughey, N.J.; Greenway, S.C.; Yacono, P.W.; Golan, D.E.; Geiger, J.D. Expression of ryanodine receptors in human embryonic kidney (HEK293) cells. *Biochem. J.* **1998**, *334 Pt 1*, 79–86. [CrossRef] [PubMed]

45. Luo, D.; Sun, H.; Xiao, R.P.; Han, Q. Caffeine induced Ca²⁺ release and capacitative Ca²⁺ entry in human embryonic kidney (HEK293) cells. *Eur. J. Pharmacol.* **2005**, *509*, 109–115. [CrossRef] [PubMed]

46. Itzhaki, I.; Rapoport, S.; Huber, I.; Mizrahi, I.; Zwi-Dantsis, L.; Arbel, G.; Schiller, J.; Gepstein, L. Calcium handling in human induced pluripotent stem cell derived cardiomyocytes. *PLoS ONE* **2011**, *6*, e18037. [CrossRef] [PubMed]

47. Blakeslee, W.W.; Wysoczynski, C.L.; Fritz, K.S.; Nyborg, J.K.; Churchill, M.E.; McKinsey, T.A. Class I HDAC inhibition stimulates cardiac protein SUMOylation through a post-translational mechanism. *Cell Signal.* **2014**, *26*, 2912–2920. [CrossRef] [PubMed]

48. Bers, D.M.; Despa, S. Cardiac myocytes Ca²⁺ and Na⁺ regulation in normal and failing hearts. *J. Pharmacol. Sci.* **2006**, *100*, 315–322. [CrossRef] [PubMed]

49. Bostjancic, E.; Zidar, N.; Glavac, D. MicroRNAs and cardiac sarcoplasmic reticulum calcium ATPase-2 in human myocardial infarction: Expression and bioinformatic analysis. *BMC Genom.* **2012**, *13*, 552. [CrossRef] [PubMed]

50. Gurha, P.; Abreu-Goodger, C.; Wang, T.; Ramirez, M.O.; Drumond, A.L.; van Dongen, S.; Chen, Y.; Bartonicek, N.; Enright, A.J.; Lee, B.; et al. Targeted deletion of microRNA-22 promotes stress-induced cardiac dilation and contractile dysfunction. *Circulation* **2012**, *125*, 2751–2761. [CrossRef] [PubMed]

51. Zhao, B.; Lucas, K.J.; Saha, T.T.; Ha, J.; Ling, L.; Kokoza, V.A.; Roy, S.; Raikhel, A.S. MicroRNA-275 targets sarco/endoplasmic reticulum Ca²⁺ adenosine triphosphatase (SERCA) to control key functions in the mosquito gut. *PLoS Genet.* **2017**, *13*, e1006943. [CrossRef] [PubMed]

52. Lee, E.M.; Shin, S.; Cha, H.J.; Yoon, Y.; Bae, S.; Jung, J.H.; Lee, S.M.; Lee, S.J.; Park, I.C.; Jin, Y.W.; et al. Suberoylanilide hydroxamic acid (SAHA) changes microRNA expression profiles in A549 human non-small cell lung cancer cells. *Int. J. Mol. Med.* **2009**, *24*, 45–50. [PubMed]

53. Yang, H.; Lan, P.; Hou, Z.; Guan, Y.; Zhang, J.; Xu, W.; Tian, Z.; Zhang, C. Histone deacetylase inhibitor SAHA epigenetically regulates miR-17-92 cluster and MCM7 to upregulate MICA expression in hepatoma. *Br. J. Cancer* **2015**, *112*, 112–121. [CrossRef] [PubMed]

54. Poddar, S.; Kesharwani, D.; Datta, M. Histone deacetylase inhibition regulates miR-449a levels in skeletal muscle cells. *Epigenetics* **2016**, *11*, 579–587. [CrossRef] [PubMed]

55. Li, S.; Li, X.; Zheng, H.; Xie, B.; Bidasee, K.R.; Rozanski, G.J. Pro-oxidant effect of transforming growth factor-beta1 mediates contractile dysfunction in rat ventricular myocytes. *Cardiovasc. Res.* **2008**, *77*, 107–117. [CrossRef] [PubMed]

56. Mufti, S.; Wenzel, S.; Euler, G.; Piper, H.M.; Schluter, K.D. Angiotensin II-dependent loss of cardiac function: Mechanisms and pharmacological targets attenuating this effect. *J. Cell. Physiol.* **2008**, *217*, 242–249. [CrossRef] [PubMed]

57. Ammanamanchi, S.; Brattain, M.G. Restoration of transforming growth factor-beta signaling through receptor RI induction by histone deacetylase activity inhibition in breast cancer cells. *J. Biol. Chem.* **2004**, *279*, 32620–32625. [CrossRef] [PubMed]

58. Khan, S.; Jena, G. Sodium butyrate, a HDAC inhibitor ameliorates eNOS, iNOS and TGF-beta1-induced fibrogenesis, apoptosis and DNA damage in the kidney of juvenile diabetic rats. *Food Chem. Toxicol.* **2014**, *73*, 127–139. [CrossRef] [PubMed]

59. Xie, L.; Santhoshkumar, P.; Reneker, L.W.; Sharma, K.K. Histone deacetylase inhibitors trichostatin A and vorinostat inhibit TGFbeta2-induced lens epithelial-to-mesenchymal cell transition. *Investig. Ophthalmol. Vis. Sci.* **2014**, *55*, 4731–4740. [CrossRef] [PubMed]

60. Huber, K.; Doyon, G.; Plaks, J.; Fyne, E.; Mellors, J.W.; Sluis-Cremer, N. Inhibitors of histone deacetylases: Correlation between isoform specificity and reactivation of HIV type 1 (HIV-1) from latently infected cells. *J. Biol. Chem.* **2011**, *286*, 22211–22218. [CrossRef] [PubMed]

61. Clocchiatti, A.; Florean, C.; Brancolini, C. Class IIa HDACs: From important roles in differentiation to possible implications in tumourigenesis. *J. Cell. Mol. Med.* **2011**, *15*, 1833–1846. [CrossRef] [PubMed]

62. Kahali, S.; Sarcar, B.; Prabhu, A.; Seto, E.; Chinnaiyan, P. Class I histone deacetylases localize to the endoplasmic reticulum and modulate the unfolded protein response. *FASEB J.* **2012**, *26*, 2437–2445. [CrossRef] [PubMed]

63. Ligeti, L.; Szenczi, O.; Prestia, C.M.; Szabo, C.; Horvath, K.; Marcsek, Z.L.; van Stiphout, R.G.; van Riel, N.A.; Op den Buijs, J.; Van der Vusse, G.J.; et al. Altered calcium handling is an early sign of streptozotocin-induced diabetic cardiomyopathy. *Int. J. Mol. Med.* **2006**, *17*, 1035–1043. [CrossRef] [PubMed]

64. Pieske, B.; Kretschmann, B.; Meyer, M.; Holubarsch, C.; Weirich, J.; Posival, H.; Minami, K.; Just, H.; Hasenfuss, G. Alterations in intracellular calcium handling associated with the inverse force-frequency relation in human dilated cardiomyopathy. *Circulation* **1995**, *92*, 1169–1178. [CrossRef] [PubMed]

65. Saini, H.K.; Dhalla, N.S. Defective calcium handling in cardiomyocytes isolated from hearts subjected to ischemia-reperfusion. *Am. J. Physiol. Heart Circ. Physiol.* **2005**, *288*, H2260–H2270. [CrossRef] [PubMed]

66. Balke, C.W.; Shorofsky, S.R. Alterations in calcium handling in cardiac hypertrophy and heart failure. *Cardiovasc. Res.* **1998**, *37*, 290–299. [CrossRef]

67. Luo, M.; Anderson, M.E. Mechanisms of altered Ca^{2+} handling in heart failure. *Circ. Res.* **2013**, *113*, 690–708. [CrossRef] [PubMed]

68. Zaniboni, M.; Pollard, A.E.; Yang, L.; Spitzer, K.W. Beat-to-beat repolarization variability in ventricular myocytes and its suppression by electrical coupling. *Am. J. Physiol. Heart Circ. Physiol.* **2000**, *278*, H677–H687. [CrossRef] [PubMed]

69. Maruyama, K.; MacLennan, D.H. Mutation of aspartic acid-351, lysine-352, and lysine-515 alters the Ca^{2+} transport activity of the Ca^{2+}-ATPase expressed in COS-1 cells. *Proc. Natl. Acad. Sci. USA* **1988**, *85*, 3314–3318. [CrossRef] [PubMed]

70. Sacchetto, R.; Testoni, S.; Gentile, A.; Damiani, E.; Rossi, M.; Liguori, R.; Drogemuller, C.; Mascarello, F. A defective SERCA1 protein is responsible for congenital pseudomyotonia in Chianina cattle. *Am. J. Pathol.* **2009**, *174*, 565–573. [CrossRef] [PubMed]

71. Kiianitsa, K.; Solinger, J.A.; Heyer, W.D. NADH-coupled microplate photometric assay for kinetic studies of ATP-hydrolyzing enzymes with low and high specific activities. *Anal. Biochem.* **2003**, *321*, 266–271. [CrossRef]

72. Bassani, J.W.; Bassani, R.A.; Bers, D.M. Relaxation in rabbit and rat cardiac cells: Species-dependent differences in cellular mechanisms. *J. Physiol.* **1994**, *476*, 279–293. [CrossRef] [PubMed]

73. Gnad, F.; Gunawardena, J.; Mann, M. PHOSIDA 2011: The posttranslational modification database. *Nucleic Acids Res.* **2011**, *39*, D253–D260. [CrossRef] [PubMed]

74. Hornbeck, P.V.; Zhang, B.; Murray, B.; Kornhauser, J.M.; Latham, V.; Skrzypek, E. PhosphoSitePlus, 2014: Mutations, PTMs and recalibrations. *Nucleic Acids Res.* **2015**, *43*, D512–D520. [CrossRef] [PubMed]

75. Marti-Renom, M.A.; Stuart, A.C.; Fiser, A.; Sanchez, R.; Melo, F.; Sali, A. Comparative protein structure modeling of genes and genomes. *Annu. Rev. Biophys. Biomol. Struct.* **2000**, *29*, 291–325. [CrossRef] [PubMed]

76. Robbins, A.K.; Horlick, R.A. Macrophage scavenger receptor confers an adherent phenotype to cells in culture. *Biotechniques* **1998**, *25*, 240–244. [PubMed]

6

Beta-Estradiol Regulates Voltage-Gated Calcium Channels and Estrogen Receptors in Telocytes from Human Myometrium

Adela Banciu [1,2,†], Daniel Dumitru Banciu [1,†], Cosmin Catalin Mustaciosu [3,4], Mihai Radu [3], Dragos Cretoiu [5,6], Junjie Xiao [7], Sanda Maria Cretoiu [5], Nicolae Suciu [6,8] and Beatrice Mihaela Radu [1,9,*]

[1] Department of Anatomy, Animal Physiology and Biophysics, Faculty of Biology, University of Bucharest, Splaiul Independentei 91-95, 050095 Bucharest, Romania; adela.banciu79@gmail.com (A.B.); danieldumitrubanciu@gmail.com (D.D.B.)

[2] Faculty of Medical Engineering, University Politehnica of Bucharest, Gheorge Polizu Street 1-7, 011061 Bucharest, Romania

[3] Department of Life and Environmental Physics, Horia Hulubei National Institute of Physics and Nuclear Engineering, Reactorului 30, P.O. Box MG-6, 077125 Magurele, Romania; cosmin@nipne.ro (C.C.M.); mradu@nipne.ro (M.R.)

[4] Faculty of Applied Chemistry and Materials Science, University Politehnica of Bucharest, 011061 Bucharest, Romania

[5] Department of Cell and Molecular Biology and Histology, Carol Davila University of Medicine and Pharmacy, 050474 Bucharest, Romania; dragos@cretoiu.ro (D.C.); sanda@cretoiu.ro (S.M.C.)

[6] Alessandrescu-Rusescu National Institute of Mother and Child Health, Fetal Medicine Excellence Research Center, 020395 Bucharest, Romania; nsuciu54@yahoo.com

[7] Cardiac Regeneration and Ageing Lab, Experimental Center of Life Sciences, School of Life Science, Shanghai University, Shanghai 200444, China; junjiexiao@shu.edu.cn

[8] Department of Obstetrics and Gynecology, Polizu Clinical Hospital, 011062 Bucharest, Romania

[9] Life, Environmental and Earth Sciences Division, Research Institute of the University of Bucharest (ICUB), 91-95 Splaiul Independenţei, 050095 Bucharest, Romania

* Correspondence: beatrice.radu@bio.unibuc.ro
† These authors contributed equally to this work.

Abstract: Voltage-gated calcium channels and estrogen receptors are essential players in uterine physiology, and their association with different calcium signaling pathways contributes to healthy and pathological conditions of the uterine myometrium. Among the properties of the various cell subtypes present in human uterine myometrium, there is increasing evidence that calcium oscillations in telocytes (TCs) contribute to contractile activity and pregnancy. Our study aimed to evaluate the effects of beta-estradiol on voltage-gated calcium channels and estrogen receptors in TCs from human uterine myometrium and to understand their role in pregnancy. For this purpose, we employed patch-clamp recordings, ratiometric Fura-2-based calcium imaging analysis, and qRT-PCR techniques for the analysis of cultured human myometrial TCs derived from pregnant and non-pregnant uterine samples. In human myometrial TCs from both non-pregnant and pregnant uterus, we evidenced by qRT-PCR the presence of genes encoding for voltage-gated calcium channels (Cav3.1, Ca3.2, Cav3.3, Cav2.1), estrogen receptors (ESR1, ESR2, GPR30), and nuclear receptor coactivator 3 (NCOA3). Pregnancy significantly upregulated Cav3.1 and downregulated Cav3.2, Cav3.3, ESR1, ESR2, and NCOA3, compared to the non-pregnant condition. Beta-estradiol treatment (24 h, 10, 100, 1000 nM) downregulated Cav3.2, Cav3.3, Cav1.2, ESR1, ESR2, GRP30, and NCOA3 in TCs from human pregnant uterine myometrium. We also confirmed the functional expression of voltage-gated calcium channels by patch-clamp recordings and calcium imaging analysis of TCs from pregnant human myometrium by perfusing with BAY K8644, which induced calcium influx through these channels. Additionally, we demonstrated that beta-estradiol (1000 nM) antagonized the effect of BAY K8644 (2.5 or 5 µM) in the same preparations. In conclusion, we evidenced the presence of voltage-gated calcium channels and estrogen receptors in TCs from non-pregnant and pregnant human uterine myometrium and their gene expression regulation by beta-estradiol in pregnant conditions. Further exploration of the calcium signaling in TCs and its modulation by estrogen hormones will contribute to the understanding of labor and pregnancy mechanisms and to the development of effective strategies to reduce the risk of premature birth.

Keywords: beta-estradiol; human uterine myometrium; telocytes; calcium signaling; voltage-gated calcium channels; estrogen receptors

---

1. Introduction

Uterine contractions represent a key point throughout pregnancy. Considered to be normal during pregnancy, uterine contractions may also be responsible for triggering premature or dysfunctional labor. During pregnancy, the myometrium suffers morphological and adaptive changes which enable it to become a forceful organ necessary for delivery. Myometrium contractility involves, among other physiological mechanisms that produce excitation in the uterus, changes in Ca^{2+} signals. Smooth myofiber contraction requires actin and myosin myofilaments and their interaction. The cross-bridge formation and contraction are mediated by elevated levels of intracellular Ca^{2+} and myosin light-chain phosphorylation.

Calcium signaling plays an essential role in uterine contractility and involves several key players, including voltage-gated calcium channels, calcium-activated chloride channels, large conductance Ca^{2+}-activated K^+ channels (BK(Ca)), calcium-sensing receptors, and transient receptor potential channels [1–6]. On the basis of their activation electrophysiological features, voltage-gated calcium channels have been classified in high-voltage-activated Ca^{2+} channels (HVA) and low-voltage-activated Ca^{2+} channels (LVA). Among the HVA channels, only the L-type Ca^{2+} channels are expressed in uterus. Their family includes Cav1.1 (CACNA1S gene), Cav1.2 (CACNA1C gene), Cav1.3 (CACNA1D gene), and Cav1.4 (CACNA1F gene) channels. Meanwhile, in uterus, the LVA channels are represented by T-type calcium channels, classified as Cav3.1 (CACNA1G gene), Cav3.2 (CACNA1H gene), and Cav3.3 channel (CACNA1I gene).

In uterine contractility, intracellular Ca^{2+} comes from two sources: entry across the sarcolemma through voltage-gated L-type Ca^{2+} channels and release from the sarcoplasmic reticulum [7]. In the uterus, the major Ca^{2+} source for contraction comes from the extracellular space through voltage-gated L-type Ca^{2+} channels [8]. Indeed, calcium influx through the L-type Ca^{2+} channels was demonstrated to be essential for labor and uterine contractility [1]. Moreover, uterine phasic contractions were abolished if the L-type Ca^{2+} channels were blocked [1]. In addition, the T-type Ca^{2+} channels might also contribute to calcium entry in smooth myocytes in human myometrium [9].

The estrogen 17beta-estradiol was demonstrated in different types of tissue (e.g., neuronal, uterine etc.) to modulate multiple components of the calcium transport pathways, including BK(Ca) [10], small conductance Ca^{2+}-activated potassium channel subtype 3 (SK3, [11]), HVA channels [12,13], LVA channels [14], Na^+/Ca^{2+} exchanger (NCX, [15]), transient receptor potential channels [16,17], etc. In particular, beta-estradiol and its receptors play an essential role in pregnancy and labor [18–20] and were shown to act on uterine myometrium by means of various calcium signaling pathways.

Telocytes (TCs) have been described by our team as a new cellular type in human myometrium [21]. Characterized by extremely long telopodes, TCs were described in the interstitial space of numerous organs, as 3-D network-forming cells by homocellular or heterocellular contacts [22,23]. It was shown that TCs release extracellular vesicles, such as exosomes and ectosomes, to regulate the functions of the surrounding cells in non-pregnant and pregnant myometrium [24]. The gene profiles, proteome and secretome features of TCs were recently revealed [25–27]. Moreover, our early studies showed that

TCs express estrogen and progesterone receptors, which indicates them as potentially responsible for myogenic contractility modulation under hormonal control [28].

There are some extensive reviews describing the main morphological features and possible functions of TCs in reproductive organs [29–32]. One of the most important aspects relates to the TCs stemness capacity which might contribute to regeneration and repair processes, as it has been previously shown [33]. In particular, a recent review highlights the contribution of calcium signaling in interstitial cells (including TCs, interstitial cells of Cajal, interstitial Cajal-like cells) to uterine, cardiac, and urinary physiology and pathology [34].

Recently, we demonstrated, by immunofluorescence and electrophysiology, the presence of the T-type calcium channels in human uterine myometrial TCs [35]. Our studies also confirmed the antagonistic pharmacological effect of mibefradil on voltage-gated calcium currents in TCs from human uterine myometrium [35] and its modulatory effect on telopodes growth after stimulation with near-infrared low-level lasers [36]. We also evidenced, by patch-clamp recordings, the presence of HVA calcium currents in TCs from human non-pregnant myometrium [35]. To date, no studies have yet quantified the mRNA levels encoding the voltage-gated calcium channels in human uterine myometrial TCs. Moreover, despite extensive studies describing the effect of beta-estradiol on human uterine myometrium, no description of its mechanism of action on TCs was done.

TCs were recently detected in human uterine leiomyoma in higher number compared with the areas of adjacent fibrotic and normal myometrium, suggesting that it may provide a scaffold for newly formed myocytes or control important downstream signaling pathways [37]. By contrast, another study claims that TCs are absent inside leiomyomas and represent ~2% of the cells in the normal myometrium [38]. Since leiomyomas are considered estrogen-dependent tumors, and leiomyoma tissue is more sensitive to estradiol (has more estrogen receptors in comparison to normal myometrium), authors suggested that the loss of the TC network might lead to myocytes taking up the hormonal sensor function, which can determine an uncontrollable proliferation of myocytes [38]. However, in our opinion, both studies have a major weakness since they do not demonstrate the presence of TCs with the aid of their most specific immunohistochemical markers CD34 and platelet-derived growth factor receptor (PDGFR) α or β [39,40], but use c-kit for the detection. TCs' c-kit positivity is also used by the authors to suggest the role of TC progenitor cells for the development of leiomyoma. Moreover, on the basis of the assumption that TCs are involved in angiogenesis [41,42], they also suggest that TCs loss may be responsible for decreased vessel formation within the myometrium and subsequent shifting from aerobic to anaerobic metabolism in smooth muscle cells [38]. The subsequent hypoxia and the decreased angiogenesis represent crucial factors for leiomyoma development [43,44].

Uterine TCs have been reported to be present in different reproductive states; while endometrial TCs have been hypothesized to be involved in glandular support and stromal cell communication, myometrial TCs were considered to be responsible for the initiation and propagation of contractile activity [45]. Calcium is a key player in uterine physiological activity, and multiple calcium signaling pathways have been evidenced to be activated in pregnancy and labor. In particular, voltage-gated calcium channels (e.g., HVA and LVA calcium channels) [46,47] or estrogen receptors [48] have been described to be involved in the spontaneous contractile activity in both pregnant and non-pregnant uterus. Therefore, our results in human myometrial TCs from pregnant and non-pregnant uterus are clinically relevant.

Our goal was to determine the level of expression of voltage-gated calcium channels and estrogen receptors in human myometrial TCs and to evidence potential differences in pregnant versus non-pregnant uterus. Additionally, we focused our attention on the modulatory effect of female sex hormones (e.g., beta-estradiol) exerted on the voltage-gated calcium channels and estrogen receptors in human myometrial TCs from pregnant uterine samples.

Our study is trying to highlight some of the calcium signaling mechanisms associated with voltage-gated calcium channels and estrogen receptors activation and to understand their involvement in the pregnancy state.

2. Results

2.1. Pregnancy Induces Changes in mRNA Levels Encoding the Voltage-Gated Calcium Channels Compared to the Non-Pregnant Condition in Human Uterine Myometrial TCs

In a previous study, we demonstrated the immunopositivity for T-type calcium channels (i.e., Cav3.1 and Cav3.2) in human myometrial TCs from pregnant and non-pregnant uterus [35]. We also evidenced the presence of T-type calcium currents and HVA currents in human myometrial TCs by employing patch-clamp recordings [35]. In the present study, we wanted to quantify the mRNA levels for voltage-gated calcium channels by qRT-PCR in both types of uterine samples. Indeed, we confirmed the presence of the followings genes: *CACNA1G* (encoding Cav3.1 T-type calcium channel), *CACNA1H* (encoding Cav3.2 T-type calcium channel) > *CACNA1I* (encoding Cav3.3 T-type calcium channel), and *CACNA1C* (encoding Cav1.2 L-type calcium channel) in TCs from both pregnant and non-pregnant uterine myometrium. The ranking of the mRNA levels encoding the voltage-gated calcium channels relative to *GAPDH* was: *CACNA1G* (Cav3.1) > *CACNA1C* (Cav1.2) > *CACNA1H* (Cav3.2) > *CACNA1I* (Cav3.3) in TCs from both pregnant (Figure 1C,D) and non-pregnant (Figure 1A,B) uterine myometrium. The same mRNA level ranking was obtained relative to the other housekeeping gene *18S rRNA*.

We also evaluated if there were any changes in the mRNA levels encoding the voltage-gated calcium channels due to the pregnancy condition (Figure 1E,F). When considering *GAPDH* as a reference gene and the non-pregnant condition as a calibrator, *CACNA1G* was upregulated 3.74-fold ($p < 0.01$), while *CACNA1H*, *CACNA1I*, and *CACNA1C* were downregulated 0.87-fold ($p < 0.05$), 0.81-fold ($p < 0.05$), and 0.56-fold (not significant), respectively, in pregnant samples versus non-pregnant samples. On the other hand, when considering *18S rRNA* as a reference gene and the non-pregnant condition as a calibrator, *CACNA1G* was upregulated 6.48-fold ($p < 0.01$), while *CACNA1H*, *CACNA1I*, and *CACNA1C* were downregulated 0.82-fold ($p < 0.05$), 0.71-fold ($p < 0.05$), and 0.29-fold (not significant), respectively, in pregnant samples versus non-pregnant samples.

2.2. Pregnancy Induces Changes in mRNA Levels Encoding the Estrogen Receptors Compared to the Non-Pregnant Condition

Estrogens (i.e., beta-estradiol) are essential hormones in uterine physiology. They may act on multiple cellular targets, including estrogen receptors (ESR1 and ESR2), G protein-coupled receptors (GPR30), and nuclear receptors (NCOA3). The role of estrogen receptors in the uterine myometrium physiology was demonstrated in different species, including human, canine, equine, rat etc. [49–53]. We might expect that these genes are present in different subtypes of cells of the human uterine myometrium, including TCs and smooth muscle cells. Therefore, we analyzed the presence of *ESR1*, *ESR2*, *GPR30*, and *NCOA3* genes in human myometrial TCs from non-pregnant and pregnant uterine samples. We confirmed the presence of all four analyzed genes, with the following ranking: *NCOA3* > *ESR1* > *GPR30* > *ESR2*. Their presence was not dependent on the pregnancy state or the used housekeeping gene (Figure 2A–D).

We also demonstrated a significant downregulation of the mRNA levels of all four genes in human myometrial TCs from pregnant uterine samples compared to non-pregnant uterine samples (Figure 2E,F). When considering *GAPDH* as a reference gene and the non-pregnant condition as a calibrator, the estrogen receptors were downregulated in pregnant samples 0.89-fold ($p < 0.05$, *ESR1*), 0.77-fold (not significant, *ESR2*), 0.67-fold ($p < 0.05$, *GPR30*), 0.86-fold ($p < 0.05$, *NCOA3*) versus non-pregnant samples. Meanwhile, when considering *18S rRNA* as a reference gene and the non-pregnant condition as a calibrator, the estrogen receptors were downregulated in pregnant samples 0.84-fold ($p < 0.05$, *ESR1*), 0.68-fold ($p < 0.05$, *ESR2*), 0.54-fold ($p < 0.01$, *GPR30*), 0.8-fold ($p < 0.05$, *NCOA3*) versus non-pregnant samples.

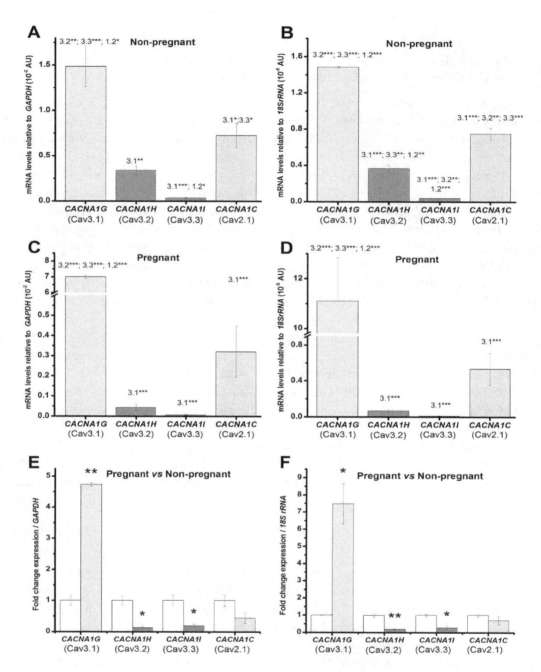

Figure 1. mRNA levels encoding the voltage-gated calcium channels in human myometrial telocytes (TCs) from non-pregnant (**A,B**) and pregnant (**C,D**) uterine samples. The mRNA levels for *CACNA1G* (Cav3.1 channel), *CACNA1H* (Cav3.2 channel), *CACNA1I* (Cav3.3 channel), and *CACNA1C* (Cav1.2 channel) were normalized against two different reference genes, *GAPDH* (**A,C**) and *18S rRNA* (**B,D**), and were plotted as mean ± SD ($N = 3$), corresponding to cell batches extracted from either pregnant or non-pregnant uterine myometrium. One-way ANOVA analysis, applied for the samples derived either from pregnant uterus or from non-pregnant uterus, indicated: (**A**) F = 78, $p < 0.001$; (**B**) F = 545, $p < 0.001$; (**C**) F = 2580, $p < 0.001$; (**D**) F = 77, $p < 0.001$. The post-hoc Bonferroni test was applied, and the statistically significant pairs are indicated by the numbers corresponding to the channels symbols, as follows: Cav3.1, 3.2, 3.3, and 1.2; (**E,F**) Fold-change for each voltage-gated calcium channel between non-pregnant and pregnant uterine samples, where non-pregnant samples were defined as calibrator, considering as housekeeping gene *GAPDH* (**E**) and *18S rRNA* (**F**), respectively. Statistical significance is indicated with asterisks (* $0.01 < p < 0.05$; ** $0.001 < p < 0.01$; *** $p < 0.001$).

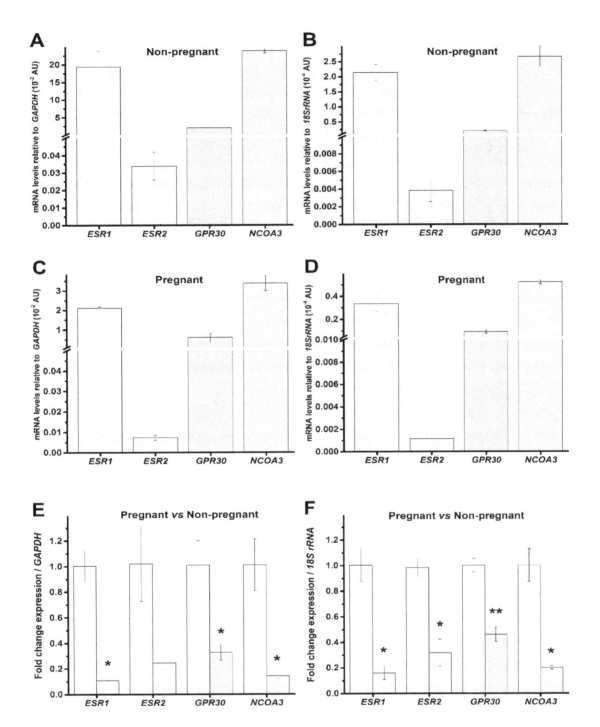

Figure 2. mRNA levels encoding the estrogen receptors and nuclear receptor co-activator in human myometrial TCs from non-pregnant (**A,B**) and pregnant (**C,D**) uterine samples. The mRNA levels of *ESR1* (estrogen receptor 1), *ESR2* (estrogen receptor 2), *GPR30* (G protein-coupled receptor 30), and *NCOA3* (nuclear receptor coactivator 3), were normalized against two different reference genes, *GAPDH* (**A,C**) and *18S rRNA* (**B,D**) and were plotted as mean ± SD ($N = 3$), corresponding to cell batches extracted from either pregnant or non-pregnant uterine myometrium. (**E,F**) Fold-change for each estrogen receptor in pregnant uterine samples versus the non-pregnant condition, where non-pregnant samples were defined as calibrator, considering as housekeeping gene *GAPDH* (**E**) and *18S rRNA* (**F**), respectively. Statistical significance is indicated with asterisks (* $0.01 < p < 0.05$; ** $0.001 < p < 0.01$).

2.3. Beta-Estradiol Downregulates Voltage-Gated Calcium Channels in TCs from Human Pregnant Uterine Myometrial Cultures

We considered only the pregnant condition for further testing the effect of beta-estradiol. Primary cultures of TCs from human pregnant myometrium were treated for 24 h with increasing concentrations of beta-estradiol (10, 100, and 1000 nM). The treatment protocol with beta-estradiol was established on the basis of previous studies [54]. We quantified by qRT-PCR the mRNA levels encoding the Cav3.1, Cav3.2, Cav3.3, and Cav1.2 channels at different doses of beta-estradiol, considering *GAPDH* or *18S rRNA* as housekeeping genes.

Beta-estradiol treatment downregulated the mRNAs encoding the voltage-gated calcium channels, with distinct patterns for each channel subtype. Considering the untreated cells as calibrator and *GAPDH* as the reference gene, *CACNA1G* (One way ANOVA, $F = 13$, $p < 0.05$), *CACNA1GH* (One way ANOVA, $F = 14$, $p < 0.05$), and *CACNA1GI* (One way ANOVA, $F = 37$, $p < 0.01$) channels were significantly downregulated (Figure 3A). The Bonferroni post-hoc analysis indicated that 100 nM beta-estradiol significantly downregulated *CACNA1G* by 0.67-fold ($p < 0.05$), *CACNA1GH* by 0.75-fold ($p < 0.05$), *CACNA1GH* by 0.86-fold ($p < 0.05$), versus non-treated samples. Considering the untreated cells as calibrator and *18S rRNA* as the reference gene, *CACNA1G* (One way ANOVA, $F = 29$, $p < 0.01$), *CACNA1GH* (One way ANOVA, $F = 12.5$, $p < 0.05$), *CACNA1GI* (One way ANOVA, $F = 24$, $p < 0.01$), and *CACNA1C* (One way ANOVA, $F = 103$, $p < 0.001$) channels were significantly downregulated (Figure 3B). The Bonferroni post-hoc analysis indicated that 100 and 1000 nM beta-estradiol significantly downregulated *CACNA1G* by 0.77-fold ($p < 0.01$) and 0.74-fold ($p < 0.05$), *CACNA1GH* by 0.73-fold ($p < 0.05$) and 0.82-fold ($p < 0.05$), *CACNA1GH* by 0.68-fold ($p < 0.05$) and 0.89-fold ($p < 0.01$), *CACNA1GC* by 0.49-fold ($p < 0.01$) and 0.36-fold ($p < 0.01$), respectively, versus the non-treated samples. It should be mentioned that the downregulation of *CACNA1GC* mRNA upon treatment with beta-estradiol was statistically significantly only relative to *18S rRNA* and not relative to *GAPDH*.

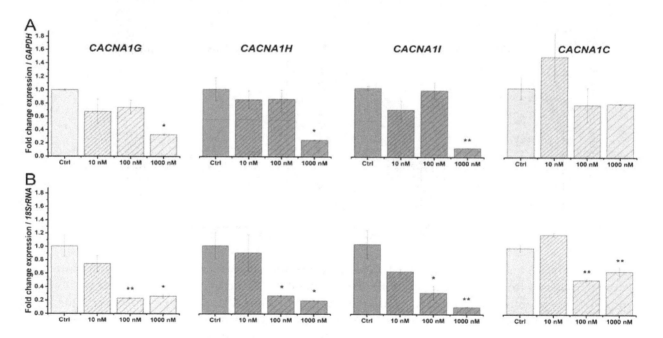

Figure 3. Effect of beta-estradiol treatment (10, 100, and 1000 nM) on the mRNA levels encoding the voltage-gated calcium channels in human myometrial TCs from pregnant uterine samples. Fold-change for each voltage-gated calcium channel upon treatment with beta-estradiol, where the non-treated samples were considered as calibrator. The data were normalized against *GAPDH* (**A**) and 18S rRNA (**B**) and were plotted as mean $\pm$ SD ($N = 3$). Statistical significance is indicated with asterisks (* $0.01 < p < 0.05$; ** $0.001 < p < 0.01$).

2.4. Beta-Estradiol Downregulates Estrogen Receptors in TCs from Human Pregnant Uterine Myometrial Cultures

Previous studies indicated that beta-estradiol was able to regulate the mRNA levels of estrogen receptors in different tissues [55–57]. Therefore, we analyzed the effect of a 24 h treatment with beta-estradiol (10,100, 1000 nM) on the mRNA levels for *ESR1* (estrogen receptor 1), *ESR2* (estrogen receptor 2), *GPR30* (G protein-coupled receptor 30), and *NCOA3* (nuclear receptor coactivator 3) in human myometrial TCs from pregnant uterus samples and normalized the data against the housekeeping genes *GAPDH* (Figure 4A) and *18S rRNA* (Figure 4B).

Considering the untreated cells as calibrator and *GAPDH* as the reference gene, beta-estradiol significantly downregulated *ESR1* (One way ANOVA, F = 31, $p < 0.01$), *ESR2* (One way ANOVA, F = 12, $p < 0.05$), *GPR30* (One way ANOVA, F = 49, $p < 0.01$), and *NCOA3* (One way ANOVA, F = 49, $p < 0.01$) (Figure 4A). The Bonferroni post-hoc analysis indicated that: 100 nM beta-estradiol significantly downregulated *ESR1* by 0.55-fold ($p < 0.05$) and *ESR2* by 0.89-fold ($p < 0.05$); 10, 100, and 1000 nM beta-estradiol significantly downregulated *GPR30* by 0.81-fold ($p < 0.01$), 0.97-fold ($p < 0.01$), and 0.72-fold ($p < 0.01$), respectively; 10 and 100 nM beta-estradiol significantly downregulated *NCAO3* by 0.89-fold ($p < 0.01$) and 0.96-fold ($p < 0.01$), respectively, versus the non-treated samples (Figure 4A).

Considering the untreated cells as calibrator and *18S rRNA* as the reference gene, *ESR1* (One way ANOVA, F = 27, $p < 0.01$), *ESR2* (One way ANOVA, F = 14, $p < 0.05$), *GPR30* (One way ANOVA, F = 46, $p < 0.01$), and *NCOA3* (One way ANOVA, F = 92, $p < 0.001$) were downregulated (Figure 4B). The Bonferroni post-hoc analysis indicated that: 100 nM beta-estradiol significantly downregulated *ESR1* by 0.37-fold ($p < 0.05$) and *ESR2* by 0.46-fold ($p < 0.05$); 10, 100, and 1000 nM beta-estradiol significantly downregulated *GPR30* by 0.8-fold ($p < 0.01$), 0.96-fold ($p < 0.01$), and 0.78-fold ($p < 0.01$), respectively; 10 and 100 nM beta-estradiol significantly downregulated *NCAO3* by 0.87-fold ($p < 0.01$) and 0.95-fold ($p < 0.01$), respectively, versus the non-treated samples (Figure 4B).

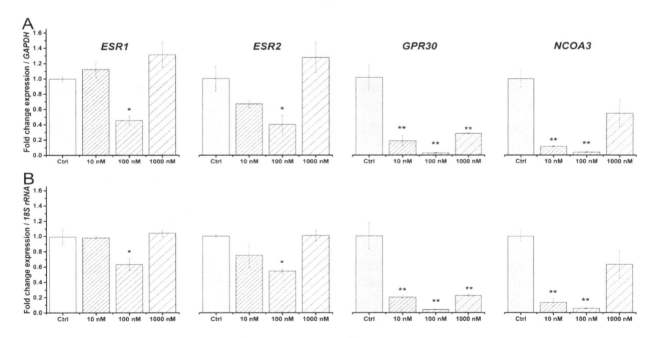

Figure 4. Effect of beta-estradiol treatment (10, 100, and 1000 nM) on the mRNA levels encoding the estrogen receptors and nuclear receptor co-activator in human myometrial TCs from pregnant uterine samples. Fold-change for each receptor upon treatment with beta-estradiol, where the non-treated samples were considered as calibrator. The data were normalized against *GAPDH* (**A**) and *18S rRNA* (**B**) and were plotted as mean ± SD ($N = 3$). Statistical significance is indicated with asterisks (* $0.01 < p < 0.05$; ** $0.001 < p < 0.01$).

2.5. Beta-Estradiol Partly Blocks Bay K8644-Induced Calcium Transients in Human Myometrial Uterine TCs

Bay K8644 was demonstrated to specifically activate HVA calcium channels [46]. We applied this agonist in order to test the functional activation of voltage-gated calcium channels in human myometrial TCs from pregnant myometrium by Fura-2AM calcium imaging (see loaded cells in Figure 5A). Consequently, we recorded calcium transients activated by 5 μM Bay K8644 (Figure 5B).

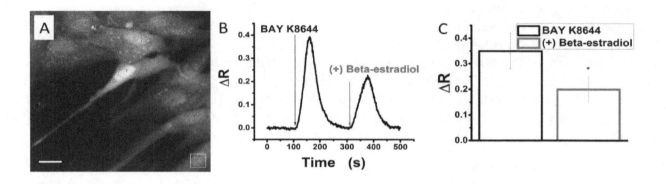

Figure 5. Beta-estradiol effect on voltage-gated calcium channels in TCs from pregnant uterine myometrium by Fura-2-based calcium imaging. (**A**) Cells in primary culture of human uterine myometrium loaded with Fura-2AM. The regions of interest (ROIs) are marked with circles corresponding to: red (reference field without cells), violet (TC), blue (other cells); Scale bar 20 μm (**B**) representative traces of fluorescence ratio (ΔR) changes induced by Bay K8644 (10 μM) in the absence and presence of beta-estradiol (1000 nM) in human myometrial TCs during calcium imaging recordings; (**C**) mean fluorescence ratio (ΔR) ± SD, $N = 5$, in the same conditions as (**B**). The statistical analysis was done by the paired Student's t test, and the level of significance is indicated with asterisks (* $p < 0.05$).

Previous studies indicated that beta-estradiol counteracts the activation exerted by Bay K8644 on cardiac calcium channels [58] and on rat cortical neurons [12,13,59]. In particular, 1000 nM beta-estradiol inhibited the HVA but not the LVA calcium currents in rat sensory neurons via a non-genomic mechanism [12,13,59]. The next step was to test the effect of beta-estradiol (1000 nM) on the calcium transients elicited by Bay K8644 (5 μM) in human myometrial TCs from pregnant uterine samples. Indeed, we evidenced the same partial inhibitory effect of beta-estradiol on calcium influx through voltage-gated calcium channels in TCs (Figure 5B). The mean fluorescence ratio of the calcium signal induced by Bay K8644 was significantly diminished from 0.35 ± 0.07 (in the absence of beta-estradiol) to 0.20 ± 0.05, $N = 5$, $p < 0.05$, paired Student's t test (in the presence of beta-estradiol) Figure 5C.

2.6. Beta-Estradiol Partly Inhibits Bay K8644-Induced Calcium Current in Human Myometrial Uterine TCs

We have also demonstrated that Bay K8644 was able to activate calcium currents in human myometrial uterine TCs (Figure 6A). The membrane capacitance had a mean of 115 ± 11 pF. Moreover, by adding beta-estradiol (1000 nM) in the presence of Bay K8644, we were able to partly inhibit Bay K8644-induced calcium current (Figure 6A). Beta-estradiol blocked Bay K8644-induced current from 85.12 ± 20.03 to 25.91 ± 10.34 pA, $N = 6$, $p < 0.05$, paired Student's t-test (Figure 6B). This inhibitory effect was present in all human myometrial uterine TCs in which Bay K8644 activated voltage-gated calcium currents.

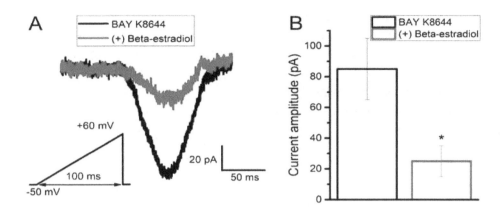

Figure 6. Beta-estradiol partly blocks voltage-gated calcium currents in TCs from pregnant uterine myometrium. (**A**) Representative patch-clamp recording in voltage-clamp mode on a human myometrial TC. Brief depolarizing ramp protocols from -50 to $+60$ mV with a duration of 100 ms were applied. HVA currents were elicited by perfusion with Bay K8644 (2.5 μM) for 1 min followed by the ramp protocol. To block the calcium current, beta-estradiol (1000 nM) was applied in the presence of Bay K8644; (**B**) amplitude of the voltage-gated calcium currents induced by Bay K8644 in the presence or absence of beta-estradiol. The statistical analysis was done by the paired Student's *t* test, and the level of significance is indicated with asterisks (* $p < 0.05$).

3. Discussion

3.1. Beta-Estradiol Regulates Voltage-Gated Calcium Channels in Human Myometrial TCs from Pregnant Uterus

We have previously evidenced HVA calcium currents induced by a brief ramp depolarization protocol in TCs from non-pregnant uterine myometrium [35]. In this study, we obtained Bay K8644-induced currents in TCs from pregnant uterine myometrium with lower amplitude compared to those recorded in TCs from non-pregnant uterine myometrium. This difference might be due to the pregnancy condition and is correlated with the lower expression of *CACNA1C* mRNA (without statistical significance) in pregnant versus non-pregnant uterine samples. While immunofluorescence indicated that the protein expression of the LVA calcium channels (e.g., Cav3.1 and Cav3.2) was upregulated in human myometrial TCs from pregnant versus non-pregnant uterine samples [35], this study showed that gene expression was upregulated for Cav3.1 and downregulated for Cav3.2 and Cav3.3.

Previous reports demonstrated that beta-estradiol regulates the mRNA expression of T-type calcium channel subunits, but the up- or downregulation was dependent on the type of tissue. To date, in the medial preoptic area and the arcuate nucleus, beta-estradiol has been shown to upregulate Cav3.1 and Cav3.2, but not Cav3.3 [60]. Meanwhile, in the pituitary, beta-estradiol has been shown to upregulate Cav3.1 and downregulate Cav3.2 and Cav3.3 [60]. By comparison, our study on human pregnant myometrial TCs indicated that chronic beta-estradiol treatment downregulated Cav3.1, Cav3.2, Cav3.3, and Cav1.2 at higher concentrations. We have also observed that the mRNA encoding Cav3.1 was upregulated in myometrial TCs in the untreated pregnant samples compared to the chronically estrogen-treated pregnant samples. Possible explanations might include the absence of some physiological factors in our preparations, including the bursting release of hormones during pregnancy and labor compared to the continuous chronic exposure (24 h) to estrogen, and the complexity and the interplay of other steroid hormones (e.g., progesterone, oxytocin etc.), prostaglandins, cytokines, nitric oxide, released in parturition [61].

Bay K8644 is a well-known agonist of the L-type calcium channels [38]. Previous studies demonstrated that beta-estradiol (1 μM) inhibited HVA, but not LVA, calcium currents in rat sensory neurons via a non-genomic mechanism [12,13,59]. This inhibition was done in a rapid, reversible and concentration-dependent manner and determined the hyperpolarization shift of the steady-state inactivation curve [12]. Our data are in accordance with these studies and demonstrate the blocking effect of beta-estradiol (1 μM) on high-voltage-activated calcium currents (Bay K8644-activated currents) in TCs from human uterine myometrium. However, 10-fold higher concentrations of beta-estradiol (10 μM) reduced not only the HVA Ba^{2+} current but also the T-type Ca^{2+} current in vascular smooth muscle cells [14]. This effect might be attributed to a non-specific antagonistic effect of beta-estradiol at higher concentrations on both types of voltage-gated calcium channels.

Interestingly, we should mention that beta-estradiol exerted different functional modulatory action on the HVA channels upon acute (1 min) or chronic (24 h) exposure. Thus, acute exposure determined a reduction in the current amplitude by partly antagonizing BAY K8644, while chronic exposure triggered gene expression downregulation.

It is of particular interest to understand if beta-estradiol regulates voltage-gated calcium channels by means of estrogen receptors activation. On this topic, there are controversial reports in the literature regarding the regulation of L-type calcium channels upon beta-estradiol treatment via estrogen receptors. To date, beta-estradiol was demonstrated to inhibit HVA calcium currents in rat cortical neurons without involving the estrogen receptors [12]. In the hypothalamus, beta-estradiol-induced upregulation in Cav3.1 was dependent on ERα, while its effect on Cav3.2 was dependent on both ERα and ERβ [60]. In the pituitary, beta-estradiol-induced effects were only dependent on the expression of ERα [60]. In our study, we did not evaluate if the beta-estradiol effect on voltage-gated calcium channels was mediated by the estrogen receptors. However, corroborating the modulatory effect of beta-estradiol exerted on voltage-gated calcium channels and on estrogen receptors, we might consider that there is an interplay between these key players that involves the activation of calcium signaling pathways. Other hormones (e.g., progesterone, oxytocin) have also been described to block voltage-gated calcium channels [62–64], and we might suppose a similar inhibition in TCs human myometrial preparations, but our study was focused only on estrogen effects.

3.2. Beta-Estradiol Regulates Estrogen Receptors in Human Myometrial TCs from Pregnant Uterus

Estrogens act specifically on estrogen receptors and mediate distinct roles in pregnancy and labor. Indeed, ERα plays a more prominent role than ERβ in mediating estrogen action in the induction of uterine oxytocin receptors before labor [65]. Alterations of the estrogen–estrogen receptors signaling have been associated with different pathologies of the reproductive system (e.g., preeclampsia, endometrial cancer, etc.). For example, low GPR30 expression levels, increased apoptosis, and reduced proliferation were associated with preeclampsia, and the pharmacological targeting of GPR30 might have clinical relevance [66]. Estrogen-induced PI3K–AKT signaling activated by GPR30 is involved in the regulation of endometrial cancer cell proliferation [67].

Previous studies on human pregnant myometrium samples indicated the expression of ERα and GPR30 mRNAs, while ERβ mRNA was not detectable [53]. By comparison, we retrieved ERα and GPR30 mRNAs expression in TCs from human pregnant myometrium samples, but we also detected low levels of ERβ mRNA. These findings are in accordance with our previous immunocytochemical studies on TCs cultured from human myometrium during and outside pregnancy [28,68]. Beside estrogen receptors, we also demonstrated, in an earlier study, the expression of progesterone receptors in TCs from human myometrial cell cultures [28], but an extensive analysis of progesterone receptors expression in pregnant and nonpregnant conditions was beyond the purpose of the present study.

Multiple studies have demonstrated that beta-estradiol regulates the mRNA levels encoding the estrogen receptors in various tissues and species, including mouse Sertoli cells [41], female rat brain [48], rat uterus [69], etc. In particular, 3 and 6 h after beta-estradiol subcutaneous injection, the downregulation of ERα and ERβ mRNA levels was demonstrated in rat uterus [69]. We showed that beta-estradiol downregulates ESR1, ESR2, GPR30, and NCOA3 in TCs from human pregnant uterine myometrial cultures. Our current data are in agreement with the in vivo downregulation of these receptors in rat uterus.

NCOA3, also known as steroid receptor coactivator-3 (src-3), was described as being associated with uterine endometrial cancer or other cancers, e.g., breast cancer [70,71]. Despite the great interest for this coregulator, little is known about its physiological role in uterine myometrium [71]. Our study documents for the first time NCOA3 expression in uterine TCs and its downregulation in pregnancy and upon beta-estradiol treatment.

Calcium was demonstrated to play an important role in the activation of estrogen receptors in different cell types. Interestingly, stimuli inducing the release of intracellular calcium determine the recruitment of estrogen receptors 1 and stimulate the expression of estrogen-responsive genes [72]. Several studies indicated that beta-estradiol regulates the expression of different receptors associated with calcium signaling mechanisms via estrogen receptors. In mouse N2A and human SK-N-SH neural cells, beta-estradiol upregulated the expression of the α and β subunits of BK channels via estrogen receptor β, in a concentration-dependent manner [10]. Alterations of calcium homeostasis may determine breast cancer progression by affecting various signaling pathways including estrogen receptors [73]. Both intra- and extracellular calcium was shown to influence ERα transcriptional activity in breast cancer cells [74].

In pregnancy, beta-estradiol exerts tissue-dependent regulatory actions. In particular, beta-estradiol and estrogens-specific agonists have been shown to differentially modulate the tone of uterine versus placental arteries, and their contribution to the regulation of human uteroplacental blood flow might be tissue-specific [75]. In rats ovariectomized on day 18 of pregnancy, estrogen treatment caused upregulation of oxytocin receptors [65].

In conclusion, our study brings novel highlights into the modulatory effects of beta-estradiol on the uterine myometrium and shows that voltage-gated calcium channels and estrogen receptors expressed in TCs are essential regulatory targets.

4. Materials and Methods

4.1. Human Uterine Samples

The myometrial tissue of non-pregnant women was obtained from a total of 10 menstruating, parous, pre-menopausal women during the proliferative phase, which were undergoing hysterectomy for benign gynecological reasons. The median age of the non-pregnant women was 42 years (range 35–49 years). The inclusion criteria used in the selection of the research subjects were indications for surgery for menometrorrhagia/dysmenorrhea as a consequence of localized or diffuse leiomyomas and no other associated medical conditions or treatment history. Hysterectomized uteri were biopsied in the pathology department where they were examined macroscopically, and apparently normal myometrial areas were chosen. The pregnant women included in the study ($n = 8$) were aged between 30 and 35 years, and the gestational age was between 38 and 40 weeks. The patients underwent cesarean surgery in labor because of dystocia and fetal distress. A small strip of myometrium was carefully dissected from the upper margin (in the midline) of the lower segment transverse incision of each patient.

All tissue samples were obtained in accordance with the protocol no. 12290 (18.08.2017) approved by the Ethics Committee of Alessandrescu-Rusescu National Institute of Mother and Child Health. The patients included in the study were enrolled from the Department of Obstetrics and Gynecology, "Polizu" Clinical Hospital, Alessandrescu-Rusescu National Institute of Mother and Child Health, Bucharest, Romania. All patients donating uterine tissue samples that were included in the study signed the informed written consent. The inclusion criteria requested that none of the patients included in the groups received any regular medication for chronic diseases.

4.2. Myometrial Cell Cultures

For the qRT-PCR experiments, primary cultures were prepared from human uterine myometrial biopsies as previously described [35]. On the basis of the immunopositivity for CD34 and platelet-derived growth factor receptor-α (PDGFRα) that was already described in human uterine myometrial TCs [35], we employed a double-labeling technique, using goat polyclonal anti-CD34 (#sc-7045, Santa Cruz

Biotechnology, Santa Cruz, CA, USA) and rabbit polyclonal anti-PDGFRα (#sc-338, Santa Cruz Biotechnology, USA) antibodies. The primary antibodies were detected by a secondary goat anti-rabbit antibody conjugated to AlexaFluor 488 and a donkey anti-goat antibody conjugated to AlexaFluor 546, from Invitrogen Molecular Probes, Eugene, OR, USA. An enriched TC culture (CD34$^+$/PDGFRα$^+$ cells) was obtained by sorting the double-labeled cells using a BD FACSCanto II - Becton Dickinson flow cytometer (BD Biosciences, Waltham, MA, USA), as previously described for cardiac TCs [69]. For the qRT-PCR experiments, the cells were chronically exposed for 24 h to beta-estradiol (#E2758, Sigma-Aldrich, St. Louis, MO, USA) at concentrations of 10, 100, and 1000 nM in serum-free medium, in order to avoid the non-specific binding of beta-estradiol to the proteins in the serum.

For the patch-clamp and calcium imaging experiments, primary cultures were prepared from human uterine myometrial biopsies as previously described [35]. The cells were used without further sorting in order to maintain the physiological environment and to prevent any alteration in the calcium signaling due to the absence of human myometrial muscle cells in the cell culture. The human myometrial cells (between first and fourth passages) were plated at a density of 5×10^4 cells/cm^2 on 24 mm Petri dishes for the patch-clamp experiments or on 24 mm glass coverslips for the calcium imaging experiments. On the basis of their morphological features (e.g., long and moniliform telopodes), only TCs were selected under the microscope for patch-clamp or calcium imaging recordings. TCs were acutely exposed for 1 min to beta-estradiol in the patch-clamp and calcium-imaging experiments, in the presence or absence of BAY K8644.

4.3. qRT-PCR

To quantify the mRNA expression levels encoding for different genes (Table 1) in primary cell cultures from human uterine myometrium, total RNA was extracted using the GenElute Mammalian Total RNA MiniPrep Kit (RTN70, Sigma), according to the manufacturer's instructions. RNA concentrations were determined by spectrophotometric measurements of absorption at 260 and 280 nm (Beckman Coulter DU 730, Carlsbad, CA, USA), and DNase I treatment was applied in order to remove contaminating genomic DNA. In agreement with the manufacturer's guidelines (Sigma-Aldrich, USA), in our experiments, the A260:A280 ratio was 2.04 ± 0.05. Reverse transcription was done using the High-Capacity cDNA Archive Kit (Applied Biosystems, Waltham, MA, USA). The human primers and TaqMan probes (Life Technologies, Carlsbad, CA, USA) used in our experiments are listed in Table 1 and were used in accordance with the manufacturer's guidelines. GAPDH and 18S rRNA were used as reference genes. The relative abundance of gene transcripts was assessed via qRT-PCR, using the TaqMan methodology and the ABI Prism 7300 Sequence Detection System (Applied Biosystems, USA). The reactions were carried out in triplicate for 50 cycles.

Table 1. Panel of primers (Life Technologies, Carlsbad, CA, USA).

Primer	Gene	Protein Encoded
Hs00167681_m1	CACNA1C	Cav1.2, L-type calcium channel
Hs00367969_m1	CACNA1G	Cav3.1, T-type calcium channel
Hs00234934_m1	CACNA1H	Cav3.2, T-type calcium channel
Hs00184168_m1	CACNA1I	Cav3.3, T-type calcium channel
Hs01922715_s1	GPR30, GPER1	G protein-coupled estrogen receptor 1
Hs01105253_m1	NCOA3	Nuclear receptor coactivator 3
Hs01100353_m1	ESR2	Estrogen receptor 2
Hs00174860_m1	ESR1	Estrogen receptor 1
Hs99999905_m1	GAPDH	Glyceraldehyde 3-phosphate dehydrogenase
Hs99999901_s1	18S rRNA	18S ribosomal RNA

4.4. Intracellular Calcium Imaging

Human myometrial TCs plated on 24 mm coverglasses were incubated for 45 min at room temperature in a dark chamber with 4 μM Fura-2 acetoxymethyl ester (Fura-2 AM; #F1221, Thermo Fisher Scientific,

Waltham, MA, USA) and 0.25% pluronic Pluronic™ F-127 (#P3000MP, Thermo Fisher Scientific, USA) in Dulbecco's Modified Eagle's Medium (DMEM; Thermo Fisher Scientific, USA). The cells were rinsed three times with Ringer solution (in mM: NaCl 126, HEPES 5, $CaCl_2$ 2, $MgCl_2$ 2, glucose 10, pH 7.4 (with Tris base), let recover for 15 min, and then imaged with an IX-71 Olympus microscope. The excitation was performed by a Xe lamp with monochromator Polychrome V (Till Photonics GmbH, Gräfelfing, Germany) at 340 ± 5 and 380 ± 5 nm, while the emission was collected by a filter at 510 ± 20 nm. The images were acquired by a cooled CCD camera iXON+ EM DU 897 (Andor, Belfast, Northern Ireland) controlled by iQ 1.8 software package, with a frequency of one pair of images per 2 s. Ca^{2+} transients were triggered by 1 min application of 5 µM ($\pm$)-Bay K8644 (#B112, Sigma-Aldrich, St. Louis, MO, USA), an L-type calcium channel agonist [46]. Beta-estradiol was applied for 1 min at 1000 nM in the presence of Bay K8644, and Ca^{2+} transients were recorded. The perfusion was performed with an MPS-2 system (World Precision Instruments, Sarasota, FL, USA). Each coverglass bearing human myometrial TCs was used for a single experimental variant.

4.5. Patch-Clamp Recordings on TCs

TCs from pregnant uterus were recorded in whole-cell configuration under the voltage-clamp mode, using an AxoPatch 200B amplifier (Molecular Devices, San Jose, CA, USA). The electrodes were pulled from borosilicate glass capillaries (GC150F; Harvard Apparatus, Edenbridge, Kent, UK) and heat polished. The final resistance of the pipette, when filled with internal solution, was 3–4 MΩ. The perfusion was performed with an MPS-2 (World Precision Instruments, Sarasota, FL, USA) system, with the tip placed at approximately 100 µm from the cell. Membrane currents were low-pass filtered at 3 kHz (−3 dB, three pole Bessel) and sampled with an Axon Digidata 1440 data acquisition system (Molecular Devices, USA), using pClamp 10 software in gap-free mode. All electrophysiological experiments were performed at room temperature (25 °C). The bath and pipette solutions were used as previously described [35,76]. We adapted the previously described protocol [35] and we applied brief depolarizing ramp protocols from −50 to +60 mV with a duration of 100 ms in order to elicit HVA currents. Uterine TCs were perfused with ($\pm$)-Bay K8644 (#B112, Sigma-Aldrich, USA) for 1 min at 2.5 µM concentration. Beta-estradiol was applied for 1 min at 1000 nM in the presence of Bay K8644.

4.6. Data Analysis

Quantitative RT-PCR data were obtained by normalizing the mRNA levels encoding for different genes to those of *GAPDH* mRNA level using the 2(-Delta C(T)) method, as previously described [77]. Statistical analysis comparing the mRNA expression for each subtype of voltage-gated calcium channel relative to the reference gene (*GAPDH* or *18S rRNA*) in pregnant and in non-pregnant myometrial TCs, was carried out by one-way ANOVA analysis followed by post-hoc Bonferroni test. The effect of pregnancy on the mRNA levels of the voltage-gated calcium channels or the estrogen receptors was analyzed by considering the unpregnant samples as calibrator and performing the one-way ANOVA analysis followed by the post-hoc Bonferroni test. The beta-estradiol effect on the mRNA levels of the voltage-gated calcium channels or the estrogen receptors was analyzed by considering the non-treated samples as calibrator and performing the one-way ANOVA analysis followed by the post-hoc Bonferroni test.

Quantitative calcium imaging results were expressed by means of emission ratio $R = I_{340}/I_{380}$, I being the emission intensity for excitation at 340 and 380 nm, which is proportional to the free cytosolic Ca^{2+} concentration ($[Ca^{2+}]_i$). The mean fluorescence ratios (ΔR) of the calcium transients induced by BAY K8644 in the absence or presence of beta-estradiol were compared by unpaired Student's *t* test.

The amplitudes of the HVA calcium currents recorded by patch-clamp in the absence or presence of beta-estradiol were compared by unpaired Student's *t* test.

All data analysis and data plotting were performed using OriginPro 8 (OriginLab Corporation, Northampton, MA, USA).

Author Contributions: S.M.C., N.S., and B.M.R. designed the study; S.M.C., D.C., and N.S. selected the patients included in the study and taken the human uterine samples; J.X. and A.D. planned the protocols for patch-clamp recordings and A.D. performed the patch-clamp experiments; A.D. and D.D.B. performed the qRT-PCR experiments; C.C.M performed the calcium-imaging experiments; A.D., M.R. and B.M.R analyzed the data; B.M.R and M.R. wrote the paper.

Acknowledgments: This work was supported by a grant of the Romanian Ministry of Research and Innovation, CCCDI - UEFISCDI, project number PN-III-P1-1.2-PCCDI-2017-0833/ 68/2018, within PNCDI III.

References

1. Wray, S.; Jones, K.; Kupittayanant, S.; Li, Y.; Matthew, A.; Monir-Bishty, E.; Noble, K.; Pierce, S.J.; Quenby, S.; Shmygol, A.V. Calcium signaling and uterine contractility. *J. Soc. Gynecol. Investig.* **2003**, *10*, 252–264. [CrossRef]

2. Herington, J.L.; Swale, D.R.; Brown, N.; Shelton, E.L.; Choi, H.; Williams, C.H.; Hong, C.C.; Paria, B.C.; Denton, J.S.; Reese, J. High-Throughput Screening of Myometrial Calcium-Mobilization to Identify Modulators of Uterine Contractility. *PLoS ONE* **2015**, *10*, e0143243. [CrossRef] [PubMed]

3. Bernstein, K.; Vink, J.Y.; Fu, X.W.; Wakita, H.; Danielsson, J.; Wapner, R.; Gallos, G. Calcium-activated chloride channels anoctamin 1 and 2 promote murine uterine smooth muscle contractility. *Am. J. Obstet. Gynecol.* **2014**, *211*, 688.e1–688.e10. [CrossRef] [PubMed]

4. Pistilli, M.J.; Petrik, J.J.; Holloway, A.C.; Crankshaw, D.J. Immunohistochemical and functional studies on calcium-sensing receptors in rat uterine smooth muscle. *Clin. Exp. Pharmacol. Physiol.* **2012**, *39*, 37–42. [CrossRef] [PubMed]

5. Ying, L.; Becard, M.; Lyell, D.; Han, X.; Shortliffe, L.; Husted, C.I.; Alvira, C.M.; Cornfield, D.N. The transient receptor potential vanilloid 4 channel modulates uterine tone during pregnancy. *Sci. Transl. Med.* **2015**, *7*, 319ra204. [CrossRef] [PubMed]

6. Wakle-Prabagaran, M.; Lorca, R.A.; Ma, X.; Stamnes, S.J.; Amazu, C.; Hsiao, J.J.; Karch, C.M.; Hyrc, K.L.; Wright, M.E.; England, S.K. BKCa channel regulates calcium oscillations induced by alpha-2-macroglobulin in human myometrial smooth muscle cells. *Proc. Natl. Acad. Sci. USA* **2016**, *113*, E2335–E2344. [CrossRef] [PubMed]

7. Wray, S. Insights from physiology into myometrial function and dysfunction. *Exp. Physiol.* **2015**, *100*, 1468–1476. [CrossRef] [PubMed]

8. Wray, S.; Burdyga, T.; Noble, D.; Noble, K.; Borysova, L.; Arrowsmith, S. Progress in understanding electro-mechanical signalling in the myometrium. *Acta Physiol.* **2015**, *213*, 417–431. [CrossRef] [PubMed]

9. Lee, S.E.; Ahn, D.S.; Lee, Y.H. Role of T-type Ca Channels in the Spontaneous Phasic Contraction of Pregnant Rat Uterine Smooth Muscle. *Korean J. Physiol. Pharmacol.* **2009**, *13*, 241–249. [CrossRef] [PubMed]

10. Li, X.T.; Qiu, X.Y. 17β-Estradiol Upregulated Expression of α and β Subunits of Larger-Conductance Calcium-Activated K(+) Channels (BK) via Estrogen Receptor β. *J. Mol. Neurosci.* **2015**, *56*, 799–807. [CrossRef] [PubMed]

11. Rahbek, M.; Nazemi, S.; Odum, L.; Gupta, S.; Poulsen, S.S.; Hay-Schmidt, A.; Klaerke, D.A. Expression of the small conductance Ca^{2+}-activated potassium channel subtype 3 (SK3) in rat uterus after stimulation with 17β-estradiol. *PLoS ONE* **2014**, *9*, e87652. [CrossRef] [PubMed]

12. Wang, Q.; Ye, Q.; Lu, R.; Cao, J.; Wang, J.; Ding, H.; Gao, R.; Xiao, H. Effects of estradiol on high-voltage-activated Ca(2+) channels in cultured rat cortical neurons. *Endocr. Res.* **2014**, *39*, 44–49. [CrossRef] [PubMed]

13. Sánchez, J.C.; López-Zapata, D.F.; Pinzón, O.A. Effects of 17beta-estradiol and IGF-1 on L-type voltage-activated and stretch-activated calcium currents in cultured rat cortical neurons. *Neuroendocrinol. Lett.* **2014**, *35*, 724–732. [PubMed]

14. Zhang, F.; Ram, J.L.; Standley, P.R.; Sowers, J.R. 17 beta-Estradiol attenuates voltage-dependent Ca^{2+} currents in A7r5 vascular smooth muscle cell line. *Am. J. Physiol.* **1994**, *266*, C975–C980. [CrossRef] [PubMed]

15. Sánchez, J.C.; López-Zapata, D.F.; Francis, L.; De Los Reyes, L. Effects of estradiol and IGF-1 on the sodium calcium exchanger in rat cultured cortical neurons. *Cell. Mol. Neurobiol.* **2011**, *31*, 619–627. [CrossRef] [PubMed]

16. Choi, Y.; Seo, H.; Kim, M.; Ka, H. Dynamic expression of calcium-regulatory molecules, TRPV6 and S100G, in the uterine endometrium during pregnancy in pigs. *Biol. Reprod.* **2009**, *81*, 1122–1130. [CrossRef] [PubMed]

17. Pohóczky, K.; Kun, J.; Szalontai, B.; Szőke, É.; Sághy, É.; Payrits, M.; Kajtár, B.; Kovács, K.; Környei, J.L.; Garai, J.; et al. Estrogen-dependent up-regulation of TRPA1 and TRPV1 receptor proteins in the rat endometrium. *J. Mol. Endocrinol.* **2016**, *56*, 135–149. [CrossRef] [PubMed]

18. Tulchinsky, D.; Korenman, S.G. The plasma estradiol as an index of fetoplacental function. *J. Clin. Investig.* **1971**, *50*, 1490–1497. [CrossRef] [PubMed]

19. Hatthachote, P.; Gillespie, J.I. Complex interactions between sex steroids and cytokines in the human pregnant myometrium: Evidence for an autocrine signaling system at term. *Endocrinology* **1999**, *140*, 2533–2540. [CrossRef] [PubMed]

20. Wu, J.J.; Geimonen, E.; Andersen, J. Increased expression of estrogen receptor beta in human uterine smooth muscle at term. *Eur. J. Endocrinol.* **2000**, *142*, 92–99. [CrossRef] [PubMed]

21. Cretoiu, S.M.; Cretoiu, D.; Popescu, L.M. Human myometrium—The ultrastructural 3D network of telocytes. *J. Cell. Mol. Med.* **2012**, *16*, 2844–2849. [CrossRef] [PubMed]

22. Cretoiu, S.M.; Popescu, L.M. Telocytes revisited. *Biomol. Concepts* **2014**, *5*, 353–369. [CrossRef] [PubMed]

23. Song, D.; Cretoiu, D.; Cretoiu, S.M.; Wang, X. Telocytes and lung disease. *Histol. Histopathol.* **2016**, *31*, 1303–1314. [PubMed]

24. Cretoiu, S.M.; Cretoiu, D.; Marin, A.; Radu, B.M.; Popescu, L.M. Telocytes: Ultrastructural, immunohistochemical and electrophysiological characteristics in human myometrium. *Reproduction* **2013**, *145*, 357–370. [CrossRef] [PubMed]

25. Song, D.; Cretoiu, D.; Zheng, M.; Qian, M.; Zhang, M.; Cretoiu, S.M.; Chen, L.; Fang, H.; Popescu, L.M.; Wang, X. Comparison of Chromosome 4 gene expression profile between lung telocytes and other local cell types. *J. Cell. Mol. Med.* **2016**, *20*, 71–80. [CrossRef] [PubMed]

26. Albulescu, R.; Tanase, C.; Codrici, E.; Popescu, D.I.; Cretoiu, S.M.; Popescu, L.M. The secretome of myocardial telocytes modulates the activity of cardiac stem cells. *J. Cell. Mol. Med.* **2015**, *19*, 1783–1794. [CrossRef] [PubMed]

27. Zheng, Y.; Cretoiu, D.; Yan, G.; Cretoiu, S.M.; Popescu, L.M.; Fang, H.; Wang, X. Protein profiling of human lung telocytes and microvascular endothelial cells using iTRAQ quantitative proteomics. *J. Cell. Mol. Med.* **2014**, *18*, 1035–1059. [CrossRef] [PubMed]

28. Cretoiu, S.M.; Cretoiu, D.; Simionescu, A.; Popescu, L.M. Telocytes in human fallopian tube and uterus express estrogen and progesterone receptors. In *Sex Steroids*; Kahn, S., Ed.; InTech: Rijeka, Croatia, 2012; pp. 91–114.

29. Roatesi, I.; Radu, B.M.; Cretoiu, D.; Cretoiu, S.M. Uterine Telocytes: A Review of Current Knowledge. *Biol. Reprod.* **2015**, *93*, 10. [CrossRef] [PubMed]

30. Cretoiu, D.; Xu, J.; Xiao, J.; Cretoiu, S.M. Telocytes and Their Extracellular Vesicles-Evidence and Hypotheses. *Int. J. Mol. Sci.* **2016**, *17*, 1322. [CrossRef] [PubMed]

31. Cretoiu, D.; Cretoiu, S.M. Telocytes in the reproductive organs: Current understanding and future challenges. *Semin. Cell Dev. Biol.* **2016**, *55*, 40–49. [CrossRef] [PubMed]

32. Cretoiu, D.; Radu, B.M.; Banciu, A.; Banciu, D.D.; Cretoiu, S.M. Telocytes heterogeneity: From cellular morphology to functional evidence. *Semin. Cell Dev. Biol.* **2017**, *64*, 26–39. [CrossRef] [PubMed]

33. Rusu, M.C.; Cretoiu, D.; Vrapciu, A.D.; Hostiuc, S.; Dermengiu, D.; Manoiu, V.S.; Cretoiu, S.M.; Mirancea, N. Telocytes of the human adult trigeminal ganglion. *Cell Biol. Toxicol.* **2016**, *32*, 199–207. [CrossRef] [PubMed]

34. Radu, B.M.; Banciu, A.; Banciu, D.D.; Radu, M.; Cretoiu, D.; Cretoiu, S.M. Calcium Signaling in Interstitial Cells: Focus on Telocytes. *Int. J. Mol. Sci.* **2017**, *18*, 397. [CrossRef] [PubMed]

35. Cretoiu, S.M.; Radu, B.M.; Banciu, A.; Banciu, D.D.; Cretoiu, D.; Ceafalan, L.C.; Popescu, L.M. Isolated human uterine telocytes: Immunocytochemistry and electrophysiology of T-type calcium channels. *Histochem. Cell Biol.* **2015**, *143*, 83–94. [CrossRef] [PubMed]

36. Campeanu, R.A.; Radu, B.M.; Cretoiu, S.M.; Banciu, D.D.; Banciu, A.; Cretoiu, D.; Popescu, L.M. Near-infrared low-level laser stimulation of telocytes from human myometrium. *Lasers Med. Sci.* **2014**, *29*, 1867–1874. [CrossRef] [PubMed]

37. Othman, E.R.; Elgamal, D.A.; Refaiy, A.M.; Abdelaal, I.I.; Abdel-Mola, A.F.; Al-Hendy, A. Identification and potential role of telocytes in human uterine leiomyoma. *Contracept. Reprod. Med.* **2016**, *20*, 12. [CrossRef] [PubMed]

38. Varga, I.; Klein, M.; Urban, L.; Danihel, L., Jr.; Polak, S.; Danihel, L., Sr. Recently discovered interstitial cells "telocytes" as players in the pathogenesis of uterine leiomyomas. *Med. Hypotheses* **2018**, *110*, 64–67. [CrossRef] [PubMed]

39. Cretoiu, S.M. Immunohistochemistry of Telocytes in the Uterus and Fallopian Tubes. *Adv. Exp. Med. Biol.* **2016**, *913*, 335–357. [PubMed]

40. Vannucchi, M.G.; Faussone-Pellegrini, M.S. The Telocyte Subtypes. *Adv. Exp. Med. Biol.* **2016**, *913*, 115–126. [PubMed]

41. Yang, J.; Li, Y.; Xue, F.; Liu, W.; Zhang, S. Exosomes derived from cardiac telocytes exert positive effects on endothelial cells. *Am. J. Transl. Res.* **2017**, *9*, 5375–5387. [PubMed]

42. Rusu, M.C.; Hostiuc, S.; Vrapciu, A.D.; Mogoantă, L.; Mănoiu, V.S.; Grigoriu, F. Subsets of telocytes: Myocardial telocytes. *Ann. Anat.* **2017**, *209*, 37–44. [CrossRef] [PubMed]

43. Uluer, E.T.; Inan, S.; Ozbilgin, K.; Karaca, F.; Dicle, N.; Sancı, M. The role of hypoxia related angiogenesis in uterine smooth muscle tumors. *Biotech. Histochem.* **2015**, *90*, 102–110. [CrossRef] [PubMed]

44. Sajewicz, M.; Konarska, M.; Wrona, A.N.; Aleksandrovych, V.; Bereza, T.; Komnata, K.; Solewski, B.; Maleszka, A.; Depukat, P.; Warchoł, Ł. Vascular density, angiogenesis and pro-angiogenic factors in uterine fibroids. *Folia Med. Cracov.* **2016**, *56*, 27–32. [PubMed]

45. Salama, N. Immunohistochemical characterization of telocytes in rat uterus in different reproductive states. *Egypt J. Histol.* **2013**, *36*, 85–194.

46. Kim, Y.H.; Chung, S.; Lee, Y.H.; Kim, E.C.; Ahn, D.S. Increase of L-type Ca^{2+} current by protease-activated receptor 2 activation contributes to augmentation of spontaneous uterine contractility in pregnant rats. *Biochem. Biophys. Res. Commun.* **2012**, *418*, 167–172. [CrossRef] [PubMed]

47. Seda, M.; Pinto, F.M.; Wray, S.; Cintado, C.G.; Noheda, P.; Buschmann, H.; Candenas, L. Functional and molecular characterization of voltage-gated sodium channels in uteri from nonpregnant rats. *Biol. Reprod.* **2007**, *77*, 855–863. [CrossRef] [PubMed]

48. Tica, A.A.; Dun, E.C.; Tica, O.S.; Gao, X.; Arterburn, J.B.; Brailoiu, G.C.; Oprea, T.I.; Brailoiu, E. G protein-coupled estrogen receptor 1-mediated effects in the rat myometrium. *Am. J. Physiol. Cell Physiol.* **2011**, *301*, C1262–C1269. [CrossRef] [PubMed]

49. Silva, E.S.; Scoggin, K.E.; Canisso, I.F.; Troedsson, M.H.; Squires, E.L.; Ball, B.A. Expression of receptors for ovarian steroids and prostaglandin E2 in the endometrium and myometrium of mares during estrus, diestrus and early pregnancy. *Anim. Reprod. Sci.* **2014**, *151*, 169–181. [CrossRef] [PubMed]

50. Kautz, E.; Gram, A.; Aslan, S.; Ay, S.S.; Selçuk, M.; Kanca, H.; Koldaş, E.; Akal, E.; Karakaş, K.; Findik, M.; et al. Expression of genes involved in the embryo-maternal interaction in the early-pregnant canine uterus. *Reproduction* **2014**, *147*, 703–717. [CrossRef] [PubMed]

51. Ilicic, M.; Butler, T.; Zakar, T.; Paul, J.W. The expression of genes involved in myometrial contractility changes during ex situ culture of pregnant human uterine smooth muscle tissue. *J. Smooth Muscle Res.* **2017**, *53*, 73–89. [CrossRef] [PubMed]

52. Vodstrcil, L.A.; Shynlova, O.; Westcott, K.; Laker, R.; Simpson, E.; Wlodek, M.E.; Parry, L.J. Progesterone withdrawal, and not increased circulating relaxin, mediates the decrease in myometrial relaxin receptor (RXFP1) expression in late gestation in rats. *Biol. Reprod.* **2010**, *83*, 825–832. [CrossRef] [PubMed]

53. Welsh, T.; Johnson, M.; Yi, L.; Tan, H.; Rahman, R.; Merlino, A.; Zakar, T.; Mesiano, S. Estrogen receptor (ER) expression and function in the pregnant human myometrium: Estradiol via ERα activates ERK1/2 signaling in term myometrium. *J. Endocrinol.* **2012**, *212*, 227–238. [CrossRef] [PubMed]

54. Chandran, S.; Cairns, M.T.; O'Brien, M.; Smith, T.J. Transcriptomic effects of estradiol treatment on cultured human uterine smooth muscle cells. *Mol. Cell. Endocrinol.* **2014**, *393*, 16–23. [CrossRef] [PubMed]

55. Lin, J.; Zhu, J.; Li, X.; Li, S.; Lan, Z.; Ko, J.; Lei, Z. Expression of genomic functional estrogen receptor 1 in mouse sertoli cells. *Reprod. Sci.* **2014**, *21*, 1411–1422. [CrossRef] [PubMed]

56. Yamaguchi, N.; Yuri, K. Estrogen-dependent changes in estrogen receptor-β mRNA expression in middle-aged female rat brain. *Brain Res.* **2014**, *1543*, 49–57. [CrossRef] [PubMed]

57. Murata, T.; Narita, K.; Ichimaru, T. Rat uterine oxytocin receptor and estrogen receptor α and β mRNA levels are regulated by estrogen through multiple estrogen receptors. *J. Reprod.* **2014**, *60*, 55–61. [CrossRef]

58. Bechem, M.; Hoffmann, H. The molecular mode of action of the Ca agonist (-) BAY K 8644 on the cardiac Ca channel. *Pflugers Arch.* **1993**, *424*, 343–353. [CrossRef] [PubMed]

59. Lee, D.Y.; Chai, Y.G.; Lee, E.B.; Kim, K.W.; Nah, S.Y.; Oh, T.H.; Rhim, H. 17Beta-estradiol inhibits high-voltage-activated calcium channel currents in rat sensory neurons via a non-genomic mechanism. *Life Sci.* **2002**, *70*, 2047–2059. [CrossRef]

60. Bosch, M.A.; Hou, J.; Fang, Y.; Kelly, M.J.; Rønnekleiv, O.K. 17Beta-estradiol regulation of the mRNA expression of T-type calcium channel subunits: Role of estrogen receptor alpha and estrogen receptor beta. *J. Comp. Neurol.* **2009**, *512*, 347–358. [CrossRef] [PubMed]

61. Ravanos, K.; Dagklis, T.; Petousis, S.; Margioula-Siarkou, C.; Prapas, Y.; Prapas, N. Factors implicated in the initiation of human parturition in term and preterm labor: A review. *Gynecol. Endocrinol.* **2015**, *31*, 679–683. [CrossRef] [PubMed]

62. Luoma, J.I.; Kelley, B.G.; Mermelstein, P.G. Progesterone inhibition of voltage-gated calcium channels is a potential neuroprotective mechanism against excitotoxicity. *Steroids* **2011**, *76*, 845–855. [CrossRef] [PubMed]

63. Sun, J.; Moenter, S.M. Progesterone treatment inhibits and dihydrotestosterone (DHT) treatment potentiates voltage-gated calcium currents in gonadotropin-releasing hormone (GnRH) neurons. *Endocrinology* **2010**, *151*, 5349–5358. [CrossRef] [PubMed]

64. Liu, B.; Hill, S.J.; Khan, R.N. Oxytocin inhibits T-type calcium current of human decidual stromal cells. *J. Clin. Endocrinol. Metab.* **2005**, *90*, 4191–4197. [CrossRef] [PubMed]

65. Murata, T.; Narita, K.; Honda, K.; Matsukawa, S.; Higuchi, T. Differential regulation of estrogen receptor alpha and beta mRNAs in the rat uterus during pregnancy and labor: Possible involvement of estrogen receptors in oxytocin receptor regulation. *Endocr. J.* **2003**, *50*, 579–587. [CrossRef] [PubMed]

66. Li, J.; Chen, Z.; Zhou, X.; Shi, S.; Qi, H.; Baker, P.N.; Zhang, H. Imbalance between proliferation and apoptosis-related impaired GPR30 expression is involved in preeclampsia. *Cell Tissue Res.* **2016**, *366*, 499–508. [CrossRef] [PubMed]

67. Wei, Y.; Zhang, Z.; Liao, H.; Wu, L.; Wu, X.; Zhou, D.; Xi, X.; Zhu, Y.; Feng, Y. Nuclear estrogen receptor-mediated Notch signaling and GPR30-mediated PI3K/AKT signaling in the regulation of endometrial cancer cell proliferation. *Oncol. Rep.* **2012**, *27*, 504–510. [PubMed]

68. Cretoiu, D.; Ciontea, S.M.; Popescu, L.M.; Ceafalan, L.; Ardeleanu, C. Interstitial Cajal-like cells (ICLC) as steroid hormone sensors in human myometrium: Immunocytochemical approach. *J. Cell. Mol. Med.* **2006**, *10*, 789–795. [CrossRef] [PubMed]

69. Li, Y.Y.; Zhang, S.; Li, Y.G.; Wang, Y. Isolation, culture, purification and ultrastructural investigation of cardiac telocytes. *Mol. Med. Rep.* **2016**, *14*, 1194–1200. [CrossRef] [PubMed]

70. Sakaguchi, H.; Fujimoto, J.; Sun, W.S.; Tamaya, T. Clinical implications of steroid receptor coactivator (SRC)-3 in uterine endometrial cancers. *J. Steroid Biochem. Mol. Biol.* **2007**, *104*, 237–240. [CrossRef] [PubMed]

71. Szwarc, M.M.; Kommagani, R.; Lessey, B.A.; Lydon, J.P. The p160/steroid receptor coactivator family: Potent arbiters of uterine physiology and dysfunction. *Biol. Reprod.* **2014**, *91*, 122. [CrossRef] [PubMed]

72. Divekar, S.D.; Storchan, G.B.; Sperle, K.; Veselik, D.J.; Johnson, E.; Dakshanamurthy, S.; Lajiminmuhip, Y.N.; Nakles, R.E.; Huang, L.; Martin, M.B. The role of calcium in the activation of estrogen receptor-alpha. *Cancer Res.* **2011**, *71*, 1658–1668. [CrossRef] [PubMed]

73. Tajbakhsh, A.; Pasdar, A.; Rezaee, M.; Fazeli, M.; Soleimanpour, S.; Hassanian, S.M.; FarshchiyanYazdi, Z.; Younesi Rad, T.; Ferns, G.A.; Avan, A. The current status and perspectives regarding the clinical implication of intracellular calcium in breast cancer. *J. Cell. Physiol.* **2018**. [CrossRef] [PubMed]

74. Leclercq, G. Calcium-induced activation of estrogen receptor alpha—New insight. *Steroids* **2012**, *77*, 924–927. [CrossRef] [PubMed]

75. Corcoran, J.J.; Nicholson, C.; Sweeney, M.; Charnock, J.C.; Robson, S.C.; Westwood, M.; Taggart, M.J. Human uterine and placental arteries exhibit tissue-specific acute responses to 17β-estradiol and estrogen-receptor-specific agonists. *Mol. Hum. Reprod.* **2014**, *20*, 433–441. [CrossRef] [PubMed]

76. Comunanza, V.; Carbone, E.; Marcantoni, A.; Sher, E.; Ursu, D. Calcium-dependent inhibition of T-type calcium channels by TRPV1 activation in rat sensory neurons. *Pflügers Arch. Eur. J. Physiol.* **2011**, *462*, 709–722. [CrossRef] [PubMed]

77. Livak, K.J.; Schmittgen, T.D. Analysis of relative gene expression data using real-time quantitative PCR and the $2^{-\Delta\Delta Ct}$. *Methods* **2001**, *25*, 402–408. [CrossRef] [PubMed]

7

The Amino-Terminal Domain of GRK5 Inhibits Cardiac Hypertrophy through the Regulation of Calcium-Calmodulin Dependent Transcription Factors

Daniela Sorriento [1], Gaetano Santulli [1,2], Michele Ciccarelli [3], Angela Serena Maione [4], Maddalena Illario [4], Bruno Trimarco [1] and Guido Iaccarino [3,*]

1. Dipartmento di "Scienze Biomediche Avanzate", Università "Federico II" di Napoli, Via Pansini 5, 80131 Napoli, Italy; danisor@libero.it (D.S.); gsantulli001@gmail.com (G.S.); trimarco@unina.it (B.T.)
2. Department of Medicine, Albert Einstein College of Medicine, Montefiore University Hospital, 1300 Morris Park Avenue, Bronx, NY 10461, USA
3. Dipartimento di Medicina, Chirurgia e Odontoiatria "Scuola Medica Salernitana"/DIPMED, Università degli Studi di Salerno, Via S. Allende, 84081 Baronissi (SA), Italy; mciccarelli@unisa.it
4. Dipartmento di "Scienze Mediche Traslazionali", Università "Federico II" di Napoli, Via Pansini 5, 80131 Napoli, Italy; mercoledi85@gmail.com (A.S.M.); illario@unina.it (M.I.)

* Correspondence: giaccarino@unisa.it

Abstract: We have recently demonstrated that the amino-terminal domain of G protein coupled receptor kinase (GRK) type 5, (GRK5-NT) inhibits NFκB activity in cardiac cells leading to a significant amelioration of LVH. Since GRK5-NT is known to bind calmodulin, this study aimed to evaluate the functional role of GRK5-NT in the regulation of calcium-calmodulin-dependent transcription factors. We found that the overexpression of GRK5-NT in cardiomyoblasts significantly reduced the activation and the nuclear translocation of NFAT and its cofactor GATA-4 in response to phenylephrine (PE). These results were confirmed in vivo in spontaneously hypertensive rats (SHR), in which intramyocardial adenovirus-mediated gene transfer of GRK5-NT reduced both wall thickness and ventricular mass by modulating NFAT and GATA-4 activity. To further verify in vitro the contribution of calmodulin in linking GRK5-NT to the NFAT/GATA-4 pathway, we examined the effects of a mutant of GRK5 (GRK5-NTPB), which is not able to bind calmodulin. When compared to GRK5-NT, GRK5-NTPB did not modify PE-induced NFAT and GATA-4 activation. In conclusion, this study identifies a double effect of GRK5-NT in the inhibition of LVH that is based on the regulation of multiple transcription factors through means of different mechanisms and proposes the amino-terminal sequence of GRK5 as a useful prototype for therapeutic purposes.

Keywords: cardiac hypertrophy; transcription factors; calmodulin; GRK

---

1. Introduction

Left ventricular hypertrophy (LVH) is an adaptive response of the heart to stress that eventually, and progressively, turns into maladaptive, evolving towards cardiac dysfunction and heart failure [1–4]. Available therapies targeting the pro-hypertrophic pathways, including angiotensin converting enzyme (ACE) inhibitors and β-adrenergic receptor (β-AR) blockers, can reduce hypertrophy significantly, but not completely [5]. During the past decade, scientific research has focused on the identification of new targets for the treatment of cardiac hypertrophy [5–7]. In particular, alterations in intracellular calcium fluxes activate intracellular pathways that are involved in the progression of LVH and myocardial remodeling [8]. Intracellular Calcium fluxes are the primary component of excitation-contraction coupling and maintain heart contractility [9,10], but can also ignite nuclear gene transcription. In particular, several transcription factors, including nuclear factor of activated T cells (NFAT), are calcium-dependent [11]. Calcium binds calmodulin (CaM) and activates the serine/threonine calcium-calmodulin phosphatase calcineurin. Activated calcineurin, in turn, dephosphorylates the transcription factor NFAT, which quickly moves from cytosol to the nucleus [12,13]. Here NFAT regulates the transcription of genes involved in the development of cardiac hypertrophy, including Atrial Natriuretic Factor, TNF-α, and Endothelin-1 [14]. NFAT works in association with other transcription factors, such as GATA-4. This latter is activated by ERK phosphorylation [15], but also Ca/CaM dependent pathway can enable it.

Previously, we demonstrated that the RH domain of G protein-coupled receptor kinase 5 (GRK5) inhibits NFκB transcriptional activity through binding of the inhibitory protein IκBα [16–18]. In vivo, the intra-myocardial injection of an adenovirus encoding for the amino-terminal domain of GRK5 (AdGRK5-NT), which comprises the RH domain, reduces LVH in spontaneously hypertensive rats (SHR) in a blood pressure-independent manner through the inhibition of NFκB transcription activity [19]. Interestingly, the same amino-terminal sequence of GRK5 that contains the RH domain also flanks a calmodulin binding site [20]. Therefore, it is possible to speculate that GRK5-NT is involved in the regulation of calcium-calmodulin dependent events of activation of transcription factors. This study aims to evaluate whether GRK5-NT can regulate the activation of transcription factors NFAT and GATA-4 through the interaction with calcium-calmodulin dependent signaling pathways.

2. Results

2.1. GRK5-NT Regulates the Activation of Calcium-Calmodulin Dependent Transcription Factors In Vitro

In cultured cardiomyoblasts, hypertrophy was induced by chronic PE stimulation. PE induced the activation of the transcription factors GATA-4, NFκB, and NFAT (Figure 1A). The overexpression of GRK5-NT inhibited the activation of these transcription factors in response to PE (Figure 1A). On the contrary, TAT-RH, which is a peptide that reproduces only the RH domain of GRK5 lacking the amino-terminal domain, did not affect GATA-4 and NFAT activation in response to PE, but was active on NFκB inhibition, as consistent with our previous findings [18]. These data suggest that GRK5-NT is able to regulate calcium-calmodulin-dependent transcription factors, through means of its amino-terminal domain and independently from the RH domain. To verify these findings, we assessed NFAT and GATA 4 nuclear translocation. Strikingly, GRK5-NT overexpression reduced their nuclear accumulation in response to PE, whereas TATRH had no effects (Figure 1B).

2.2. GRK5-NT Inhibits NFAT Activation by Competing for Binding to Calmodulin

To further clarify in vitro the molecular mechanisms by which GRK5-NT regulates the activation of calcium-calmodulin-dependent transcription factors, we tested the hypothesis that GRK5-NT regulates these factors by sequestrating calmodulin. To this aim, cardiomyoblasts were transfected with a plasmid encoding a mutated form of GRK5 in the calmodulin binding site (GRK5-NTPB), which enables the kinase to bind calmodulin [21]. Figure 2A shows that, when compared to GRK5-NT, GRK5-NTPB had no significant effect on PE-induced nuclear translocation of GATA-4 and NFAT, strongly suggesting that GRK5-NT regulates GATA-4 and NFAT activation by competing with calcineurin for binding to calmodulin. Therefore, we evaluated the effects of GRK5-NT and GRK5-NTPB on calmodulin/calcineurin interaction via immunoprecipitation and western blot. We observed that PE induces such interaction, which is not affected by GRK5-NTPB, but is markedly reduced by GRK5-NT (Figure 2B).

A

Whole extracts

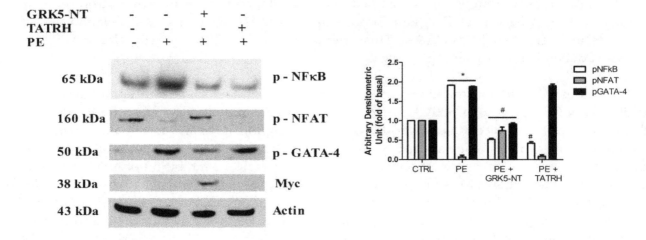

B

Nuclear extracts

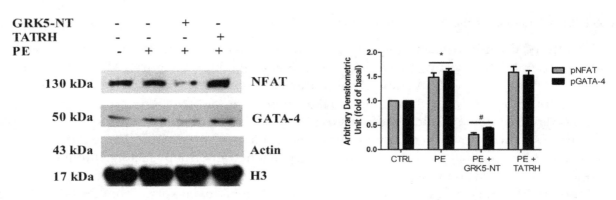

Figure 1. The amino-terminal domain of the G Protein Coupled Receptor Kinase 5 (GRK5-NT) regulates the activation of calcium-calmodulin dependent transcription factors in vitro. (**A**) In cultured cardiomyoblasts H9C2, hypertrophy was induced by chronic stimulation with Phenylephrine (PE) and the activation of NFκB, nuclear factor of activated T cells (NFAT), and GATA-4 in response to PE was evaluated by western blot. PE triggered the phosphorylation of GATA-4, NFκB and the dephosphorylation of NFAT. The overexpression of GRK5-NT reduces such phenomenon while TAT-RH, which has only the RH domain of GRK5, did not modify PE-induced GATA-4 and NFAT activation albeit being effective on NFκB activation. Images are representative of three independent experiments. Densitometric analysis is shown in bar graph as mean ± SD; * $p < 0.05$ vs. Control and # $p < 0.05$ vs. PE; (**B**) nuclear accumulation of NFAT and GATA-4 was evaluated by western blot in nuclear extracts from H9C2 cells. GRK5-NT reduced the PE-dependent nuclear translocation of these factors while TATRH is not able to exert the same effect. Actin was used as control of nuclear extracts purity. Images are representative of three independent experiments. Densitometric analysis is shown in bar graph as mean ± SD; * $p < 0.05$ vs. Control and # $p < 0.05$ vs. PE.

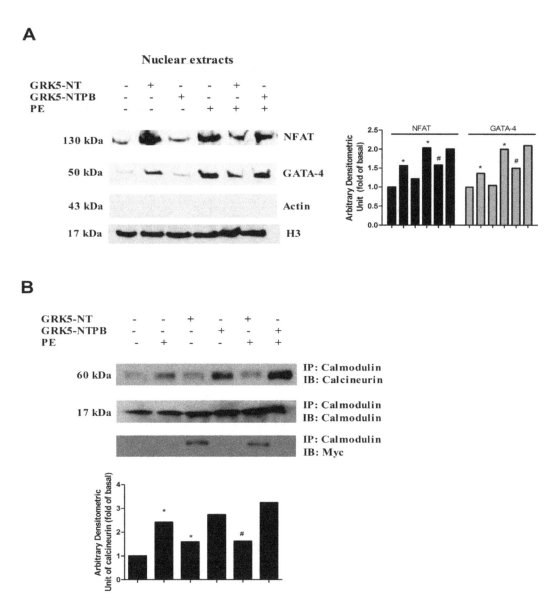

Figure 2. The amino-terminal domain of the G Protein Coupled Receptor Kinase 5 (GRK5-NT) inhibits NFAT activation by sequestrating calmodulin. (**A**) Cardiomyoblasts were transfected with a plasmid encoding GRK5-NT or GRK5-NTPB and NFAT and GATA-4 nuclear translocation was evaluated by western blot in presence and absence of phenylephrine (PE). GRK5-NTPB had no significant effect on nuclear translocation and activation of NFAT and GATA4; instead GRK5-NT inhibited such phenomenon. Actin was used as control of nuclear extracts purity and histone 3 (H3) was used as loading control. Densitometric analysis is shown in bar graph as mean $\pm$ SD, * $p < 0.05$ vs. Control and # $p < 0.05$ vs. PE; (**B**) cardiomyoblasts were transfected with a plasmid encoding GRK5-NT or GRK5-NTPB and calmodulin was precipitated in whole lysates from these cells. Calcineurin was evaluated by western blot. GRK5-NT reduced calmodulin/calcineurin interaction. GRK5-NTPB had no effect on such interaction. All images are representative of three independent experiments. Densitometric analysis is shown in bar graph as mean $\pm$ SD, * $p < 0.05$ vs. Control and # $p < 0.05$ vs. PE.

2.3. GRK5-NT Regulates the Activation of Calcium-Calmodulin Dependent Transcription Factors In Vivo

To confirm our in vitro data, we evaluated the effects of GRK5-NT on NFAT and GATA-4 activation in an animal model of hypertrophy. Spontaneously hypertensive rats (SHR) underwent intra-myocardial injections of an adenovirus encoding for GRK5-NT (AdGRK5-NT) or Lac-Z as control (AdLac-Z), as previously described [19]. Rats were monitored for three weeks by CUS to assess the effect of such treatment on LVH. After 21 days, LVH was significantly reduced, as underlined by the reduction of IVS (Figure 3A) and LVM/BW (Figure 3B). Moreover, cardiac function was recovered in treated rats (Figure 3C). Hearts were then collected and the nuclear translocation of NFAT and GATA-4 was assessed by immunoblot. In hypertrophic SHR rats, there was a significant accumulation of NFAT and GATA-4 in the nucleus as compared with WKY (Figure 4A). The treatment with AdGRK5-NT significantly inhibited such nuclear translocation (Figure 4A) thereby indicating that GRK5-NT can reduce NFAT and GATA-4 activation in response to hypertrophy in vivo. Importantly, this finding was confirmed by EMSA assay, which shows that GRK5-NT significantly reduces both NFAT (Figure 4B) and GATA-4 (Figure 4C) ability to bind DNA in an established model of cardiac hypertrophy (i.e., SHR).

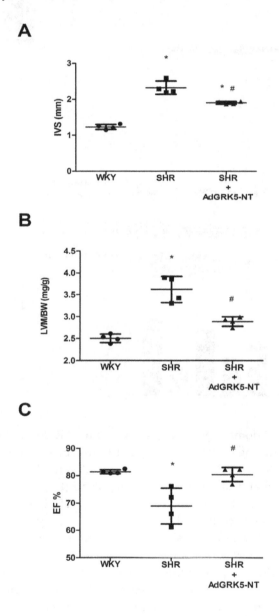

Figure 3. The amino-terminal domain of the G Protein Coupled Receptor Kinase 5 (GRK5-NT) inhibits calcium-calmodulin dependent transcription factors in vivo. (**A–C**) Spontaneously hypertensive and hypertrophic rats (SHR) were treated with an intra-myocardial injection of an adenovirus encoding for GRK5-NT (AdGRK5-NT) or Lac-Z (AdLac-Z), as described in methods, and cardiac hypertrophy was evaluated by echocardiography. 21 days after injection, a reduction of IVS (**A**) and LVM/BW (**B**), was found in AdGRK5-NT versus SHR controls. Cardiac function was recovered in treated SHR vs. SHR (**C**) * $p < 0.05$ vs. WKY and # $p < 0.05$ vs. SHR + AdLac-Z.

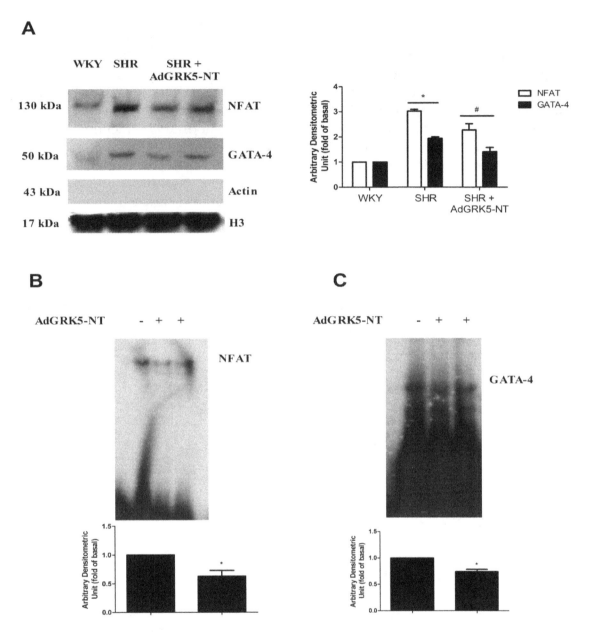

Figure 4. The amino-terminal domain of the G Protein Coupled Receptor Kinase 5 (GRK5-NT) regulates NFAT and GATA-4 activity in vivo. (**A**) The nuclear translocation of NFAT and GATA-4 was evaluated by immunoblot in rat hearts. In hypertrophic SHR rats, there was a significant accumulation of NFAT and GATA-4 in nuclear extracts respect to WKY. The treatment with AdGRK5-NT significantly reduced nuclear translocation of these factors. Densitometric analysis is shown in bar graph as mean ± SD. Actin was used as control of nuclear extracts purity; * $p < 0.05$ vs. WKY and # $p < 0.05$ vs. SHR + AdLac-Z; (**B,C**) to confirm the inhibition of transcription factors activity, we analyzed NFAT and GATA-4 ability to bind DNA by EMSA in control and treated SHR. GRK5-NT reduced both NFAT (**B**) and GATA-4 (**C**) activity in response to hypertrophy. All of the images are representative of at least three independent experiments. Densitometric analysis is shown in bar graph as mean ± SD; * $p < 0.05$ vs. SHR + AdLac-Z.

3. Discussion

Several transcription factors are involved in the regulation and development of LVH [11,22,23] by regulating the expression of critical hypertrophic genes in response to specific stimuli [24,25]. We have recently demonstrated that the treatment with GRK5-NT inhibits NFκB activity in hypertrophied hearts [19]. Here, we describe a novel level of inhibition of LVH by using the ability of the GRK5 sequence to regulate CaM signaling and to prevent NFAT and GATA-4 activation. Indeed, we show that GRK5-NT regulates the activation of the calcium-calmodulin-dependent transcription factor, NFAT, and its cofactor GATA-4, through binding to calmodulin (Figure 5). Other calcium-calmodulin dependent pathways are involved in the development of cardiac hypertrophy, such as CaMKs signaling, and we cannot exclude the possibility that they could be affected by GRK5-NT. However, in this study, we focused on NFAT and its co-factor GATA-4, since they are among the main cardiac transcription factors whose activation is strictly dependent on calcium-calmodulin interactions. Hence, our data indicate that within the GRK5-NT sequence, there are two regions with the potentiality to regulate cardiac hypertrophy in two different ways: by inhibiting NFκB through the binding of the RH domain to IκBα [19] and by inhibiting NFAT through the binding and sequestration of calmodulin in the amino-terminal domain. GRK5-NT can simultaneously affect these intracellular signaling pathways since our data show that it can inhibit both NFκB and NFAT when compared with TATRH.

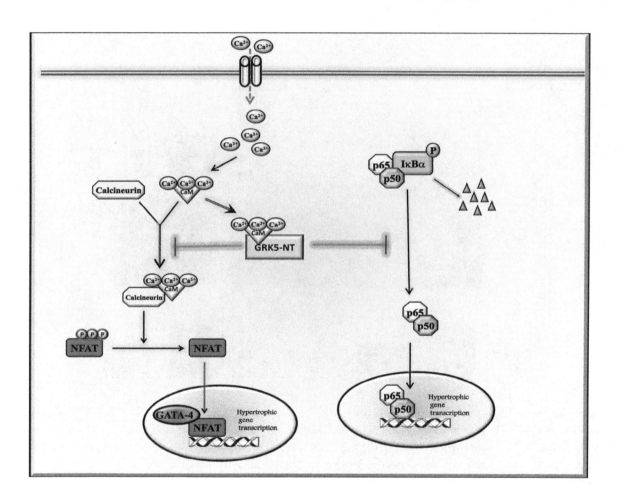

Figure 5. Schematic representation of the molecular mechanism. GRK5-NT regulates GATA-4 and NFAT activation by competing with calcineurin for binding to calcium-calmodulin. PE induces an increase of intracellular calcium levels. Calcium binds calmodulin and activates calcium-calmodulin dependent factors such as calcineurin. Calcineurin, in turn, dephosphorylates NFAT leading to NFAT nuclear translocation and activation of gene transcription. GRK5-NT binds calcium-calmodulin, through the calmodulin binding site (CBS), thus inhibiting its binding to calcineurin. Inactivated calcineurin is not anymore able to dephosphorylate NFAT and this causes the inhibition of NFAT dependent gene transcription and the inhibition of NFAT dependent GATA-4 activation. Besides this effect of GRK5-NT, the peptide is also able to inhibit IκB degradation, thus preventing NFκB transcription activity.

A direct association between NFAT and NFκB activity in cardiac myocytes has been shown to promote cardiac hypertrophy and ventricular remodeling [26]. In particular, since the inhibition of NFκB with IκBαM or the dominant negative of IKKβ reduces NFAT activity both in vitro and in vivo [26], it is likely that the inhibitory effect of GRK5-NT on NFAT activity could be a consequence of GRK5-NT dependent inhibition of NFκB. Actually, in our model, we did not find a mechanistic association between NFκB and NFAT activity. Indeed, while GRK5-NT regulates both NFκB and NFAT activity, TAT-RH—which specifically inhibits NFκB signaling—does not modify NFAT or GATA-4 activation. Such discrepancy could be attributable to the different ways of inhibition of NFκB transcriptional activity. Indeed, the inhibition of NFκB that was obtained via overexpressing IκBαM or a dominant negative IKKβ is mainly based on the inhibition of IκB phosphorylation. In our model, the effects of GRK5-NT are instead based on a protein-protein interaction without interfering with IκB phosphorylation [16]. Such binding leads to the generation of a macromolecular complex that could prevent the binding of NFAT. Furthermore, GRK5-NT regulates NFAT in different manner respect to NFκB that is based on the sequestration of calmodulin through the amino-terminal domain. These findings suggest that GRK5-NT dependent inhibition of NFAT is not due to NFκB regulation, but the peptide is able to regulate both intracellular signalings simultaneously.

Here, we show that GRK5-NT also regulates GATA-4, which is not itself a calcium-calmodulin dependent transcription factor. It has been shown that its activation induced by PE stimulation is coupled with serine phosphorylation by Extracellular signal–regulated kinase 2 (ERK2) [15]. However, besides phosphorylation, the transcriptional activity of GATA4 is also regulated through interaction with other cofactors such as p300, MEF2, SRF, and NFAT [11,24]. Among them, the interaction with NFAT is noteworthy since NFAT plays a critical role in activating the hypertrophic gene program, and this activity is partly dependent on its interaction with GATA4 [24]. These findings suggest the functional importance of calcium signaling also in the activation of GATA4 and support our finding on GRK5-NT dependent GATA-4 inhibition through the modulation of calcium-calmodulin signaling.

Previous reports show that GRK5 also induces cardiac hypertrophy [27,28] by activating NFAT-dependent gene transcription [29]. In this context, GRK5-NT exerts opposite effects compared with the full-length sequence of GRK5. This is because GRK5-NT includes the amino-terminal domain of GRK5 (first 170 aminoacidic sequence) and lacks the catalytic domain that is instead responsible for GRK5 effects on gene transcription. Indeed, GRK5 exacerbates cardiac hypertrophy by phosphorylation of different substrates (plasma membrane receptors and transcription factors) [27–30]. On the contrary, GRK5-NT acts by protein-protein interaction within the RH domain and calmodulin binding within the amino-terminal sequence. However, we cannot exclude the possibility that GRK5-NT could interfere with endogenous GRK5 signaling. Literature is quite discordant on the effects of calmodulin binding on GRK5 signaling. Indeed, some reports show that CaM binding to GRK5 inhibits its catalytic activity [20], while others suggest that this interaction inhibits the binding of GRK5 to plasma membrane favoring its nuclear translocation to induce HDAC phosphorylation [27]. Pitcher et al. which generate the mutant plasmid of GRK5 (GRK5-NTPB) showed that the binding of nuclear Calcium-Calmodulin to the amino-terminal CaM-binding site of GRK5 is required for nuclear export [21]. Thus, further studies are needed to better clarify this issue. Our data show that GRK5-NT in basal condition exerts a pro-hypertrophic effect inducing NFAT nuclear translocation, and this appears to be in contrast with the effect of the peptide in response to PE. Hypertrophic stimuli increase the intracellular calcium levels and enhance the affinity of calmodulin to its substrates. Thus the effect of GRK5-NT to compete with calcineurin for calmodulin binding could take place mainly in response to stimuli. In resting conditions, GRK5-NT may interfere with other physiological pathways and indirectly regulate NFAT nuclear translocation. Taken together, our data demonstrate the ability of GRK5-NT to control several fundamental cardiac transcription factors in response to hypertrophy by different mechanisms of action which involve different domains of GRK5. Specifically, here we demonstrate the ability of GRK5-NT to regulate calcium-calmodulin dependent transcription factors, proposing GRK5 as a potential regulator of calcium signaling through its calmodulin binding site. In conclusion, our data confirm the usefulness of GRK5-NT for the treatment of LVH, suggesting the GRK5-NT sequence as a prototype for the generation of small molecules that are to be used for therapeutic applications.

4. Materials and Methods

4.1. Cell Culture

A cell line of cardiac myoblasts (H9C2) was maintained in culture in Dulbecco Modified Eagle Medium (DMEM) supplemented with 10% FBS at 37 °C in 95% air-5% CO_2.

4.2. Plasmids

p-GRK5-NT is a pcDNA3.1 myc/his plasmid encoding the amino-terminal domain of GRK5, including the RH domain (aa 1-176), and was described previously [16]; p-GRK5-NTPB, which is a plasmid encoding a mutant form of GRK5 which lacks the ability to bind calmodulin, was a kind gift of Julie Pitcher (University College London, London, UK) and was used as template for cloning p-GRK5-NT mutant (GRK5-NT mut), as previously described [16]; TAT-RH was described previously [18]. All experiments were performed using TAT alone and empty pcDNA3.1 as controls. Transient transfection of the plasmids was performed, as previously described [31], using Lipofectamine 2000 from Invitrogen (Thermo Fisher Scientific, Waltham, MA, USA) in 70% confluent H9C2, accordingly to manufacturer instructions.

4.3. Western Blot

The experiments were performed as described previously [16,32]. H9C2 were treated with 10^{-7} M phenylephrine (PE, Sigma-Aldrich Corporation, St. Louis, MO, USA) for 24 h. In some experiments, cells were treated with the synthetic peptide TAT-RH, which only reproduce the RH domain of GRK5, as described previously [18] or transfected with plasmids encoding for GRK5-NT or its mutant in the amino-terminal calmodulin binding. At the end of the treatment, cells were lysed in RIPA/SDS buffer, and protein concentration was determined by using Pierce BCA assay kit (Thermo Fisher Scientific, Waltham, MA, USA) [33,34]. Total extracts were electrophoresed by SDS/PAGE and transferred to nitrocellulose [35]. The antibodies anti-NFATc4 (B-2) (SC-271597), GATA-4 (G-4) (sc-25310), p-NFATc4 (80.S168/170) (sc-135770), p-GATA-4 (H-4) (sc-377543), β-actin (C-4) (sc-47778), and histone H3 (FL-136) (sc-10809) were from Santa Cruz Biotechnology (Santa Cruz Biotechnology, Inc, Dallas, TX, USA). In some experiments, nuclear proteins were isolated from heart samples as previously described [16]. Densitometric analysis was performed using Image Quant 5.2 software (Molecular Dynamics Inc., Caesarea, Israel). Images are representative of at least three independent experiments quantified and corrected for appropriate loading control.

4.4. In Vivo Study

Experiments were carried out accordingly with the Federico II University Ethical Committee on 12-week–old spontaneously hypertensive male rats SHR (AdLac-Z n = 6, and AdGRK5-NT n = 6) and 4 normotensive Wistar-Kyoto (WKY) rats as control. The animals were obtained from Charles River (Wilmington, MA, USA) and had access to water and food ad libitum. Anesthesia was obtained through isoflurane (4%). After the induction of anesthesia, rats were orotracheally intubated, the inhaled concentration of isoflurane was reduced to 1.8%, and the lungs were mechanically ventilated (New England Medical Instruments Scientific, Inc., Chelmsford, MA, USA). The chest was opened under sterile conditions through a right parasternal mini-thoracotomy to expose the heart. Then, we performed four injections (50 μL each) of AdGRK5-NT (10^{10} pfu/mL) or AdLac-Z (10^{10} pfu/mL) as control, into the cardiac wall (anterior, lateral, posterior, and apical), as previously validated [19]. Finally, the chest wall was quickly closed in layers, and animals were observed and monitored until recovery.

4.5. Cardiac Ultrasounds (CUS)

Transthoracic CUS was performed at days 0, 7, 14, and 21 after surgery using a dedicated small-animal high-resolution imaging system (VeVo 770, Visualsonics, Inc., Amsterdam, The Netherlands). The rats were anesthetized with isoflurane (4%) inhalation and maintained by mask ventilation, as described above. The chest was shaved with a depilatory cream (Veet, Reckitt-Benckiser, Milan, Italy). Left ventricular (LV) end-diastolic and LV end-systolic diameters (LVEDD and LVESD, respectively) were measured at the level of the papillary muscles from the parasternal short-axis view [36,37]. Intraventricular septal (IVS) and LV posterior wall thickness (PW) were measured at the end of the diastolic phase. LV mass (LVM) was obtained, as described and corrected by body weight [19,36]. All of the measurements were averaged on at least five consecutive cardiac cycles and were analyzed by investigators that were blinded to treatment.

4.6. EMSA

Electrophoretic mobility shift assay (EMSA) was performed on lysates from treated hearts as previously described [16].

4.7. Statistical Analysis

All data are presented as mean ± SEM. Two-way ANOVA with Bonferroni post hoc test was performed to compare the different parameters between groups. A p value < 0.05 was considered significant. Statistical analysis was performed using GraphPad Prism version 5.01 (GraphPad Software, Inc., San Diego, CA, USA).

5. Conclusions

This study identifies a double effect of GRK5-NT in the inhibition of LVH that is based on the regulation of multiple transcription factors through means of different mechanisms. Indeed, GRK5-NT is able to inhibit NFκB activity through the interaction of RH domain with IkBα and to inhibit NFAT activity by means of calcium-calmodulin sequestration. In conclusion, here we propose the amino-terminal sequence of GRK5 as a useful prototype for therapeutic purposes.

Acknowledgments: We are grateful to Julie Pitcher (University College London, London, UK) for providing GRK5-NTPB plasmid. Research was funded by "BEYONDSILOS, FP7-ICT-PCP" and "Campania Bioscience, PON03PE_00060_8" Grants to Guido Iaccarino.

Author Contributions: Daniela Sorriento and Guido Iaccarino conceived and designed the work; Daniela Sorriento, Gaetano Santulli, Michele Ciccarelli and Angela Serena Maione performed the experiments; Daniela Sorriento, Gaetano Santulli, Michele Ciccarelli, Maddalena Illario, Bruno Trimarco and Guido Iaccarino analyzed data and wrote the paper.

References

1.   Devereux, R.B.; Roman, M.J. Left ventricular hypertrophy in hypertension: Stimuli, patterns, and consequences. *Hypertens. Res.* **1999**, *22*, 1–9. [CrossRef] [PubMed]

2.   Schmieder, R.E.; Messerli, F.H. Hypertension and the heart. *J. Hum. Hypertens.* **2000**, *14*, 597–604. [CrossRef] [PubMed]

3.   Santulli, G.; Iaccarino, G. Adrenergic signaling in heart failure and cardiovascular aging. *Maturitas* **2016**, *93*, 65–72. [CrossRef] [PubMed]

4.   Vakili, B.A.; Okin, P.M.; Devereux, R.B. Prognostic implications of left ventricular hypertrophy. *Am. Heart J.* **2001**, *141*, 334–341. [CrossRef] [PubMed]

5.   Hardt, S.E.; Sadoshima, J. Negative regulators of cardiac hypertrophy. *Cardiovasc. Res.* **2004**, *63*, 500–509. [CrossRef] [PubMed]

6.   Sadoshima, J.; Izumo, S. The cellular and molecular response of cardiac myocytes to mechanical stress. *Annu. Rev. Physiol.* **1997**, *59*, 551–571. [CrossRef] [PubMed]

7.   Molkentin, J.D.; Dorn, G.W., II. Cytoplasmic signaling pathways that regulate cardiac hypertrophy. *Annu. Rev. Physiol.* **2001**, *63*, 391–426. [CrossRef] [PubMed]

8.   Heineke, J.; Molkentin, J.D. Regulation of cardiac hypertrophy by intracellular signalling pathways. *Nat. Rev. Mol. Cell Biol.* **2006**, *7*, 589–600. [CrossRef] [PubMed]

9.   Gambardella, J.; Trimarco, B.; Iaccarino, G.; Santulli, G. New Insights in Cardiac Calcium Handling and Excitation-Contraction Coupling. *Adv. Exp. Med. Biol.* **2017**. [CrossRef]

10. Santulli, G.; Nakashima, R.; Yuan, Q.; Marks, A.R. Intracellular calcium release channels: An update. *J. Physiol.* **2017**, *595*, 3041–3051. [CrossRef] [PubMed]

11. Akazawa, H.; Komuro, I. Roles of cardiac transcription factors in cardiac hypertrophy. *Circ. Res.* **2003**, *92*, 1079–1088. [CrossRef] [PubMed]

12. Schulz, R.A.; Yutzey, K.E. Calcineurin signaling and NFAT activation in cardiovascular and skeletal muscle development. *Dev. Biol.* **2004**, *266*, 1–16. [CrossRef] [PubMed]

13. Yuan, Q.; Yang, J.; Santulli, G.; Reiken, S.R.; Wronska, A.; Kim, M.M.; Osborne, B.W.; Lacampagne, A.; Yin, Y.; Marks, A.R. Maintenance of normal blood pressure is dependent on IP3R1-mediated regulation of eNOS. *Proc. Nat. Acad. Sci. USA* **2016**, *113*, 8532–8537. [CrossRef] [PubMed]

14. Hogan, P.G.; Chen, L.; Nardone, J.; Rao, A. Transcriptional regulation by calcium, calcineurin, and NFAT. *Genes Dev.* **2003**, *17*, 2205–2232. [CrossRef] [PubMed]

15. Liang, Q.; Wiese, R.J.; Bueno, O.F.; Dai, Y.S.; Markham, B.E.; Molkentin, J.D. The transcription factor GATA4 is activated by extracellular signal-regulated kinase 1- and 2-mediated phosphorylation of serine 105 in cardiomyocytes. *Mol. Cell Biol.* **2001**, *21*, 7460–7469. [CrossRef] [PubMed]

16. Sorriento, D.; Ciccarelli, M.; Santulli, G.; Campanile, A.; Altobelli, G.G.; Cimini, V.; Galasso, G.; Astone, D.; Piscione, F.; Pastore, L.; et al. The G-protein-coupled receptor kinase 5 inhibits NFκB transcriptional activity by inducing nuclear accumulation of IκBα. *Proc. Nat. Acad. Sci. USA* **2008**, *105*, 17818–17823. [CrossRef] [PubMed]

17. Sorriento, D.; Illario, M.; Finelli, R.; Iaccarino, G. To NFκB or not to NFκB: The Dilemma on How to Inhibit a Cancer Cell Fate Regulator. *Transl. Med. UniSa* **2012**, *4*, 73–85. [PubMed]

18. Sorriento, D.; Campanile, A.; Santulli, G.; Leggiero, E.; Pastore, L.; Trimarco, B.; Iaccarino, G. A new synthetic protein, TAT-RH, inhibits tumor growth through the regulation of NFκB activity. *Mol. Cancer* **2009**, *8*, 97. [CrossRef] [PubMed]

19. Sorriento, D.; Santulli, G.; Fusco, A.; Anastasio, A.; Trimarco, B.; Iaccarino, G. Intracardiac injection of AdGRK5-NT reduces left ventricular hypertrophy by inhibiting NFκB-dependent hypertrophic gene expression. *Hypertension* **2010**, *56*, 696–704. [CrossRef] [PubMed]

20. Pronin, A.N.; Satpaev, D.K.; Slepak, V.Z.; Benovic, J.L. Regulation of G protein-coupled receptor kinases by calmodulin and localization of the calmodulin binding domain. *J. Biol. Chem.* **1997**, *272*, 18273–18280. [CrossRef] [PubMed]

21. Johnson, L.R.; Scott, M.G.; Pitcher, J.A. G protein-coupled receptor kinase 5 contains a DNA-binding nuclear localization sequence. *Mol. Cell Biol.* **2004**, *24*, 10169–10179. [CrossRef] [PubMed]

22. Frey, N.; Olson, E.N. Cardiac hypertrophy: The good, the bad, and the ugly. *Annu. Rev. Physiol.* **2003**, *65*, 45–79. [CrossRef] [PubMed]

23. Russell, B.; Motlagh, D.; Ashley, W.W. Form follows function: How muscle shape is regulated by work. *J. Appl. Physiol.* **2000**, *88*, 1127–1132. [CrossRef] [PubMed]

24. Molkentin, J.D.; Lu, J.R.; Antos, C.L.; Markham, B.; Richardson, J.; Robbins, J.; Grant, S.R.; Olson, E.N. A calcineurin-dependent transcriptional pathway for cardiac hypertrophy. *Cell* **1998**, *93*, 215–228. [CrossRef]

25. Gordon, J.W.; Shaw, J.A.; Kirshenbaum, L.A. Multiple facets of NF-κB in the heart: To be or not to NF-κB. *Circ. Res.* **2011**, *108*, 1122–1132. [CrossRef] [PubMed]

26. Liu, Q.; Chen, Y.; Auger-Messier, M.; Molkentin, J.D. Interaction between NFκB and NFAT coordinates cardiac hypertrophy and pathological remodeling. *Circ. Res.* **2012**, *110*, 1077–1086. [CrossRef] [PubMed]

27. Martini, J.S.; Raake, P.; Vinge, L.E.; DeGeorge, B.R., Jr.; Chuprun, J.K.; Harris, D.M.; Gao, E.; Eckhart, A.D.; Pitcher, J.A.; Koch, W.J. Uncovering G protein-coupled receptor kinase-5 as a histone deacetylase kinase in the nucleus of cardiomyocytes. *Proc. Nat. Acad. Sci USA* **2008**, *105*, 12457–12462. [CrossRef] [PubMed]

28. Belmonte, S.L.; Blaxall, B.C. G protein-coupled receptor kinase 5: Exploring its hype in cardiac hypertrophy. *Circ. Res.* **2012**, *111*, 957–958. [CrossRef] [PubMed]

29. Hullmann, J.E.; Grisanti, L.A.; Makarewich, C.A.; Gao, E.; Gold, J.I.; Chuprun, J.K.; Tilley, D.G.; Houser, S.R.; Koch, W.J. GRK5-mediated exacerbation of pathological cardiac hypertrophy involves facilitation of nuclear NFAT activity. *Circ. Res.* **2014**, *115*, 976–985. [CrossRef] [PubMed]

30. Dzimiri, N.; Muiya, P.; Andres, E.; Al-Halees, Z. Differential functional expression of human myocardial G protein receptor kinases in left ventricular cardiac diseases. *Eur. J. Pharmacol.* **2004**, *489*, 167–177. [CrossRef] [PubMed]

31. Sorriento, D.; Santulli, G.; Del Giudice, C.; Anastasio, A.; Trimarco, B.; Iaccarino, G. Endothelial cells are able to synthesize and release catecholamines both in vitro and in vivo. *Hypertension* **2012**, *60*, 129–136. [CrossRef] [PubMed]

32. Santulli, G.; Campanile, A.; Spinelli, L.; Assante di Panzillo, E.; Ciccarelli, M.; Trimarco, B.; Iaccarino, G. G protein-coupled receptor kinase 2 in patients with acute myocardial infarction. *Am. J. Cardiol.* **2011**, *107*, 1125–1130. [CrossRef] [PubMed]

33. Ciccarelli, M.; Sorriento, D.; Cipolletta, E.; Santulli, G.; Fusco, A.; Zhou, R.H.; Eckhart, A.D.; Peppel, K.; Koch, W.J.; Trimarco, B.; et al. Impaired neoangiogenesis in β2-adrenoceptor gene-deficient mice: Restoration by intravascular human β2-adrenoceptor gene transfer and role of NFκB and CREB transcription factors. *Br. J. Pharmacol.* **2011**, *162*, 712–721. [CrossRef] [PubMed]

34. Santulli, G.; Basilicata, M.F.; De Simone, M.; Del Giudice, C.; Anastasio, A.; Sorriento, D.; Saviano, M.; Del Gatto, A.; Trimarco, B.; Pedone, C.; et al. Evaluation of the anti-angiogenic properties of the new selective αVβ3 integrin antagonist RGDechiHCit. *J. Transl. Med.* **2011**, *9*, 7. [CrossRef] [PubMed]

35. Iaccarino, G.; Izzo, R.; Trimarco, V.; Cipolletta, E.; Lanni, F.; Sorriento, D.; Iovino, G.L.; Rozza, F.; De Luca, N.; Priante, O.; et al. β2-adrenergic receptor polymorphisms and treatment-induced regression of left ventricular hypertrophy in hypertension. *Clin. Pharmacol. Ther.* **2006**, *80*, 633–645. [CrossRef] [PubMed]

36. Santulli, G.; Cipolletta, E.; Sorriento, D.; Del Giudice, C.; Anastasio, A.; Monaco, S.; Maione, A.S.; Condorelli, G.; Puca, A.; Trimarco, B.; et al. CaMK4 Gene Deletion Induces Hypertension. *J. Am. Heart Assoc.* **2012**, *1*, e001081. [CrossRef] [PubMed]

37. Santulli, G.; Xie, W.; Reiken, S.R.; Marks, A.R. Mitochondrial calcium overload is a key determinant in heart failure. *Proc. Nat. Acad. Sci. USA* **2015**, *112*, 11389–11394. [CrossRef] [PubMed]

8

Role of KCa3.1 Channels in Modulating Ca^{2+} Oscillations during Glioblastoma Cell Migration and Invasion

Luigi Catacuzzeno * and Fabio Franciolini *

Department of Chemistry, Biology and Biotechnology, University of Perugia, 06134 Perugia, Italy
* Correspondence: luigi.catacuzzeno@unipg.it (L.C.); fabio.franciolini@unipg.it (F.F)

Abstract: Cell migration and invasion in glioblastoma (GBM), the most lethal form of primary brain tumors, are critically dependent on Ca^{2+} signaling. Increases of [Ca^{2+}]$_i$ in GBM cells often result from Ca^{2+} release from the endoplasmic reticulum (ER), promoted by a variety of agents present in the tumor microenvironment and able to activate the phospholipase C/inositol 1,4,5-trisphosphate PLC/IP$_3$ pathway. The Ca^{2+} signaling is further strengthened by the Ca^{2+} influx from the extracellular space through Ca^{2+} release-activated Ca^{2+} (CRAC) currents sustained by Orai/STIM channels, meant to replenish the partially depleted ER. Notably, the elevated cytosolic [Ca^{2+}]$_i$ activates the intermediate conductance Ca^{2+}-activated K (KCa3.1) channels highly expressed in the plasma membrane of GBM cells, and the resulting K$^+$ efflux hyperpolarizes the cell membrane. This translates to an enhancement of Ca^{2+} entry through Orai/STIM channels as a result of the increased electromotive (driving) force on Ca^{2+} influx, ending with the establishment of a recurrent cycle reinforcing the Ca^{2+} signal. Ca^{2+} signaling in migrating GBM cells often emerges in the form of intracellular Ca^{2+} oscillations, instrumental to promote key processes in the migratory cycle. This has suggested that KCa3.1 channels may promote GBM cell migration by inducing or modulating the shape of Ca^{2+} oscillations. In accordance, we recently built a theoretical model of Ca^{2+} oscillations incorporating the KCa3.1 channel-dependent dynamics of the membrane potential, and found that the KCa3.1 channel activity could significantly affect the IP$_3$ driven Ca^{2+} oscillations. Here we review our new theoretical model of Ca^{2+} oscillations in GBM, upgraded in the light of better knowledge of the KCa3.1 channel kinetics and Ca^{2+} sensitivity, the dynamics of the Orai/STIM channel modulation, the migration and invasion mechanisms of GBM cells, and their regulation by Ca^{2+} signals.

Keywords: KCa3.1 channels; glioblastoma; cell migration; calcium oscillations; mathematical model

1. The Glioblastoma

The large majority (more than 90%) of cancer deaths are due not to the primary tumor per se, but to relapses arising from new foci established in distant organs via metastasis [1]. Glioblastoma (GBM), the most common and aggressive form of primary brain tumors, is no exception, though it does not metastasize in the classical way (that is, by colonizing other tissues via the bloodstream), but invades brain parenchyma by detaching from the original tumor mass and infiltrating into the healthy tissue by degrading the extracellular matrix or squeezing through the brain interstitial spaces. The urgency of tackling the migration and invasion issues of GBM tumors is clear.

The 2016 World Health Organization classification of the various types of brain tumors regards the presence of isocitrate dehydrogenase gene (*IDH1/2*) mutations as one of the most critical biomarkers. Accordingly, *IDH1/2* wildtype gliomas are categorized as glioblastoma (formerly primary glioblastoma) and *IDH1/2*-mutated gliomas (including formerly classified secondary glioblastoma) as astrocytic glioma and oligodendroglioma. GBMs are further subdivided into four groups (Proneural,

Neural, Classical, and Mesenchymal), mainly based on the abnormally high levels of mutated genes (i.e., *EGFR* is highly upregulated in >98% of Classical GBM, whereas *TP53* (p53), which is most frequently mutated in Proneural GBM (50–60% of patients) is rarely mutated in Classical GBM). In spite of the intensive basic and clinical studies carried out over the past decades, and modern diagnostics and treatments, the average life expectancy for GBM patients is still only around 15 months. The major obstacle with GBM remains its high migratory and invasive potential into healthy brain parenchyma, which prevents complete surgical removal of tumor cells. Even with full clinical treatment (temozolomide-based chemotherapy and radiation therapy), tumors normally recur at some distance from the site of resection, establishing new tumor lesions that are by far the primary cause of mortality in GBM patients. Arguably, at the time of surgery, large numbers of cells have already detached from the original tumor mass and invaded normal brain tissue. Although GBM cell migration and invasion have been deeply investigated, many aspects of these processes are still poorly understood.

GBM cell migration is a highly regulated multistep process that initiates with GBM cells losing adhesion with surrounding elements, avoiding the cell death often associated with extracellular matrix (ECM) disconnection, and acquiring a highly migratory phenotype, which is a critical feature of the invasive process. The basic mechanisms underlying migration of GBM cells are common to most types of migratory cells. Migration is a property of many non-tumor cells, although it is often restricted to specific developmental stages or environmental conditions; the migration of tumor cells could be viewed as the result of mutation-induced dysregulation of specific biochemical pathways that in healthy tissue keep cell migration dormant.

2. Glioblastoma Cell Migration and Ca^{2+} Signaling

2.1. Cell Migration

The basic mechanisms of cell locomotion are now fairly well established. Locomotion can be described as the cyclical repeating of two main processes: (i) protrusion of the cell front due to local gain of cell volume mostly generated by active $Na^+/K^+/2Cl^-$ cotransport accompanied by isoosmotically obliged water, and actin polymerization, with formation of pseudopods; (ii) retraction of the rear cell body in the direction of motion, due to forces produced by actomyosin contraction, accompanied by loss of cell volume generated by passive ion (mainly K^+ and Cl^-) fluxes and osmotic water [2,3]. These two processes involve the coordinated and localized formation of integrin-dependent cell adhesions at the leading edge, and their disassembly at the cell rear [4,5].

Protrusion of the cell front is sustained by localized polymerization of submembrane actin-based cytoskeleton that generates the pushing force and forms flat lamellipodia or needle-like filipodia. A large variety of signaling molecules have been shown to play a leading role in these processes, including the Rho GTPases family (that act as molecular switches to control downstream transduction pathways), and their effector proteins CDC42, RAC1, and RhoA. PI3 kinases have also been deeply implicated in controlling actin polymerization and lamellipodium extension. Activation of PI3 kinase by the pro-invasive signal molecules present in the tumor microenvironment functions as the trigger of the process, in that its activation initiates actin polymerization and generates membrane protrusion [5].

The retraction of the cell rear depends on the contractile forces generated by the activation of the myosin motors along crosslinked actin filaments. The myosin molecule is formed by two heavy chains that make up the two heads and the coiled-coil tail, and four—two essential and two regulatory—myosin light chains (MLCs). The myosin motor is primarily activated by a Ca^{2+}-dependent cascade whereby a cytosolic Ca^{2+} increase activates a MLC kinase (MLCK) that leads to the regulatory MLC phosphorylation, which allows the myosin motor to crosslink with actin and produce tension to pull the rear end of the cell (however, some studies claim that the motor activity of myosin is not required for cell migration [6,7], playing instead a role in the establishment of cell polarity and in the coordination between different cell domains [7,8]). Clearly, to make the rear cell effectively move, the focal adhesions that anchor the actin cytoskeleton to the extracellular matrix (ECM) must be disassembled, normally through a proteolytic process that involves calpain, but also actin microtubules and focal adhesion kinase (FAK)-recruited dynamin that internalizes the integrins. Increasing actomyosin tension disrupts the possible residual resistances of focal adhesions at the cell rear [2]. Myosin can also be modulated by other pathways, including a Rho-dependent cascade (Rho $\rightarrow$ ROCK $\rightarrow$ MLC phosphorylation, or MLC phosphatase activation) that does not depend on cytosolic Ca^{2+}.

This locomotion cycle can be initiated by a variety of external stimuli or directional cues which are sensed and decoded by specialized receptors on the plasma membrane and transferred to the cell interior via distinct signaling pathways to actuate the mechanical response. Most common migration stimuli rely on cells sensing (by specific G protein-coupled receptors) local gradients in the concentration of chemical factors (chemoattractants) [9] present in the tumor microenvironment (for instance chemokines IL-8 or CXCL12), or certain growth factors, including the platelet-derived growth factor (PDGF) that plays an important role in tumor metastasis [10]. Cancer cells can also be guided by gradients of bound ligands. Generally attached to the ECM, these ligands are recognized by specific integrins that form oriented focal adhesion to direct the formation of invadopodia, thus the direction of movement. Common is also the observation of cancer cells being guided by mechanical stimuli (mechanotaxis), namely the stiffness of the extracellular matrix they try to penetrate [11]. It is important to note that in vivo cells can be simultaneously subjected to multiple types of cues that need be evaluated as a whole in order to provide appropriate responses.

2.2. Cell Volume Changes Associated with Migration

It is important to recognize that these two steps—cell front protrusion and cell rear retraction—normally occur in succession, implying that cells are subjected to major changes in cell volume, especially the cell rear retraction. This requires compensatory ion fluxes (and osmotic water flow) across the plasma membrane in order to maintain proper cytosol osmolarity. Along the line proposed by [12] with regard to cell migration in general, during the protrusion of the cell front the maintenance of a stable osmolarity is sustained by cotransporters and ion exchangers, for instance $Na^+/K^+/2Cl^-$ cotransport, followed by osmotically driven water. By contrast, the control of osmolarity during cell rear retraction critically depends on the activation of K channels and ensuing efflux of K^+ ions (accompanied by Cl^- fluxes, to which the membrane has become highly permeant, and osmotic water). Following the opening of K channels the membrane potential hyperpolarizes to a value between the equilibrium potentials of Cl^- and K^+ (E_{Cl} and E_K), a condition that allows a significant efflux of KCl with consequent loss of water and cell volume decrease that facilitates cell body retraction. This view can explain the slowing of cell locomotion following the inhibition of K and Cl channels [13]. A number of papers indicate that the channel types primarily involved in osmolarity control are the intermediate conductance Ca^{2+}-activated K channel (KCa3.1) and the Cl channel ClC3 [13,14].

2.3. Ca^{2+} Oscillations

Much work shows that cell migration is strictly regulated by Ca^{2+}, which is no surprise given that many proteins involved in migration such as myosin, myosin light chain kinase (MLCK), Ca^{2+}/calmodulin-dependent protein kinase II are Ca^{2+} sensitive [15]. In the resting cell Ca^{2+} concentration is kept very low (<100 nM) to prevent activation of Ca^{2+}-sensitive proteins that act as Ca^{2+} sensor and initiation of unwanted biological processes. Ca^{2+} concentration can easily increase more than ten-fold (well over 1 µM) following several types of cell stimulation, and this would bring many of the major cytoplasmic Ca^{2+}-sensitive proteins (calmodulin, PKA, Ca^{2+}-activated channels, etc.) to activate. This opens the question of how Ca^{2+} signaling, under these conditions, can selectively activate a specific Ca^{2+} sensor protein and the associated biochemical cascade. Several strategies have evolved in this respect, but specificity is mostly attained by confining the signal at sub-cellular level, or by regulating kinetics and magnitude of the Ca^{2+} signal. Since in the cytoplasm Ca^{2+} diffusion is extremely low and buffering high, the opening of Ca^{2+} permeant channels results in a very localized Ca^{2+} increase (the Ca^{2+} microdomain), attaining spatial discrimination of the target proteins and linked downstream pathways. Another strategy relies on the temporal features of Ca^{2+} signals (rising rate, duration, repetition of spikes, etc.) that biological systems are capable to interpret.

In non-excitable cells, including GBM, typical Ca^{2+} signals induced experimentally by robust chemical (hormone) stimulation are slow, large and generally sustained. Hormones binding to their specific G-protein-coupled receptors (GPCRs) often results in the PLC-dependent synthesis of inositol 1,4,5-trisphosphate (IP_3), and consequent Ca^{2+} release from the endoplasmic reticulum (ER) via IP_3 receptors. Eventually, the sustained Ca^{2+} increase subsides as result of the activation of Ca^{2+} pumps placed on the sarco/endoplasmic reticulum (SERCA), or the plasma membrane (PMCA) that transfer Ca^{2+} ions from the cytosol to intracellular stores or the extracellular space.

A physiologically more meaningful type of Ca^{2+} signal, especially in relation to cell migration, where critical Ca^{2+}-dependent processes are cyclical, comes in the form of Ca^{2+} oscillations. It has recently become clear that this type of oscillatory Ca^{2+} increases is what normally occurs when using physiological agonist stimulations. The mechanisms underlying Ca^{2+} oscillations are not fully understood, and often depend on the cell model and agonist used. In some systems Ca^{2+} oscillations are secondary to oscillations of the cytoplasmic IP_3 level, which may be due to various types of feedback control that for brief intervals uncouple the GTP-coupled receptor from PLC. For instance, the reported negative feedback of the PLC products, IP_3 and diacylglycerol (DAG), on PLC itself, or upstream on the GPCR, would result in IP_3 oscillations [16], which would in turn generate cyclical release of Ca^{2+} from intracellular stores, and the Ca^{2+} oscillations [17].

More commonly, however, Ca^{2+} oscillations can be observed in the presence of a constant level of IP_3. The prevailing view is that, under these conditions, the repetitive Ca^{2+} oscillations secondary to moderate (physiological) hormone or agonist activation of membrane PLC and production of a constant amount of IP_3 arises from the biphasic effects of cytosolic Ca^{2+} on the IP_3 receptor gating, with the first phase being embodied by the establishment of the Ca^{2+}-induced Ca^{2+} release (CICR) mechanism, whereby a moderate increase of $[Ca^{2+}]$ in the cytosol via IP_3 receptor causes a positive feedback activation of the IP_3 receptor that results in a higher Ca^{2+} released from internal stores ([18]; Figure 1A).

The second phase occurs when the cytosolic $[Ca^{2+}]$ reaches significantly higher levels, shifting the previous positive feedback on IP_3 receptors into a delayed negative feedback that results in their closing and $[Ca^{2+}]_i$ returning to resting level by the action of both SERCA and PMCA Ca^{2+} pumps (Figure 1A). This view finds support from the observation that at constant levels of IP_3, the activity of the IP_3 receptor and the cytosolic Ca^{2+} concentration display a bell-shaped function, whereby low Ca^{2+} concentrations activate the receptor, whereas high Ca^{2+} concentrations inhibit it [19–22]. Ca^{2+} oscillations generated according to this scheme (i.e., Figure 1A) would however show a rapid decrease of spike amplitude with time, because part of the Ca^{2+} released from the ER during each spike will be pumped out of the cell by PMCA, and be no longer available for refilling the ER and contributing to the next Ca^{2+} spike. This Ca^{2+} oscillations time course is observed experimentally upon removal of external Ca^{2+}, or blockade of Ca^{2+} influx from the extracellular space. To have sustained (or slow decaying) Ca^{2+} oscillations, external Ca^{2+} pumped out of the cell by PMCAs must be allowed to re-enter the cell. This is most commonly accomplished through the mechanism of store-operated Ca^{2+} entry (SOCE).

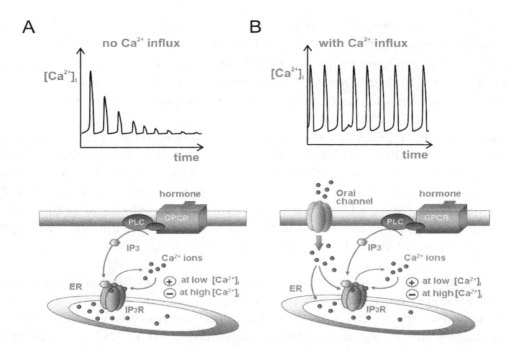

Figure 1. Ca^{2+} oscillations in response to inositol thriphosphate (IP$_3$) increase, with and without Ca^{2+} influx from extracellular space. (**A**) Bottom, drawing illustrating the hormone-based production of IP$_3$ that activates the IP$_3$ receptor to release Ca^{2+} from endoplasmic reticulum (ER). The biphasic effects of cytosolic Ca^{2+} on IP$_3$ receptor gating (the basic mechanism for Ca^{2+} oscillations), whereby Ca^{2+} modulates positively the receptor at low [Ca^{2+}] but negatively at high [Ca^{2+}], is also illustrated. Top, Ca^{2+} oscillations as produced from the schematics below. Note the decaying trend of Ca^{2+} spikes due to the absence of Ca^{2+} influx from extracellular space; (**B**) Here the drawing has been enriched with a Ca^{2+} influx apparatus from extracellular space through ER-depletion activated Orai channels on the plasma membrane (bottom), which generates sustained Ca^{2+} oscillations (top). For clarity, SERCA and PMCA Ca^{2+} pumps have not been sketched in the drawing, although their activity has always been taken into account. For the same reason, we omitted to draw STIM protein of the ER.

2.4. Store-Operated Ca^{2+} Entry (SOCE)

When Ca^{2+} is released from the ER, i.e., following hormone stimulation, the Ca^{2+} concentration within the organelle decreases. It is this decrease to trigger the process (known as SOCE) whereby Ca^{2+} channels in the plasma membrane are activated, allowing entry of Ca^{2+} into the cell. Although the process of SOCE was described some 30 years ago, its mechanisms and molecular counterparts—the STIM and Orai proteins—have been identified much more recently [23–25]. The stromal interaction molecules (STIMs) are a family of one-passage integral proteins of the ER that act as sensors of the Ca^{2+} concentration within the ER. Orai proteins form instead a group of highly selective Ca^{2+} channels placed in the plasma membrane. Orai channels are neither activated by voltage, nor by agonists (in the classical sense), and are normally silent when the cell is at rest and the ER filled with Ca^{2+} ions. Orai Ca^{2+} channels are opened by the interaction with activated STIM proteins, which occurs following depletion of Ca^{2+} in the ER, and their opening results in Ca^{2+} ions flowing inside the cell (the mechanism has been recently reviewed in depth on neurons by [26]). This Ca^{2+} entry into the cytoplasm enables the refilling of Ca^{2+} depleted ER through the SERCA pumps, and make the Ca^{2+} oscillations amplitude to be kept stable with time (Figure 1B). In the context of our review topic, this feedback mechanism of store-operated Ca^{2+} entry gains further relevance for two more reasons: it contributes to further modulating the IP$_3$ receptor (this modulation is also dependent on the cell type—for instance, in U87-MG human GBM cells Ca^{2+} oscillations are affected only marginally by removal of Ca^{2+} influx [27], whereas in rat C6 GBM cells this maneuver fully suppresses Ca^{2+}

oscillations [28]), and it represents the target on which the KCa3.1 channel exerts its modulation of Ca^{2+} oscillations, as we will see later.

It needs to be recalled that the store-operated Ca^{2+} entry is not the only suggested mechanism—although it is by far the most widely accepted—that links hormone stimulation to increased plasma membrane Ca^{2+} entry. For instance, heteromers of Orai1 and Orai3 can be activated by arachidonic acid [26], and the Ca^{2+}-permeant TRPC6 by receptor-induced PLC activation via direct action of DAG. Notably, neither pathway involves IP_3 receptor or Ca^{2+} depletion from the ER.

2.5. Ca^{2+} Oscillations and Cell Migration

Numerous studies have shown that migrating GBM cells display evident Ca^{2+} oscillations having a period of several tens of seconds, whose presence appears to be essential to the process of migration. For example, in U87-MG cells Ca^{2+} oscillations may be induced by the pro-migratory fetal calf serum (FCS), and trigger focal adhesion disassembly during cell rear retraction [29,30]. Similarly, in D54-MG cells prolonged exposure to bradykinin induces Ca^{2+} oscillations, which significantly enhance cell motility [31]. Mechanistically, the Ca^{2+} signal may translate into actomyosin contraction and rear cell retraction by first binding to calmodulin (CaM). The activated complex, Ca^{2+}/CaM, then activates MLCK, the kinase that phosphorylates the regulatory myosin light chains (MLC), thereby triggering cross-bridge movements and actomyosin contraction. It has been shown that MLCK can be modulated by several other protein kinases including PKA, PKC, and RhoK, indicating that this critical enzyme for cell migration is under the influence and control of many extracellular signals and cytoplasmic pathways. Although the consensus on this mechanism and the role of MLCK is widespread, a little caution is needed as most studies used a pharmacological approach, but really specific MLCK inhibitors were not available [32]. A recent report showed, for instance, that knockout cells for MLCK maintained unaltered (or increased) their migratory ability [33]. This shows that if the critical role of Ca^{2+} in cell motility is undisputed, uncertainties may remain as to how Ca^{2+} is linked to the cell migratory machinery. The role of the Ca^{2+}-activated K channels in cell volume changes and the dynamics of Ca^{2+} signals during migration may be a valid alternative.

3. The Intermediate Conductance Ca^{2+}-Activated K Channel

The intermediate conductance Ca^{2+}-activated K channel, KCa3.1, belongs to the Ca^{2+}-activated K channel (KCa) family, which comprises the large- (KCa1.1), intermediate (KCa3.1), and small conductance (KCa2.1-3) K channels, as originally classified according to their single-channel conductance. Genetic relationship and Ca^{2+} activation mechanisms have later shown that these channels form two well-defined and distantly related groups. One, including only the large-conductance KCa1.1 channel, is gated by the cooperative action of membrane depolarization and $[Ca^{2+}]_i$, while the other group includes both the intermediate conductance KCa3.1 and small-conductance KCa2.1-3 channels, gated solely by cytosolic Ca^{2+} increase.

3.1. Biophysics, Pharmacology, and Gating of the KCa3.1 Channel

The biophysical properties of the KCa3.1 channel have been investigated using various experimental approaches and cell models. Work on cultured human glioblastoma cells shows high K^+ selectivity, moderate inward rectification, and single-channel conductance of 20–60 pS in symmetrical 150 mM K^+. The KCa3.1 channel is voltage-insensitive, but highly sensitive to $[Ca^{2+}]_i$ showing an EC_{50} of <200 nM. Ca^{2+} ions activate the KCa3.1 channel via the Ca^{2+}-binding protein CaM, which is constitutively bound to the membrane-proximal region of the intracellular C terminus of the channel. Binding of Ca^{2+} to CaM results in conformational changes of the channel, and its opening. In addition to serving as a Ca^{2+} sensor for channel opening, CaM also regulates the assembly and trafficking of the channel protein to the plasma membrane [34,35].

The KCa3.1 channel can be blocked by peptidic toxins isolated from various scorpions, such as charybdotoxin (ChTx) and maurotoxin (MTx), which display high affinity (IC_{50} in the nM range).

Small synthetic molecules have also been developed, many derived from the clotrimazole template, the classical KCa3.1 channel blocker. The most widely used TRAM-34, developed by Wulff's group, inhibits KCa3.1 channels with an $IC_{50} < 20$ nM and displays high selectivity over the other KCa channels. As for KCa3.1 channel activators, we recall the benzimidazolone 1-EBIO that activates the channels with an $EC_{50} < 30$ μM and DC-EBIO that exhibits 10-fold higher potency. Two structurally similar and still more potent molecules are the oxime NS309 and the benzothiazole SKA-31, both showing an EC_{50} for KCa3.1 channels < 20 nM. We finally mention riluzole, another potent activator of KCa3.1 channels, which is the only FDA-approved drug for treatment of amyotrophic lateral sclerosis (ALS).

3.2. KCa3.1 Channel Expression and Impact on GBM Migration and Invasion

Initial biochemical and electrophysiological studies showed that KCa3.1 channels were diffusely expressed in virtually all cell types investigated. Notably, the KCa3.1 channel was not found in the brain (later the KCa3.1 channel was described in microglia [7], several types of brain neurons [8,36], and the nodes of Ranvier of cerebellar Purkinje neurons [37]), although it was highly expressed in established cell lines from brain tumors, as well as in brain tumors in situ. Moreover, KCa3.1 channel expression was reported in many other tumor types, and implicated in malignancy (increased cell growth, migration, invasion, apoptosis evasion). The KCa3.1 mRNA and protein expression were also found to be significantly enhanced in cancer stem cells derived from both the established cell line U87-MG and primary cell line FCN9 [38]. The REMBRANDT database shows that the gene KCNN4, which encodes the KCa3.1 channel, is overexpressed in more than 30% of gliomas, and its expression is associated with a poor prognosis.

To test the role of the KCa3.1 channel in GBM cells' infiltration into brain parenchyma, human GL-15 GBM cells were xenografted into the brains of SCID mice and later treated with the specific KCa3.1 blocker TRAM-34. Immunofluorescence analyses of cerebral slices after a five-week treatment revealed a significant reduction of tumor infiltration, compared with TRAM-34 untreated mice [39]. Reduction of tumor infiltration was also observed in the brain of mice transplanted with KCa3.1-silenced GL-15 cells, indicating a direct role of KCa3.1 channels in GBM tumor infiltration [40]. Similarly, KCa3.1 channel block with TRAM-34 or silencing by short hairpin RNA (shRNA) completely abolished CXCL12-induced GL-15 cell migration [40]. A strong correlation of KCa3.1 channel expression with migration was reported by Calogero's group in GBM-derived cancer stem cells (CSC) [38]. Blockage of the KCa3.1 channel with TRAM-34 was found to have a much greater impact on the motility of CSCs (75% reduction) that express a high level of KCa3.1 channel than on the FCN9 parental population (32% reduction), where the KCa3.1 channel is expressed at much lower levels. Similar results were also observed with the CSCs derived from U87-MG [38] and with mesenchymal glioblastoma stem cells [41]. On the same line stand the results later obtained by Sontheimer's group showing that pharmacological inhibition of the KCa3.1 channel (with TRAM-34) in U251 glioma cells, or silencing it with inducible siRNA, resulted in a significant reduction of tumor cells' migration in vitro and invasion into surrounding brain parenchyma of SCID mice [42]. On a retrospective study of a patient genomic database, they further showed that KCa3.1 channel expression was inversely correlated with patient survival.

3.3. Basic Functions of KCa3.1 Channel: Regulation of Cell Ca^{2+} Signaling

KCa3.1 channels control a number of basic cellular processes involved in the modulation of several higher-order biological functions critical to brain tumors' malignancy, including migration and invasion. The most relevant and common basic cellular processes controlled by KCa3.1 channels are the modulation of Ca^{2+} signaling and the control of cell volume. Here we will concentrate on the KCa3.1 channel modulation of Ca^{2+} signaling, the focus of this review.

In virtually all cells a robust stimulation of PLC-coupled membrane receptors triggers an initial IP_3-mediated release of Ca^{2+} from intracellular stores, followed by Ca^{2+} influx through store-operated Orai Ca^{2+} channels, which are activated in response to Ca^{2+} depletion of the ER.

One consequence of this Ca^{2+} influx, besides the activation of Ca^{2+}-dependent target proteins (ion channels included), is membrane depolarization, which, if left unchecked, would more and more strongly inhibit further Ca^{2+} influx due to the reduced electrochemical driving force for Ca^{2+} ions. It has been shown that the efflux of K^+ ions following KCa3.1 channels' activation by incoming Ca^{2+} hyperpolarizes the membrane towards E_K, increasing the electromotive force on incoming Ca^{2+} ions [43,44]. This represents an indirect modulatory mechanism of Ca^{2+} entry and, as a result, of cell Ca^{2+} signaling. The Ca^{2+}-dependent and voltage-independent gating of KCa3.1 channels appears to be particularly well suited for this function, since the coupling of Ca^{2+} influx and activation of K^+ channels will create a positive feedback loop whereby more Ca^{2+} influx will increase the K^+ efflux, hyperpolarize the plasma membrane, and further stimulate Ca^{2+} entry, thus amplifying the signal transduction.

This paradigm was first demonstrated in human macrophages, where KCa3.1 channels have been shown to hyperpolarize the membrane and increase the driving force for Ca^{2+} ions during store-operated Ca^{2+} entry [43], and in T cells, where KCa3.1 channels are rapidly upregulated following cell activation, and supposedly used to maximize Ca^{2+} influx during the reactivation of memory T cells [45,46]. Activated T cells isolated from KCa3.1$^{-/-}$ mice show a defective Ca^{2+} response to T cell receptor activation [47]. Also, mast cells appear to use KCa3.1 channels to hyperpolarize the membrane and increase the Ca^{2+} influx following antigen-mediated stimulation. In this case KCa3.1 channels have been found to physically interact with the Orai1 subunit, suggesting that they may be activated by the Ca^{2+} microdomain that forms close to the store-operated Ca^{2+} channel [28,48]. A similar conclusion has been reached for rat microglial cells, where the Orai1-KCa3.1 functional coupling could be interrupted by BAPTA but not by EGTA [49]. Since the two Ca^{2+} chelators have similar Ca^{2+} affinity but different Ca^{2+} binding rates (10 times higher for BAPTA), only a physical coupling between Orai1 and KCa3.1, with a separating distance of few nanometers, would explain these results.

Evidence of cross-talk between the KCa3.1 channel and store depletion-activated Ca^{2+} influx has also been reported for GBM cells. An increase of $[Ca^{2+}]_i$, consisting of a fast peak due to IP$_3$-dependent Ca^{2+} release from intracellular stores followed by a sustained phase of Ca^{2+} influx across the plasma membrane, supposedly activated by intracellular stores' depletion, was found upon prolonged application of histamine on GL-15 GBM cells [44]. The enhancing role of the KCa3.1 channel in sustaining the protracted influx of external Ca^{2+} was shown by the marked reduction of the sustained histamine-induced $[Ca^{2+}]_i$ increase following application of TRAM-34. This observation could be significant with regard to the KCa3.1 channels' contribution to GBM cell migration exerted through modulation of Ca^{2+} signals.

In another GBM cell line—U87-MG—we also observed that the promigratory FCS is able to promote Ca^{2+} oscillations, and these oscillations cyclically activate the KCa3.1 channels during cell migration [13]. Using a modeling approach, we also found that a channel activity with the properties of KCa3.1 channels could modulate these Ca^{2+} oscillations (it increased the amplitude, duration, and frequency of each Ca^{2+} spike [50]). The results we obtained, which are illustrated in the following section, were unexpectedly interesting, and also able to explain old experiments showing that the KCa3.1 channel inhibition by ChTx abolishes the bradykinin-induced Ca^{2+} oscillations in C6 glioma cells [51].

3.4. Modulation of Ca^{2+} Oscillations by KCa3.1 Channel Activity

We saw earlier that the mechanism underlying Ca^{2+} oscillations is essentially based on the biphasic effects of Ca^{2+} on IP$_3$ receptor gating—activatory at low and inhibitory at high concentrations. We also discussed the modulation that the KCa3.1 channel could exert on Ca^{2+} oscillations. This modulation relies on the high Ca^{2+} sensitivity and voltage independence of the KCa3.1 channel, and on the output of its activity, that is, the hyperpolarization of the cell membrane potential (V_m) as result of the K^+ efflux. As KCa3.1 channels are activated by Ca^{2+} concentrations well within the range observed during hormone-induced Ca^{2+} oscillations (150–600 nM; cf. above), they are expected to cyclically open (during the Ca^{2+} spikes) and cause cyclic membrane hyperpolarizations in phase with the Ca^{2+} oscillations.

We have observed experimentally such KCa3.1 channel oscillatory activity and associated membrane potential oscillations in U87-MG cells in response to FCS [13], an agent known to induce Ca^{2+} oscillations in these cells [29,30]. Notably, because of their voltage dependence and much lower Ca^{2+} sensitivity, the KCa1.1 channels, also highly expressed in GBM cells, could not be activated under these conditions, even at the peak Ca^{2+} concentration of the oscillations ([13], but see [52] for a case in which the KCa1.1 channel activity may control Ca^{2+} influx). In conclusion, since V_m controls the driving force for external Ca^{2+} entry, KCa3.1-dependent V_m oscillations are expected to cause oscillations in the amplitude of Ca^{2+} influx through the hormone-activated plasma membrane Ca^{2+} channels, which will in turn feedback onto, and modulate the Ca^{2+} oscillations themselves.

We recently implemented basic theoretical models of Ca^{2+} oscillations with the oscillating membrane V_m, as produced by the cyclic activation of KCa3.1 channels. The resulting model was used to predict how the hormone-induced Ca^{2+} oscillations would be influenced by the KCa3.1 channels-induced V_m oscillations in phase with Ca^{2+} oscillations. We found that the cyclic activation of KCa3.1 channels by Ca^{2+} oscillations induces V_m fluctuations that in turn determine oscillations in the Ca^{2+} influx from the extracellular medium, as a result of the oscillating V_m-dependent changes in the driving force for Ca^{2+} ions influx. Since Ca^{2+} influx peaks are in phase with the Ca^{2+} oscillations, the Ca^{2+} spikes will also be increased by the in-phase Ca^{2+} influx. Our model calculations in fact show that KCa3.1-induced V_m oscillations strengthen the hormone-induced $[Ca^{2+}]_i$ signals by increasing the amplitude, duration, and oscillatory frequency of Ca^{2+} spikes (Figure 2; see also [50]).

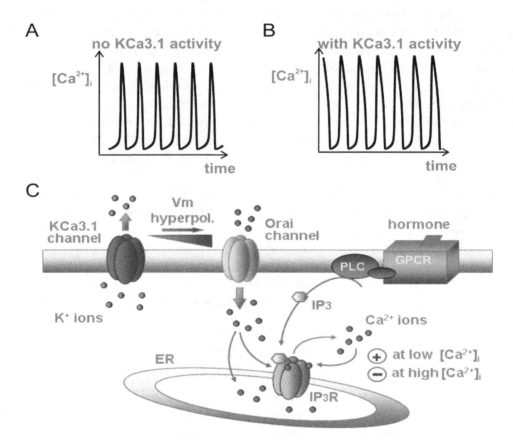

Figure 2. Modulation of Ca^{2+} oscillations by the KCa3.1 channel. Drawing (**C**) illustrating the main ion fluxes generating or modulating the Ca^{2+} oscillations in our model when the KCa3.1 channel is cut off (**A**) or introduced into the system (**B**). Please notice the much longer duration (width) and slightly higher amplitude of the Ca^{2+} oscillations in the presence of KCa3.1 channels.

3.5. KCa3.1 Channels Switch Ca^{2+} Oscillations On and Off

Arguably, the KCa3.1 channels can do more than just modulate the amplitude and frequency of Ca^{2+} oscillations: they can trigger or suppress them. This has been an unexpected prediction of our model that we later found to have been experimentally observed. As experimentally found, our model predicts that Ca^{2+} oscillations can be generated only within a specific range of IP$_3$ concentrations, while a stable, non-oscillating Ca^{2+} level is present for both lower and higher IP$_3$ concentrations (Figure 3A). Surprisingly, it was found that when the KCa3.1 conductance is changed in the system, the resulting alteration of the KCa3.1 channels-induced V_m oscillations significantly change the IP$_3$ concentrations needed to produce Ca^{2+} oscillations. In particular, increasing the activity of KCa3.1 channels shifts leftward the IP$_3$ range within which Ca^{2+} oscillations are produced. This implies that there are ranges of IP$_3$ levels where the presence or absence of Ca^{2+} oscillations are only dictated by the activity (conductance) of KCa3.1 channels (Figure 3B bottom, shaded areas).

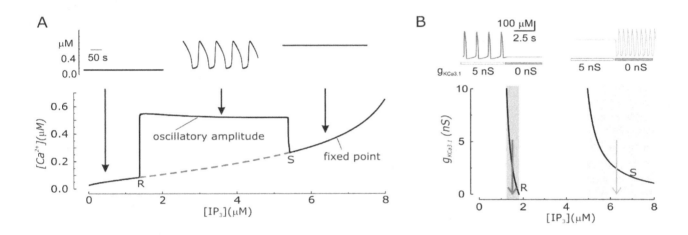

Figure 3. Ca^{2+} oscillation development depends on the IP$_3$ concentration and the activity of KCa3.1 channels. (**A**) Bottom. Bifurcation diagram showing that IP$_3$ concentration determines the establishment of the Ca^{2+} oscillations. The red dashed line represents the unstable equilibrium [Ca^{2+}]$_i$. The computation was performed with g$_{KCa3.1}$ = 5 nS. R and S indicate the Hopf bifurcations where Ca^{2+} oscillations appear and disappear, respectively, upon increasing the IP$_3$ levels. Top. The traces show the temporal changes of the [Ca^{2+}]$_i$ for three different concentrations of IP$_3$ (indicated by arrows in the bottom part). (**B**) Bottom. Plot of the two Hopf bifurcations (R and S) as a function of the IP$_3$ concentration and g$_{KCa3.1}$. Three different ranges of IP$_3$ concentrations, where the KCa3.1-induced V_m oscillations appear to have different effects in the modulation of Ca^{2+} oscillations are evidenced (shaded areas; see text). More specifically in the pink region the presence of KCa3.1 channels is necessary for the existence of Ca^{2+} oscillations, in the light blue region KCa3.1 channels prevents Ca^{2+} oscillations, and finally in the white region in between KCa3.1 channels modulate the shape and duration of oscillation. Top. Simulated Ca^{2+} oscillations obtained with two IP$_3$ concentrations (1.5 left and 6.2 μM right, as indicated by the red and blue arrows below) within the regions where removing KCa3.1 channels make Ca^{2+} oscillations disappear or appear. (Modified from [48].)

These observations could explain a number of experimental results where KCa3.1 channel modulation was shown to have an effect on Ca^{2+} oscillations. In both activated T lymphocytes and C6 glioma cells, inhibition of V_m oscillations by blocking KCa3.1 channels is able to suppress Ca^{2+} oscillations [51,53]. Moreover *ras*-induced transformation of fibroblasts, which has been reported to upregulate KCa3.1 channels, has also been shown to trigger Ca^{2+} oscillations [54,55]. Finally, KCa3.1 inhibition leads to induction of Ca^{2+} oscillations in about half of the tested glioblastoma cells [52].

3.6. Glioblastoma KCa3.1 Channels Are Activated by Serum-Induced Ca²⁺ Oscillations and Participate to Cell Migration

Glioblastoma cells in vivo are exposed to a variety of tumor microenvironment components and soluble factors that can markedly affect their migratory ability. Among them there are unknown serum components that infiltrate into the tumor area of high-grade gliomas as result of the blood brain barrier breakdown [6,56]. As already shown, FCS enhances migration of U87-MG glioblastoma cells, and does so by inducing Ca²⁺ oscillations that are critically involved in the detachment of focal adhesions (by activating the focal adhesion kinase), and subsequent retraction of the cell rear [29,30]. On the same cell model we further reported that FCS, besides Ca²⁺ oscillations, induces an oscillatory activity of KCa3.1 channels, and this KCa3.1 channel activity is a necessary step to promote U87-MG cell migration by FCS [13]. In the same study, we also found a stable activation of Cl⁻ currents upon FCS stimulation. Altogether these observations sketch a coherent picture of the cell migratory process, in that, the retraction of the cell rear after the detachment of focal adhesions involves a major cell shape rearrangement and volume reduction. In these instances, the (oscillatory) KCa3.1 channel activity and the consequent cyclical K⁺ efflux, in combination with Cl⁻ efflux (and essential osmotic water), form part of the whole machinery for achieving the cyclical volume reduction needed.

Cell locomotion is thought to be promoted by the cycling replication of four main steps: (i) protrusion of the cell front, associated to asymmetric polymerization of the actin-based cytoskeleton; (ii) adhesion of the protruded cell front to the substratum, mediated by the binding of focal adhesions to the extracellular polymers in contact with the cell; (iii) unbinding of the focal adhesions at the cell rear; (iv) retraction of the rear cell body in the direction of motion (cf. Figure 4). Two of these steps, namely cell front protrusion and cell rear retraction, involve major changes in cell volume and thus require compensatory ion fluxes (and osmotic water flow) across the membrane. Along the lines proposed by [12], during the protrusion of the cell front the maintenance of a stable osmolarity is sustained by cotransporters and ion exchangers—for instance, Na⁺/K⁺/2Cl⁻ cotransport—followed by osmotically driven water. During this phase [Ca²⁺]ᵢ is relatively low (red segment in the Ca²⁺ oscillation, Figure 4a), a condition in which KCa3.1 channels are closed and the resting membrane potential is near E_Cl. This condition prevents or strongly limits the loss of KCl via K and Cl channels that would nullify the osmotic work of the Na⁺/K⁺/2Cl⁻ cotransport.

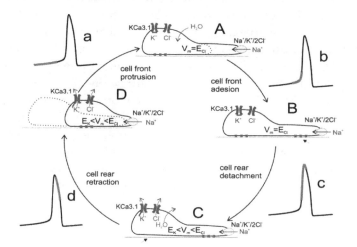

Figure 4. Schematic diagram of cell migratory cycle. The classic view of the cell migration process can be split down into four main cyclical steps. The cycle begins with the cell front protrusion, due to the activity of Na⁺/K⁺/2Cl⁻ cotransport (yellow) (**A**) and establishment of adhesion structures (**B**). The elongated cell then removes/weakens the rear adhesions (**C**) so that ensuing contraction can pull the rear cell portion forward (**D**). The concomitant values of [Ca²⁺]ᵢ is indicated by the red portion on the associated Ca²⁺ oscillation. V_m, ion and water fluxes are also illustrated (see text for details).

By contrast, the control of osmolarity during cell rear retraction critically depends on the activation of KCa3.1 channels and ensuing efflux of K^+ ions (accompanied by Cl^- fluxes, to which membrane has become highly permeant, and osmotic water). Following the increase in $[Ca^{2+}]_i$ towards the peak of the Ca^{2+} oscillation (Figure 4c,d), KCa3.1 channels open and the membrane potential hyperpolarizes to a value between E_{Cl} and E_K, a condition that would allow a significant efflux of KCl with consequent loss of water and cell volume decrease (Figure 4C,D). This would facilitate the process of cell body retraction. This view can explain the slowing of cell locomotion following the inhibition of K and Cl channels [13].

4. Conclusions

Substantial evidence suggests that the oscillatory activity of the KCa3.1 channel during Ca^{2+} oscillations may have both direct and indirect modulatory effects on GBM cell migration. As has long been recognized, the K^+ efflux through KCa3.1 channels directly participates in the cell volume changes during cell rear retraction. The same oscillatory K^+ efflux, however, will also determine the membrane potential oscillations that alter the driving force for Ca^{2+} influx and indirectly modulate the properties and even the presence of Ca^{2+} oscillations, and thus the timing of the migratory machinery. Both modulatory roles make KCa3.1 channels pivotal in GBM cell migration, and thus potential pharmacological targets for this deadly tumor. Although this review concentrates on the role of KCa3.1 and STIM/Orai channels in modulating Ca^{2+} oscillations and cell migration, it needs to be said that other channels including TRPM8 and KCa1.1, also abundantly expressed in glioblastoma cells, may also be important in these processes. Accordingly, TRPM8 [41] and KCa1.1 channel signaling [57] have been demonstrated to be required for basal and radiation-induced migration, respectively, and certainly contribute to Ca^{2+} oscillations.

Author Contributions: Both L.C. and F.F. performed literature search and wrote the paper.

References

1. Hanahan, D.; Weinberg, R.A. Hallmarks of cancer: The next generation. *Cell* **2011**, *144*, 646–674. [CrossRef] [PubMed]
2. Ridley, A.J.; Schwartz, M.A.; Burridge, K.; Firtel, R.A.; Ginsberg, M.H.; Borisy, G.; Parsons, J.T.; Horwitz, A.R. Cell migration: Integrating signals from front to back. *Science* **2003**, *302*, 1704–1709. [CrossRef] [PubMed]
3. Ridley, A.J. Life at the leading edge. *Cell* **2011**, *145*, 1012–1022. [CrossRef] [PubMed]
4. Carragher, N.O.; Frame, M.C. Focal adhesion and actin dynamics: A place where kinases and proteases meet to promote invasion. *Trends Cell Biol.* **2004**, *14*, 241–249. [CrossRef] [PubMed]
5. Parsons, J.T.; Horwitz, A.R.; Schwartz, M.A. Cell adhesion: Integrating cytoskeletal dynamics and cellular tension. *Nat. Rev. Mol. Cell Biol.* **2010**, *11*, 633–643. [CrossRef] [PubMed]
6. Lund, C.V.; Nguyen, M.T.; Owens, G.C.; Pakchoian, A.J.; Shaterian, A.; Kruse, C.A.; Eliceiri, B.P. Reduced glioma infiltration in Src-deficient mice. *J. Neurooncol.* **2006**, *78*, 19–29. [CrossRef] [PubMed]
7. Wessels, D.; Soll, D.R.; Knecht, D.; Loomis, W.F.; De Lozanne, A.; Spudich, J. Cell motility and chemotaxis in Dictyostelium amebae lacking myosin heavy chain. *Dev. Biol.* **1988**, *128*, 164–177. [CrossRef]
8. Lombardi, M.L.; Knecht, D.A.; Dembo, M.; Lee, J. Traction force microscopy in Dictyostelium reveals distinct roles for myosin II motor and actin-crosslinking activity in polarized cell movement. *J. Cell Sci.* **2007**, *120 Pt 9*, 1624–1634. [CrossRef]
9. Roca-Cusachs, P.; Sunyer, R.; Trepat, X. Mechanical guidance of cell migration: Lessons from chemotaxis. *Curr. Opin. Cell Biol.* **2013**, *25*, 543–549. [CrossRef] [PubMed]
10. Ostman, A.; Heldin, C.H. PDGF receptors as targets in tumor treatment. *Adv. Cancer Res.* **2007**, *97*, 247–274. [PubMed]

11. Lo, C.M.; Wang, H.B.; Dembo, M.; Wang, Y.L. Cell movement is guided by the rigidity of the substrate. *Biophys. J.* **2000**, *79*, 144–152. [CrossRef]

12. Schwab, A.; Fabian, A.; Hanley, P.J.; Stock, C. Role of ion channels and transporters in cell migration. *Physiol. Rev.* **2012**, *92*, 1865–1913. [CrossRef] [PubMed]

13. Catacuzzeno, L.; Aiello, F.; Fioretti, B.; Sforna, L.; Castigli, E.; Ruggieri, P.; Tata, A.M.; Calogero, A.; Franciolini, F. Serum-activated K and Cl currents underlay U87-MG glioblastoma cell migration. *J. Cell Physiol.* **2011**, *226*, 1926–1933. [CrossRef] [PubMed]

14. Cuddapah, V.A.; Habela, C.W.; Watkins, S.; Moore, L.S.; Barclay, T.T.; Sontheimer, H. Kinase activation of ClC-3 accelerates cytoplasmic condensation during mitotic cell rounding. *Am. J. Physiol. Cell Physiol.* **2012**, *302*, C527–C538. [CrossRef] [PubMed]

15. Wei, C.; Wang, X.; Zheng, M.; Cheng, H. Calcium gradients underlying cell migration. *Curr. Opin. Cell Biol.* **2012**, *24*, 254–261. [CrossRef] [PubMed]

16. Cuthbertson, K.S.; Chay, T.R. Modelling receptor-controlled intracellular calcium oscillators. *Cell Calcium* **1991**, *12*, 97–109. [CrossRef]

17. Gaspers, L.D.; Bartlett, P.J.; Politi, A.; Burnett, P.; Metzger, W.; Johnston, J.; Joseph, S.K.; Höfer, T.; Thomas, A.P. Hormone-induced calcium oscillations depend on cross-coupling with inositol 1,4,5-trisphosphate oscillations. *Cell Rep.* **2014**, *9*, 1209–1218. [CrossRef] [PubMed]

18. Dupont, G.; Combettes, L.; Bird, G.S.; Putney, J.W. Calcium oscillations. *Cold Spring Harb. Perspect. Biol.* **2011**, *3*, a004226. [CrossRef] [PubMed]

19. Iino, M. Biphasic Ca^{2+} dependence of inositol 1,4,5-trisphosphate-induced Ca release in smooth muscle cells of the guinea pig taenia caeci. *J. Gen. Physiol.* **1990**, *95*, 1103–1122. [CrossRef] [PubMed]

20. Bezprozvanny, I.; Watras, J.; Ehrlich, B.E. Bell-shaped calcium-response curves of Ins(1,4,5)P3- and calcium-gated channels from endoplasmic reticulum of cerebellum. *Nature* **1991**, *351*, 751–754. [CrossRef] [PubMed]

21. Finch, E.A.; Turner, T.J.; Goldin, S.M. Calcium as a coagonist of inositol 1,4,5-trisphosphate-induced calcium release. *Science* **1991**, *252*, 443–446. [CrossRef] [PubMed]

22. Kaznacheyeva, E.; Lupu, V.D.; Bezprozvanny, I. Single-channel properties of inositol (1,4,5)-trisphosphate receptor heterologously expressed in HEK-293 cells. *J. Gen. Physiol.* **1998**, *111*, 847–856. [CrossRef] [PubMed]

23. Liou, J.; Kim, M.L.; Heo, W.D.; Jones, J.T.; Myers, J.W.; Ferrell, J.E., Jr.; Meyer, T. STIM is a Ca^{2+} sensor essential for Ca^{2+}-store-depletion-triggered Ca^{2+} influx. *Curr. Biol.* **2005**, *15*, 1235–1241. [CrossRef] [PubMed]

24. Roos, J.; DiGregorio, P.J.; Yeromin, A.V.; Ohlsen, K.; Lioudyno, M.; Zhang, S.; Safrina, O.; Kozak, J.A.; Wagner, S.L.; Cahalan, M.D.; et al. STIM1, an essential and conserved component of store-operated Ca^{2+} channel function. *J. Cell Biol.* **2005**, *169*, 435–445. [CrossRef] [PubMed]

25. Peinelt, C.; Vig, M.; Koomoa, D.L.; Beck, A.; Nadler, M.J.; Koblan-Huberson, M.; Lis, A.; Fleig, A.; Penner, R.; Kinet, J.P. Amplification of CRAC current by STIM1 and CRACM1 (Orai1). *Nat. Cell Biol.* **2006**, *8*, 771–773. [CrossRef] [PubMed]

26. Moccia, F.; Ruffinatti, F.A.; Zuccolo, E. Intracellular Ca^{2+} Signals to Reconstruct a Broken Heart: Still a Theoretical Approach? *Curr. Drug Targets* **2015**, *6*, 793–815. [CrossRef]

27. Dubois, C.; Vanden Abeele, F.; Lehen'kyi, V.; Gkika, D.; Guarmit, B.; Lepage, G.; Slomianny, C.; Borowiec, A.S.; Bidaux, G.; Benahmed, M.; et al. Remodeling of channel-forming ORAI proteins determines an oncogenic switch in prostate cancer. *Cancer Cell* **2014**, *26*, 19–32. [CrossRef] [PubMed]

28. Duffy, S.M.; Ashmole, I.; Smallwood, D.T.; Leyland, M.L.; Bradding, P. Orai/CRACM1 and KCa3.1 ion channels interact in the human lung mast cell plasma membrane. *Cell Commun. Signal.* **2015**, *13*, 32. [CrossRef] [PubMed]

29. Rondé, P.; Giannone, G.; Gerasymova, I.; Stoeckel, H.; Takeda, K.; Haiech, J. Mechanism of calcium oscillations in migrating human astrocytoma cells. *Biochim. Biophys. Acta* **2000**, *1498*, 273–280. [CrossRef]

30. Giannone, G.; Rondé, P.; Gaire, M.; Haiech, J.; Takeda, K. Calcium oscillations trigger focal adhesion disassembly in human U87 astrocytoma cells. *J. Biol. Chem.* **2002**, *277*, 26364–26371. [CrossRef] [PubMed]

31. Montana, V.; Sontheimer, H. Bradykinin promotes the chemotactic invasion of primary brain tumors. *J. Neurosci.* **2011**, *31*, 4858–4867. [CrossRef] [PubMed]

32. Totsukawa, G.; Wu, Y.; Sasaki, Y.; Hartshorne, D.J.; Yamakita, Y.; Yamashiro, S.; Matsumura, F. Distinct roles of MLCK and ROCK in the regulation of membrane protrusions and focal adhesion dynamics during cell migration of fibroblasts. *J. Cell Biol.* **2004**, *164*, 427–439. [CrossRef] [PubMed]

33. Chen, C.; Tao, T.; Wen, C.; He, W.Q.; Qiao, Y.N.; Gao, Y.Q.; Chen, X.; Wang, P.; Chen, C.P.; Zhao, W.; et al. Myosin light chain kinase (MLCK) regulates cell migration in a myosin regulatory light chain phosphorylation-independent mechanism. *J. Biol. Chem.* **2014**, *289*, 28478–28488. [CrossRef] [PubMed]

34. Sforna, L.; Megaro, A.; Pessia, M.; Franciolini, F.; Catacuzzeno, L. Structure, Gating and Basic Functions of the Ca^{2+}-activated K Channel of Intermediate Conductance. *Curr. Neuropharmacol.* **2018**, *16*, 608–617. [CrossRef] [PubMed]

35. Catacuzzeno, L.; Franciolini, F. Editorial: The Role of Ca^{2+}-activated K^+ Channels of Intermediate Conductance in Glioblastoma Malignancy. *Curr. Neuropharmacol.* **2018**, *16*, 607. [CrossRef] [PubMed]

36. Vicente-Manzanares, M.; Koach, M.A.; Whitmore, L.; Lamers, M.L.; Horwitz, A.F. Segregation and activation of myosin IIB creates a rear in migrating cells. *J. Cell Biol.* **2008**, *183*, 543–554. [CrossRef] [PubMed]

37. Kaushal, V.; Koeberle, P.D.; Wang, Y.; Schlichter, L.C. The Ca^{2+}-activated K^+ channel KCNN4/KCa3.1 contributes to microglia activation and nitric oxide-dependent neurodegeneration. *J. Neurosci.* **2007**, *27*, 234–244. [CrossRef] [PubMed]

38. Ruggieri, P.; Mangino, G.; Fioretti, B.; Catacuzzeno, L.; Puca, R.; Ponti, D.; Miscusi, M.; Franciolini, F.; Ragona, G.; Calogero, A. The inhibition of KCa3.1 channels activity reduces cell motility in glioblastoma derived cancer stem cells. *PLoS ONE* **2012**, *7*, e47825. [CrossRef] [PubMed]

39. D'Alessandro, G.; Catalano, M.; Sciaccaluga, M.; Chece, G.; Cipriani, R.; Rosito, M.; Grimaldi, A.; Lauro, C.; Cantore, G.; Santoro, A.; et al. KCa3.1 channels are involved in the infiltrative behavior of glioblastoma in vivo. *Cell Death Dis.* **2013**, *4*, e773. [CrossRef] [PubMed]

40. Sciaccaluga, M.; Fioretti, B.; Catacuzzeno, L.; Pagani, F.; Bertollini, C.; Rosito, M.; Catalano, M.; D'Alessandro, G.; Santoro, A.; Cantore, G.; et al. CXCL12-induced glioblastoma cell migration requires intermediate conductance Ca^{2+}-activated K^+ channel activity. *Am. J. Physiol. Cell Physiol.* **2010**, *299*, C175–C184. [CrossRef] [PubMed]

41. Klumpp, L.; Sezgin, E.C.; Skardelly, M.; Eckert, F.; Huber, S.M. KCa3.1 Channels and Glioblastoma: In Vitro Studies. *Curr. Neuropharmacol.* **2018**, *16*, 627–635. [CrossRef] [PubMed]

42. Turner, K.L.; Honasoge, A.; Robert, S.M.; McFerrin, M.M.; Sontheimer, H. A proinvasive role for the Ca(2+)-activated K(+) channel KCa3.1 in malignant glioma. *Glia* **2014**, *62*, 971–981. [CrossRef] [PubMed]

43. Gao, Y.D.; Hanley, P.J.; Rinné, S.; Zuzarte, M.; Daut, J. Calcium-activated K^+ channel ($K_{Ca}3.1$) activity during Ca^{2+} store depletion and store-operated Ca^{2+} entry in human macrophages. *Cell Calcium* **2010**, *48*, 19–27. [CrossRef] [PubMed]

44. Fioretti, B.; Catacuzzeno, L.; Sforna, L.; Aiello, F.; Pagani, F.; Ragozzino, D.; Castigli, E.; Franciolini, F. Histamine hyperpolarizes human glioblastoma cells by activating the intermediate-conductance Ca^{2+}-activated K^+ channel. *Am. J. Physiol. Cell Physiol.* **2009**, *297*, C102–C110. [CrossRef] [PubMed]

45. Ghanshani, S.; Wulff, H.; Miller, M.J.; Rohm, H.; Neben, A.; Gutman, G.A.; Cahalan, M.D.; Chandy, K.G. Up-regulation of the IKCa1 potassium channel during T-cell activation. Molecular mechanism and functional consequences. *J. Biol. Chem.* **2000**, *275*, 37137–37149. [CrossRef] [PubMed]

46. Wulff, H.; Beeton, C.; Chandy, K.G. Potassium channels as therapeutic targets for autoimmune disorders. *Curr. Opin. Drug Discov. Dev.* **2003**, *6*, 640–647.

47. Di, L.; Srivastava, S.; Zhdanova, O.; Ding, Y.; Li, Z.; Wulff, H.; Lafaille, M.; Skolnik, E.Y. Inhibition of the K^+ channel KCa3.1 ameliorates T cell-mediated colitis. *Proc. Natl. Acad. Sci. USA* **2010**, *107*, 1541–1546. [CrossRef] [PubMed]

48. Duffy, M.S.; Berger, P.; Cruse, G.; Yang, W.; Bolton, S.J.; Bradding, P. The K^+ channel iKCA1 potentiates Ca^{2+} influx and degranulation in human lung mast cells. *J. Allergy Clin. Immunol.* **2004**, *114*, 66–72. [CrossRef] [PubMed]

49. Ferreira, R.; Schlichter, L.C. Selective activation of KCa3.1 and CRAC channels by P2Y2 receptors promotes Ca(2+) signaling, store refilling and migration of rat microglial cells. *PLoS ONE* **2013**, *8*, e62345. [CrossRef] [PubMed]

50. Catacuzzeno, L.; Fioretti, B.; Franciolini, F. A theoretical study on the role of Ca^{2+}-activated K^+ channels in the regulation of hormone-induced Ca^{2+} oscillations and their synchronization in adjacent cells. *J. Theor. Biol.* **2012**, *309*, 103–112. [CrossRef] [PubMed]

51. Reetz, G.; Reiser, G. [Ca2+]i oscillations induced by bradykinin in rat glioma cells associated with Ca^{2+} store-dependent Ca^{2+} influx are controlled by cell volume and by membrane potential. *Cell Calcium* **1996**, *19*, 143–156. [CrossRef]

52. Stegen, B.; Klumpp, L.; Misovic, M.; Edalat, L.; Eckert, M.; Klumpp, D.; Ruth, P.; Huber, S.M. K^+ channel signaling in irradiated tumor cells. *Eur. Biophys. J.* **2016**, *45*, 585–598. [CrossRef] [PubMed]

53. Verheugen, J.A.; Vijverberg, H.P. Intracellular Ca^{2+} oscillations and membrane potential fluctuations in intact human T lymphocytes: Role of K^+ channels in Ca^{2+} signaling. *Cell Calcium* **1995**, *17*, 287–300. [CrossRef]

54. Hashii, M.; Nozawa, Y.; Higashida, H. Bradykinin-induced cytosolic Ca^{2+} oscillations and inositol tetrakisphosphate-induced Ca^{2+} influx in voltage-clamped ras-transformed NIH/3T3 fibroblasts. *J. Biol. Chem.* **1993**, *268*, 19403–19410. [PubMed]

55. Rane, S.G. A Ca2(+)-activated K^+ current in ras-transformed fibroblasts is absent from nontransformed cells. *Am. J. Physiol.* **1991**, *260*, C104–C112. [CrossRef] [PubMed]

56. Seitz, R.J.; Wechsler, W. Immunohistochemical demonstration of serum proteins in human cerebral gliomas. *Acta Neuropathol.* **1987**, *73*, 145–152. [CrossRef] [PubMed]

57. Steinle, M.; Palme, D.; Misovic, M.; Rudner, J.; Dittmann, K.; Lukowski, R.; Ruth, P.; Huber, S.M. Ionizing radiation induces migration of glioblastoma cells by activating BK K^+ channels. *Radiother. Oncol.* **2011**, *101*, 122–126. [CrossRef] [PubMed]

9

Endothelial Ca^{2+} Signaling and the Resistance to Anticancer Treatments: Partners in Crime

Francesco Moccia

Laboratory of General Physiology, Department of Biology and Biotechnology "L. Spallanzani", University of Pavia, I-27100 Pavia, Italy; francesco.moccia@unipv.it

Abstract: Intracellular Ca^{2+} signaling drives angiogenesis and vasculogenesis by stimulating proliferation, migration, and tube formation in both vascular endothelial cells and endothelial colony forming cells (ECFCs), which represent the only endothelial precursor truly belonging to the endothelial phenotype. In addition, local Ca^{2+} signals at the endoplasmic reticulum (ER)–mitochondria interface regulate endothelial cell fate by stimulating survival or apoptosis depending on the extent of the mitochondrial Ca^{2+} increase. The present article aims at describing how remodeling of the endothelial Ca^{2+} toolkit contributes to establish intrinsic or acquired resistance to standard anti-cancer therapies. The endothelial Ca^{2+} toolkit undergoes a major alteration in tumor endothelial cells and tumor-associated ECFCs. These include changes in TRPV4 expression and increase in the expression of P2X7 receptors, Piezo2, Stim1, Orai1, TRPC1, TRPC5, Connexin 40 and dysregulation of the ER Ca^{2+} handling machinery. Additionally, remodeling of the endothelial Ca^{2+} toolkit could involve nicotinic acetylcholine receptors, gasotransmitters-gated channels, two-pore channels and Na$^+$/H$^+$ exchanger. Targeting the endothelial Ca^{2+} toolkit could represent an alternative adjuvant therapy to circumvent patients' resistance to current anti-cancer treatments.

Keywords: Ca^{2+} signaling; tumor; endothelial cells; endothelial progenitor cells; endothelial colony forming cells; anticancer therapies; VEGF; resistance to apoptosis

---

1. Introduction

An increase in intracellular Ca^{2+} concentration ([Ca^{2+}]$_i$) has long been known to play a crucial role in angiogenesis and arterial remodeling [1–5]. Accordingly, growth factors and cytokines, such as vascular endothelial growth factor (VEGF), epidermal growth factor (EGF), basic fibroblast growth factor (bFGF), insulin-like growth factor-1 (IGF-1), angiopoietin and stromal derived factor-1α (SDF-1α), trigger robust Ca^{2+} signals in vascular endothelial cells [6–12], which recruit a number of downstream Ca^{2+}-dependent pro-angiogenic decoders. These include, but are not limited to, the transcription factors, Nuclear factor of activated T-cells (NFAT), Nuclear factor-kappaB (NF-κB) and cAMP responsive element binding protein (CREB) [8,13,14], myosin light chain kinase (MLCK) and myosin 2 [8,15], endothelial nitric oxide synthase (eNOS) [16,17], extracellular signal–regulated kinases $\frac{1}{2}$ (ERK 1/2) [18,19] and Akt [19,20]. Not surprisingly, therefore, subsequent studies clearly revealed that endothelial Ca^{2+} signals may also drive tumor angiogenesis, growth and metastasis [3,21–24]. However, the process of tumor vascularization is far more complex than originally envisaged [25]. Accordingly, the angiogenic switch, which is the initial step in the multistep process that ensures cancer cells with an adequate supply of oxygen and nutrients and provides them with an escape route to enter peripheral circulation, is triggered by the recruitment of bone marrow-derived endothelial progenitor cells (EPCs), according to a process termed vasculogenesis [26–28]. Similar to mature endothelial cells, EPCs require an increase in [Ca^{2+}]$_i$ to proliferate, assembly into capillary-like tubular networks in vitro and form patent neovessels in vivo [29–31]. Of note, intracellular Ca^{2+} signals finely regulate proliferation and in vitro tubulogenesis also in tumor-derived EPCs (T-EPCs) [23,32,33]. An established tenet of neoplastic transformation is the remodeling of the Ca^{2+} machinery in malignant cells, which contributes to the distinct hallmarks of cancer described by Hanahan and Weinberg [34–36]. Tumor endothelial cells (T-ECs) and T-EPCs do not derive from the malignant clone, but they display a dramatic dysregulation of their Ca^{2+} signaling toolkit [29,32,37]. The present article surveys the most recent updates on the remodeling of endothelial Ca^{2+} signals during tumor vascularization. In particular, it has been outlined which Ca^{2+}-permeable channels and Ca^{2+}-transporting systems are up- or down-regulated in T-ECs and T-EPCs and how they impact on neovessel formation and/or apoptosis resistance in the presence of anti-cancer drugs. Finally, the hypothesis that the remodeling of endothelial Ca^{2+} signals may be deeply involved in tumor resistance to standard therapeutic treatments, including chemotherapy, radiotherapy and anti-angiogenic therapy is widely discussed.

2. Ca^{2+} Signaling in Normal Endothelial Cells: A Brief Introduction

The resting $[Ca^{2+}]_i$ in vascular endothelial cells is set at around 100–200 nM by the concerted interaction of three Ca^{2+}-transporting systems, which extrude Ca^{2+} across the plasma membrane, such as the Plasma-Membrane Ca^{2+}-ATPase and the Na^+/Ca^{2+} exchanger (NCX), or sequester cytosolic Ca^{2+} into the endoplasmic reticulum (ER), the largest intracellular Ca^{2+} reservoir [2,38–40], such as the SarcoEndoplasmic Reticulum Ca^{2+}-ATPase (SERCA). Endothelial cells lie at the interface between the vascular wall and the underlying tissue; therefore, they are continuously exposed to a myriad of low levels soluble factors, including growth factors, hormones and transmitters, which may induce highly localized events of inositol-1,4,5-trisphosphate (InsP$_3$)-dependent Ca^{2+} release from the ER even in the absence of global cytosolic elevations in $[Ca^{2+}]_i$ [41–45]. These spontaneous InsP$_3$-dependent Ca^{2+} microdomains are redirected towards the mitochondrial matrix through the direct physical association specific components of the outer mitochondrial membrane (OMM) with specialized ER regions, which are known as mitochondrial-associated membranes (MAMs) [46]. This constitutive ER-to-mitochondria Ca^{2+} shuttle drives cellular bioenergetics by activating intramitochondrial Ca^{2+}-dependent dehydrogenases, such as pyruvate dehydrogenase, NAD-isocitrate dehydrogenase and oxoglutarate dehydrogenase [47–49]. This pro-survival Ca^{2+} transfer may be switched into a pro-death Ca^{2+} signal by various apoptotic stimuli [46,47,50]. For instance, hydrogen peroxide (H_2O_2), menadione, resveratrol, ceramide, and etoposide boost the InsP$_3$-dependent ER-to-mitochondria Ca^{2+} communication, thereby causing a massive increase in mitochondrial Ca^{2+} concentration ($[Ca^{2+}]_{mit}$), which ultimately results in the opening of mitochondrial permeability transition pore and in the release of pro-apoptotic factors into the cytosol [46,51–53]. The hypoxic microenvironment of a growing tumor may then trigger an oxygen (O_2)-sensitive transcriptional program in tumor cells by activating two basic helix-loop-helix transcription factors, i.e., the hypoxia-inducible factors HIF-1 and HIF-2, which drive the expression of a myriad of growth factors and cytokines [54]. These include, but are not limited to, VEGF, EGF, bFGF, IGF-1, angiopoietin and SDF-1α [27,54], which are liberated into peripheral circulation according to a concentration gradient, which delivers a strong pro-angiogenic signal to vascular endothelial cells residing in close proximity to the primary tumor site [3,33]. Growth factors bind to their specific tyrosine kinase receptors (TKRs), such as VEGFR-2 (KDR/Flk-1), EGFR (ErbB-1), and IGF-1R, thereby stimulating phospholipase Cγ (PLCγ) to cleave phosphatidylinositol 4,5-bisphosphate (PIP$_2$) into the two intracellular second messengers, InsP$_3$ and diacylglycerol (DAG) [1,2,55]. The following increase in cytosolic InsP$_3$ levels further stimulates ER-dependent Ca^{2+} release through InsP$_3$ receptors (InsP$_3$Rs), which can be amplified by the recruitment of adjoining ryanodine receptors (RyRs) through the process of Ca^{2+}-induced Ca^{2+} release (CICR) [1,2]. The following drop in ER Ca^{2+} concentration ($[Ca^{2+}]_{ER}$) is detected by Stromal Interaction Molecule 1 (Stim1), a sensor of ER Ca^{2+} levels, which is prompted to aggregate into oligomers and relocate towards ER-plasma membrane junctions, known as puncta and positioned in close vicinity to the plasma membrane (10–20 nm). Herein, Stim1 interacts with and gates the Ca^{2+}-permeable channel, Orai1, thereby triggering the so-called store-operated Ca^{2+} entry (SOCE), the most important Ca^{2+} entry route in endothelial cells [42,56–59]. In addition, Stim1 may recruit additional Ca^{2+}-permeable channels, which belong to the Canonical Transient Receptor Potential (TRPC) sub-family [2,59]. Accordingly, the TRP superfamily of cation channels comprises 28 members, subdivided into six sub-families: TRPC, TRPV (Vanilloid), TRPM (Melastatin), TRPP (Polycystin), TRPML (Mucolipin) and TRPA (Ankyrin) based on the homology of their amino acid sequences [60]. More specifically, endothelial SOCE could involve TRPC1 and TRPC4, which are recruited by Stim1 into a supermolecular heteromeric complex [61], whose Ca^{2+} selectivity is determined by Orai1 [62,63]. Moreover, TRPC3 and TRPC6 may mediate DAG-induced Ca^{2+} entry in several types of endothelial cells [64,65]. This toolkit of Ca^{2+} release/entry channels may be differently exploited by growth factors to stimulate angiogenesis by eliciting diverse patterns of Ca^{2+} signals depending on the vascular bed. For instance, VEGF triggers a biphasic increase in $[Ca^{2+}]_i$ in human umbilical vein endothelial cells (HUVECs), which consists in an initial $InsP_3$-dependent Ca^{2+} peak followed by a plateau phase of intermediate amplitude due to SOCE activation [57,58]. Likewise, $InsP_3$ and SOCE shape VEGF- and EGF-induced intracellular Ca^{2+} oscillations in sheep uterine artery endothelial cells [66] and in rat microvascular endothelial cells (CMECs) [7], respectively. VEGF-induced Ca^{2+} influx in HUVECs may, however, be sustained by TRPC3, which causes Na^+ accumulation beneath the plasma membrane and stimulates the forward (i.e., Ca^{2+} entry) mode of NCX [18]. Moreover, the DAG-gated channel, TRPC6, underlies the monotonic increase in $[Ca^{2+}]_i$ induced by VEGF in human dermal microvascular endothelial cells (HDMECs) [67]. Finally, TRPC1 is engaged by bFGF to mediate Ca^{2+} entry in HDMECs [68]. These data have been recently confirmed by directly monitoring angiogenesis in developing zebrafish; this model showed that VEGF stimulated biphasic Ca^{2+} signals to drive migration in stalk cells and intracellular Ca^{2+} oscillations to promote proliferation in tip cells [8]. Besides growth factors-activated channels, vascular endothelial cells dispose of a larger toolkit of plasmalemmal Ca^{2+}-permeable channels that can be recruited by a multitude of chemical and physical stimuli [2]. For instance, endothelial Ca^{2+} entry may be mediated by additional intracellular second messengers, such as arachidonic acid (AA) and AA metabolites, i.e., epoxyeicosatrienoic acids (EETs) and 2-arachidonoylglycerol, which activate TRPV4 [69,70]; NO, which gates TRPC5 [71]; adenosine 5′-diphosphoribose (ADPR) and low micromolorar doses of H_2O_2, which converge on TRPM2 activation [72]; and cyclic nucleotides [73]. Moreover, vascular endothelial cells are endowed with several Ca^{2+}-permeable ionotropic receptors, including ATP-sensitive P_{2X} receptors [74], acetylcholine-sensitive nicotinic receptors [75], and N-methyl-D-aspartate (NMDA) receptors [76]. Finally, mechanical stimuli (e.g., laminar shear stress, pulsatile stretch, and changes in the local osmotic pressure) elicit Ca^{2+} influx by recruiting a variety of mechano-sensitive channels, such as TRPP2 [77], heteromeric TRPC1-TRPP2 [78], TRPV4 [79], TRPC1-TRPP2-TRPV4 [80], and Piezo1 [81]. Recently, the Ca^{2+} toolkit has also been explored in human EPCs [29]; most of the work has been carried out in endothelial colony forming cells (ECFCs), which represent the only EPC subset truly belonging to the endothelial, rather than the myeloid, lineage [82]. VEGF triggers pro-angiogenic intracellular Ca^{2+} oscillations in ECFCs by triggering the interaction between $InsP_3$-dependent Ca^{2+} release and SOCE, which is mediated by Stim1, Orai1 and TRPC1 [83,84]. Conversely, RyRs and the DAG-sensitive channels, TRPC3 and TRPC6, are absent and do not contribute to Ca^{2+} signaling [84–86]. Of note, AA may promote proliferation by directly activating TRPV4 and inducing NO release in the presence of extracellular growth factors and cytokines [86]. Finally, the $InsP_3$-dependent ER-to-mitochondria Ca^{2+} shuttle is at work and finely regulates the sensitivity to apoptotic stimuli in ECFCs, too [87].

Herein, the mechanisms whereby the remodeling of the endothelial transportome, i.e., the specific arsenal of ion channels and transporters expressed by vascular endothelial cells and ECFCs, confers resistance to anti-cancer therapies have been subdivided into two main categories: (1) enhanced neovascularization, which attenuates the therapeutic outcome of anticancer treatments by nourishing cancer cells with O_2 and nutrients and removing their catabolic waste, and further provides them with a direct access to peripheral circulation, thereby favoring metastasis (Figure 1 and Table 1); and (2) resistance to apoptosis, which hampers the cellular stress induced by chemo- and radiotherapy on tumor endothelial cells and interferes with the dismantling of cancer vasculature (Figure 2 and Table 2).

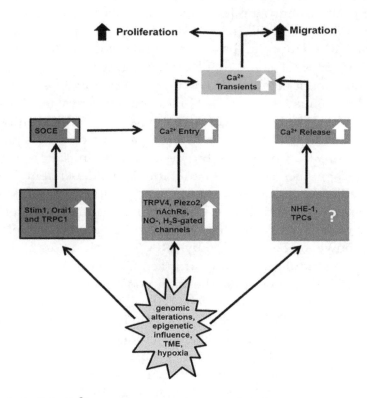

Figure 1. The endothelial Ca^{2+} transportome is remodeled to sustain tumor vascularization. The equence of events is illustrated by the black arrows. Upward arrows indicate the over-expression of a specific Ca^{2+}-permeable channel or transporter and the stimulation of a precise cellular process. See the text for further details.

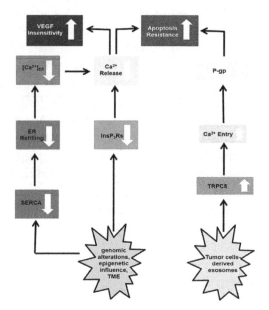

Figure 2. The endothelial Ca^{2+} transportome is remodeled to promote tumor endothelial cell resistance to apoptosis. The sequence of events is illustrated by the black arrows. Downward arrows indicate the down-regulation of a specific Ca^{2+}-permeable channel/transporter or of a precise cellular process. Upward arrows indicate the over-expression of a specific Ca^{2+}-permeable channel or the stimulation of a precise cellular process.

Table 1. Channels and transporters directly supporting tumor vascularization.

Channel/ Transporter	Tumor and Cell Type (T-EC, T-ECFC T-EPC)	Expression Levels (Transcripts and/or Proteins)	Effect on Tumor Vascularization	Strategy to Target Tumor Vascularization	Ref.
TRPV4	Breast Cancer: T-ECs	↑	Stimulates B-TEC proliferation, migration and in vitro tubulogenesis	Channel blockade with shTRPV4 or with CAI (0.1–10 μM)	[88,89]
TRPV4	Lewis Lung Carcinoma: T-ECs (isolated from prostate adenocarcinoma)	↓	Inhibits T-EC mechanosensation, proliferation and migration in vitro and promotes the formation of a malfunctioning, leaky and exceedingly expanded vascular network in vivo	Injection of TRPV4 agonist GSK (10 μg/kg) to normalize tumor vasculature and favor cisplatin-induced tumor regression	[90–92]
Piezo2 proteins	Glioma: T-ECs	↑	Regulates tumor angiogenesis, vascular leakage and permeability	Blockade with siPiezo2	[93]
P2X7Rs	Breast cancer: T-ECs	↑	Inhibits B-TEC migration and normalizes B-TECs-derived vessels in vitro	Activated by BzATP (50 μM)	[94]
Stim1, Orai1, TRPC1	Renal cellular carcinoma: T-ECFCs	↑	Stimulate T-EPC proliferation and in vitro tubulogenesis	Blockade with siStim1 and siOrai1 and with YM-58483/BTP2 (20 μM), La^{3+} (10 μM), Gd^{3+} (10 μM), CAI (2–10 μM), 2-APB (50 μM), and genistein (50 μM)	[95]
Stim1, Orai1, TRPC1	Breast cancer: T-ECFCs	=	Control T-ECFC proliferation and in vitro tubulogenesis	Blockade with YM-58483/BTP2 (20 μM), La^{3+} (10 μM), and CAI (10 μM)	[96]
Stim1, Orai1, TRPC1	Infantile hemangioma: T-ECFCs	↑	Control T-ECFCs proliferation in vitro	Blockade with with YM-58483/BTP2 (20 μM), La^{3+} (10 μM), and Pyr6 (10 μM)	[97]
α7-nAchRs	Lewis lung carcinoma: T-ECs and T-EPCs	Not determined	Controls tumor growth and angiogenesis in vivo	Blockade with mecamylamine (1.0 μg/kg) or hexamethonium (1.0 μg/kg)	[98,99]
			Stimulates EPC proliferation, migration and tubulogenesis in vitro and EPC recruitment in vivo	Blockade in vitro with mecamylamine (1 μM) and α-bungarotoxin (10 nM) and in vivo with mecamylamine (0.24 mg/kg per day)	[100,101]
Connexin40	Melanoma and urogenital cancers: T-EC	↑	Stimulates tumor angiogenesis and growth in vivo	Blockade in vivo with ^{40}Gap^{27} peptide (100 μg)	[102]
NHE-1	Breast cancer: TECs	Not determined	Stimulates B-TEC migration in vitro	Blocked with siNHE-1 and with cariporide (50 μM)	[103]

The generic term EPC, in this context, refers to circulating pro-angiogenic cells which cannot be grouped into the ECFC sub-family and are likely to belong to the myeloid lineage.

Table 2. Components of the endothelial Ca^{2+} toolkit that determine endothelial cell resistance to chemotherapeutic drugs.

Channel/ Transporter	Tumor and Cell Type (T-EC and T-EPC)	Expression Levels	Effect on Tumor Vascularization	Strategy to Target Tumor Vascularization	Ref.
TRPC5	Breast Cancer: T-ECs	↑	Stimulates endothelial resistance to adriamycin	Channel blockade with the specific blocking antibody T5E3 (concentration not reported)	[104]
InsP$_3$Rs	RCC: T-ECFCs	↓	Favor T-ECFC resistance to rapamycin	Preventing InsP$_3$-dependent ER-mitochondria Ca^{2+} shuttle with selective InsP$_3$R inhibitors or cytosolic Ca^{2+} buffers (e.g., BAPTA)	[87]

3. Enhanced Neovascularization

3.1. Vanilloid Transient Receptor Potential 4 (TRPV4)

TRPV4 has been the first endothelial Ca^{2+}-permeable channel to be clearly involved in tumor vascularization [88]. TRPV4 is gated by an array of chemical and physical cues and represents, therefore, the archetypal of polymodal TRP channels [60]. For instance, TRPV4 may be activated by physiological stimuli, including AA and its cytochrome P450-derived metabolites mediators, i.e., EETs, acidic pH, hypotonic swelling, mechanical deformation, heat (>17–24 °C), and dimethylallyl pyrophosphate (DMAPP) [105,106]. Furthermore, TRPV4-mediated Ca^{2+} entry is elicited by manifold synthetic compounds, including the α-phorbol esters, phorbol 12-myristate 13-acetate (PMA) and 4α-phorbol 12,13-didecanoate (4α PDD), and the small molecule drugs, GSK1016790A (GSK) and JNc-440 [60,107]. TRPV4 has long been known to stimulate angiogenesis and arteriogenesis [4,5,108] by stimulating endothelial cell proliferation [5,109] and migration [110]. TRPV4-mediated Ca^{2+} entry is translated into a pro-angiogenic signal by several decoders, such as the Ca^{2+}-dependent transcription factors NFAT cytoplasmic 1 (NFATc1), myocyte enhancer factor 2C (MEF2C), and Kv channel interacting protein 3, calsenilin (KCNIP3/CSEN/DREAM), which drive endothelial cell proliferation, [4], β1-integrin and phosphatidylinositol 3-kinase (PI3-K), which promote endothelial cell motility [111]. The opening of only few TRPV4 channels, that tend to assemble into a four-channel cluster, results in spatially-restricted cytosolic Ca^{2+} microdomains, known as Ca^{2+} sparklets, which selectively recruit the downstream Ca^{2+}-dependent effectors [112,113]. A recent study revealed that TRPV4 was dramatically up-regulated in breast tumor-derived endothelial cells (B-TECs) and that TRPV4-mediated Ca^{2+} entry significantly increased the rate of cell migration as compared to control cells [88]. TRPV4 promoted B-TEC motility by eliciting local Ca^{2+} pulses at the leading edge of migrating cells [88], which were reminiscent of TRPV4-dependent Ca^{2+} sparklets [112]. TRPV4 was physiologically gated by AA [89], which is quite abundant in breast cancer microenvironment [114]. Likewise, cytosolic phospholipase A2 (PLA2), which cleaves AA from membrane phospholipids in response to physiological stimuli [115] is up-regulated and promotes cancer development by stimulating angiogenesis in several types of tumors, including breast cancer [116]. Therefore, TRPV4 might represent a novel and specific target to treat breast cancer as it is only barely expressed and does not drive migration in healthy endothelial cells [88].

Subsequently, the role of TRPV4 was investigated in prostate adenocarcinoma-derived endothelial cells (A-TECs). Unlike B-TECs, TRPV4 was down-regulated in A-TECs, which increased their sensitivity towards extracellular matrix stiffness, boosted their migration rate and favored the development of an aberrant (i.e., non-uniform, abnormally dilated and leaky) tumor vascular network [90]. This feature gains therapeutic relevance as the resultant hostile (i.e., low extracellular pH, hypoxia, and high interstitial pressure) microenvironment fuels tumor progression and hampers the efficacy of chemotherapy, radiation therapy, anti-angiogenic therapy immunotherapy [117,118]. Accordingly, overexpression or pharmacological activation of TRPV4 with GSK restored A-TEC mechanosensitivity and normalized their abnormal tube formation in vitro by inhibiting enhanced basal Rho activity [91]. Moreover, the daily intraperitoneal injection of GSK was able to normalize tumor vasculature in a xenograft mouse model of Lewis Lung Carcinoma (LLC), thereby improving cisplatin delivery and causing significant tumor shrinkage [91]. In addition, TRPV4-mediated Ca^{2+} entry reduced A-TEC proliferation in vitro by inhibiting the extracellular signal-regulated kinases 1/2 [92]. This mechanism further contributes to GSK-induced dismantling of LLC vasculature in vivo [92]. Therefore, remodeling of TRPV4-mediated Ca^{2+} entry may be used to effectively target tumor vascularization, although the most effective approach may depend on the tumor type. Accordingly, TRPV4 should be inhibited to halt tumor vascularization in breast cancer, while it must be stimulated to normalize tumor vasculature in LLC [23].

3.2. Piezo Proteins

Piezo1 and Piezo2 proteins are two recently identified non-selective cation channels that mediate mechanosensory transduction in mammalian cells [119,120]. Piezo proteins are gigantic homotetrameric complexes endowed with one or four ion-conducting pores: each subunit comprises over 2500 amino acids and presents 24–40 predicted transmembrane domains [119]. Piezo channels are Ca^{2+}-permeable and, therefore, lead to robust Ca^{2+} entry in response to mechanical deformation of the plasma membrane; unlike TRPV4 channels [105], Piezo-mediated Ca^{2+} entry is directly activated by tension within the lipid bilayer of the plasma membrane rather than by physical coupling to the sub-membranal cytoskeleton or intracellular second messengers [120]. A recent study demonstrated that the endothelial Piezo1 was activated by laminal shear stress to drive embryonic vascular development [81]. Piezo1 promoted vascular endothelial cell migration, alignment and re-alignment along the direction of blood flow by engaging the Ca^{2+}-dependent decoders, eNOS and calpain [81,121]. More recently, Piezo2 was found to be up-regulated in T-ECs from mouse xenografted with GL261 glioma cells [93]. Knocking down Piezo2 with a selective small interfering RNA (siRNA) reduced glioma angiogenesis and normalized tumor neovessels [93]. Moreover, suppressing Piezo2 expression decreased VEGF- and interleukin-1β-induced angiogenesis in the mouse corneal neovascularization model [93]. Finally, Piezo2-mediated Ca^{2+} entry elicited the Ca^{2+}-dependent transcription of Wnt11 and, consequently, the nuclear translocation of β-catenin in HUVECs, thereby promoting their angiogenic activity in vitro [93]. Although this mechanism remains to be confirmed in T-ECs, Piezo2 stands out as a crucial regulator of tumor angiogenesis and should be probed as a novel target for more effective anti-cancer treatments.

3.3. P2X7 Receptors

ATP and its metabolite, adenosine, are major constituents of tumor microenvironment and may differently affect tumor growth, immune cells and tumor-host interaction by activating a wealth of metabotropic (i.e., P2Y1, P2Y2, P2Y4, P2Y6, P2Y11, P2Y12, P2Y13, and P2Y14) and ionotropic (P2X1–2X8,) receptors [122]. Of note, ATP has long been known to stimulate angiogenesis though metabotropic P2y receptors [123]. Nevertheless, a recent investigation demonstrated that P2X7 stimulates tumor angiogenesis in vivo. Two different tumor cell lines, i.e., HEK293 and CT26 colon carcinoma cells, were transfected with P2X7 receptors and subsequently xenografted into immunodeficient or immunocompetent BALB/c mice, respectively. Tumor growth and angiogenesis were significantly enhanced by P_{2X7} expression; consequently, pharmacological inhibition (with AZ10606120) or genetic silencing of P_{2X7} decreased tumor growth and dramatically reduced vascular density [124]. This study further confirmed that P2X7 receptors were significantly up-regulated in several types of cancer cells, including those from breast cancer, and stimulated angiogenesis by promoting VEGF release [124]. More recently, it was found that P2X7 receptors were over-expressed also in B-TECs [94]. This study revealed that the activation of these purinergic receptors with high doses of ATP (>20 μM) and BzATP, a selective P_{2X7} agonist, inhibited B-TEC, but not HDMEC, migration in vitro. The anti-angiogenic effect of P_{2X7} was mediated by the Ca^{2+}-sensitive adenylate cyclase 10 (AC10), which increased cyclic adenosine monophosphate (cAMP) and recruited EPAC-1 to dampen cell migration by inducing cytoskeletal remodeling [94]. Moreover, P2X7 receptors-induced cAMP production stabilized bidimensional tumor vessels by favoring pericyte attraction towards B-TECs and reducing endothelial permeability [94]. Intriguingly, hypoxia prevented the anti-angiogenic ability of P_{2X7} receptors by likely reducing their expression [94,125]. These data, therefore, strongly suggest that stimulating P2X7 receptors could provide an efficient strategy to normalize tumor vasculature, thereby enhancing the delivery of cytotoxic drugs and of O_2 for radiotherapy. In this context, it should be pointed out that P2X7 receptors target hematopoietic EPCs to glioblastoma [126]. Although this investigation was conducted on healthy cells, and remains therefore to be validated in T-EPCs, it suggests that ATP may differently affect tumor endothelial cells and T-EPCs. Alternatively, the effect exerted by P2X7 con T-EPC fate could be cancer-dependent and needs to be further investigated.

3.4. Stim1, Orai1 and Canonical Transient Receptor Potential 1 (TRPC1)

SOCE represents the most important Ca^{2+} entry pathway supporting the pro-angiogenic activity of human ECFCs [29,33,56]. Accordingly, TRPV4 boosted ECFC proliferation rate only when accompanied by the administration of a robust dose of growth factors [86,127], whereas TRPV1 stimulated ECFC proliferation and tubulogenesis by mediating the intracellular intake of anandamide in a Ca^{2+}-independent manner [128]. SOCE is activated by the pharmacological (by blocking SERCA-mediated Ca^{2+} sequestration) or physiological (by stimulating InsP$_3$Rs) depletion of the ER Ca^{2+} stores and is mediated by the dynamic interplay between Stim1, Orai1 and TRPC1 [29,56,58,84]. It is, however, still unknown whether Orai1 and TRPC1 form two independent Stim1-gated Ca^{2+}-permeable routes [129] or assemble into a unique heteromeric supermolecular complex in ECFCs [130]. A recent series of studies revealed that SOCE maintained VEGF-induced intracellular Ca^{2+} oscillations and promoted ECFC proliferation and in vitro tubulogenesis by recruiting the Ca^{2+}-dependent transcription factor, NF-κB [58,83]. Of note, SOCE was significantly enhanced in metastatic renal cellular carcinoma (RCC)-derived ECFCs (RCC-ECFCs) due to the up-regulation of Stim1, Orai1 and TRPC1 [95]. Similar to normal cells, the pharmacological blockade of SOCE with YM-58483/BTP2 or with low micromolar doses of lanthanides prevented proliferation and tube formation in RCC-ECFCs [95]. This finding strongly suggests that SOCE could provide an alternative target for the treatment of metastatic RCC [32,131], which develops either intrinsic or acquired refractoriness towards conventional treatments, such as anti-VEGF inhibitors and anti-mammalian target of rapamycin (mTOR) blockers [132]. As more extensively discussed below, the overall remodeling of the intracellular Ca^{2+} toolkit in T-ECFCs could indeed be responsible for the relative or complete failure of standard therapies in RCC patients. Conversely, SOCE was not significantly up-regulated in breast cancer-derived ECFCs (BC-ECFCs) [96]. Accordingly, Orai1 and TRPC1 expression were not significantly altered, while Stim1 was significantly more abundant as compared to control cells. Nevertheless, a tight stoichiometric ratio between Stim1, Orai1 and TRPC1 is required for SOCE to be activated [133]. If all Orai1 and TRPC1 channel proteins are gated by the physiological levels of Stim1, any increase in Stim1 expression will not be sufficient to enhance SOCE as there will be no further channels available on the plasma membrane. Similar to RCC-ECFCs, however, the pharmacological inhibition of SOCE abrogated BC-ECFC proliferation and tube formation, thereby confirming that Orai1 and TRPC1 could serve as reliable targets to interfere with tumor vascularization, although this hypothesis remains to be validated in vivo [134,135]. The strict requirement of Stim1 for tumor vascularization is further suggested by the recent finding that Stim1 transcription in hypoxic tumors is finely regulated by HIF-1 [136]. SOCE, in turn, was found to stimulate HIF-1 accumulation in hypoxic cancer cells by engaging Ca^{2+}/calmodulin-dependent protein kinase II and p300 [136]. Therefore, targeting SOCE could also affect the expression of the primary transcription factor responsible for RCC and breast cancer growth and metastasis [137,138]. Of note, HIF-1 has been shown to control also TRPC1 expression [139], although it is still unclear whether this regulation also occurs in tumor microenvironment and, if so, why TRPC1 is up-regulated in RCC-ECFCs, but not in RCC-ECFCs.

The role played by SOCE in tumor vascularization has, finally, been uncovered also in infantile hemangioma (IH), the most common childhood malignancy which may cause disfigurement, ulceration and obstruction and, if not treated, ultimately leads to patients' death [140]. IH is a vascular tumor that arises as a consequence of dysregulation of angiogenesis and vasculogenesis [140]. The clonal expansion of an endothelial progenitor/stem cell population, which is closely reminiscent of ECFCs, is deeply involved in IH vascularization [141,142]. A recent investigation provided the evidence that Stim1, Orai1 and TRPC1 drive the higher rate of IH-derived ECFC (IH-ECFC) growth as compared to control cells [97]. Stim1, Orai1 and TRPC1 were not up-regulated in IH-ECFCs; however, the ER Ca^{2+} store was depleted to such an extent that Stim1 was basally activated and gated the constitutive activation of Orai1 and TRPC1 [97]. Stim2 displays a lower Ca^{2+} affinity as respect to Stim1 and supports basal Ca^{2+} entry in HUVECs [143]. Nevertheless, the pharmacological abrogation of Stim2 silencing did not affect constitutive SOCE in IH-ECFCs [97]. Constitutive SOCE boosted IH-ECFC proliferation by enhancing NO release [97], thereby emerging as an alternative target to treat IH in propranolol-resistant patients [144].

3.5. Neuronal Nicotinic Receptors (nAchRs)

nAchRs belong to a super-family of Cys-loop ligand-gated non-selective cation channels that are physiologically activated by acetylcholine, mediate fast synaptic transmission in neurons and, by virtue of their resolvable Ca^{2+}-permeability, control a number of Ca^{2+}-dependent processes, including neurotransmitter release and synaptic plasticity [145,146]. However, nAchRs are also largely expressed in non-neuronal brain cells, such as astrocytes, in epithelial cells and in several types of vascular cells, including smooth muscle cells and endothelial cells [75,147–149]. It has been established that $\alpha 7$ homomeric nAchRs ($\alpha 7$-nAchRs) promote endothelial cell proliferation, migration and tube formation both in vitro and in vivo by recruiting an array of Ca^{2+}-dependent effectors [98,99]. These include eNOS, mitogen-activated protein kinase, phosphoinositide 3-kinase (PI3K), NF-κB, matrix metalloproteinase-2 and -9 [98,99,150]. In addition, $\alpha 7$-nAChRs were shown to induce the JAK2/STAT3 signaling cascade to promote endothelial cell survival [151]. Intriguingly, $\alpha 7$-nAchRs possess the highest Ca^{2+}-permeability among the known nAchR subtypes [152]. These pieces of evidence ignited the hypothesis that nicotine accelerated tumor growth by stimulating endothelial $\alpha 7$-nAchRs, thereby promoting angiogenesis and tumor vascularization [147,149]. In support of this model, nicotine induced tumor growth in a mouse model of LLC by stimulating endothelial cell proliferation and tube formation. Nicotine-induced tumor vascularization was significantly reduced by pharmacological blockade (with mecamylamine or hexamethonium) as well as by genetic silencing of $\alpha 7$-nAChRs. The signaling pathways recruited by $\alpha 7$-nAChRs to sustain tumor angiogenesis were not deeply investigated, but nicotine stimulated endothelial cells to release NO, prostacyclin and VEGF [98,99]. It should, however, be pointed out that the expression and role of $\alpha 7$-nAChRs in T-ECs has not been investigated, yet. Nevertheless, hypoxia has been shown to increase $\alpha 7$-nAChRs expression in a mouse model of hindlimb ischemia [98], whereas $\alpha 7$-nAChRs may stimulate HIF-1αtranscription [153]. These observations support the hypothesis that $\alpha 7$-nAChRs are actually over-expressed in T-ECs.

In addition to promoting angiogenesis, nicotine could recruit $\alpha 7$-nAChRs to boost vasculogenesis. A recent study revealed that nicotine induced proliferation, migration and tube formation also in ECFCs and that this effect was inhibited by mecamylamine or α-bungarotoxin [100]. Moreover, nicotine triggered EPC mobilization from bone marrow in a cohort of mice xenografted with colorectal cancer cells, thereby fostering tumor growth and vascularization [101]. Lastly, exposure to second hand smoke stimulated tumor angiogenesis and increased the number of circulating EPCs in a mouse model of LLC by enhancing VEGF release: mecamylamine, however, halted VEGF release, thereby reducing tumor size and capillary density. The pro-angiogenic effect of nicotine was, therefore, likely to be mediated by nAchRs [154]. We are yet to know whether and how $\alpha 7$-nAChRs are altered in T-ECs and T-EPCs. Nevertheless, these ionotropic receptors could be regarded as a promising target for alternative anti-angiogenic therapies.

3.6. Gasotransmitters-Activated Ca^{2+}-Permeable Channels

Gaseous mediators or gasotransmitters are endogenous signaling messengers that, although being toxic at high concentrations, regulate a multitude of physiological processes, ranging from the regulation of vascular tone to synaptic plasticity and mitochondrial bioenergetics [155–158]. The gasotransmitters NO and hydrogen sulphide (H_2S) have recently been shown to stimulate endothelial cells through an increase in $[Ca^{2+}]_i$ [6,156,159], while the role of CO in angiogenesis is less clear [160]. NO promotes angiogenesis and disease progression in several types of malignancies [161,162], including breast cancer [163]. The administration of two structurally unrelated NO donors, i.e., S-nitroso-N-acetylpenicillamine (SNAP) or sodium nitroprusside (SNP), was recently found to trigger Ca^{2+} influx and migration in B-TECs [164]. These effects were mimicked by elevating endogenous NO release with L-arginine [164], which is the physiological substrate for eNOS [156]. Of note, AA-induced TRPV4 activation in B-TECs was inhibited by preventing NO production with N^G-nitro-L-arginine methyl ester (L-NAME) [88,164]; moreover, AA- and NO-induced Ca^{2+} entry were both sensitive to protein kinase A (PKA) inhibition [164]. It is, therefore, likely that NO elicits Ca^{2+} entry in B-TECs by gating TRPV4. In agreement with this hypothesis, TRPV4 may be activated by NO through direct S-nitrosylation [71] and is phosphorylated by PKA upon AA stimulation in vascular endothelial cells [70]. Finally, NO-induced Ca^{2+} entry and migration were dramatically reduced in HDMECs [164], in which TRPV4 expression was significantly down-regulated [88]. Besides TRPV4, however, NO is able to recruit multiple TRP channels, such as TRPC1, TRPC4, TRPC5, TRPV1, and TRPV3 [71], some of which are up-regulated in T-ECFCs [95,165]. Unfortunately, it is still unclear whether NO elicits intracellular Ca^{2+} entry in these cells. Although future work is mandatory to understand whether NO stimulates TRP channels, as well as other Ca^{2+}-permeable channels, to promote tumor vascularization, endothelial Ca^{2+} signaling is emerging as an attractive target to prevent its pro-tumorigenic effect.

H_2S has also been shown to promote angiogenesis in a Ca^{2+}-dependent manner. For instance, H_2S mediated VEGF-induced Ea.hy926 cell proliferation and migration by inducing $InsP_3$-dependent ER Ca^{2+} release without the contribution of extracellular Ca^{2+} entry [6]. The components of the endothelial Ca^{2+} toolkit recruited by H_2S may, however, vary depending on the vascular bed [166]. H_2S induced ER-dependent Ca^{2+} release through $InsP_3Rs$ and RyRs followed by a sustained SOCE in primary cultures of human saphenous vein endothelial cells [167], whereas it recruited the reverse mode of NCX by gating a Na^+- and Ca^{2+}-permeable pathway in rat aortic endothelial cells [168] and HDMECs [21]. Conversely, NaHS did not elicit any resolvable elevation in $[Ca^{2+}]_i$ in ECFCs [6] and its role in neovasculogenesis in vivo operated by truly endothelial precursors remains to be elucidated [169]. H_2S-induced Ca^{2+} signals may be translated into a pro-angiogenic signal by multiple Ca^{2+}-dependent decoders, including the PI3K/Akt and the ERK/p38 signaling pathways [155,156]. Growing evidence demonstrated that H_2S drove disease progression and angiogenesis in several types of tumor, such as RCC and colorectal cancer [169]. Intriguingly, sodium hydrosulfide (NaHS), a widely employed H_2S donor, induced intracellular Ca^{2+} signals in both B-TECs and HDMECs; however, NaHS-elicited Ca^{2+} signals were enhanced and arose within a significantly lower range (nanomolar vs. micromolar) in B-TECs [21]. Consequently, NaHS promoted proliferation and migration in B-TECs, but not in control endothelial cells [21]. The Ca^{2+} response to H_2S was mediated by a Ca^{2+}-permeable non-selective cation channel and was sustained by membrane hyperpolarization through the activation of a K^+ conductance [21], likely an ATP-sensitive K^+ channel [157]. The molecular nature of this Ca^{2+}-permeable route is yet to be identified [166]. Nevertheless, H_2S is able to stimulate TRPV3 and TRPV6 in bone marrow-derived mesenchymal cells by direct sulfhydration of some of Cys residues within their protein structure [170]. Moreover, H_2S activated TRPA1 in RIN14B cells [171]. Deciphering the molecular target of H_2S in tumor endothelium is, therefore, mandatory to devise alternative anti-cancer treatments. In addition, both eNOS and cys-tathionine gamma lyase (CSE), the enzyme which catalyzes H_2S production in vascular endothelial cells, are Ca^{2+}-sensitive [16,42,172]. Therefore, targeting a Ca^{2+} entry/release pathway tightly coupled to either eNOS (i.e., Orai1, [173]) or CSE (yet to be identified) has the potential to interfere with multiple pro-angiogenic pathways and, therefore, exert a more profound anti-tumor effect.

3.7. Connexin 40 (Cx40)

Connexin (Cx) hemichannels, also termed connexons, have long been known to provide the building blocks of gap junctions, thereby enabling the transfer of small solutes, ions and signaling molecules, such as Ca^{2+} and $InsP_3$, between adjacent cells [174]. Three diverse Cx isoform exist in vascular endothelial cells, i.e., Cx37, Cx40, and Cx43, and synchronize robust NO release induced by extracellular autacoids by mediating intercellular Ca^{2+} communication [175,176]. In addition, unopposed Cx hemichannels were found to mediate extracellular Ca^{2+} entry and NO release in endothelial cells from different vascular beds [16,74,177–179]. Earlier work suggested that Cxs served as tumor suppressors and were down-regulated in cancer, thereby affecting vascular integrity and reducing vascular leakage [180–182]. However, a recent study challenged this dogma by showing that Cx40 was over-expressed in T-ECs and promoted disease progression and angiogenesis in melanoma and urogenital cancers [102]. Cx40 stimulated tumor growth by inducing eNOS recruitment, which strongly suggest that intracellular Ca^{2+} levels increased during the angiogenic process [102]. Intriguingly, targeting Cx40 function with [40]Gap27, a peptide that has long been use to inhibit Cx40-mediated intercellular communication and extracellular Ca^{2+} entry [1,16,178], normalized tumor vasculature and enhanced the efficacy of the chemotherapeutic drug, cyclophosphamide [102]. Therefore, although these findings remain to be confirmed in other tumor types, and the role served by Ca^{2+} is still unclear, Cx40 deserves careful consideration for the design of new anticancer drugs.

3.8. Na^+/H^+ Exchanger-1 (NHE-1)

The Na^+/H^+ exchanger NHE-1 is a reversible electroneutral antiporter that maintains cytosolic pH by expelling H^+ at expense of the inwardly directed Na^+ electrochemical gradient with a 1:1 stoichiometric ratio [183]. NHE-1 induces endothelial cell proliferation, migration and tube formation by means of several Ca^{2+}-dependent effectors, such as calpain [184], eNOS [185], and ERK 1/2 [186]. Accordingly, thrombin-induced NHE-1 activation was able to increase sub-membranal Na^+ levels, thereby switching NCX into the reverse mode and mediating extracellular Ca^{2+} entry in HUVECs [187,188]. Moreover, NHE-1-induced cytosolic alkalinization triggered ER-dependent Ca^{2+} release through InsP3Rs in bovine aortic endothelial cells and human pulmonary artery endothelial cells [189,190]. NHE1 is constitutively activated in cancer cells to favor extracellular acidification and stimulate metastasis and invasion by facilitating protease-mediated degradation of the extracellular matrix [191]. In addition, NHE-1 is transcriptionally regulated by HIF-1 and is up-regulated in a multitude of carcinomas [191,192]. A recent series of studies demonstrated that NHE-1 was over-expressed in endothelial cells exposed to tumor microenvironment [193] and was able to boost vascularization, invasion and metastasis in several types of tumors, including breast cancer [191,194]. Accordingly, aldosterone-induced NHE-1 activation promoted B-TEC proliferation, migration and cytosolic alkalinization [103]. Further work is required to assess whether NHE-1 activation stimulates tumor vascularization through an increase in $[Ca^{2+}]_i$. However, NHE-1 blockers, including cariporide and the more specific 3-methyl-4-flouro analog of 5-aryl-4(4-(5-methyl-14-imidazol-4-yl) piperidin-1-yl)pyrimidine (Compound 9t), have been put forward as alternative anti-cancer drugs [195].

3.9. Two-Pore Channels (TPCs)

The ER is the largest endogenous Ca^{2+} store in vascular endothelial cells by accounting for ≈75% of the total Ca^{2+} storage capacity [2]. The remainder 25% of the total stored Ca^{2+} is located within the mitochondria and the acidic Ca^{2+} stores of the endolysosomal (EL) system [2]. As more widely illustrated in [196], the EL Ca^{2+} store releases Ca^{2+} through many Ca^{2+}-permeable channels, including Mucolipin TRP 1 (TRPML1), Melastatin TRPM 2 (TRPM2) and two-pore channels 1 and 2 (TPC1–2) [197,198]. The newly discovered second messenger, nicotinic acid adenine dinucleotide phosphate (NAADP), is the physiological stimulus that gates TPC1–2 in response to extracellular stimulation [196,199,200]. NAADP-induced EL Ca^{2+} release is, in turn, amplified by juxtaposed ER-embedded InsP3Rs and RyRs through the CICR process, thereby initiating a regenerative Ca^{2+} wave [196,200,201]. NAADP-gated TPC2 channels are also expressed in vascular ECs [202,203], whereas N-ECFCs display larger amounts of TPC1 [86,204]. A recent study demonstrated that NAADP-induced Ca^{2+} signals promoted tumor vascularization and metastasis in murine models xenografted with B16 melanoma cells [205]. Of note, the pharmacological blockade of NAADP-induced Ca^{2+} release with Ned-19 dampened melanoma growth, vascularization and lung metastasis [205]. Future work will have to assess whether TPC2 channels are up-regulated in T-ECs and whether

NAADP-induced Ca^{2+} signaling also drive T-ECFC incorporation into tumor neovessels. However, TPCs stand out as promising targets to develop alternative anti-angiogenic treatments.

4. Resistance to Apoptosis

4.1. Canonical Transient Receptor Potential 5 (TRPC)

TRPC5 forms a homotetrameric Ca^{2+}-permeable channel that is gated upon PLCβ activation by Gq/11-coupled membrane receptors through a yet to be identified signaling cascade [2,206,207]. Accordingly, although some studies reported that TRPC5 is recruited in a store-dependent manner by Stim1 [208], it has been proposed that TRPC5 activation by PLCβ does not involve ER store depletion [209]. In addition, TRPC5-mediated Ca^{2+} entry is elicited by several physiological messengers, including reduced thioredoxin, protons, sphingosine-1-phosphate, lysophospholipids, NO and Ca^{2+} itself [207]. Finally, TRPC5 presents a spontaneous activity that is increased by lanthanides, cold temperatures (47 °C to 25 °C) and membrane stretch; consequently, TRPC5 serves as a cold sensor in the peripheral nervous system [210]. Of note, TRPC5 may establish physical associations with a multitude of molecular partners, including TRPC1, TRPC4 and TRPC6, which regulate its membrane localization and biophysical properties [207]. TRPC5 differently tunes angiogenesis depending on the vascular bed. For instance, TRPC5 promoted proliferation and tube formation by inducing intracellular Ca^{2+} oscillations in EA.hy926 cells [211]. Conversely, a TRPC6-TRPC5 channel interaction inhibited angiogenesis by decreasing the rate of migration in bovine aortic ECs (BOECs). In this context, TRPC6-mediated Ca^{2+} entry triggered an ERK-mediated phosphorylation cascade that leads to MLCK activation and TRPC5 externalization on the plasma membrane [212,213]. It has recently been shown that endothelial TRPC5 could underlie the development of chemoresistance to anticancer drugs in both breast cancer [104,214,215] and colorectal carcinoma [216]. P-glycoprotein (P-gp), also termed multidrug resistance protein 1 (MDR1), is a multidrug efflux transporter that expels xenobiotics out from the cytoplasm into the extracellular milieu [217]. P-gp overexpression, therefore, confers resistance to malignant cells, which become insensitive to a wide range of cancer chemotherapeutics, including adriamycin, vincristine, taxol, and anthracyclines [217]. Earlier evidence demonstrated that TRPC5 was up-regulated and induced P-gp overexpression by hyper-stimulating the Ca^{2+}-dependent transcription factor, NFATc3 in chemoresistant MCF-7 breast cancer cells [214]. In agreement with this observation, microRNA 320a (miR-320a), which is able to associate with and degrade TRPC5 and NFATc3 transcripts in normal cells, was down-regulated in chemoresistant breast cancer cells due to the hypermethylation of its promoter sequence [218]. TRPC5 up-regulation induced resistance to adriamycin, paclixatel, epirubicin, mitoxantrone and vincristine [214]. Additionally, TRPC5-mediated Ca^{2+} entry promoted transcription of HIF-1α gene, thereby boosting VEGF release and enhancing tumor angiogenesis [219]. Remarkably, TRPC5-based chemoresistance could be shuttled to tumor endothelial via intercellular communication. Adriamycin-resistant MCF-7 cells could pack the up-regulated TRPC5 channels into mobile extracellular vesicles (EVs), which are released in tumor microenvironment and transferred their signaling content to surrounding endothelial cells. This scenario is supported by the observations that HDMECs exposed to TRPC5-containing EVs, which were collected from adriamycin MCF-7 breast cancer cells, over-expressed the TRPC3-NFATc3-P-gp signaling pathway and developed resistance to adriamycin-induced apoptosis [104]. Moreover, TRPC5-containing vesicles were identified in peripheral blood of breast cancer patients receiving chemotherapy and of nude mice bearing adriamycin-resistant MCF-7 tumor xenografts [215]. Furthermore, P-gp production was enriched in tumor endothelium of adriamycin-resistant MCF-7 xenografts than in other sites and was sensitive to TRPC5 inhibition with a specific siRNA (siTRPC5) [104]. These data, therefore, suggest that TRPC5 provide a promising target to design alternative adjuvant anticancer treatments [220]. Accordingly, a blocking TRPC5 antibody reduced P-gp expression, retarded cancer growth and boosted paclitaxel-induced tumor regression in chemoresistant breast cancer in vivo [104,214,215]. The endothelial effects of TRPC5 in breast cancer are seemingly limited to T-ECs, as BC-ECFCs do not express this channel [96]. Future work will have to assess whether, besides conferring B-TECs with the resistance to chemotherapeutic drugs, TRPC5 up-regulation accelerates breast cancer angiogenesis.

4.2. Inositol-1,4,5-Trisphosphate (InsP₃) Receptors (InsP₃Rs)

InsP$_3$Rs are non-selective cation channels which mediate ER-dependent Ca^{2+} release, thereby controlling multiple endothelial cell functions, including bioenergetics, apoptosis, angiogenesis and vasculogenesis (see Paragraph 2. Ca^{2+} signaling in normal endothelial cells: a brief introduction). Three distinct InsP$_3$R isoforms exist in both vascular endothelial cells and ECFCs [2,29], i.e., InsP$_3$R1, InsP$_3$R2 and InsP$_3$R3, which may associate into homo- or hetero-tetrameric ER-embedded channels [2]. It has recently been shown that InsP$_3$Rs were dramatically down-regulated in RCC-ECFCs, thereby preventing the onset of VEGF-induced intracellular Ca^{2+} oscillations, proliferation and in vitro tubulogenesis [95] (Figure 3). More specifically, RCC-ECFCs only expressed InsP$_3$R1, while InsP$_3$R2 and InsP$_3$R3 were absent [95]. This result was surprising as InsP$_3$R1 was transcriptionally regulated by HIF-2 in human RCC cancer cell lines [221]. The failure of the pro-angiogenic Ca^{2+} response to VEGF also involves the chronic reduction in the ER Ca^{2+} concentration ([Ca^{2+}]$_{ER}$) in RCC-ECFCs, as monitored by using an ER-targeted aequorin Ca^{2+} indicator [87,222]. Therefore, in contrast with the widely accepted belief that VEGF sustains the angiogenic switch [223], VEGF does not stimulate ECFC-dependent neovessel formation in RCC patients [27,32]. This observation shed novel light on the refractoriness to anti-VEGF therapies in individuals suffering from RCC [224–226]. It has long been known that humanized monoclonal anti-VEGF antibodies, such as bevacizumab, or small molecule tyrosine kinase inhibitors, such as sorafenib and sunitinib, did not increase the overall survival of RCC patients, who ultimately developed secondary (acquired) resistance and succumbed because of tumor relapse and metastasis. In addition, targeting VEGF-dependent pathway proved to be ineffective in a large cohort of subjects, who displayed intrinsic refractoriness to these anti-VEGF drugs and did not show any improvement in their progression free survival [227,228]. ECFCs play a key role during the early phases of the angiogenic switch that supports tumor vascularization and metastasis [27,28,229]. If tumor vasculature is dismantled by anti-VEGF drugs, the following drop in P$_{O2}$ will release in circulation a cytokine storm that attracts ECFCs from their bone marrow and/or vascular niches. ECFCs will home to the shrunk tumor, but, being insensitive to VEGF, will not be affected by the presence of anti-VEGF drugs. Consequently, they will proliferate in response to the mixture of growth factors liberated in tumor microenvironment and will restore blood supply to cancer cells [32]. Remodeling of the Ca^{2+} toolkit in RCC-ECFCs could, therefore, underlie the resistance to anti-angiogenic therapies in RCC patients. Similar data were obtained in BC-ECFCs, in which the significant reduction in [Ca^{2+}]$_{ER}$ prevented VEGF from triggering robust intracellular Ca^{2+} oscillations, proliferation and tube formation, although the pattern of InsP$_3$R expression remained unchanged [37,96]. Again, this result is consistent with notion that also breast cancer patients present intrinsic or secondary refractoriness to anti-VEGF therapies [54,230]. The reduction in [Ca^{2+}]$_{ER}$ observed in several types of tumor-associated ECFCs, including IH-ECFCs [95–97], was likely to reflect the down-regulation of SERCA2B activity [87,95]. Accordingly, ATP-induced InsP$_3$-dependent ER Ca^{2+} release in RCC-ECFCs decayed to resting Ca^{2+} levels with slower kinetics as compared to normal ECFCs [95]. Intriguingly, the gene expression profile of RCC- and BC-ECFCs resulted to be dramatically different with respect to normal cells [37]: BC-ECFCs and RCC-ECFCs presented, respectively, 382 and 71 differently expressed genes (DEGs) as compared to healthy cells, including TMTC1 [37]. TMTC1 is a tetratricopeptide repeat-containing adapter protein, which binds to and inhibits SERCA2B, thereby reducing ER Ca^{2+} levels and dampening agonist-induced intracellular Ca^{2+} release [231]. It is conceivable that TMTC1 up-regulation in T-ECFCs contributes to the chronic underfilling of their ER Ca^{2+} reservoir. In further agreement with this observation, electron microscopy revealed that both RCC- and BC-ECFCs presented dramatic ultrastructural differences as compared to control cells [87,96]. In particular, T-ECFCs presented a remarkable expansion of ER volume, whereas mitochondria were more abundant and very often elongated as compared to N-ECFCs [87,96]. This ultrastructural remodeling is consistent with the ER stress caused by the chronic reduction in $[Ca^{2+}]_{ER}$ [232,233].

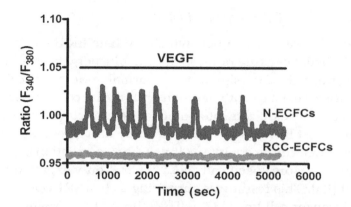

Figure 3. VEGF does not trigger pro-angiogenic Ca^{2+} oscillations in tumor-derived endothelial colony forming cells. VEGF (10 ng/mL) triggers intracellular Ca^{2+} oscillations in N-ECFCs, but not in RCC-ECFCs. Adapted from [95].

5. Targeting the Endothelial Ca^{2+} Toolkit to Circumvent the Resistance to Anticancer Treatments

Remodeling of the Ca^{2+} toolkit in tumor cells led many authors to search for alternative strategies to treat cancer. Intracellular Ca^{2+} signaling controls most, if not all, the so-called cancer hallmarks and could, therefore, be targeted to inhibit or, at least, retard tumor growth and metastasis [35,36,234–240]. As shown above, the Ca^{2+} transportome is also altered in stromal cancer cells [32,241,242], including endothelial cells and ECFCs. Remodeling of the endothelial Ca^{2+} toolkit could play a crucial role in the refractoriness to anticancer treatments, by supporting tumor vascularization and decreasing the susceptibility to pro-apoptotic stimuli. Therefore, the endothelial Ca^{2+} transportome might provide an efficient target for adjuvant therapies to conventional anti-cancer treatments. Three strategies could be pursued to improve the therapeutic outcome of standard therapies by interfering with the endothelial Ca^{2+} machinery: (1) blocking Ca^{2+} signaling to dampen angiogenesis and/or vasculogenesis; (2) stimulating Ca^{2+} entry to normalize tumor vessels, thereby improving the delivery and efficacy of chemo-, radio- and immunotherapy; and (3) manipulating Ca^{2+} signaling to endothelial cell apoptosis and dismantle tumor vasculature.

SOCE is, perhaps, the most suitable target to affect tumor vasculature by inhibiting both angiogenesis and vasculogenesis. Although there is no report of SOCE expression in T-ECs, Stim1 and Orai1 control proliferation and tube formation in normal endothelial cells, such as rat CMECs [7], bovine brain capillary endothelial cells [243], mouse lymphatic endothelial cells [244], and HUVEC [57,58]. Moreover, the pharmacological blockade of SOCE attenuates the rate of cell growth and abrogates in vitro tubulogenesis in RCC-, IH- and BC-derived ECFCs [95–97]. In addition, SOCE controls proliferation, migration and metastasis in a multitude of different cancer cell lines [134,135], which expands the cellular targets of SOCE inhibitors to the whole tumor microenvironment. We [131,135] and others [133,245] have recently described the Orai1 and TRPC1 inhibitors, some of which have been listed in Table 1, that could serve as a molecular template to design novel anticancer drugs. Unfortunately, none of these drugs have reached the milestone of being approved by US Food and Drug Administration (FDA) due to their scarce selectivity and high toxicity. For instance, carboxyamidotriazole (CAI), a non-selective blocker of Ca^{2+} signaling, was originally used to inhibit angiogenesis in vitro and tumor vascularization in vivo [131,246–248]. Depending on the cell type, CAI was able to block SOCE by occluding the mitochondrial Ca^{2+} uniporter [249,250] or reducing $InsP_3$ synthesis, which in turn prevents $InsP_3$-dependent Ca^{2+} release and Stim activation [56,251,252]. Intriguingly, CAI inhibited proliferation and tube formation also in RCC- and BC-ECFCs by preventing

InsP$_3$-dependent ER Ca^{2+} release [95,96]. Additionally, CAI was found to block growth and motility in several types of cancer cell lines [252,253]. Therefore, phase I-III clinical trials were launched to assess CAI toxicity and tolerability in patients suffering from several types of malignancies, including RCC, breast cancer, ovarian cancer, melanoma, non-small cell lung carcinoma, and gastrointestinal (stomach and pancreas) adenocarcinomas [56,248,254,255]. As discussed elsewhere [131], this drug caused disease stabilization when administrated alone or as adjuvant of chemo- or radio-therapy, and induced well tolerable side effects in most patients, such diarrhea, nausea and/or vomiting, fatigue and constipation. The therapy was discontinued only in RCC patients, who underwent disease progression and suffered from unacceptable toxicities, such as neuropsychiatric difficulties and asthenia [256]. As mentioned earlier, however, the effect of CAI is not directed towards the SOCE machinery, but is indirect. In addition to SOCE, CAI may also block TRPV4 and ER leakage channels [89,95,96,131]. A recent investigation, however, screened a library of >1800 FDA-approved drugs to search for specific SOCE blockers and identified five novel compounds, i.e., leflunomide, teriflunomide, lansoprazole, tolvaptan and roflumilast, that could be successfully used in therapy (leflunomide and terifluonomide) or provide the template to design more selective Orai1 inhibitors (i.e., lansoprazole, tolvaptan and roflumilast) [257].

An alternative strategy consists in stimulating endothelial Ca^{2+} signaling to induce tumor normalization by activating distinct Ca^{2+} entry routes depending on the tumor type. For instance, TRPV4-mediated Ca^{2+} entry drives tumor normalization in LLC [90,91], whereas P$_{2X7}$ receptors could be targeted to normalize tumor vasculature in breast cancer [94]. Tumor normalization, in turn, represents a promising adjuvant approach to facilitate cancer therapy by increasing the diffusion of chemotherapeutic drugs, improving radiotherapy efficiency and favoring the recruitment of tumor-killing immune cells [118,258]. Several synthetic agonists may selectively induce TRPV4 opening, such as 4αPDD derivatives, RN-1747, and JNc-440. Moreover, GlaxoSmithKline commercialized several patent applications of small molecule TRPV4 activators, the most famous of which is GSK [105,259]. Likewise, BzATP is regarded as the most potent P$_{2X7}$ receptor agonist, while 2-meSATP and ATPγS are only partial agonists and $\alpha\beta$-meATP and $\beta\gamma$-meATP exert a rather weaker on activation [260]. Clearly, further studies are required to uncover additional components of the endothelial Ca^{2+} toolkit potentially implicated in tumor normalization. Nevertheless, a recent investigation reported that angiopoietins, which induce vessel maturation by regulating the interaction between luminal endothelial cells and mural cells, such as vascular smooth muscle cells and pericytes, stimulate HUVEC migration by promoting ER-dependent Ca^{2+} release through InsP$_3$Rs and RyRs [261]. These findings lend further support to the hypothesis that targeting the endothelial Ca^{2+} signaling provides a suitable means to accelerate the dismantling of tumor vasculature by standard anticancer therapies.

Finally, the endothelial Ca^{2+} machinery could be properly manipulated to enhance the pro-apoptotic outcome of chemo- and radiation-therapy. For instance, TRPC5-mediated Ca^{2+} entry could be inhibited in B-TECs by taking advantage of a battery of novel small molecule inhibitors, such as Pico145 [262], 3,5,7-trihydroxy-2-(2-bromophenyl)-4H-chromen-4-one (AM12) [263], 2-aminobenzimidazole derivatives [264], ML204 [265], and neuroactive steroids [266]. Alternatively, the [Ca^{2+}]$_{ER}$ could be augmented to such an extent to induce the pro-apoptotic InsP$_3$-driven ER-to-mitochondria Ca^{2+} transfer. Pinton's group demonstrated that phototherapy induces a p53-dependent increase in [Ca^{2+}]$_{mit}$, which leads to tumor disruption in vivo [239,267]. Moreover, cytotoxic ER-dependent Ca^{2+} mobilization could be promoted by conjugating thapsigargin, a selective SERCA inhibitor, with a protease-specific peptide carrier, which is cleaved by the prostate-specific membrane antigen (PMSA) [268]. PMSA is widely expressed in the endothelium of many solid tumors [269,270], including RCC, thereby selectively favoring thapsigargin release in TME and inducing cancer and stromal cell apoptosis [268,271]. This prodrug has been termed mipsagargin or prodrug G202 and has recently been probed in a phase I clinical trials in patients suffering from refractory, advanced or metastatic solid tumors [272]. We do not know yet whether [Ca^{2+}]$_{ER}$ is also decreased in the endothelium of tumor neovessels, as ECFCs are likely to be replaced/diluted by local endothelial cells after the angiogenic switch [27]. Nevertheless, mipsagargin is likely to cause pro-apoptotic Ca^{2+} release in all stromal cells, including T-ECs.

As outlined elsewhere [23,32,135], caution is warranted when targeting a ubiquitous intracellular second messenger, such as Ca^{2+}. It should, however, be pointed out that several inhibitors of voltage-gated Ca^{2+} channels, such as verapamil, nifedipine and nitrendipin, are routinely employed in clinical practice to treat severe cardiovascular disorders, including hypertension, arrhythmia, acute myocardial infarction-induced heart failure and chronic stable angina [135,273]. In agreement with this observation, a phase I clinical trial is currently assessing the therapeutic outcome of Ca^{2+} electroporation on cutaneous metastases of solid tumors as compared to standard electrochemotherapy with bleomycin (https://clinicaltrials.gov/ct2/show/NCT01941901). Ca^{2+} electroporation is predicted to enhance the rate of cancer cell death by resulting in cytotoxic Ca^{2+} accumulation in the cytosol and in mitochondria [36]. Finally, the pharmacological inhibition of intracellular Ca^{2+} signaling did not elicit any intolerable side effects, such as immune depression, bleeding or neuropathic disorders, in least three distinct models of human cancer xenografts [24,205,274].

6. Conclusions

The present article discussed how remodeling of the endothelial Ca^{2+} toolkit (or transportome) could contribute to the resistance to anti-cancer treatments, which hampers from the very beginning their therapeutic outcome (intrinsic resistance) or leads to tumor relapse (acquired resistance) and patients' death. The intimate relationship between endothelial Ca^{2+} signaling and refractoriness to anti-cancer treatments cannot be fully appreciated by studying normal/healthy endothelial cells [32]. For instance, the role of VEGF in promoting tumor neovascularization has been extensively acknowledged based upon the observation that VEGF triggers pro-angiogenic Ca^{2+} signals in normal endothelial cells [6,58,251] and ECFCs. Nevertheless, VEGF does not stimulate proliferation and tube formation in T-ECFCs, which play a crucial role in sustaining the angiogenic switch and are likely to restore tumor vasculature prior to recurrence of disease progression. Therefore, to be effective in the patients, a strategy aiming at targeting the Ca^{2+} toolkit must be first probed on tumor-associated endothelial cells and ECFCs in vitro, as the their Ca^{2+} machinery could be different from that of naïve cells. The protocol to isolate ECFCs from peripheral blood does not require an unreasonable volume of blood ($\approx$40 mL), but it takes no less than three weeks [31] due to lack of ECFCs-specific membrane antigens. It will be imperative to speed up this procedure to accelerate the therapeutic translation of the findings generated by basic research. Isolating T-ECs represents a more technically demanding challenge, but several strategies were designed to collect and expand T-ECs from several types of solid cancers [275]. Further work on patients-derived T-ECs or T-ECFCs is mandatory to identify novel components of the endothelial Ca^{2+} toolkit involved in the refractoriness to anti-cancer therapies. Most of the attention is, of course, currently paid to Stim1 and Orai1 [33,131] and to the multiple TRP channel subfamilies that drive physiological angiogenesis [2,131,276,277]. Additional components of the endothelial Ca^{2+} transportome deserve careful investigation. For instance, Orai3 was found to up-regulated in several types of T-ECFCs [95,165], but its role in tumor vascularization is currently unknown. Of note, Orai3 may replace Orai1 as the pore-forming subunit of store-operated channels in cancer cells and could, therefore, emerge as a promising target for anti-cancer therapies [278]. NCX provides another unconventional Ca^{2+} entry that regulates proliferation and tube formation in healthy endothelial cells [18,279], but has been scarcely investigated in tumor neovessels. Finally, nAchRs are not the only ionotropic receptors expressed in vascular endothelial cells. N-methyl-D-aspartate receptors (NMDARs) are widely expressed in brain microvascular endothelial cells [280], in which they recruit eNOS and stimulate NO release in response to synaptic activity [281]. Aberrant glutamate signaling has been associated to glioma growth [282] and NMDARs-mediated Ca^{2+} entry could engage Ca^{2+}-dependent decoders other than eNOS in brain endothelium. Therefore, the expression and role of endothelial NMDARs in glioblastoma should be carefully evaluated.

Acknowledgments: The author gratefully acknowledges the contribution and support of all the colleagues and students who participated in the studies described in the present article. In particular, I would like to thank Vittorio Rosti and Germano Guerra. I am also grateful to Teresa Soda for her continuous support and comprehension.

References

1. Moccia, F.; Tanzi, F.; Munaron, L. Endothelial remodelling and intracellular calcium machinery. *Curr. Mol. Med.* **2014**, *14*, 457–480. [CrossRef] [PubMed]

2. Moccia, F.; Berra-Romani, R.; Tanzi, F. Update on vascular endothelial Ca^{2+} signalling: A tale of ion channels, pumps and transporters. *World J. Biol. Chem.* **2012**, *3*, 127–158. [CrossRef] [PubMed]

3. Munaron, L.; Pla, A.F. Endothelial Calcium Machinery and Angiogenesis: Understanding Physiology to Interfere with Pathology. *Curr. Med. Chem.* **2009**, *16*, 4691–4703. [CrossRef] [PubMed]

4. Troidl, C.; Nef, H.; Voss, S.; Schilp, A.; Kostin, S.; Troidl, K.; Szardien, S.; Rolf, A.; Schmitz-Rixen, T.; Schaper, W.; et al. Calcium-dependent signalling is essential during collateral growth in the pig hind limb-ischemia model. *J. Mol. Cell. Cardiol.* **2010**, *49*, 142–151. [CrossRef] [PubMed]

5. Troidl, C.; Troidl, K.; Schierling, W.; Cai, W.J.; Nef, H.; Mollmann, H.; Kostin, S.; Schimanski, S.; Hammer, L.; Elsasser, A.; et al. Trpv4 induces collateral vessel growth during regeneration of the arterial circulation. *J. Cell. Mol. Med.* **2009**, *13*, 2613–2621. [CrossRef] [PubMed]

6. Potenza, D.M.; Guerra, G.; Avanzato, D.; Poletto, V.; Pareek, S.; Guido, D.; Gallanti, A.; Rosti, V.; Munaron, L.; Tanzi, F.; et al. Hydrogen sulphide triggers VEGF-induced intracellular Ca^{2+} signals in human endothelial cells but not in their immature progenitors. *Cell Calcium* **2014**, *56*, 225–236. [CrossRef] [PubMed]

7. Moccia, F.; Berra-Romani, R.; Tritto, S.; Signorelli, S.; Taglietti, V.; Tanzi, F. Epidermal growth factor induces intracellular Ca^{2+} oscillations in microvascular endothelial cells. *J. Cell. Physiol.* **2003**, *194*, 139–150. [CrossRef] [PubMed]

8. Noren, D.P.; Chou, W.H.; Lee, S.H.; Qutub, A.A.; Warmflash, A.; Wagner, D.S.; Popel, A.S.; Levchenko, A. Endothelial cells decode VEGF-mediated Ca^{2+} signaling patterns to produce distinct functional responses. *Sci. Signal.* **2016**, *9*, ra20. [CrossRef] [PubMed]

9. Munaron, L.; Fiorio Pla, A. Calcium influx induced by activation of tyrosine kinase receptors in cultured bovine aortic endothelial cells. *J. Cell. Physiol.* **2000**, *185*, 454–463. [CrossRef]

10. Gupta, S.K.; Lysko, P.G.; Pillarisetti, K.; Ohlstein, E.; Stadel, J.M. Chemokine receptors in human endothelial cells. Functional expression of CXCR4 and its transcriptional regulation by inflammatory cytokines. *J. Biol. Chem.* **1998**, *273*, 4282–4287. [CrossRef] [PubMed]

11. Moccia, F.; Bonetti, E.; Dragoni, S.; Fontana, J.; Lodola, F.; Romani, R.B.; Laforenza, U.; Rosti, V.; Tanzi, F. Hematopoietic progenitor and stem cells circulate by surfing on intracellular Ca^{2+} waves: A novel target for cell-based therapy and anti-cancer treatment? *Curr. Signal Trans. Ther.* **2012**, *7*, 161–176. [CrossRef]

12. Yang, C.; Ohk, J.; Lee, J.Y.; Kim, E.J.; Kim, J.; Han, S.; Park, D.; Jung, H.; Kim, C. Calmodulin Mediates Ca^{2+}-Dependent Inhibition of Tie2 Signaling and Acts as a Developmental Brake During Embryonic Angiogenesis. *Arterioscler. Thromb. Vasc. Biol.* **2016**, *36*, 1406–1416. [CrossRef] [PubMed]

13. Zhu, L.P.; Luo, Y.G.; Chen, T.X.; Chen, F.R.; Wang, T.; Hu, Q. Ca^{2+} oscillation frequency regulates agonist-stimulated gene expression in vascular endothelial cells. *J. Cell Sci.* **2008**, *121*, 2511–2518. [CrossRef] [PubMed]

14. Chen, F.; Zhu, L.; Cai, L.; Zhang, J.; Zeng, X.; Li, J.; Su, Y.; Hu, Q. A stromal interaction molecule 1 variant up-regulates matrix metalloproteinase-2 expression by strengthening nucleoplasmic Ca^{2+} signaling. *Biochim. Biophys. Acta* **2016**, *1863*, 617–629. [CrossRef] [PubMed]

15. Tsai, F.C.; Seki, A.; Yang, H.W.; Hayer, A.; Carrasco, S.; Malmersjo, S.; Meyer, T. A polarized Ca^{2+}, diacylglycerol and STIM1 signalling system regulates directed cell migration. *Nat. Cell Biol.* **2014**, *16*, 133–144. [CrossRef] [PubMed]

16. Berra-Romani, R.; Avelino-Cruz, J.E.; Raqeeb, A.; Della Corte, A.; Cinelli, M.; Montagnani, S.; Guerra, G.; Moccia, F.; Tanzi, F. Ca^{2+}-dependent nitric oxide release in the injured endothelium of excised rat aorta: A promising mechanism applying in vascular prosthetic devices in aging patients. *BMC Surg.* **2013**, *13* (Suppl. 2), S40. [CrossRef] [PubMed]

17. Charoensin, S.; Eroglu, E.; Opelt, M.; Bischof, H.; Madreiter-Sokolowski, C.T.; Kirsch, A.; Depaoli, M.R.; Frank, S.; Schrammel, A.; Mayer, B.; et al. Intact mitochondrial Ca^{2+} uniport is essential for agonist-induced activation of endothelial nitric oxide synthase (eNOS). *Free Radic. Biol. Med.* **2017**, *102*, 248–259. [CrossRef] [PubMed]

18. Andrikopoulos, P.; Eccles, S.A.; Yaqoob, M.M. Coupling between the TRPC3 ion channel and the NCX1 transporter contributed to VEGF-induced ERK1/2 activation and angiogenesis in human primary endothelial cells. *Cell. Signal.* **2017**, *37*, 12–30. [CrossRef] [PubMed]

19. Lyubchenko, T.; Woodward, H.; Veo, K.D.; Burns, N.; Nijmeh, H.; Liubchenko, G.A.; Stenmark, K.R.; Gerasimovskaya, E.V. P2Y1 and P2Y13 purinergic receptors mediate Ca^{2+} signaling and proliferative responses in pulmonary artery vasa vasorum endothelial cells. *Am. J. Physiol. Cell Physiol.* **2011**, *300*, C266–C275. [CrossRef] [PubMed]

20. Sameermahmood, Z.; Balasubramanyam, M.; Saravanan, T.; Rema, M. Curcumin modulates SDF-1alpha/CXCR4-induced migration of human retinal endothelial cells (HRECs). *Investig. Ophthalmol. Vis. Sci.* **2008**, *49*, 3305–3311. [CrossRef] [PubMed]

21. Pupo, E.; Pla, A.F.; Avanzato, D.; Moccia, F.; Cruz, J.E.; Tanzi, F.; Merlino, A.; Mancardi, D.; Munaron, L. Hydrogen sulfide promotes calcium signals and migration in tumor-derived endothelial cells. *Free Radic. Biol. Med.* **2011**, *51*, 1765–1773. [CrossRef] [PubMed]

22. Fiorio Pla, A.; Munaron, L. Functional properties of ion channels and transporters in tumour vascularization. *Philos. Trans. R. Soc. Lond. B Biol. Sci.* **2014**, *369*, 20130103. [CrossRef] [PubMed]

23. Moccia, F. Remodelling of the Ca^{2+} Toolkit in Tumor Endothelium as a Crucial Responsible for the Resistance to Anticancer Therapies. *Curr. Signal Trans. Ther.* **2017**, *12*. [CrossRef]

24. Chen, Y.F.; Chiu, W.T.; Chen, Y.T.; Lin, P.Y.; Huang, H.J.; Chou, C.Y.; Chang, H.C.; Tang, M.J.; Shen, M.R. Calcium store sensor stromal-interaction molecule 1-dependent signaling plays an important role in cervical cancer growth, migration, and angiogenesis. *Proc. Natl. Acad. Sci. USA* **2011**, *108*, 15225–15230. [CrossRef] [PubMed]

25. Jain, R.K.; Carmeliet, P. SnapShot: Tumor angiogenesis. *Cell* **2012**, *149*, 1408.e1. [CrossRef] [PubMed]

26. Gao, D.C.; Nolan, D.; McDonnell, K.; Vahdat, L.; Benezra, R.; Altorki, N.; Mittal, V. Bone marrow-derived endothelial progenitor cells contribute to the angiogenic switch in tumor growth and metastatic progression. *Biochim. Biophys. Acta* **2009**, *1796*, 33–40. [CrossRef] [PubMed]

27. Moccia, F.; Zuccolo, E.; Poletto, V.; Cinelli, M.; Bonetti, E.; Guerra, G.; Rosti, V. Endothelial progenitor cells support tumour growth and metastatisation: Implications for the resistance to anti-angiogenic therapy. *Tumour Biol.* **2015**, *36*, 6603–6614. [CrossRef] [PubMed]

28. Yoder, M.C.; Ingram, D.A. The definition of EPCs and other bone marrow cells contributing to neoangiogenesis and tumor growth: Is there common ground for understanding the roles of numerous marrow-derived cells in the neoangiogenic process? *Biochim. Biophys. Acta* **2009**, *1796*, 50–54. [CrossRef] [PubMed]

29. Moccia, F.; Guerra, G. Ca^{2+} Signalling in Endothelial Progenitor Cells: Friend or Foe? *J. Cell. Physiol.* **2016**, *231*, 314–327. [CrossRef] [PubMed]

30. Maeng, Y.S.; Choi, H.J.; Kwon, J.Y.; Park, Y.W.; Choi, K.S.; Min, J.K.; Kim, Y.H.; Suh, P.G.; Kang, K.S.; Won, M.H.; et al. Endothelial progenitor cell homing: Prominent role of the IGF2-IGF2R-PLCbeta2 axis. *Blood* **2009**, *113*, 233–243. [CrossRef] [PubMed]

31. Moccia, F.; Ruffinatti, F.A.; Zuccolo, E. Intracellular Ca^{2+} Signals to Reconstruct A Broken Heart: Still A Theoretical Approach? *Curr. Drug Targets* **2015**, *16*, 793–815. [CrossRef] [PubMed]

32. Moccia, F.; Poletto, V. May the remodeling of the Ca^{2+} toolkit in endothelial progenitor cells derived from cancer patients suggest alternative targets for anti-angiogenic treatment? *Biochim. Biophys. Acta* **2015**, *1853*, 1958–1973. [CrossRef] [PubMed]

33. Moccia, F.; Lodola, F.; Dragoni, S.; Bonetti, E.; Bottino, C.; Guerra, G.; Laforenza, U.; Rosti, V.; Tanzi, F. Ca^{2+} signalling in endothelial progenitor cells: A novel means to improve cell-based therapy and impair tumour vascularisation. *Curr. Vasc. Pharmacol.* **2014**, *12*, 87–105. [CrossRef] [PubMed]

34. Bergers, G.; Hanahan, D. Modes of resistance to anti-angiogenic therapy. *Nat. Rev. Cancer* **2008**, *8*, 592–603. [CrossRef] [PubMed]

35. Prevarskaya, N.; Ouadid-Ahidouch, H.; Skryma, R.; Shuba, Y. Remodelling of Ca^{2+} transport in cancer: How it contributes to cancer hallmarks? *Philos. Trans. R. Soc. Lond. B Biol. Sci.* **2014**, *369*, 20130097. [CrossRef] [PubMed]

36. Monteith, G.R.; Prevarskaya, N.; Roberts-Thomson, S.J. The calcium-cancer signalling nexus. *Nat. Rev. Cancer* **2017**, *17*, 367–380. [CrossRef] [PubMed]

37. Moccia, F.; Fotia, V.; Tancredi, R.; Della Porta, M.G.; Rosti, V.; Bonetti, E.; Poletto, V.; Marchini, S.; Beltrame, L.; Gallizzi, G.; et al. Breast and renal cancer-Derived endothelial colony forming cells share a common gene signature. *Eur. J. Cancer* **2017**, *77*, 155–164. [CrossRef] [PubMed]

38. Moccia, F.; Berra-Romani, R.; Baruffi, S.; Spaggiari, S.; Signorelli, S.; Castelli, L.; Magistretti, J.; Taglietti, V.; Tanzi, F. Ca^{2+} uptake by the endoplasmic reticulum Ca^{2+}-ATPase in rat microvascular endothelial cells. *Biochem. J.* **2002**, *364 Pt 1*, 235–244. [CrossRef] [PubMed]

39. Berra-Romani, R.; Raqeeb, A.; Guzman-Silva, A.; Torres-Jacome, J.; Tanzi, F.; Moccia, F. Na^+-Ca^{2+} exchanger contributes to Ca^{2+} extrusion in ATP-stimulated endothelium of intact rat aorta. *Biochem. Biophys. Res. Commun.* **2010**, *395*, 126–130. [CrossRef] [PubMed]

40. Paszty, K.; Caride, A.J.; Bajzer, Z.; Offord, C.P.; Padanyi, R.; Hegedus, L.; Varga, K.; Strehler, E.E.; Enyedi, A. Plasma membrane Ca^{2+}-ATPases can shape the pattern of Ca^{2+} transients induced by store-operated Ca^{2+} entry. *Sci. Signal.* **2015**, *8*, ra19. [CrossRef] [PubMed]

41. Cardenas, C.; Foskett, J.K. Mitochondrial Ca^{2+} signals in autophagy. *Cell Calcium* **2012**, *52*, 44–51. [CrossRef] [PubMed]

42. Zuccolo, E.; Lim, D.; Kheder, D.A.; Perna, A.; Catarsi, P.; Botta, L.; Rosti, V.; Riboni, L.; Sancini, G.; Tanzi, F.; et al. Acetylcholine induces intracellular Ca^{2+} oscillations and nitric oxide release in mouse brain endothelial cells. *Cell Calcium* **2017**, *66*, 33–47. [CrossRef] [PubMed]

43. Huang, T.Y.; Chu, T.F.; Chen, H.I.; Jen, C.J. Heterogeneity of $[Ca^{2+}](i)$ signaling in intact rat aortic endothelium. *FASEB J.* **2000**, *14*, 797–804. [PubMed]

44. Duza, T.; Sarelius, I.H. Localized transient increases in endothelial cell Ca^{2+} in arterioles in situ: Implications for coordination of vascular function. *Am. J. Physiol. Heart Circ. Physiol.* **2004**, *286*, H2322–H2331. [CrossRef] [PubMed]

45. Kansui, Y.; Garland, C.J.; Dora, K.A. Enhanced spontaneous Ca^{2+} events in endothelial cells reflect signalling through myoendothelial gap junctions in pressurized mesenteric arteries. *Cell Calcium* **2008**, *44*, 135–146. [CrossRef] [PubMed]

46. Pedriali, G.; Rimessi, A.; Sbano, L.; Giorgi, C.; Wieckowski, M.R.; Previati, M.; Pinton, P. Regulation of Endoplasmic Reticulum-Mitochondria Ca^{2+} Transfer and Its Importance for Anti-Cancer Therapies. *Front. Oncol.* **2017**, *7*, 180. [CrossRef] [PubMed]

47. De Stefani, D.; Rizzuto, R.; Pozzan, T. Enjoy the Trip: Calcium in Mitochondria Back and Forth. *Annu. Rev. Biochem.* **2016**, *85*, 161–192. [CrossRef] [PubMed]

48. Marcu, R.; Wiczer, B.M.; Neeley, C.K.; Hawkins, B.J. Mitochondrial matrix Ca^{2+} accumulation regulates cytosolic NAD(+)/NADH metabolism, protein acetylation, and sirtuin expression. *Mol. Cell. Biol.* **2014**, *34*, 2890–2902. [CrossRef] [PubMed]

49. Dong, Z.; Shanmughapriya, S.; Tomar, D.; Siddiqui, N.; Lynch, S.; Nemani, N.; Breves, S.L.; Zhang, X.; Tripathi, A.; Palaniappan, P.; et al. Mitochondrial Ca^{2+} Uniporter Is a Mitochondrial Luminal Redox Sensor that Augments MCU Channel Activity. *Mol. Cell* **2017**, *65*, 1014.e7–1028.e7. [CrossRef] [PubMed]

50. Bittremieux, M.; Parys, J.B.; Pinton, P.; Bultynck, G. ER functions of oncogenes and tumor suppressors: Modulators of intracellular Ca^{2+} signaling. *Biochim. Biophys. Acta* **2016**, *1863 Pt B*, 1364–1378. [CrossRef] [PubMed]

51. Pinton, P.; Giorgi, C.; Pandolfi, P.P. The role of PML in the control of apoptotic cell fate: A new key player at ER-mitochondria sites. *Cell Death Differ.* **2011**, *18*, 1450–1456. [CrossRef] [PubMed]

52. Zhu, H.; Jin, Q.; Li, Y.; Ma, Q.; Wang, J.; Li, D.; Zhou, H.; Chen, Y. Melatonin protected cardiac microvascular endothelial cells against oxidative stress injury via suppression of IP3R-$[Ca^{2+}]c$/VDAC-$[Ca^{2+}]m$ axis by activation of MAPK/ERK signaling pathway. *Cell Stress Chaperones* **2018**, *23*, 101–113. [CrossRef] [PubMed]

53. Madreiter-Sokolowski, C.T.; Gottschalk, B.; Parichatikanond, W.; Eroglu, E.; Klec, C.; Waldeck-Weiermair, M.; Malli, R.; Graier, W.F. Resveratrol Specifically Kills Cancer Cells by a Devastating Increase in the Ca^{2+} Coupling Between the Greatly Tethered Endoplasmic Reticulum and Mitochondria. *Cell. Physiol. Biochem.* **2016**, *39*, 1404–1420. [CrossRef] [PubMed]

54. Carmeliet, P.; Jain, R.K. Molecular mechanisms and clinical applications of angiogenesis. *Nature* **2011**, *473*, 298–307. [CrossRef] [PubMed]

55. Berridge, M.J.; Bootman, M.D.; Roderick, H.L. Calcium signalling: Dynamics, homeostasis and remodelling. *Nat. Rev. Mol. Cell Biol.* **2003**, *4*, 517–529. [CrossRef] [PubMed]

56. Moccia, F.; Dragoni, S.; Lodola, F.; Bonetti, E.; Bottino, C.; Guerra, G.; Laforenza, U.; Rosti, V.; Tanzi, F. Store-dependent Ca^{2+} entry in endothelial progenitor cells as a perspective tool to enhance cell-based therapy and adverse tumour vascularization. *Curr. Med. Chem.* **2012**, *19*, 5802–5818. [CrossRef] [PubMed]

57. Abdullaev, I.F.; Bisaillon, J.M.; Potier, M.; Gonzalez, J.C.; Motiani, R.K.; Trebak, M. Stim1 and Orai1 mediate CRAC currents and store-operated calcium entry important for endothelial cell proliferation. *Circ. Res.* **2008**, *103*, 1289–1299. [CrossRef] [PubMed]

58. Li, J.; Cubbon, R.M.; Wilson, L.A.; Amer, M.S.; McKeown, L.; Hou, B.; Majeed, Y.; Tumova, S.; Seymour, V.A.L.; Taylor, H.; et al. Orai1 and CRAC channel dependence of VEGF-activated Ca^{2+} entry and endothelial tube formation. *Circ. Res.* **2011**, *108*, 1190–1198. [CrossRef] [PubMed]

59. Blatter, L.A. Tissue Specificity: SOCE: Implications for Ca^{2+} Handling in Endothelial Cells. *Adv. Exp. Med. Biol.* **2017**, *993*, 343–361. [PubMed]

60. Gees, M.; Colsoul, B.; Nilius, B. The role of transient receptor potential cation channels in Ca^{2+} signaling. *Cold Spring Harb. Perspect. Biol.* **2010**, *2*, a003962. [CrossRef] [PubMed]

61. Sundivakkam, P.C.; Freichel, M.; Singh, V.; Yuan, J.P.; Vogel, S.M.; Flockerzi, V.; Malik, A.B.; Tiruppathi, C. The Ca^{2+} sensor stromal interaction molecule 1 (STIM1) is necessary and sufficient for the store-operated Ca^{2+} entry function of transient receptor potential canonical (TRPC) 1 and 4 channels in endothelial cells. *Mol. Pharmacol.* **2012**, *81*, 510–526. [CrossRef] [PubMed]

62. Cioffi, D.L.; Wu, S.; Chen, H.; Alexeyev, M.; St Croix, C.M.; Pitt, B.R.; Uhlig, S.; Stevens, T. Orai1 determines calcium selectivity of an endogenous TRPC heterotetramer channel. *Circ. Res.* **2012**, *110*, 1435–1444. [CrossRef] [PubMed]

63. Xu, N.; Cioffi, D.L.; Alexeyev, M.; Rich, T.C.; Stevens, T. Sodium entry through endothelial store-operated calcium entry channels: Regulation by Orai1. *Am. J. Physiol. Cell Physiol.* **2015**, *308*, C277–C288. [CrossRef] [PubMed]

64. Antigny, F.; Jousset, H.; Konig, S.; Frieden, M. Thapsigargin activates Ca^{2+} entry both by store-dependent, STIM1/Orai1-mediated, and store-independent, TRPC3/PLC/PKC-mediated pathways in human endothelial cells. *Cell Calcium* **2011**, *49*, 115–127. [CrossRef] [PubMed]

65. Weissmann, N.; Sydykov, A.; Kalwa, H.; Storch, U.; Fuchs, B.; Mederos y Schnitzler, M.; Brandes, R.P.; Grimminger, F.; Meissner, M.; Freichel, M.; et al. Activation of TRPC6 channels is essential for lung ischaemia-reperfusion induced oedema in mice. *Nat. Commun.* **2012**, *3*, 649. [CrossRef] [PubMed]

66. Boeldt, D.S.; Grummer, M.A.; Magness, R.R.; Bird, I.M. Altered VEGF-stimulated Ca^{2+} signaling in part underlies pregnancy-adapted eNOS activity in UAEC. *J. Endocrinol.* **2014**, *223*, 1–11. [CrossRef] [PubMed]

67. Hamdollah Zadeh, M.A.; Glass, C.A.; Magnussen, A.; Hancox, J.C.; Bates, D.O. VEGF-mediated elevated intracellular calcium and angiogenesis in human microvascular endothelial cells in vitro are inhibited by dominant negative TRPC6. *Microcirculation* **2008**, *15*, 605–614. [CrossRef] [PubMed]

68. Antoniotti, S.; Lovisolo, D.; Fiorio Pla, A.; Munaron, L. Expression and functional role of bTRPC1 channels in native endothelial cells. *FEBS Lett.* **2002**, *510*, 189–195. [CrossRef]

69. Ho, W.S.; Zheng, X.; Zhang, D.X. Role of endothelial TRPV4 channels in vascular actions of the endocannabinoid, 2-arachidonoylglycerol. *Br. J. Pharmacol.* **2015**, *172*, 5251–5264. [CrossRef] [PubMed]

70. Zheng, X.; Zinkevich, N.S.; Gebremedhin, D.; Gauthier, K.M.; Nishijima, Y.; Fang, J.; Wilcox, D.A.; Campbell, W.B.; Gutterman, D.D.; Zhang, D.X. Arachidonic acid-induced dilation in human coronary arterioles: Convergence of signaling mechanisms on endothelial TRPV4-mediated Ca^{2+} entry. *J. Am. Heart Assoc.* **2013**, *2*, e000080. [CrossRef] [PubMed]

71. Yoshida, T.; Inoue, R.; Morii, T.; Takahashi, N.; Yamamoto, S.; Hara, Y.; Tominaga, M.; Shimizu, S.; Sato, Y.; Mori, Y. Nitric oxide activates TRP channels by cysteine S-nitrosylation. *Nat. Chem. Biol.* **2006**, *2*, 596–607. [CrossRef] [PubMed]

72. Dietrich, A.; Gudermann, T. Another TRP to endothelial dysfunction: TRPM2 and endothelial permeability. *Circ. Res.* **2008**, *102*, 275–277. [CrossRef] [PubMed]

73. Kwan, H.Y.; Cheng, K.T.; Ma, Y.; Huang, Y.; Tang, N.L.; Yu, S.; Yao, X. CNGA2 contributes to ATP-induced noncapacitative Ca^{2+} influx in vascular endothelial cells. *J. Vasc. Res.* **2010**, *47*, 148–156. [CrossRef] [PubMed]

74. Berra-Romani, R.; Raqeeb, A.; Avelino-Cruz, J.E.; Moccia, F.; Oldani, A.; Speroni, F.; Taglietti, V.; Tanzi, F. Ca^{2+} signaling in injured in situ endothelium of rat aorta. *Cell Calcium* **2008**, *44*, 298–309. [CrossRef] [PubMed]

75. Moccia, F.; Frost, C.; Berra-Romani, R.; Tanzi, F.; Adams, D.J. Expression and function of neuronal nicotinic ACh receptors in rat microvascular endothelial cells. *Am. J. Physiol. Heart Circ. Physiol.* **2004**, *286*, H486–H491. [CrossRef] [PubMed]

76. LeMaistre, J.L.; Sanders, S.A.; Stobart, M.J.; Lu, L.; Knox, J.D.; Anderson, H.D.; Anderson, C.M. Coactivation of NMDA receptors by glutamate and D-serine induces dilation of isolated middle cerebral arteries. *J. Cereb. Blood Flow Metab.* **2012**, *32*, 537–547. [CrossRef] [PubMed]

77. AbouAlaiwi, W.A.; Takahashi, M.; Mell, B.R.; Jones, T.J.; Ratnam, S.; Kolb, R.J.; Nauli, S.M. Ciliary polycystin-2 is a mechanosensitive calcium channel involved in nitric oxide signaling cascades. *Circ. Res.* **2009**, *104*, 860–869. [CrossRef] [PubMed]

78. Berrout, J.; Jin, M.; O'Neil, R.G. Critical role of TRPP2 and TRPC1 channels in stretch-induced injury of blood-brain barrier endothelial cells. *Brain Res.* **2012**, *1436*, 1–12. [CrossRef] [PubMed]

79. Filosa, J.A.; Yao, X.; Rath, G. TRPV4 and the regulation of vascular tone. *J. Cardiovasc. Pharmacol.* **2013**, *61*, 113–119. [CrossRef] [PubMed]

80. Du, J.; Ma, X.; Shen, B.; Huang, Y.; Birnbaumer, L.; Yao, X. TRPV4, TRPC1, and TRPP2 assemble to form a flow-sensitive heteromeric channel. *FASEB J.* **2014**, *28*, 4677–4685. [CrossRef] [PubMed]

81. Li, J.; Hou, B.; Tumova, S.; Muraki, K.; Bruns, A.; Ludlow, M.J.; Sedo, A.; Hyman, A.J.; McKeown, L.; Young, R.S.; et al. Piezo1 integration of vascular architecture with physiological force. *Nature* **2014**, *515*, 279–282. [CrossRef] [PubMed]

82. Medina, R.J.; Barber, C.L.; Sabatier, F.; Dignat-George, F.; Melero-Martin, J.M.; Khosrotehrani, K.; Ohneda, O.; Randi, A.M.; Chan, J.K.Y.; Yamaguchi, T.; et al. Endothelial Progenitors: A Consensus Statement on Nomenclature. *Stem Cells Transl. Med.* **2017**, *6*, 1316–1320. [CrossRef] [PubMed]

83. Dragoni, S.; Laforenza, U.; Bonetti, E.; Lodola, F.; Bottino, C.; Berra-Romani, R.; Carlo Bongio, G.; Cinelli, M.P.; Guerra, G.; Pedrazzoli, P.; et al. Vascular endothelial growth factor stimulates endothelial colony forming cells proliferation and tubulogenesis by inducing oscillations in intracellular Ca^{2+} concentration. *Stem Cells* **2011**, *29*, 1898–1907. [CrossRef] [PubMed]

84. Sanchez-Hernandez, Y.; Laforenza, U.; Bonetti, E.; Fontana, J.; Dragoni, S.; Russo, M.; Avelino-Cruz, J.E.; Schinelli, S.; Testa, D.; Guerra, G.; et al. Store-operated Ca^{2+} entry is expressed in human endothelial progenitor cells. *Stem Cells Dev.* **2010**, *19*, 1967–1981. [CrossRef] [PubMed]

85. Dragoni, S.; Laforenza, U.; Bonetti, E.; Lodola, F.; Bottino, C.; Guerra, G.; Borghesi, A.; Stronati, M.; Rosti, V.; Tanzi, F.; et al. Canonical transient receptor potential 3 channel triggers vascular endothelial growth factor-induced intracellular Ca^{2+} oscillations in endothelial progenitor cells isolated from umbilical cord blood. *Stem Cells Dev.* **2013**, *22*, 2561–2580. [CrossRef] [PubMed]

86. Zuccolo, E.; Dragoni, S.; Poletto, V.; Catarsi, P.; Guido, D.; Rappa, A.; Reforgiato, M.; Lodola, F.; Lim, D.; Rosti, V.; et al. Arachidonic acid-evoked Ca^{2+} signals promote nitric oxide release and proliferation in human endothelial colony forming cells. *Vascul. Pharmacol.* **2016**, *87*, 159–171. [CrossRef] [PubMed]

87. Poletto, V.; Dragoni, S.; Lim, D.; Biggiogera, M.; Aronica, A.; Cinelli, M.; De Luca, A.; Rosti, V.; Porta, C.; Guerra, G.; et al. Endoplasmic Reticulum Ca^{2+} Handling and Apoptotic Resistance in Tumor-Derived Endothelial Colony Forming Cells. *J. Cell. Biochem.* **2016**, *117*, 2260–2271. [CrossRef] [PubMed]

88. Pla, A.F.; Ong, H.L.; Cheng, K.T.; Brossa, A.; Bussolati, B.; Lockwich, T.; Paria, B.; Munaron, L.; Ambudkar, I.S. TRPV4 mediates tumor-derived endothelial cell migration via arachidonic acid-activated actin remodeling. *Oncogene* **2012**, *31*, 200–212.

89. Fiorio Pla, A.; Grange, C.; Antoniotti, S.; Tomatis, C.; Merlino, A.; Bussolati, B.; Munaron, L. Arachidonic acid-induced Ca^{2+} entry is involved in early steps of tumor angiogenesis. *Mol. Cancer Res.* **2008**, *6*, 535–545. [PubMed]

90. Adapala, R.K.; Thoppil, R.J.; Ghosh, K.; Cappelli, H.C.; Dudley, A.C.; Paruchuri, S.; Keshamouni, V.; Klagsbrun, M.; Meszaros, J.G.; Chilian, W.M.; et al. Activation of mechanosensitive ion channel TRPV4 normalizes tumor vasculature and improves cancer therapy. *Oncogene* **2016**, *35*, 314–322. [CrossRef] [PubMed]

91. Thoppil, R.J.; Cappelli, H.C.; Adapala, R.K.; Kanugula, A.K.; Paruchuri, S.; Thodeti, C.K. TRPV4 channels regulate tumor angiogenesis via modulation of Rho/Rho kinase pathway. *Oncotarget* **2016**, *7*, 25849–25861. [CrossRef] [PubMed]

92.   Thoppil, R.J.; Adapala, R.K.; Cappelli, H.C.; Kondeti, V.; Dudley, A.C.; Gary Meszaros, J.; Paruchuri, S.;
       Thodeti, C.K. TRPV4 channel activation selectively inhibits tumor endothelial cell proliferation. *Sci. Rep.*
       2015, *5*, 14257. [CrossRef] [PubMed]
93.   Yang, H.; Liu, C.; Zhou, R.M.; Yao, J.; Li, X.M.; Shen, Y.; Cheng, H.; Yuan, J.; Yan, B.; Jiang, Q. Piezo2 protein:
       A novel regulator of tumor angiogenesis and hyperpermeability. *Oncotarget* **2016**, *7*, 44630–44643. [CrossRef]
       [PubMed]
94.   Avanzato, D.; Genova, T.; Fiorio Pla, A.; Bernardini, M.; Bianco, S.; Bussolati, B.; Mancardi, D.; Giraudo, E.;
       Maione, F.; Cassoni, P.; et al. Activation of P2X7 and P2Y11 purinergic receptors inhibits migration and
       normalizes tumor-derived endothelial cells via cAMP signaling. *Sci. Rep.* **2016**, *6*, 32602. [CrossRef]
       [PubMed]
95.   Lodola, F.; Laforenza, U.; Bonetti, E.; Lim, D.; Dragoni, S.; Bottino, C.; Ong, H.L.; Guerra, G.; Ganini, C.;
       Massa, M.; et al. Store-operated Ca^{2+} entry is remodelled and controls in vitro angiogenesis in endothelial
       progenitor cells isolated from tumoral patients. *PLoS ONE* **2012**, *7*, e42541. [CrossRef] [PubMed]
96.   Lodola, F.; Laforenza, U.; Cattaneo, F.; Ruffinatti, F.A.; Poletto, V.; Massa, M.; Tancredi, R.; Zuccolo, E.;
       Khdar, A.D.; Riccardi, A.; et al. VEGF-induced intracellular Ca^{2+} oscillations are down-regulated and do
       not stimulate angiogenesis in breast cancer-derived endothelial colony forming cells. *Oncotarget* **2017**, *8*,
       95223–95246. [PubMed]
97.   Zuccolo, E.; Bottino, C.; Diofano, F.; Poletto, V.; Codazzi, A.C.; Mannarino, S.; Campanelli, R.; Fois, G.;
       Marseglia, G.L.; Guerra, G.; et al. Constitutive Store-Operated Ca^{2+} Entry Leads to Enhanced Nitric
       Oxide Production and Proliferation in Infantile Hemangioma-Derived Endothelial Colony-Forming Cells.
       Stem Cells Dev. **2016**, *25*, 301–319. [CrossRef] [PubMed]
98.   Heeschen, C.; Weis, M.; Aicher, A.; Dimmeler, S.; Cooke, J.P. A novel angiogenic pathway mediated by
       non-neuronal nicotinic acetylcholine receptors. *J. Clin. Investig.* **2002**, *110*, 527–536. [CrossRef] [PubMed]
99.   Heeschen, C.; Jang, J.J.; Weis, M.; Pathak, A.; Kaji, S.; Hu, R.S.; Tsao, P.S.; Johnson, F.L.; Cooke, J.P. Nicotine
       stimulates angiogenesis and promotes tumor growth and atherosclerosis. *Nat. Med.* **2001**, *7*, 833–839.
       [CrossRef] [PubMed]
100.   Yu, M.; Liu, Q.; Sun, J.; Yi, K.; Wu, L.; Tan, X. Nicotine improves the functional activity of late endothelial
       progenitor cells via nicotinic acetylcholine receptors. *Biochem. Cell Biol.* **2011**, *89*, 405–410. [CrossRef]
       [PubMed]
101.   Natori, T.; Sata, M.; Washida, M.; Hirata, Y.; Nagai, R.; Makuuchi, M. Nicotine enhances neovascularization
       and promotes tumor growth. *Mol. Cells* **2003**, *16*, 143–146. [PubMed]
102.   Alonso, F.; Domingos-Pereira, S.; Le Gal, L.; Derre, L.; Meda, P.; Jichlinski, P.; Nardelli-Haefliger, D.;
       Haefliger, J.A. Targeting endothelial connexin40 inhibits tumor growth by reducing angiogenesis and
       improving vessel perfusion. *Oncotarget* **2016**, *7*, 14015–14028. [CrossRef] [PubMed]
103.   Rigiracciolo, D.C.; Scarpelli, A.; Lappano, R.; Pisano, A.; Santolla, M.F.; Avino, S.; De Marco, P.; Bussolati, B.;
       Maggiolini, M.; De Francesco, E.M. GPER is involved in the stimulatory effects of aldosterone in breast
       cancer cells and breast tumor-derived endothelial cells. *Oncotarget* **2016**, *7*, 94–111. [CrossRef] [PubMed]
104.   Dong, Y.; Pan, Q.; Jiang, L.; Chen, Z.; Zhang, F.; Liu, Y.; Xing, H.; Shi, M.; Li, J.; Li, X.; et al. Tumor endothelial
       expression of P-glycoprotein upon microvesicular transfer of TrpC5 derived from adriamycin-resistant breast
       cancer cells. *Biochem. Biophys. Res. Commun.* **2014**, *446*, 85–90. [CrossRef] [PubMed]
105.   White, J.P.; Cibelli, M.; Urban, L.; Nilius, B.; McGeown, J.G.; Nagy, I. TRPV4: Molecular Conductor of a
       Diverse Orchestra. *Physiol. Rev.* **2016**, *96*, 911–973. [CrossRef] [PubMed]
106.   Everaerts, W.; Nilius, B.; Owsianik, G. The vanilloid transient receptor potential channel TRPV4: From
       structure to disease. *Prog. Biophys. Mol. Biol.* **2010**, *103*, 2–17. [CrossRef] [PubMed]
107.   He, D.; Pan, Q.; Chen, Z.; Sun, C.; Zhang, P.; Mao, A.; Zhu, Y.; Li, H.; Lu, C.; Xie, M.; et al. Treatment of
       hypertension by increasing impaired endothelial TRPV4-KCa2.3 interaction. *EMBO Mol. Med.* **2017**, *9*,
       1491–1503. [CrossRef] [PubMed]
108.   Chen, C.K.; Hsu, P.Y.; Wang, T.M.; Miao, Z.F.; Lin, R.T.; Juo, S.H. TRPV4 Activation Contributes Functional
       Recovery from Ischemic Stroke via Angiogenesis and Neurogenesis. *Mol. Neurobiol.* **2017**. [CrossRef]
       [PubMed]
109.   Hatano, N.; Suzuki, H.; Itoh, Y.; Muraki, K. TRPV4 partially participates in proliferation of human brain
       capillary endothelial cells. *Life Sci.* **2013**, *92*, 317–324. [CrossRef] [PubMed]

110. Matthews, B.D.; Thodeti, C.K.; Tytell, J.D.; Mammoto, A.; Overby, D.R.; Ingber, D.E. Ultra-rapid activation of TRPV4 ion channels by mechanical forces applied to cell surface beta1 integrins. *Integr. Biol. (Camb.)* **2010**, *2*, 435–442. [CrossRef] [PubMed]

111. Thodeti, C.K.; Matthews, B.; Ravi, A.; Mammoto, A.; Ghosh, K.; Bracha, A.L.; Ingber, D.E. TRPV4 Channels Mediate Cyclic Strain-Induced Endothelial Cell Reorientation Through Integrin-to-Integrin Signaling. *Circ. Res.* **2009**, *104*, 1123–1130. [CrossRef] [PubMed]

112. Sonkusare, S.K.; Bonev, A.D.; Ledoux, J.; Liedtke, W.; Kotlikoff, M.I.; Heppner, T.J.; Hill-Eubanks, D.C.; Nelson, M.T. Elementary Ca^{2+} signals through endothelial TRPV4 channels regulate vascular function. *Science* **2012**, *336*, 597–601. [CrossRef] [PubMed]

113. Zhao, L.; Sullivan, M.N.; Chase, M.; Gonzales, A.L.; Earley, S. Calcineurin/nuclear factor of activated T cells-coupled vanilliod transient receptor potential channel 4 Ca^{2+} sparklets stimulate airway smooth muscle cell proliferation. *Am. J. Respir. Cell Mol. Biol.* **2014**, *50*, 1064–1075. [CrossRef] [PubMed]

114. Wen, Z.H.; Su, Y.C.; Lai, P.L.; Zhang, Y.; Xu, Y.F.; Zhao, A.; Yao, G.Y.; Jia, C.H.; Lin, J.; Xu, S.; et al. Critical role of arachidonic acid-activated mTOR signaling in breast carcinogenesis and angiogenesis. *Oncogene* **2013**, *32*, 160–170. [CrossRef] [PubMed]

115. Munaron, L. Shuffling the cards in signal transduction: Calcium, arachidonic acid and mechanosensitivity. *World J. Biol. Chem.* **2011**, *2*, 59–66. [CrossRef] [PubMed]

116. Kim, E.; Tunset, H.M.; Cebulla, J.; Vettukattil, R.; Helgesen, H.; Feuerherm, A.J.; Engebraten, O.; Maelandsmo, G.M.; Johansen, B.; Moestue, S.A. Anti-vascular effects of the cytosolic phospholipase A2 inhibitor AVX235 in a patient-derived basal-like breast cancer model. *BMC Cancer* **2016**, *16*, 191. [CrossRef] [PubMed]

117. Carmeliet, P.; Jain, R.K. Principles and mechanisms of vessel normalization for cancer and other angiogenic diseases. *Nat. Rev. Drug Discov.* **2011**, *10*, 417–427. [CrossRef] [PubMed]

118. Goel, S.; Duda, D.G.; Xu, L.; Munn, L.L.; Boucher, Y.; Fukumura, D.; Jain, R.K. Normalization of the vasculature for treatment of cancer and other diseases. *Physiol. Rev.* **2011**, *91*, 1071–1121. [CrossRef] [PubMed]

119. Bagriantsev, S.N.; Gracheva, E.O.; Gallagher, P.G. Piezo proteins: Regulators of mechanosensation and other cellular processes. *J. Biol. Chem.* **2014**, *289*, 31673–31681. [CrossRef] [PubMed]

120. Honore, E.; Martins, J.R.; Penton, D.; Patel, A.; Demolombe, S. The Piezo Mechanosensitive Ion Channels: May the Force Be with You! *Rev. Physiol. Biochem. Pharmacol.* **2015**, *169*, 25–41. [PubMed]

121. Ranade, S.S.; Qiu, Z.; Woo, S.H.; Hur, S.S.; Murthy, S.E.; Cahalan, S.M.; Xu, J.; Mathur, J.; Bandell, M.; Coste, B.; et al. Piezo1, a mechanically activated ion channel, is required for vascular development in mice. *Proc. Natl. Acad. Sci. USA* **2014**, *111*, 10347–10352. [CrossRef] [PubMed]

122. Di Virgilio, F.; Adinolfi, E. Extracellular purines, purinergic receptors and tumor growth. *Oncogene* **2017**, *36*, 293–303. [CrossRef] [PubMed]

123. Burnstock, G. Purinergic Signaling in the Cardiovascular System. *Circ. Res.* **2017**, *120*, 207–228. [CrossRef] [PubMed]

124. Adinolfi, E.; Raffaghello, L.; Giuliani, A.L.; Cavazzini, L.; Capece, M.; Chiozzi, P.; Bianchi, G.; Kroemer, G.; Pistoia, V.; Di Virgilio, F. Expression of P2X7 receptor increases in vivo tumor growth. *Cancer Res.* **2012**, *72*, 2957–2969. [CrossRef] [PubMed]

125. Azimi, I.; Beilby, H.; Davis, F.M.; Marcial, D.L.; Kenny, P.A.; Thompson, E.W.; Roberts-Thomson, S.J.; Monteith, G.R. Altered purinergic receptor-Ca^{2+} signaling associated with hypoxia-induced epithelial-mesenchymal transition in breast cancer cells. *Mol. Oncol.* **2016**, *10*, 166–178. [CrossRef] [PubMed]

126. Fang, J.; Chen, X.; Wang, S.; Xie, T.; Du, X.; Liu, H.; Li, X.; Chen, J.; Zhang, B.; Liang, H.; et al. The expression of P2X(7) receptors in EPCs and their potential role in the targeting of EPCs to brain gliomas. *Cancer Biol. Ther.* **2015**, *16*, 498–510. [CrossRef] [PubMed]

127. Dragoni, S.; Guerra, G.; Fiorio Pla, A.; Bertoni, G.; Rappa, A.; Poletto, V.; Bottino, C.; Aronica, A.; Lodola, F.; Cinelli, M.P.; et al. A functional Transient Receptor Potential Vanilloid 4 (TRPV4) channel is expressed in human endothelial progenitor cells. *J. Cell. Physiol.* **2015**, *230*, 95–104. [CrossRef] [PubMed]

128. Hofmann, N.A.; Barth, S.; Waldeck-Weiermair, M.; Klec, C.; Strunk, D.; Malli, R.; Graier, W.F. TRPV1 mediates cellular uptake of anandamide and thus promotes endothelial cell proliferation and network-formation. *Biol. Open* **2014**, *3*, 1164–1172. [CrossRef] [PubMed]

129. Ong, H.L.; Jang, S.I.; Ambudkar, I.S. Distinct contributions of Orai1 and TRPC1 to agonist-induced $[Ca^{2+}](i)$ signals determine specificity of Ca^{2+}-dependent gene expression. *PLoS ONE* **2012**, *7*, e47146. [CrossRef] [PubMed]

130. Gueguinou, M.; Harnois, T.; Crottes, D.; Uguen, A.; Deliot, N.; Gambade, A.; Chantome, A.; Haelters, J.P.; Jaffres, P.A.; Jourdan, M.L.; et al. SK3/TRPC1/Orai1 complex regulates SOCE-dependent colon cancer cell migration: A novel opportunity to modulate anti-EGFR mAb action by the alkyl-lipid Ohmline. *Oncotarget* **2016**, *7*, 36168–36184. [CrossRef] [PubMed]

131. Moccia, F.; Dragoni, S.; Poletto, V.; Rosti, V.; Tanzi, F.; Ganini, C.; Porta, C. Orai1 and Transient Receptor Potential Channels as novel molecular targets to impair tumor neovascularisation in renal cell carcinoma and other malignancies. *Anticancer Agents Med. Chem.* **2014**, *14*, 296–312. [CrossRef] [PubMed]

132. Porta, C.; Giglione, P.; Paglino, C. Targeted therapy for renal cell carcinoma: Focus on 2nd and 3rd line. *Expert Opin. Pharmacother.* **2016**, *17*, 643–655. [CrossRef] [PubMed]

133. Prakriya, M.; Lewis, R.S. Store-Operated Calcium Channels. *Physiol. Rev.* **2015**, *95*, 1383–1436. [CrossRef] [PubMed]

134. Vashisht, A.; Trebak, M.; Motiani, R.K. STIM and Orai proteins as novel targets for cancer therapy. A Review in the Theme: Cell and Molecular Processes in Cancer Metastasis. *Am. J. Physiol. Cell Physiol.* **2015**, *309*, C457–C469. [CrossRef] [PubMed]

135. Moccia, F.; Zuccolo, E.; Poletto, V.; Turin, I.; Guerra, G.; Pedrazzoli, P.; Rosti, V.; Porta, C.; Montagna, D. Targeting Stim and Orai Proteins as an Alternative Approach in Anticancer Therapy. *Curr. Med. Chem.* **2016**, *23*, 3450–3480. [CrossRef] [PubMed]

136. Li, Y.; Guo, B.; Xie, Q.; Ye, D.; Zhang, D.; Zhu, Y.; Chen, H.; Zhu, B. STIM1 Mediates Hypoxia-Driven Hepatocarcinogenesis via Interaction with HIF-1. *Cell Rep.* **2015**, *12*, 388–395. [CrossRef] [PubMed]

137. Gudas, L.J.; Fu, L.; Minton, D.R.; Mongan, N.P.; Nanus, D.M. The role of HIF1alpha in renal cell carcinoma tumorigenesis. *J. Mol. Med. (Berl.)* **2014**, *92*, 825–836. [CrossRef] [PubMed]

138. Semenza, G.L. Regulation of the breast cancer stem cell phenotype by hypoxia-inducible factors. *Clin. Sci. (Lond.)* **2015**, *129*, 1037–1045. [CrossRef] [PubMed]

139. Wang, J.; Weigand, L.; Lu, W.; Sylvester, J.T.; Semenza, G.L.; Shimoda, L.A. Hypoxia inducible factor 1 mediates hypoxia-induced TRPC expression and elevated intracellular Ca^{2+} in pulmonary arterial smooth muscle cells. *Circ. Res.* **2006**, *98*, 1528–1537. [CrossRef] [PubMed]

140. Leaute-Labreze, C.; Harper, J.I.; Hoeger, P.H. Infantile haemangioma. *Lancet* **2017**, *390*, 85–94. [CrossRef]

141. Bischoff, J. Progenitor cells in infantile hemangioma. *J. Craniofac. Surg.* **2009**, *20* (Suppl. 1), 695–697. [CrossRef] [PubMed]

142. Khan, Z.A.; Melero-Martin, J.M.; Wu, X.; Paruchuri, S.; Boscolo, E.; Mulliken, J.B.; Bischoff, J. Endothelial progenitor cells from infantile hemangioma and umbilical cord blood display unique cellular responses to endostatin. *Blood* **2006**, *108*, 915–921. [CrossRef] [PubMed]

143. Brandman, O.; Liou, J.; Park, W.S.; Meyer, T. STIM2 is a feedback regulator that stabilizes basal cytosolic and endoplasmic reticulum Ca^{2+} levels. *Cell* **2007**, *131*, 1327–1339. [CrossRef] [PubMed]

144. Greenberger, S.; Bischoff, J. Infantile hemangioma-mechanism(s) of drug action on a vascular tumor. *Cold Spring Harb. Perspect. Med.* **2011**, *1*, a006460. [CrossRef] [PubMed]

145. Zoli, M.; Pucci, S.; Vilella, A.; Gotti, C. Neuronal and extraneuronal nicotinic acetylcholine receptors. *Curr. Neuropharmacol.* **2017**. [CrossRef]

146. Yakel, J.L. Nicotinic ACh receptors in the hippocampal circuit; functional expression and role in synaptic plasticity. *J. Physiol.* **2014**, *592*, 4147–4153. [CrossRef] [PubMed]

147. Egleton, R.D.; Brown, K.C.; Dasgupta, P. Nicotinic acetylcholine receptors in cancer: Multiple roles in proliferation and inhibition of apoptosis. *Trends Pharmacol. Sci.* **2008**, *29*, 151–158. [CrossRef] [PubMed]

148. Cooke, J.P.; Ghebremariam, Y.T. Endothelial nicotinic acetylcholine receptors and angiogenesis. *Trends Cardiovasc. Med.* **2008**, *18*, 247–253. [CrossRef] [PubMed]

149. Egleton, R.D.; Brown, K.C.; Dasgupta, P. Angiogenic activity of nicotinic acetylcholine receptors: Implications in tobacco-related vascular diseases. *Pharmacol. Ther.* **2009**, *121*, 205–223. [CrossRef] [PubMed]

150. Dom, A.M.; Buckley, A.W.; Brown, K.C.; Egleton, R.D.; Marcelo, A.J.; Proper, N.A.; Weller, D.E.; Shah, Y.H.; Lau, J.K.; Dasgupta, P. The alpha7-nicotinic acetylcholine receptor and MMP-2/-9 pathway mediate the proangiogenic effect of nicotine in human retinal endothelial cells. *Investig. Ophthalmol. Vis. Sci.* **2011**, *52*, 4428–4438. [CrossRef] [PubMed]

151. Smedlund, K.; Tano, J.Y.; Margiotta, J.; Vazquez, G. Evidence for operation of nicotinic and muscarinic acetylcholine receptor-dependent survival pathways in human coronary artery endothelial cells. *J. Cell Biochem.* **2011**, *112*, 1978–1984. [CrossRef] [PubMed]

152. Fucile, S. Ca^{2+} permeability of nicotinic acetylcholine receptors. *Cell Calcium* **2004**, *35*, 1–8. [CrossRef] [PubMed]

153. Shi, D.; Guo, W.; Chen, W.; Fu, L.; Wang, J.; Tian, Y.; Xiao, X.; Kang, T.; Huang, W.; Deng, W. Nicotine promotes proliferation of human nasopharyngeal carcinoma cells by regulating alpha7AChR, ERK, HIF-1alpha and VEGF/PEDF signaling. *PLoS ONE* **2012**, *7*, e43898.

154. Zhu, B.Q.; Heeschen, C.; Sievers, R.E.; Karliner, J.S.; Parmley, W.W.; Glantz, S.A.; Cooke, J.P. Second hand smoke stimulates tumor angiogenesis and growth. *Cancer Cell* **2003**, *4*, 191–196. [CrossRef]

155. Altaany, Z.; Moccia, F.; Munaron, L.; Mancardi, D.; Wang, R. Hydrogen sulfide and endothelial dysfunction: Relationship with nitric oxide. *Curr. Med. Chem.* **2014**, *21*, 3646–3661. [CrossRef] [PubMed]

156. Mancardi, D.; Pla, A.F.; Moccia, F.; Tanzi, F.; Munaron, L. Old and new gasotransmitters in the cardiovascular system: Focus on the role of nitric oxide and hydrogen sulfide in endothelial cells and cardiomyocytes. *Curr. Pharm. Biotechnol.* **2011**, *12*, 1406–1415. [CrossRef] [PubMed]

157. Wang, R. Physiological implications of hydrogen sulfide: A whiff exploration that blossomed. *Physiol. Rev.* **2012**, *92*, 791–896. [CrossRef] [PubMed]

158. Hartmann, C.; Nussbaum, B.; Calzia, E.; Radermacher, P.; Wepler, M. Gaseous Mediators and Mitochondrial Function: The Future of Pharmacologically Induced Suspended Animation? *Front. Physiol.* **2017**, *8*, 691. [CrossRef] [PubMed]

159. Munaron, L. Intracellular calcium, endothelial cells and angiogenesis. *Recent Pat. Anticancer Drug Discov.* **2006**, *1*, 105–119. [CrossRef] [PubMed]

160. Loboda, A.; Jozkowicz, A.; Dulak, J. Carbon monoxide: Pro- or anti-angiogenic agent? Comment on Ahmad et al. (Thromb Haemost 2015; 113: 329–337). *Thromb. Haemost.* **2015**, *114*, 432–433. [CrossRef] [PubMed]

161. Tran, A.N.; Boyd, N.H.; Walker, K.; Hjelmeland, A.B. NOS Expression and NO Function in Glioma and Implications for Patient Therapies. *Antioxid. Redox Signal.* **2017**, *26*, 986–999. [CrossRef] [PubMed]

162. Mocellin, S. Nitric oxide: Cancer target or anticancer agent? *Curr. Cancer Drug Targets* **2009**, *9*, 214–236. [CrossRef] [PubMed]

163. Basudhar, D.; Somasundaram, V.; de Oliveira, G.A.; Kesarwala, A.; Heinecke, J.L.; Cheng, R.Y.; Glynn, S.A.; Ambs, S.; Wink, D.A.; Ridnour, L.A. Nitric Oxide Synthase-2-Derived Nitric Oxide Drives Multiple Pathways of Breast Cancer Progression. *Antioxid. Redox Signal.* **2017**, *26*, 1044–1058. [CrossRef] [PubMed]

164. Fiorio Pla, A.; Genova, T.; Pupo, E.; Tomatis, C.; Genazzani, A.; Zaninetti, R.; Munaron, L. Multiple roles of protein kinase a in arachidonic acid-mediated Ca^{2+} entry and tumor-derived human endothelial cell migration. *Mol. Cancer Res.* **2010**, *8*, 1466–1476. [CrossRef] [PubMed]

165. Dragoni, S.; Laforenza, U.; Bonetti, E.; Reforgiato, M.; Poletto, V.; Lodola, F.; Bottino, C.; Guido, D.; Rappa, A.; Pareek, S.; et al. Enhanced expression of Stim, Orai, and TRPC transcripts and proteins in endothelial progenitor cells isolated from patients with primary myelofibrosis. *PLoS ONE* **2014**, *9*, e91099. [CrossRef] [PubMed]

166. Munaron, L.; Avanzato, D.; Moccia, F.; Mancardi, D. Hydrogen sulfide as a regulator of calcium channels. *Cell Calcium* **2013**, *53*, 77–84. [CrossRef] [PubMed]

167. Bauer, C.C.; Boyle, J.P.; Porter, K.E.; Peers, C. Modulation of Ca^{2+} signalling in human vascular endothelial cells by hydrogen sulfide. *Atherosclerosis* **2010**, *209*, 374–380. [CrossRef] [PubMed]

168. Moccia, F.; Bertoni, G.; Pla, A.F.; Dragoni, S.; Pupo, E.; Merlino, A.; Mancardi, D.; Munaron, L.; Tanzi, F. Hydrogen sulfide regulates intracellular Ca^{2+} concentration in endothelial cells from excised rat aorta. *Curr. Pharm. Biotechnol.* **2011**, *12*, 1416–1426. [CrossRef] [PubMed]

169. Katsouda, A.; Bibli, S.I.; Pyriochou, A.; Szabo, C.; Papapetropoulos, A. Regulation and role of endogenously produced hydrogen sulfide in angiogenesis. *Pharmacol. Res.* **2016**, *113 Pt A*, 175–185. [CrossRef] [PubMed]

170. Liu, Y.; Yang, R.; Liu, X.; Zhou, Y.; Qu, C.; Kikuiri, T.; Wang, S.; Zandi, E.; Du, J.; Ambudkar, I.S.; et al. Hydrogen Sulfide Maintains Mesenchymal Stem Cell Function and Bone Homeostasis via Regulation of Ca^{2+} Channel Sulfhydration. *Cell Stem Cell* **2014**, *15*, 66–78. [CrossRef] [PubMed]

171. Ujike, A.; Otsuguro, K.; Miyamoto, R.; Yamaguchi, S.; Ito, S. Bidirectional effects of hydrogen sulfide via ATP-sensitive K(+) channels and transient receptor potential A1 channels in RIN14B cells. *Eur. J. Pharmacol.* **2015**, *764*, 463–470. [CrossRef] [PubMed]

172. Yang, G.; Wu, L.; Jiang, B.; Yang, W.; Qi, J.; Cao, K.; Meng, Q.; Mustafa, A.K.; Mu, W.; Zhang, S.; et al. H2S as a physiologic vasorelaxant: Hypertension in mice with deletion of cystathionine gamma-lyase. *Science* **2008**, *322*, 587–590. [CrossRef] [PubMed]

173. Dedkova, E.N.; Blatter, L.A. Nitric oxide inhibits capacitative Ca^{2+} entry and enhances endoplasmic reticulum Ca^{2+} uptake in bovine vascular endothelial cells. *J. Physiol.* **2002**, *539 Pt 1*, 77–91. [CrossRef] [PubMed]

174. Saez, J.C.; Leybaert, L. Hunting for connexin hemichannels. *FEBS Lett.* **2014**, *588*, 1205–1211. [CrossRef] [PubMed]

175. Boeldt, D.S.; Krupp, J.; Yi, F.X.; Khurshid, N.; Shah, D.M.; Bird, I.M. Positive versus negative effects of VEGF165 on Ca^{2+} signaling and NO production in human endothelial cells. *Am. J. Physiol. Heart Circ. Physiol.* **2017**, *312*, H173–H181. [CrossRef] [PubMed]

176. Boittin, F.X.; Alonso, F.; Le Gal, L.; Allagnat, F.; Beny, J.L.; Haefliger, J.A. Connexins and M3 muscarinic receptors contribute to heterogeneous Ca^{2+} signaling in mouse aortic endothelium. *Cell. Physiol. Biochem.* **2013**, *31*, 166–178. [CrossRef] [PubMed]

177. Berra-Romani, R.; Raqeeb, A.; Torres-Jácome, J.; Guzman-Silva, A.; Guerra, G.; Tanzi, F.; Moccia, F. The mechanism of injury-induced intracellular calcium concentration oscillations in the endothelium of excised rat aorta. *J. Vasc. Res.* **2012**, *49*, 65–76. [CrossRef] [PubMed]

178. De Bock, M.; Culot, M.; Wang, N.; Bol, M.; Decrock, E.; De Vuyst, E.; da Costa, A.; Dauwe, I.; Vinken, M.; Simon, A.M.; et al. Connexin channels provide a target to manipulate brain endothelial calcium dynamics and blood-brain barrier permeability. *J. Cereb. Blood Flow Metab.* **2011**, *31*, 1942–1957. [CrossRef] [PubMed]

179. De Bock, M.; Wang, N.; Decrock, E.; Bol, M.; Gadicherla, A.K.; Culot, M.; Cecchelli, R.; Bultynck, G.; Leybaert, L. Endothelial calcium dynamics, connexin channels and blood-brain barrier function. *Prog. Neurobiol.* **2013**, *108*, 1–20. [CrossRef] [PubMed]

180. McLachlan, E.; Shao, Q.; Wang, H.L.; Langlois, S.; Laird, D.W. Connexins act as tumor suppressors in three-dimensional mammary cell organoids by regulating differentiation and angiogenesis. *Cancer Res.* **2006**, *66*, 9886–9894. [CrossRef] [PubMed]

181. Trosko, J.E.; Ruch, R.J. Gap junctions as targets for cancer chemoprevention and chemotherapy. *Curr. Drug Targets* **2002**, *3*, 465–482. [CrossRef] [PubMed]

182. Zhang, J.; O'Carroll, S.J.; Henare, K.; Ching, L.M.; Ormonde, S.; Nicholson, L.F.; Danesh-Meyer, H.V.; Green, C.R. Connexin hemichannel induced vascular leak suggests a new paradigm for cancer therapy. *FEBS Lett.* **2014**, *588*, 1365–1371. [CrossRef] [PubMed]

183. Counillon, L.; Bouret, Y.; Marchiq, I.; Pouyssegur, J. Na(+)/H(+) antiporter (NHE1) and lactate/H(+) symporters (MCTs) in pH homeostasis and cancer metabolism. *Biochim. Biophys. Acta* **2016**, *1863*, 2465–2480. [CrossRef] [PubMed]

184. Mo, X.G.; Chen, Q.W.; Li, X.S.; Zheng, M.M.; Ke, D.Z.; Deng, W.; Li, G.Q.; Jiang, J.; Wu, Z.Q.; Wang, L.; et al. Suppression of NHE1 by small interfering RNA inhibits HIF-1alpha-induced angiogenesis in vitro via modulation of calpain activity. *Microvasc. Res.* **2011**, *81*, 160–168. [CrossRef] [PubMed]

185. Ayajiki, K.; Kindermann, M.; Hecker, M.; Fleming, I.; Busse, R. Intracellular pH and tyrosine phosphorylation but not calcium determine shear stress-induced nitric oxide production in native endothelial cells. *Circ. Res.* **1996**, *78*, 750–758. [CrossRef] [PubMed]

186. Yuen, N.; Lam, T.I.; Wallace, B.K.; Klug, N.R.; Anderson, S.E.; O'Donnell, M.E. Ischemic factor-induced increases in cerebral microvascular endothelial cell Na/H exchange activity and abundance: Evidence for involvement of ERK1/2 MAP kinase. *Am. J. Physiol. Cell Physiol.* **2014**, *306*, C931–C942. [CrossRef] [PubMed]

187. Ghigo, D.; Bussolino, F.; Garbarino, G.; Heller, R.; Turrini, F.; Pescarmona, G.; Cragoe, E.J., Jr.; Pegoraro, L.; Bosia, A. Role of Na^+/H^+ exchange in thrombin-induced platelet-activating factor production by human endothelial cells. *J. Biol. Chem.* **1988**, *263*, 19437–19446. [PubMed]

188. Siffert, W.; Akkerman, J.W. Na^+/H^+ exchange and Ca^{2+} influx. *FEBS Lett.* **1989**, *259*, 1–4. [CrossRef]

189. Danthuluri, N.R.; Kim, D.; Brock, T.A. Intracellular alkalinization leads to Ca^{2+} mobilization from agonist-sensitive pools in bovine aortic endothelial cells. *J. Biol. Chem.* **1990**, *265*, 19071–19076. [PubMed]

190. Nishio, K.; Suzuki, Y.; Takeshita, K.; Aoki, T.; Kudo, H.; Sato, N.; Naoki, K.; Miyao, N.; Ishii, M.; Yamaguchi, K. Effects of hypercapnia and hypocapnia on $[Ca^{2+}]i$ mobilization in human pulmonary artery endothelial cells. *J. Appl. Physiol. (1985)* **2001**, *90*, 2094–2100. [CrossRef] [PubMed]

191. Amith, S.R.; Fliegel, L. Regulation of the Na^+/H^+ Exchanger (NHE1) in Breast Cancer Metastasis. *Cancer Res.* **2013**, *73*, 1259–1264. [CrossRef] [PubMed]

192. Reshkin, S.J.; Greco, M.R.; Cardone, R.A. Role of pHi, and proton transporters in oncogene-driven neoplastic transformation. *Philos. Trans. R. Soc. Lond. B Biol. Sci.* **2014**, *369*, 20130100. [CrossRef] [PubMed]

193. Pedersen, A.K.; Mendes Lopes de Melo, J.; Morup, N.; Tritsaris, K.; Pedersen, S.F. Tumor microenvironment conditions alter Akt and Na^+/H^+ exchanger NHE1 expression in endothelial cells more than hypoxia alone: Implications for endothelial cell function in cancer. *BMC Cancer* **2017**, *17*, 542. [CrossRef] [PubMed]

194. Orive, G.; Reshkin, S.J.; Harguindey, S.; Pedraz, J.L. Hydrogen ion dynamics and the Na^+/H^+ exchanger in cancer angiogenesis and antiangiogenesis. *Br. J. Cancer* **2003**, *89*, 1395–1399. [CrossRef] [PubMed]

195. Spugnini, E.P.; Sonveaux, P.; Stock, C.; Perez-Sayans, M.; De Milito, A.; Avnet, S.; Garcia, A.G.; Harguindey, S.; Fais, S. Proton channels and exchangers in cancer. *Biochim. Biophys. Acta* **2015**, *1848 Pt B*, 2715–2726. [CrossRef] [PubMed]

196. Morgan, A.J.; Platt, F.M.; Lloyd-Evans, E.; Galione, A. Molecular mechanisms of endolysosomal Ca^{2+} signalling in health and disease. *Biochem. J.* **2011**, *439*, 349–374. [CrossRef] [PubMed]

197. Jha, A.; Brailoiu, E.; Muallem, S. How does NAADP release lysosomal Ca^{2+}? *Channels* **2014**, *8*, 174–175. [CrossRef] [PubMed]

198. Patel, S.; Docampo, R. Acidic calcium stores open for business: Expanding the potential for intracellular Ca^{2+} signaling. *Trends Cell Biol.* **2010**, *20*, 277–286. [CrossRef] [PubMed]

199. Cosker, F.; Cheviron, N.; Yamasaki, M.; Menteyne, A.; Lund, F.E.; Moutin, M.J.; Galione, A.; Cancela, J.M. The ecto-enzyme CD38 is a nicotinic acid adenine dinucleotide phosphate (NAADP) synthase that couples receptor activation to Ca^{2+} mobilization from lysosomes in pancreatic acinar cells. *J. Biol. Chem.* **2010**, *285*, 38251–38259. [CrossRef] [PubMed]

200. Galione, A. A primer of NAADP-mediated Ca^{2+} signalling: From sea urchin eggs to mammalian cells. *Cell Calcium* **2015**, *58*, 27–47. [CrossRef] [PubMed]

201. Moccia, F.; Nusco, G.A.; Lim, D.; Kyozuka, K.; Santella, L. NAADP and InsP3 play distinct roles at fertilization in starfish oocytes. *Dev. Biol.* **2006**, *294*, 24–38. [CrossRef] [PubMed]

202. Favia, A.; Desideri, M.; Gambara, G.; D'Alessio, A.; Ruas, M.; Esposito, B.; Del Bufalo, D.; Parrington, J.; Ziparo, E.; Palombi, F.; et al. VEGF-induced neoangiogenesis is mediated by NAADP and two-pore channel-2-dependent Ca^{2+} signaling. *Proc. Natl. Acad. Sci. USA* **2014**, *111*, E4706–E4715. [CrossRef] [PubMed]

203. Brailoiu, G.C.; Gurzu, B.; Gao, X.; Parkesh, R.; Aley, P.K.; Trifa, D.I.; Galione, A.; Dun, N.J.; Madesh, M.; Patel, S.; et al. Acidic NAADP-sensitive calcium stores in the endothelium: Agonist-specific recruitment and role in regulating blood pressure. *J. Biol. Chem.* **2010**, *285*, 37133–37137. [CrossRef] [PubMed]

204. Di Nezza, F.; Zuccolo, E.; Poletto, V.; Rosti, V.; De Luca, A.; Moccia, F.; Guerra, G.; Ambrosone, L. Liposomes as a Putative Tool to Investigate NAADP Signaling in Vasculogenesis. *J. Cell. Biochem.* **2017**, *118*, 3722–3729. [CrossRef] [PubMed]

205. Favia, A.; Pafumi, I.; Desideri, M.; Padula, F.; Montesano, C.; Passeri, D.; Nicoletti, C.; Orlandi, A.; Del Bufalo, D.; Sergi, M.; et al. NAADP-Dependent Ca^{2+} Signaling Controls Melanoma Progression, Metastatic Dissemination and Neoangiogenesis. *Sci. Rep.* **2016**, *6*, 18925. [CrossRef] [PubMed]

206. Schaefer, M.; Plant, T.D.; Obukhov, A.G.; Hofmann, T.; Gudermann, T.; Schultz, G. Receptor-mediated regulation of the nonselective cation channels TRPC4 and TRPC5. *J. Biol. Chem.* **2000**, *275*, 17517–17526. [CrossRef] [PubMed]

207. Zholos, A.V. TRPC5. *Handb. Exp. Pharmacol.* **2014**, *222*, 129–156. [PubMed]

208. Yuan, J.P.; Zeng, W.; Huang, G.N.; Worley, P.F.; Muallem, S. STIM1 heteromultimerizes TRPC channels to determine their function as store-operated channels. *Nat. Cell Biol.* **2007**, *9*, 636–645. [CrossRef] [PubMed]

209. DeHaven, W.I.; Jones, B.F.; Petranka, J.G.; Smyth, J.T.; Tomita, T.; Bird, G.S.; Putney, J.W., Jr. TRPC channels function independently of STIM1 and Orai1. *J. Physiol.* **2009**, *587 Pt 10*, 2275–2298. [CrossRef] [PubMed]

210. Zimmermann, K.; Lennerz, J.K.; Hein, A.; Link, A.S.; Kaczmarek, J.S.; Delling, M.; Uysal, S.; Pfeifer, J.D.; Riccio, A.; Clapham, D.E. Transient receptor potential cation channel, subfamily C, member 5 (TRPC5) is a cold-transducer in the peripheral nervous system. *Proc. Natl. Acad. Sci. USA* **2011**, *108*, 18114–18119. [CrossRef] [PubMed]

211. Antigny, F.; Girardin, N.; Frieden, M. Transient Receptor Potential Canonical Channels Are Required for in Vitro Endothelial Tube Formation. *J. Biol. Chem.* **2012**, *287*, 5917–5927. [CrossRef] [PubMed]

212. Chaudhuri, P.; Colles, S.M.; Bhat, M.; Van Wagoner, D.R.; Birnbaumer, L.; Graham, L.M. Elucidation of a TRPC6-TRPC5 channel cascade that restricts endothelial cell movement. *Mol. Biol. Cell* **2008**, *19*, 3203–3211. [CrossRef] [PubMed]

213. Chaudhuri, P.; Rosenbaum, M.A.; Birnbaumer, L.; Graham, L.M. Integration of TRPC6 and NADPH oxidase activation in lysophosphatidylcholine-induced TRPC5 externalization. *Am. J. Physiol. Cell Physiol.* **2017**, *313*, C541–C555. [CrossRef] [PubMed]

214. Ma, X.; Cai, Y.; He, D.; Zou, C.; Zhang, P.; Lo, C.Y.; Xu, Z.; Chan, F.L.; Yu, S.; Chen, Y.; et al. Transient receptor potential channel TRPC5 is essential for P-glycoprotein induction in drug-resistant cancer cells. *Proc. Natl. Acad. Sci. USA* **2012**, *109*, 16282–16287. [CrossRef] [PubMed]

215. Ma, X.; Chen, Z.; Hua, D.; He, D.; Wang, L.; Zhang, P.; Wang, J.; Cai, Y.; Gao, C.; Zhang, X.; et al. Essential role for TrpC5-containing extracellular vesicles in breast cancer with chemotherapeutic resistance. *Proc. Natl. Acad. Sci. USA* **2014**, *111*, 6389–6394. [CrossRef] [PubMed]

216. Wang, T.; Chen, Z.; Zhu, Y.; Pan, Q.; Liu, Y.; Qi, X.; Jin, L.; Jin, J.; Ma, X.; Hua, D. Inhibition of transient receptor potential channel 5 reverses 5-Fluorouracil resistance in human colorectal cancer cells. *J. Biol. Chem.* **2015**, *290*, 448–456. [CrossRef] [PubMed]

217. Pokharel, D.; Roseblade, A.; Oenarto, V.; Lu, J.F.; Bebawy, M. Proteins regulating the intercellular transfer and function of P-glycoprotein in multidrug-resistant cancer. *Ecancermedicalscience* **2017**, *11*, 768. [CrossRef] [PubMed]

218. He, D.X.; Gu, X.T.; Jiang, L.; Jin, J.; Ma, X. A methylation-based regulatory network for microRNA 320a in chemoresistant breast cancer. *Mol. Pharmacol.* **2014**, *86*, 536–547. [CrossRef] [PubMed]

219. Zhu, Y.; Pan, Q.; Meng, H.; Jiang, Y.; Mao, A.; Wang, T.; Hua, D.; Yao, X.; Jin, J.; Ma, X. Enhancement of vascular endothelial growth factor release in long-term drug-treated breast cancer via transient receptor potential channel 5-Ca^{2+}-hypoxia-inducible factor 1alpha pathway. *Pharmacol. Res.* **2015**, *93*, 36–42. [CrossRef] [PubMed]

220. He, D.X.; Ma, X. Transient receptor potential channel C5 in cancer chemoresistance. *Acta Pharmacol. Sin.* **2016**, *37*, 19–24. [CrossRef] [PubMed]

221. Messai, Y.; Noman, M.Z.; Hasmim, M.; Janji, B.; Tittarelli, A.; Boutet, M.; Baud, V.; Viry, E.; Billot, K.; Nanbakhsh, A.; et al. ITPR1 protects renal cancer cells against natural killer cells by inducing autophagy. *Cancer Res.* **2014**, *74*, 6820–6832. [CrossRef] [PubMed]

222. Lim, D.; Bertoli, A.; Sorgato, M.C.; Moccia, F. Generation and usage of aequorin lentiviral vectors for Ca^{2+} measurement in sub-cellular compartments of hard-to-transfect cells. *Cell Calcium* **2016**, *59*, 228–239. [CrossRef] [PubMed]

223. De la Puente, P.; Muz, B.; Azab, F.; Azab, A.K. Cell trafficking of endothelial progenitor cells in tumor progression. *Clin. Cancer. Res.* **2013**, *19*, 3360–3368. [CrossRef] [PubMed]

224. Escudier, B.; Szczylik, C.; Porta, C.; Gore, M. Treatment selection in metastatic renal cell carcinoma: Expert consensus. *Nat. Rev. Clin. Oncol.* **2012**, *9*, 327–337. [CrossRef] [PubMed]

225. Porta, C.; Paglino, C.; Imarisio, I.; Canipari, C.; Chen, K.; Neary, M.; Duh, M.S. Safety and treatment patterns of multikinase inhibitors in patients with metastatic renal cell carcinoma at a tertiary oncology center in Italy. *BMC Cancer* **2011**, *11*. [CrossRef] [PubMed]

226. Porta, C.; Paglino, C.; Imarisio, I.; Ganini, C.; Sacchi, L.; Quaglini, S.; Giunta, V.; De Amici, M. Changes in circulating pro-angiogenic cytokines, other than VEGF, before progression to sunitinib therapy in advanced renal cell carcinoma patients. *Oncology* **2013**, *84*, 115–122. [CrossRef] [PubMed]

227. Ribatti, D. Tumor refractoriness to anti-VEGF therapy. *Oncotarget* **2016**, *7*, 46668–46677. [CrossRef] [PubMed]

228. Hiles, J.J.; Kolesar, J.M. Role of sunitinib and sorafenib in the treatment of metastatic renal cell carcinoma. *Am. J. Health Syst. Pharm.* **2008**, *65*, 123–131. [CrossRef] [PubMed]

229. Naito, H.; Wakabayashi, T.; Kidoya, H.; Muramatsu, F.; Takara, K.; Eino, D.; Yamane, K.; Iba, T.; Takakura, N. Endothelial Side Population Cells Contribute to Tumor Angiogenesis and Antiangiogenic Drug Resistance. *Cancer Res.* **2016**, *76*, 3200–3210. [CrossRef] [PubMed]

230. Loges, S.; Schmidt, T.; Carmeliet, P. Mechanisms of resistance to anti-angiogenic therapy and development of third-generation anti-angiogenic drug candidates. *Genes Cancer* **2010**, *1*, 12–25. [CrossRef] [PubMed]

231. Sunryd, J.C.; Cheon, B.; Graham, J.B.; Giorda, K.M.; Fissore, R.A.; Hebert, D.N. TMTC1 and TMTC2 are novel endoplasmic reticulum tetratricopeptide repeat-containing adapter proteins involved in calcium homeostasis. *J. Biol. Chem.* **2014**, *289*, 16085–16099. [CrossRef] [PubMed]

232. Sammels, E.; Parys, J.B.; Missiaen, L.; De Smedt, H.; Bultynck, G. Intracellular Ca^{2+} storage in health and disease: A dynamic equilibrium. *Cell Calcium* **2010**, *47*, 297–314. [CrossRef] [PubMed]

233. Schuck, S.; Prinz, W.A.; Thorn, K.S.; Voss, C.; Walter, P. Membrane expansion alleviates endoplasmic reticulum stress independently of the unfolded protein response. *J. Cell Biol.* **2009**, *187*, 525–536. [CrossRef] [PubMed]

234. Dubois, C.; Vanden Abeele, F.; Prevarskaya, N. Targeting apoptosis by the remodelling of calcium-transporting proteins in cancerogenesis. *FEBS J.* **2013**, *280*, 5500–5510. [CrossRef] [PubMed]

235. Prevarskaya, N.; Skryma, R.; Shuba, Y. Ion channels and the hallmarks of cancer. *Trends Mol. Med.* **2010**, *16*, 107–121. [CrossRef] [PubMed]

236. Prevarskaya, N.; Skryma, R.; Shuba, Y. Targeting Ca^{2+} transport in cancer: Close reality or long perspective? *Expert Opin. Ther. Targets* **2013**, *17*, 225–241. [CrossRef] [PubMed]

237. Vanoverberghe, K.; Vanden Abeele, F.; Mariot, P.; Lepage, G.; Roudbaraki, M.; Bonnal, J.L.; Mauroy, B.; Shuba, Y.; Skryma, R.; Prevarskaya, N. Ca^{2+} homeostasis and apoptotic resistance of neuroendocrine-differentiated prostate cancer cells. *Cell Death Differ.* **2004**, *11*, 321–330. [CrossRef] [PubMed]

238. Cui, C.; Merritt, R.; Fu, L.; Pan, Z. Targeting calcium signaling in cancer therapy. *Acta Pharm. Sin. B* **2017**, *7*, 3–17. [CrossRef] [PubMed]

239. Bonora, M.; Giorgi, C.; Pinton, P. Novel frontiers in calcium signaling: A possible target for chemotherapy. *Pharmacol. Res.* **2015**, *99*, 82–85. [CrossRef] [PubMed]

240. Leanza, L.; Manago, A.; Zoratti, M.; Gulbins, E.; Szabo, I. Pharmacological targeting of ion channels for cancer therapy: In vivo evidences. *Biochim. Biophys. Acta* **2016**, *1863 Pt B*, 1385–1397. [CrossRef] [PubMed]

241. Nielsen, N.; Lindemann, O.; Schwab, A. TRP channels and STIM/ORAI proteins: Sensors and effectors of cancer and stroma cell migration. *Br. J. Pharmacol.* **2014**, *171*, 5524–5540. [CrossRef] [PubMed]

242. Munaron, L. Systems biology of ion channels and transporters in tumor angiogenesis: An omics view. *Biochim. Biophys. Acta* **2015**, *1848 Pt B*, 2647–2656. [CrossRef] [PubMed]

243. Kito, H.; Yamamura, H.; Suzuki, Y.; Yamamura, H.; Ohya, S.; Asai, K.; Imaizumi, Y. Regulation of store-operated Ca^{2+} entry activity by cell cycle dependent up-regulation of Orai2 in brain capillary endothelial cells. *Biochem. Biophys. Res. Commun.* **2015**, *459*, 457–462. [CrossRef] [PubMed]

244. Choi, D.; Park, E.; Jung, E.; Seong, Y.J.; Hong, M.; Lee, S.; Burford, J.; Gyarmati, G.; Peti-Peterdi, J.; Srikanth, S.; et al. ORAI1 Activates Proliferation of Lymphatic Endothelial Cells in Response to Laminar Flow Through Kruppel-Like Factors 2 and 4. *Circ. Res.* **2017**, *120*, 1426–1439. [CrossRef] [PubMed]

245. Tian, C.; Du, L.; Zhou, Y.; Li, M. Store-operated CRAC channel inhibitors: Opportunities and challenges. *Future Med. Chem.* **2016**, *8*, 817–832. [CrossRef] [PubMed]

246. Luzzi, K.J.; Varghese, H.J.; MacDonald, I.C.; Schmidt, E.E.; Kohn, E.C.; Morris, V.L.; Marshall, K.E.; Chambers, A.F.; Groom, A.C. Inhibition of angiogenesis in liver metastases by carboxyamidotriazole (CAI). *Angiogenesis* **1998**, *2*, 373–379. [CrossRef] [PubMed]

247. Oliver, V.K.; Patton, A.M.; Desai, S.; Lorang, D.; Libutti, S.K.; Kohn, E.C. Regulation of the pro-angiogenic microenvironment by carboxyamido-triazole. *J. Cell. Physiol.* **2003**, *197*, 139–148. [CrossRef] [PubMed]

248. Patton, A.M.; Kassis, J.; Doong, H.; Kohn, E.C. Calcium as a molecular target in angiogenesis. *Curr. Pharm. Des.* **2003**, *9*, 543–551. [CrossRef] [PubMed]

249. Mignen, O.; Brink, C.; Enfissi, A.; Nadkarni, A.; Shuttleworth, T.J.; Giovannucci, D.R.; Capiod, T. Carboxyamidotriazole-induced inhibition of mitochondrial calcium import blocks capacitative calcium entry and cell proliferation in HEK-293 cells. *J. Cell Sci.* **2005**, *118 Pt 23*, 5615–5623. [CrossRef] [PubMed]

250. Enfissi, A.; Prigent, S.; Colosetti, P.; Capiod, T. The blocking of capacitative calcium entry by 2-aminoethyl diphenylborate (2-APB) and carboxyamidotriazole (CAI) inhibits proliferation in Hep G2 and Huh-7 human hepatoma cells. *Cell Calcium* **2004**, *36*, 459–467. [CrossRef] [PubMed]

251. Faehling, M.; Kroll, J.; Fohr, K.J.; Fellbrich, G.; Mayr, U.; Trischler, G.; Waltenberger, J. Essential role of calcium in vascular endothelial growth factor A-induced signaling: Mechanism of the antiangiogenic effect of carboxyamidotriazole. *FASEB J.* **2002**, *16*, 1805–1807. [CrossRef] [PubMed]

252. Wu, Y.; Palad, A.J.; Wasilenko, W.J.; Blackmore, P.F.; Pincus, W.A.; Schechter, G.L.; Spoonster, J.R.; Kohn, E.C.; Somers, K.D. Inhibition of head and neck squamous cell carcinoma growth and invasion by the calcium influx inhibitor carboxyamido-triazole. *Clin. Cancer. Res.* **1997**, *3*, 1915–1921. [PubMed]

253. Moody, T.W.; Chiles, J.; Moody, E.; Sieczkiewicz, G.J.; Kohn, E.C. CAI inhibits the growth of small cell lung cancer cells. *Lung Cancer* **2003**, *39*, 279–288. [CrossRef]

254. Hussain, M.M.; Kotz, H.; Minasian, L.; Premkumar, A.; Sarosy, G.; Reed, E.; Zhai, S.; Steinberg, S.M.; Raggio, M.; Oliver, V.K.; et al. Phase II trial of carboxyamidotriazole in patients with relapsed epithelial ovarian cancer. *J. Clin. Oncol.* **2003**, *21*, 4356–4363. [CrossRef] [PubMed]
255. Griffioen, A.W.; Molema, G. Angiogenesis: Potentials for pharmacologic intervention in the treatment of cancer, cardiovascular diseases, and chronic inflammation. *Pharmacol. Rev.* **2000**, *52*, 237–268. [PubMed]
256. Stadler, W.M.; Rosner, G.; Small, E.; Hollis, D.; Rini, B.; Zaentz, S.D.; Mahoney, J.; Ratain, M.J. Successful implementation of the randomized discontinuation trial design: An application to the study of the putative antiangiogenic agent carboxyaminoimidazole in renal cell carcinoma—CALGB 69901. *J. Clin. Oncol.* **2005**, *23*, 3726–3732. [CrossRef] [PubMed]
257. Rahman, S.; Rahman, T. Unveiling some FDA-approved drugs as inhibitors of the store-operated Ca^{2+} entry pathway. *Sci. Rep.* **2017**, *7*, 12881. [CrossRef] [PubMed]
258. Viallard, C.; Larrivee, B. Tumor angiogenesis and vascular normalization: Alternative therapeutic targets. *Angiogenesis* **2017**, *20*, 409–426. [CrossRef] [PubMed]
259. Vincent, F.; Duncton, M.A. TRPV4 agonists and antagonists. *Curr. Top. Med. Chem.* **2011**, *11*, 2216–2226. [CrossRef] [PubMed]
260. Coddou, C.; Yan, Z.; Obsil, T.; Huidobro-Toro, J.P.; Stojilkovic, S.S. Activation and regulation of purinergic P2X receptor channels. *Pharmacol. Rev.* **2011**, *63*, 641–683. [CrossRef] [PubMed]
261. Pafumi, I.; Favia, A.; Gambara, G.; Papacci, F.; Ziparo, E.; Palombi, F.; Filippini, A. Regulation of Angiogenic Functions by Angiopoietins through Calcium-Dependent Signaling Pathways. *BioMed Res. Int.* **2015**, *2015*, 965271. [CrossRef] [PubMed]
262. Rubaiy, H.N.; Ludlow, M.J.; Bon, R.S.; Beech, D.J. Pico145—Powerful new tool for TRPC1/4/5 channels. *Channels* **2017**, *11*, 362–364. [CrossRef] [PubMed]
263. Naylor, J.; Minard, A.; Gaunt, H.J.; Amer, M.S.; Wilson, L.A.; Migliore, M.; Cheung, S.Y.; Rubaiy, H.N.; Blythe, N.M.; Musialowski, K.E.; et al. Natural and synthetic flavonoid modulation of TRPC5 channels. *Br. J. Pharmacol.* **2016**, *173*, 562–574. [CrossRef] [PubMed]
264. Zhu, Y.; Lu, Y.; Qu, C.; Miller, M.; Tian, J.; Thakur, D.P.; Zhu, J.; Deng, Z.; Hu, X.; Wu, M.; et al. Identification and optimization of 2-aminobenzimidazole derivatives as novel inhibitors of TRPC4 and TRPC5 channels. *Br. J. Pharmacol.* **2015**, *172*, 3495–3509. [CrossRef] [PubMed]
265. Miller, M.; Shi, J.; Zhu, Y.; Kustov, M.; Tian, J.B.; Stevens, A.; Wu, M.; Xu, J.; Long, S.; Yang, P.; et al. Identification of ML204, a novel potent antagonist that selectively modulates native TRPC4/C5 ion channels. *J. Biol. Chem.* **2011**, *286*, 33436–33446. [CrossRef] [PubMed]
266. Majeed, Y.; Amer, M.S.; Agarwal, A.K.; McKeown, L.; Porter, K.E.; O'Regan, D.J.; Naylor, J.; Fishwick, C.W.; Muraki, K.; Beech, D.J. Stereo-selective inhibition of transient receptor potential TRPC5 cation channels by neuroactive steroids. *Br. J. Pharmacol.* **2011**, *162*, 1509–1520. [CrossRef] [PubMed]
267. Giorgi, C.; Bonora, M.; Missiroli, S.; Poletti, F.; Ramirez, F.G.; Morciano, G.; Morganti, C.; Pandolfi, P.P.; Mammano, F.; Pinton, P. Intravital imaging reveals p53-dependent cancer cell death induced by phototherapy via calcium signaling. *Oncotarget* **2015**, *6*, 1435–1445. [CrossRef] [PubMed]
268. Doan, N.T.; Paulsen, E.S.; Sehgal, P.; Moller, J.V.; Nissen, P.; Denmeade, S.R.; Isaacs, J.T.; Dionne, C.A.; Christensen, S.B. Targeting thapsigargin towards tumors. *Steroids* **2015**, *97*, 2–7. [CrossRef] [PubMed]
269. Denmeade, S.R.; Mhaka, A.M.; Rosen, D.M.; Brennen, W.N.; Dalrymple, S.; Dach, I.; Olesen, C.; Gurel, B.; Demarzo, A.M.; Wilding, G.; et al. Engineering a prostate-specific membrane antigen-activated tumor endothelial cell prodrug for cancer therapy. *Sci. Transl. Med.* **2012**, *4*, 140ra86. [CrossRef] [PubMed]
270. Liu, H.; Moy, P.; Kim, S.; Xia, Y.; Rajasekaran, A.; Navarro, V.; Knudsen, B.; Bander, N.H. Monoclonal antibodies to the extracellular domain of prostate-specific membrane antigen also react with tumor vascular endothelium. *Cancer Res.* **1997**, *57*, 3629–3634. [PubMed]
271. Quynh Doan, N.T.; Christensen, S.B. Thapsigargin, Origin, Chemistry, Structure-Activity Relationships and Prodrug Development. *Curr. Pharm. Des.* **2015**, *21*, 5501–5517. [CrossRef] [PubMed]
272. Mahalingam, D.; Wilding, G.; Denmeade, S.; Sarantopoulas, J.; Cosgrove, D.; Cetnar, J.; Azad, N.; Bruce, J.; Kurman, M.; Allgood, V.E.; et al. Mipsagargin, a novel thapsigargin-based PSMA-activated prodrug: Results of a first-in-man phase I clinical trial in patients with refractory, advanced or metastatic solid tumours. *Br. J. Cancer* **2016**, *114*, 986–994. [CrossRef] [PubMed]
273. Monteith, G.R.; McAndrew, D.; Faddy, H.M.; Roberts-Thomson, S.J. Calcium and cancer: Targeting Ca^{2+} transport. *Nat. Rev. Cancer* **2007**, *7*, 519–530. [CrossRef] [PubMed]

274. Zhu, H.; Zhang, H.; Jin, F.; Fang, M.; Huang, M.; Yang, C.S.; Chen, T.; Fu, L.; Pan, Z. Elevated Orai1 expression mediates tumor-promoting intracellular Ca^{2+} oscillations in human esophageal squamous cell carcinoma. *Oncotarget* **2014**, *5*, 3455–3471. [CrossRef] [PubMed]

275. Dudley, A.C. Tumor endothelial cells. *Cold Spring Harb. Perspect. Med.* **2012**, *2*, a006536. [CrossRef] [PubMed]

276. Fiorio Pla, A.; Gkika, D. Emerging role of TRP channels in cell migration: From tumor vascularization to metastasis. *Front. Physiol.* **2013**, *4*, 311. [CrossRef] [PubMed]

277. Munaron, L.; Genova, T.; Avanzato, D.; Antoniotti, S.; Fiorio Pla, A. Targeting calcium channels to block tumor vascularization. *Recent Pat. Anticancer Drug Discov.* **2013**, *8*, 27–37. [CrossRef] [PubMed]

278. Motiani, R.K.; Abdullaev, I.F.; Trebak, M. A novel native store-operated calcium channel encoded by Orai3: Selective requirement of Orai3 versus Orai1 in estrogen receptor-positive versus estrogen receptor-negative breast cancer cells. *J. Biol. Chem.* **2010**, *285*, 19173–19183. [CrossRef] [PubMed]

279. Andrikopoulos, P.; Baba, A.; Matsuda, T.; Djamgoz, M.B.; Yaqoob, M.M.; Eccles, S.A. Ca^{2+} influx through reverse mode Na^+/Ca^{2+} exchange is critical for vascular endothelial growth factor-mediated extracellular signal-regulated kinase (ERK) 1/2 activation and angiogenic functions of human endothelial cells. *J. Biol. Chem.* **2011**, *286*, 37919–37931. [CrossRef] [PubMed]

280. Hogan-Cann, A.D.; Anderson, C.M. Physiological Roles of Non-Neuronal NMDA Receptors. *Trends Pharmacol. Sci.* **2016**, *37*, 750–767. [CrossRef] [PubMed]

281. Stobart, J.L.; Lu, L.; Anderson, H.D.; Mori, H.; Anderson, C.M. Astrocyte-induced cortical vasodilation is mediated by D-serine and endothelial nitric oxide synthase. *Proc. Natl. Acad. Sci. USA* **2013**, *110*, 3149–3154. [CrossRef] [PubMed]

282. Takano, T.; Lin, J.H.; Arcuino, G.; Gao, Q.; Yang, J.; Nedergaard, M. Glutamate release promotes growth of malignant gliomas. *Nat. Med.* **2001**, *7*, 1010–1015. [CrossRef] [PubMed]

10

Calcium Ion Channels: Roles in Infection and Sepsis Mechanisms of Calcium Channel Blocker Benefits in Immunocompromised Patients at Risk for Infection

John A. D'Elia and Larry A. Weinrauch *

E P Joslin Research Laboratory, Kidney and Hypertension Section, Joslin Diabetes Center, Department of Medicine, Mount Auburn Hospital, Harvard Medical School, Boston and Cambridge, 521 Mount Auburn Street Watertown, MA 02472, USA; jd'elia@joslin.harvard.edu
* Correspondence: lweinrauch@hms.harvard.edu

Abstract: Immunosuppression may occur for a number of reasons related to an individual's frailty, debility, disease or from therapeutic iatrogenic intervention or misadventure. A large percentage of morbidity and mortality in immunodeficient populations is related to an inadequate response to infectious agents with slow response to antibiotics, enhancements of antibiotic resistance in populations, and markedly increased prevalence of acute inflammatory response, septic and infection related death. Given known relationships between intracellular calcium ion concentrations and cytotoxicity and cellular death, we looked at currently available data linking blockade of calcium ion channels and potential decrease in expression of sepsis among immunosuppressed patients. Notable are relationships between calcium, calcium channel, vitamin D mechanisms associated with sepsis and demonstration of antibiotic-resistant pathogens that may utilize channels sensitive to calcium channel blocker. We note that sepsis shock syndrome represents loss of regulation of inflammatory response to infection and that vitamin D, parathyroid hormone, fibroblast growth factor, and klotho interact with sepsis defense mechanisms in which movement of calcium and phosphorus are part of the process. Given these observations we consider that further investigation of the effect of relatively inexpensive calcium channel blockade agents of infections in immunosuppressed populations might be worthwhile.

Keywords: calcium ion channels; calcium channel blockade; sepsis; infection; immunosuppression

1. Introduction

Immunosuppression may occur for a number of reasons related to an individual's frailty, debility, disease or from therapeutic iatrogenic intervention or misadventure. A large percentage of morbidity and mortality in immunodeficient populations is related to an inadequate response to infectious agents with slow response to antibiotics, enhancements of antibiotic resistance and markedly increased prevalence of acute inflammatory response, septic and infection related death. Given the known relationships between intracellular calcium ion concentrations, cytotoxicity and cellular death, we review currently available data linking blockade of calcium ion channels and potential decrease in expression of sepsis among immunosuppressed patients. We consider possibilities for therapeutic interference with calcium ion channels that may alter immune responses to invading organisms in immunocompromised populations (Table 1).

Table 1. Pathophysiologic Interactions of calcium ion channels with the production of pathogen mediated sepsis syndrome: Potential for therapeutic interference with calcium ion channels.

Pathogenic Organism	Calcium Ion Channel Effect
Organ specific toxicity	Envelope protein increases toxic level of cytosolic calcium in host cell
Cell membrane or wall	Antibiotic efflux may be diminished by calcium channel blockade
Organism replication	Interference with calcium dependent RNA transcription
Cytoplasm (mitochondria)	Potential to interrupt intracellular calcium shifts and interrupt calcium efflux
Host Defense	**Calcium Ion Channel Effect**
Immunocompetence	Calcium activated potassium channels regulate 　Lymphocyte activation 　Mitogenesis 　Cell volume
Cellular Immunity 　Permeability, necrosis, apoptosis	Release of intracellular calcium leads to mitochondrial permeability and influx of extracellular calcium and permeability, necrosis, apoptosis
Humoral Immunity	Calcium controls antibody formation
Inflammasome	Calcium has a role in the production of TNF alpha, IL-1 beta

2. Calcium, Calcium Channel, Vitamin D Mechanisms Associated with Sepsis

In a survey of the United Kingdom General Practice database (Table 2) over 500,000 patients were listed as hypertensive. In this survey, use of angiotensin converting enzyme inhibitors was associated with a significantly higher rate of hospitalization for sepsis as well as mortality at thirty days when compared to use of angiotensin receptor blockers or calcium channel blockers [1]. One retrospective cohort study revealed that when 387 patients hospitalized for pneumonia who had not been treated with calcium channel blockers, were compared to 387 similarly hospitalized patients with pneumonia who had received calcium channel blockers there was a significantly higher incidence of bacteremia, respiratory insufficiency, and transfer to intensive care unit compared among those not so treated [2]. One retrospective evaluation of immunosuppressed recipients of kidney allografts from the pre-angiotensin receptor blocker era noted significantly higher incidence of sepsis and shorter survivals of allograft function for 33 patients who had not received calcium channel blockers as opposed to 36 who did receive calcium channel blockers [3].

Table 2. Calcium Ion Channels and hypothetical role in expression of infections.

I. Relationship of clinical infections to calcium channels (retrospective studies)
　　A.　Pneumonia: acute infection without antibiotic resistance

　　　　1.　　General population
　　　　2.　　Recipients of Kidney Transplant allografts
　　　　a.　　Positive outcome results with calcium channel blockers
　　　　b.　　Negative outcome results with calcium channel blockers

　　B.　Sepsis with or without antibiotic-resistant pathogens

　　　　1.　　Potential benefit of calcium channel blockers in combination with more expensive drugs in an attempt to reduce costs while minimizing side effects of higher doses
　　　　a.　　quinolone-resistant streptococcus
　　　　b.　　rifampicin-resistant mycobacterium
　　　　c.　　quinine-resistant plasmodium
　　　　d.　　praziquantel-resistant schistosome
　　　　e.　　amphotericin-resistant leishmania
　　　　f.　　eflornithine-resistant trypanosome

II. Role of calcium movement in white blood cell defense against pathogens
　　A.　Neutrophils and macrophages may benefit from calcium channel blockers

　　　　1.　　Restoration of capacity to attack pathogen
　　　　2.　　Limitation of capacity of pathogen to extrude antibiotic

Table 2. *Cont.*

III. Sepsis with shock following trauma

    A.   Intracellular calcium movement from storage sites may be stabilized by calcium channel -blockers, limiting cell injury B. Capillary leak during sepsis1.

        1.   Angiopoietin 2 as a factor in sepsis-related capillary leaking
        2.   Flunarizine, which blocks both calcium influx and calcium movement from intracellular stores, in prevention of angiopoietin 2—related capillary leaking

IV. Pathogen colony growth mechanisms may or may not involve calcium

    A.   Generation of anti-oxidants (catalase, dismutase)

        1.   *Bortadella pertussis*
        2.   *Pseudomonas aeruginosa*

    B.   Efflux of calcium from host cells

        1.   *Bacillus anthracis*
        2.   *Clostridium perfringens*
        3.   *Streptococcus pneumoniae*

    C.   Colony growth mechanisms inhibited by calcium channel blockers

        1.   *Aspergillus fumigatus*
        2.   *Saccharomyces cerevisiae*
        3.   *Candida albicans*
        4.   *Cryptococcus neoformans*

V. Mechanisms of Calcium balance in kidney failure

    A.   Vitamin D (and cathelicidin) deficiency association with infection risk

        1.   Promotes availability of calcium through intestinal absorption
        2.   Promotes bone calcification at the growing front of osteoid

    B.   Parathyroid hormone excess promotes lysis of calcified bone with excess calcium/phosphorus that may deposit in soft tissue, including blood vessels with skin cellulitis (calciphylaxis).

        1.   Inhibits resorption of phosphate in kidney proximal tubule.
        2.   Promotes activity of 1-alpha hydroxylase in renal proximal tubule, which converts 25(OH) Vitamin D to its most active form $1,25(OH)_2$ vitamin D.

    C.   Fibroblast Growth Factor 23 minimizes the accumulation of phosphate which is associated with an increased mortality rate.

        1.   Cooperates with parathyroid hormone in the inhibition of resorption of phosphate in kidney proximal tubule.
        2.   Competes with parathyroid hormone's action on 1-alpha hydroxylase in the proximal tubule to generate the most active form of Vitamin D.

    D.   Klotho, the antiaging gene, cooperates with parathyroid hormone and fibroblast growth factor 23 by inhibiting phosphate resorption in the kidney proximal tubule.

Since the patients reviewed in studies of angiotensin-active medications and calcium channel blockers would have had hypertension, a precise mechanism for protection from sepsis would have to include protection from injury to blood vessels supplying skin, bronchus, urinary bladder unless there had been a blood vessel indwelling catheter-line, suggesting the vascular protection hypothesis might be too narrow. Consequently, researchers have developed other approaches. An experimental model for testing the impact of calcium channel blockers on sepsis involving ligation of the cecum with puncture of the wall of the intestine was one such example. This septic shock model was used to [4] demonstrate a lesser accumulation of oxygen radicals [5] as well as longer survival if diltiazem were injected prior to the onset of septic shock.

2.1. Positive Results

Several studies have addressed the use of dihydropyridine and non-dihydropyridine calcium channel blockers in the peri-operative period and in longer-term follow up (Table 2), demonstrating improved allograft function [6,7] that did not consistently appear to be secondary to blood pressure control [8,9]. One study did demonstrate stable serum creatinine associated with a fall in renal vascular resistance calculated from mean arterial pressure + renal blood flow [10]. A unique study found the occurrence of acute graft dysfunction (acute tubular necrosis) by biopsy to be significantly lower with verapamil vs. a non-calcium channel medication [11].

Rapid calcium influx into cytoplasm is associated with cell death [12,13] Clinical use for medications affecting calcium channels may involve control of aberrant cardiac rhythm while moderating vascular spasm [14]. As multiple types of calcium ion channels (N-type, L type and T type voltage-dependent calcium channels) became recognizable and multiple pharmaceutical agents with differing action profiles became available, further studies to find a role for these agents were performed.

The fact that some of these agents could directly influence biosynthesis of aldosterone in human adrenocortical cells [15] led to an early interest in their use in hypertension. Animal studies indicated an inhibition of the mineralocorticoid receptor as an additional property of dihydropyridine calcium channel blockers with felodipine being twice as powerful as amlodipine [16]. The other classes of calcium channel blockers (diltiazem, verapamil) were not effective here. Medium-sized blood vessels experience pulsatile flow as pressure is conveyed down to points of resistance and reflected backwards. Arterioles experience non-pulsatile constant flow, which can be increased with vasoconstriction or decreased with vasodilatation. Nitric oxide regulates arteriole flow while calcium-activated potassium channels regulate conduit artery flow [17]. Pathological concentrations of the vasoconstrictor angiotensin can cause damage to the muscular layer of larger blood vessels and to the endothelium of smaller blood vessels [18].

Uncontrolled hypertension may result in loss of vascular integrity with impaired endothelial resistance to sepsis [19,20] through activation of inflammation cascades which inhibit expression of nitric oxide synthase as well as the loss of control of glucose disposal such that oxidative stress disrupts calcium/potassium vascular physiology. These same mechanisms are operative in chronic loss of kidney function through injury to blood vessels and interstitial matrix, which diminishes production of $1,25(OH)_2$ Vitamin D (calcitriol) in renal proximal tubules. Since calcitriol has been shown to be additive to the inotropic effects of norepinephrine and vasopressin [21] and since receptors for $1,25(OH)_2$ D are found in increased amounts in hypertrophied heart muscle [22], the fact that calcitriol synthesis in the proximal tubule decreases as kidney function is lost may be seen as protective from accelerated hypertension. Since vitamin D is useful in controlling excessive secretion of parathyroid hormone in kidney failure, replacement therapy is routine in hemodialysis units. But a further benefit may be protection from sepsis as found in experimental model studies [23–26].

Initial enthusiasm for calcium channel blockers in renal transplantation related to their role in control of hypertension as well as the possibility that calcium channel blockers might be organ protective from intracellular calcium infusion related cell death (as seen in the necrotic myocardium during acute coronary occlusion). Then chronic myocardial stress in a rat hypertension model demonstrated protection by a calcium channel blocker (mibefradil) from myocardial scarring with an increase in interstitial collagen thought to be the result of increased fibroblast collagen production [27]. But since outcome studies had not shown long-term benefits of vasodilation following myocardial infarction, there was concern for perioperative hypotension with acute kidney allograft injury, which eventually became an impediment to the study of calcium channel blockers in renal transplant centers, particularly as the potential treatment population has aged from less than 55 to greater than 75 years of age under certain circumstances.

Thus, observations connecting calcium channel blockers with prolonged survival of kidney transplant allografts may have been due to both preservation of circulation [8–10] to the allograft at risk for ischemia due to the vasoconstrictive effect of the immunosuppressive agent, cyclosporine, but perhaps also to inhibition of mechanisms in the immunological rejection process [6,7]. An unanswered question in multiple centers from several countries is the method of preservation of donor kidneys at a point in time after nephrectomy when they are most susceptible to acute injury. A powerful vasodilator, the calcium channel blocker, lidoflazine, was shown to protect Lewis rat donor kidneys preserved in University of Wisconsin perfusate [28]. Additive effects of cyclosporine and verapamil on suppression of cytotoxic T lymphocyte cell proliferation [29] are considered to be operating via the calcium/calmodulin/calcineurin pathway to nuclear direction of the expression of antibodies through nuclear factor kappa Beta. Additive effects of cyclosporine and verapamil may also be found in suppression of the generation of cytokines by lymphocytes through inhibition of the incorporation of thymidine into DNA or uridine into RNA or leucine into protein.

Eventually the possibility that calcium channel blockers have targets other than on slow calcium channels at the cell borders has been extended to include effects on calcium movement into and out of the cytosol from storage areas within the cell (endoplasmic reticulum, sarcoplasmic reticulum, mitochondria) or on the calcium extrusion technique of pathogen survival in a host whose levels of calcium could be fatal to the pathogen [30]. Our experience with infection in diabetic patients with nephropathy has been limited to end-stage renal disease. Use of peritoneal dialysis for urgent removal of fluid to prevent pulmonary edema [31] or for chronic removal of uremic toxins had been associated with frequent instances of fever, abdominal pain and cloudy peritoneal effluent. When this triad occurred together, the diagnosis of peritonitis with a positive culture of effluent fluid would be expected. However, if only one of the diagnostic pointers were present, there might be some time before a second or third would appear. Our prospective study to determine whether frequent culture of peritoneal dialysis effluent fluid might yield an early warning prior to the clinical syndrome did not produce positive culture results worthy of the extensive expense [32]. At the height of an incursion of illicit distribution of cocaine into the New England sector of the USA, we found the availability of clean needles did not protect diabetic dialysis patients from life-threatening cellulitis, abscess formation, and sepsis [33] All the instances of visceral infection occurred in the setting of cocaine intake. Bacterial infection was detected four times as often in this target population while evidence of spread of hepatitis virus was ten-fold greater amongst users of cocaine. In our studies of cardiovascular end-points in the setting of diabetes associated with cardio-renal complications, there were serendipitous findings in the field of kidney transplantation ten years after the introduction of cyclosporine, which had marked a time of significant improvement in prevention of allograft rejection.

But when our retrospective review of 70 consecutive diabetic recipients of kidney allografts looking for cardiovascular end-points did confirm outcome benefits, it was purely by serendipity that the benefits were mainly derived through protection from attacks of sepsis [3]. Of the 70 study subjects we reviewed from this center, data on 69 was available for analysis. There were 36 individuals who had received calcium channel blockers and 33 who had not received calcium channel blockers. Of the 36, 21 were treated with diltiazem, 10 with nifedipine, and 5 with verapamil. One study subject who initially received nifedipine was subsequently switched to diltiazem. There were no significant outcome differences in cardiovascular events which were found to be less common than sepsis-related events. So, sepsis became the focus of outcome analysis in our retrospective review. Of the 19 individuals with a prior history of cardiovascular events, six who received calcium channel blockers had two sepsis-related events while 13 who had not received calcium channel blockers had 22 sepsis-related events. Of the 50 individuals without a prior history of cardiovascular events, 30 who had received calcium channel blockers had 13 sepsis-related events while 20 individuals who had not received calcium channel blockers experienced 39 sepsis-related events. While these unadjusted differences are highly significant, a weakness of the retrospective analysis would be unequal follow-up times. On the other hand, Kaplan-Meier survival curves were able to demonstrate 1.6 ($p < 0.02$) and 1.5 ($p < 0.05$)-fold greater survivals for patients and allografts, respectively for calcium channel blocker-treated vs non-treated subgroups at 24 months. When both calcium channel blocker and beta blocker medications were used, diabetic non-living allograft kidney recipients having been rigorously evaluated by experienced cardiologists [34] were documented to have durations of survival equivalent to those of non-diabetic recipients of living donor allografts using data from the UNOS Registry [35] and the US Renal Data System [36].

2.2. *Negative Results*

More recently we had the opportunity to perform a post-hoc evaluation of kidney transplantation outcomes in the International Folic Acid for Vascular Outcomes Reduction in Transplant (FAVORIT) trial, finding the incidence of infection risks to be much higher in diabetic recipients of allografts from living or non-living donors [37]. Since follow-up in this study began on average three years post-transplantation surgery, years four to ten were the observation period when most of the acute rejection and cardiovascular issues would probably have been resolved. Thus, the FAVORIT study was useful in documenting the importance of infection as the most serious outcome measure under the unique conditions involved. Of the 4110 individuals recruited into the FAVORIT trial, there were 2447 non-diabetic study subjects of whom 199 (8.2%) died during the study; 166 Type 1 diabetic study subjects of whom 44 (26.5%) died during the study; and 1497 Type 2 diabetic study subjects of whom 250 (16.7%) died during the study. Thus, there were 493 deaths with 191(38.7%) from cardiovascular causes and 286 (58.0%) from non-cardiovascular sources. Of the 286 non-cardiovascular deaths, 113 (59.2%) were due to infection while 76 (39.8%) were due to malignancy associated with immune-suppression.

Our results confirmed those from the UK National Health Service which had reported a 70% greater infection-related mortality for diabetic compared to non-diabetic kidney transplant recipients [38]. Between 2002 and 2007 when FAVORIT trial recruited 4110 renal transplant recipients, 4009 were followed to completion of the portion reviewed herein. Of the 4009, 2323 had received a non-living donor allograft while 1868 had received a living donor organ. Of the 2323 non-living allograft recipients, 871 (37.5%) were taking a calcium channel blocker while 1452(62.4%) were not taking a calcium channel blocker. Of the 1686 living donor organ recipients, 531 (31.5%) were taking a calcium channel blocker while 1155 (68.5%) were not taking a calcium channel blocker. Kaplan-Meier survival curves demonstrated no benefit for patient or allograft survivals over six years of follow-up. This is approximately nine years post-transplant surgery, a point at which both patient and allograft survival rates were not less than 75% for several sub-groups [39]. We have not analyzed this data according to the impact of statin medication upon groups with or without use of calcium channel blockers. Statins appear to inhibit a process by which excess growth of vascular smooth muscle cells pathologically associated with a subendothelial collection of matrix metalloprotein and LDL cholesterol. Studies also indicate that statins activate an L-type calcium that can be antagonized by calcium channel blockers, thus amlodipine cooperates in this statin end-point [40].

Rescue of neutrophilic phagocytic function through use of nifedipine among hemodialysis patients has been reported [41]. When calcium channel blockers were tested for protection from infections in the intra-venous tubing used as temporary or permanent access for hemodialysis, there was no clinical benefit reported [42]. Another instance of failure to find benefit from calcium channel blockers in the infection risk situation, occurred with use of verapamil for the first ten post-operative days following kidney transplantation [43]. Among 152 renal transplant recipients 16 of 77 (21%) who received 240 mg of verapamil for the first ten post-operative days developed a significant infection over the next 2–6 months compared with 4 of 75 (5%) who did not. 20 patients were hospitalized for infection in this brief follow up, 4 died (20%), including 3 of the 16 (19%) who had received verapamil vs. 1 of 4 (25%) who did not receive verapamil. However, episodes of infection occurred 6–20 weeks after verapamil, leave some biologic doubt as to causality. Despite levels of cyclosporine being higher during verapamil treatment, the occurrence of biopsy -proven allograft rejection was not significantly different between the two groups: 19/77 (25%) for the verapamil group vs. 22/75 (29%) for the no-verapamil group. The incidence of acute graft dysfunction, however, from a non-rejection cause (acute tubular necrosis) was lower in the verapamil group: 9/77 (12%) vs. 31/75 (41%) in the no-verapamil group. Since episodes of acute renal failure occurred in the immediate post-operative period, a statistical relationship with plausible biologic validity to verapamil is noted. And post-hoc evaluation of infection risk studies in populations with multiple risk factors are difficult to control for analysis. In a post-hoc evaluation of the FAVORIT study, which had identified infection as the single greatest long-term risk factor to survival of allograft recipients [37], there was no evidence for protection from serious infection for allograft recipients who had received calcium channel blockers vs. those who had not received calcium channel blockers [39].

3. Antibiotic-Resistant Pathogens May Utilize Channels Sensitive to Calcium Channel Blockers

Specific mechanisms for protection from sepsis by calcium channel blockers may include control of concentration of cytosolic calcium of mono- and poly-morphonuclear cells for efficient chemotaxis, migration, adhesion, phagocytosis [44,45] while limiting excessive cytokine response with massive capillary leaking found in the adult respiratory distress syndrome [46,47].

3.1. Treatment Resistance through Antibiotic Extrusion by the Pathogen or Host Cell

Ultimately, an effect on invading pathogens to limit natural selection of antibiotic-resistant strains would be a cost-effect efficiency goal (Table 2). Calcium channel blockers have been studied in quinolone-resistant pneumococcal pneumonia [48], rifampicin-resistant tuberculosis [49], quinine-resistant *Plasmodium falciparum* malaria [50,51], and praziquantel-resistant *Schistosoma mansoni* infestation [52,53], amphotericin-resistant *Leishmania donovani* [54], and eflornithine-resistant *Trypanosoma cruzi* [55]. Benidipine may be useful for treatment of *Toxoplasma gondii* infestation for individuals allergic to sulfonamides like sulfadiazine or sulfamethoxazole [56].

Pathogens able to efficiently extrude antibiotics quickly become resistant since they are being treated with progressively higher doses. Calcium channel blockers have been useful through closing the channel through which antibiotics are extruded. *Mycobacterium tuberculosis* has the capacity to extrude antibiotics which have entered through its tough outer membrane. The efflux mechanism [57] can be demonstrated following engulfing of the pathogen by macrophage. Verapamil inhibits the efflux pumping of antibiotics by closing the critical channel [58–60], which is not a typical calcium channel as shown by comparison with isomers and metabolites that are very effective at the extrusion channel, but ineffective at the calcium channel.

Since treatment of resistant mycobacteria with new medications is costly, studies have utilized verapamil to achieve in vitro and in vivo eradication by bedaquiline at lower effective dose [61,62]. One of the dangerous side effects of bedaquiline, an ATP synthase inhibitor, is prolongation of QTc interval, increasing the risk for ventricular arrhythmias. Verapamil suppresses this side effect. The calcium channel blocker, nimodipine used in slow, continuous infusion has been found effective in normalizing the electrocardiographic changes found in experimental cerebral malaria [54]. Resistance to rifampicin and other antibiotics like ethambutol is a major impediment to recovery from pulmonary tuberculosis for individuals who are immune-suppressed by malnutrition, uncontrolled glycemia, uremia, cancer chemotherapy. In mice, the use of verapamil has been shown to reduce drug resistance by decreasing the extrusion of rifampicin. A similar mechanism for prevention of uncontrolled infestation with Plasmodium falciparum due to drug resistance has been described with effective use of a calcium channel blocker.

Two biologic defensive mechanisms are responsible for the antibiotic resistance of *Mycobacterium tuberculosis* and *M. smegmatis*. The first is the antibiotic efflux process which has now been demonstrated on the cell membrane where a protein is described that forms the structure of a channel with a preference for cations [63]. The second is the finding that the same protein channel which extrudes cations and antibiotics also produces a toxin which causes sufficient necrosis to allow the pathogen to escape from the macrophage which had recently engulfed it. This same channel also allows the pathogen to take up nutrients before it escapes [64]. Several gram negative and gram-positive bacteria have been found to have the potential for antibiotic efflux. Among the more well-known gram-negative bacteria are *Bacteroides, Brucella, Campylobacter, Enterobacter, haemophilus, Neisseria, Pseudomonas, Vibrio*. Among the more well-known gram-positive bacteria are *Bacillus, Clostridium, Listeria, Mycobacterium, Staphylococcus*. A rare *Mycobacterium, M. abscessus* is an important pathogen in cystic fibrosis, an inherited disorder with an increased risk of pneumonia associated with airway obstruction. Although this pathogen is fast-growing, it is frequently antibiotic-resistant [65]. Data on calcium channel blocker impact in this rare form of tuberculosis may be forth-coming from current studies, particularly in India. It is now known that antibiotic-resistant *M. tuberculosis* not only works by extrusion of antibiotics, but also by preventing their entry in the first place.

3.2. Other Mechanisms of Pathogen Resistance Involving Calcium Ion Channels

The normal concentration of calcium in the plasma is in milli-molar range (2 mMol/L) while that of the cytosol of both host cells [66] and invading pathogens [67] is in micromoles (0.1 micromol/L), a thousand-fold difference. In addition to this initial environmental shock, the pathogen may then be attacked by a sudden burst of intracellular calcium at an even higher concentration as the host cell attempts eradication (Table 2). Two protective responses have been demonstrated in pathogens. The first adaptive mechanism seeks to protect from an impending state of oxidative stress through generation of anti-oxidant proteins, such as catalase and superoxide dismutase. Catalase may be responsible for survival of antibiotic-resistant *P. aeruginosa* [68], but for *Bordetella pertussis*, dismutase is more important than catalase for survival within polymorphonuclear leukocytes [69]. The second adaptive mechanism involves influx of cytosolic calcium across the plasma membrane of the host [30] by means of energy generated by ATPase, which mechanism has been observed in *Bacillus anthracis*, *Clostridium perfringens*, and *Streptococcus pneumoniae* (Table 2). It would appear the first of these mechanisms would be of relatively short-term benefit to the pathogen. The second of these two would allow for a more interference with host cellular defenses. Thus, there are attempts to develop drugs that inhibit the calcium efflux transport function of the pathogen [57].

Calcium-related mechanisms have been identified in certain fungi which may be invasive to humans. *Aspergillus fumigatus*, an allergen and an invasive pulmonary pathogen, utilizes the calcium/calmodulin/calcineurin pathway [70] in colony growth through extension and branching of hyphae. Cyclosporine is a growth inhibitor. Calcium pathways are actively-employed by *Saccharomyces cerevisiae*, *Candida allicins* and *Cryptococcus neoformans*. Inhibition of growth of *C. neoformans* by the calcineurin inhibitor, tacrolimus, has been detected [71]. Inhibition of development of hyphae of *Candida albicans* (Table 2) has been reported for verapamil [72].

Measurement of cytosolic calcium of the Sprague-Dawley rat [73] or human [74] hepatocyte by the fluorescent indicator, Fura 4f, indicated a level of ≈ 1 micro-molar could be achieved by ATP stimulation. Thus, the normal concentration of 0.1 micro-molar would be elevated ten-fold during hepatitis virus cell invasion through a mobilization of calcium. But in a counter move, hepatitis virus pathogens may direct cytosolic calcium levels above 1 micro-molar into temporary stores in mitochondria. Since the leading intra-cellular store of calcium is endoplasmic reticulum, the first phase calcium release would occur from this store. High concentration of this cation might lead to host cell injury. An example of unregulated calcium entry into cytosol has been studied in retinal cells which are destroyed in rodents exposed to HIV-1 virus [75].

Therefore, in a second phase while a short-term bolus injection of calcium might eradicate an invading pathogen, to protect the host cell from prolonged excess calcium this cation must be removed from cytosol to either intra-cellular stores by sarco-endoplasmic reticulum ATPase pumping or to extra-cellular environment by means of plasma membrane ATPase pumping. As extra-cellular calcium intake proceeds to replenish endoplasmic reticulum stores, a temporary high-plateau of calcium concentration may be followed by a lower resting concentration (≈ 0.1 micro-molar).

3.3. Store Operated Calcium Entry (SOCE)

With the advent of CCB it became apparent that cytotoxic levels of calcium entry across the plasma membrane could be blocked with beneficial results. Entry of calcium into the cytosol from intracellular storage zones in the endoplasmic or sarcoplasmic reticulum (Store operated calcium entry, SOCE) may have therapeutic implications [76]. The juxtaposition of the sarcoplasmic endoplasmic reticulum allows for calcium signaling to control entry for calcium into the cytosol through the plasma membrane. Studies of defense mechanisms in liver cells infected with hepatitis B virus identify SOCE as being therapeutic for a short period of time. SOCE mechanisms may be activated during times of cytogenetic, inflammatory and hemodynamic instability.

In the instance of the hepatitis B patient or the victim of an attack by poliovirus [77], calcium transfer most likely occurred in the setting of a single pathogen in an uncompromised host.

This pathogenesis is more clearly delineated than that found in a trauma victim with multi-organism sepsis and shock due to capillary leak, sometimes referred to as adult respiratory distress syndrome (ARDS) during which SOCE can be sudden and to a massive degree [46,47]. An instance of SOCE that can occur without invasion by a pathogen is that of cancer metastasis [78]. Inhibition of proliferation of human leiomyoma by means of calcium release from intracellular stores via voltage gated calcium channels has been shown to be enhanced by simvastatin [66].

3.4. Hyperglycemia and Calcium Ion Channels

A related series of observations involves activity of the cardiac ATPase enzyme that may demonstrate increased expression following exposure to the diabetes medication liraglutide, a member of the glucagon-like peptide class of anti-hyperglycemia agents [79]. Increased cardiomyocyte plasma membrane ATPase calcium pumping has been linked to initiation and acceleration of congestive heart failure symptoms thought to be due to elevated intra-cellular calcium concentration. On the other hand, several studies have shown improved heart function with use of the diabetes medications empagliflozin, canagliflozin, or dapagliflozin, members of the sodium/glucose transporter 2 (SGLT2) inhibition class of antihyperglycemic agents [80]. Reversal of elevated cardiomyocyte calcium concentration may explain the beneficial result in heart failure reported with these medications. This mechanism could apply as well to heart failure patients who do not have the problem of hyperglycemia at least on a short-term basis [81]. Infection outcome results with SGLT2 inhibition would certainly be of interest, particularly in the realm of antibiotic-resistant pathogens.

Restoration in the capacity for phagocytosis when hyperglycemia is reversed with anti-hyperglycemic agents is reproducible and confirmed. The mechanism of disarming of the neutrophil appears to have been an increase in calcium concentration in the cytosol. If the source of calcium were the external medium of the neutrophil, then the effect of nifedipine/amlodipine would have to be exerted at the plasma membrane. But if the source of cytosolic calcium were stores in the endoplasmic reticulum, then the mechanism would be the target of future research agents. Which-ever source of calcium were involved, it would appear the pathological increase in calcium concentration in the cytoplasm was brought under control by means of a calcium channel blocker. Dihydropyridine calcium channel blockers like amlodipine or nifedipine in hypertension or with sulfonylurea medications, like glyburide or glipizide, in hyperglycemia has been found in rescue of neutrophil phagocytic function [44,45]. While the untested hypothesis that use of statin for cholesterol control may lower cost of expensive antibiotics has some rational grounds due to calcium channel blocking activity it should be remembered that for the type 2 diabetes population it is this same mechanism which may be responsible for hyperglycemia due to inhibition of beta cell insulin secretion [82].

4. Loss of Regulation of Inflammatory Response to Infection: Mechanisms Associated with the Sepsis Shock Syndrome

An experimental model for testing the impact of calcium channel blockers on sepsis involves ligation of the cecum with puncture of the wall of the intestine. This model has demonstrated relatively longer survival if diltiazem were injected prior to the onset of septic shock [4] in association with decreased formation of oxygen radicals [5]. Specific mechanisms for protection from sepsis by calcium channel blockers include: a decrease in cytosolic calcium of inflammation mediating cells, thereby limiting excessive cytokine responses, such as occurs in the adult respiratory distress syndrome [46]; an improved capacity to combat pathogens (chemotaxis, movement, adhesion, phagocytosis) through an increase in cytosolic calcium of polymorphonuclear cells and macrophages by release from intracellular stores (endoplasmic reticulum, sarcoplasmic reticulum, mitochondria) through alternate channels while the slow L channels from the exterior are blocked [47]; and an effect on invading pathogens to limit their capacity to select strains capable of rapid development of resistance to antibiotics. More recent reports of protection from sepsis have emerged from studies of antibiotic-resistant *Myocobacteria tuberculosis*. Along these lines, one analog of a calcium channel blocker (verapamil) may be more effective than another in assisting rifampicin in its battle with *M. tuberculosis* for reasons not easily understood in terms of L-type calcium channels, but rather by efficiency of docking at a site critical to the rifampin efflux mechanism. It is entirely possible that no available calcium blocker can reverse capillary leak as a complication of infection. Newer analogs however may become available [83]. It is now known that antibiotic-resistant *M. tuberculosis* bacilli inhibit entry and accelerate extrusion of antibiotics. Control of cytosolic calcium entry or transport from internal sources has been found critical for macrophage and neutrophil defense functions. Channels from endo- and sarcoplasmic reticula may interact with trauma/infection stresses for release of calcium from relevant storage areas. The endoplasmic reticulum utilizes inositol-3-phosphate receptors while the sarcoplasmic reticulum utilizes ryanodine receptors.

Sepsis Leads to Capillary Leaking

Understanding of the of the sepsis syndrome has shifted focus from that of an overwhelming inflammation cascade to an emphasis upon leaking of capillary fluid, resulting in hemoconcentration with diminished blood flow to vital organs and eventually into a state of shock due to hypovolemic hypotension (Table 2). This capillary leak results from endothelial dysfunction secondary to changes in the angiopoietin system. Ordinarily, angiopoietin 1 maintains the vascular barrier, but the effect of bacterial invasion is to enhance the concentration of angiopoietin 2 through its release from storage sites in endothelial cells mediated by tumor necrosis factor alpha (TNF α). A higher concentration of angiopoietin 1 vs. angiopoietin 2 normally operates the vascular barrier through a growth factor receptor, Tie2, which is inhibited when angiopoietin 2 concentration increases. So, a marker for the severity of the state of sepsis would be the rising level of circulating angiopoietin 2, a for-warning of imminent hypovolemia due to interstitial space accumulation of electrolyte/protein-containing fluid with strong osmolar capacity. Although the local presence of angiopoietin 2 at the portal of entry may be beneficial as a way of washing out the invasive pathogen, the target of sepsis-related research would be counter-measures by which a prolonged elevated level of circulating angiopoietin 2 might be brought into balance [84,85]. Calcium entry into endothelial cells under stress appears to occur via T-type calcium channels rather than L-type. Thus, amlodipine might have no effect upon capillary leak while flunarizine might be highly effective. In addition, flunarizine has been shown to inhibit vasoconstriction by norepinephrine and to diminish the risk of small vessel thrombosis [86]. Angiotensin caused vasoconstriction can damage muscular layers of larger and endothelium of smaller blood vessels [18]. The result may be a loss of resistance to sepsis [19,20].

5. Vitamin D, Parathyroid Hormone, Fibroblast Growth Factor, and Klotho Interact with Sepsis Defense Mechanisms in Which Movement of Calcium and Phosphorus Are Part of the Process

Given various findings of calcium movement inside host defense cells (polymorphonuclears, monocytes, macrophages), the issue of interaction between Vitamin D treatment for protection is an obvious area of sepsis study since lack of vitamin D is associated with risk of pneumonia [87,88]. In this connection relevant roles of vitamin D consist of assisting parathyroid hormone in the absorption of calcium in the intestinal tract as well as an independent capacity in the absorption of the companion anion phosphate which does not appear to be a target for parathyroid hormone. This increased prevalence of upper respiratory tract infections has been noted in patients with low vitamin D levels without [87,88] or with [89,90] advanced kidney failure. Intensive Care Unit patients who are septic have been found to have significantly lower concentrations of Vitamin D binding protein than those who were not septic. And there was also a positive correlation between Vitamin D and the toll-like receptors needed for expression of cathelicidin LL-37 levels [25]. Cathelicidins and defensins are antimicrobial peptides which cooperate in protection from infection. Since these agents are found in bronchopulmonary sites under attack from pathogens, such as *Mycobacterium tuberculosis*, they can be expected to offer a measure of protection in the lung.

Genes are being identified which code for peptides from pathogens, setting in motion expression of receptors (called Toll-like) which assist in generation of cathelicidins [25]. Susceptibility to tuberculosis is associated with vitamin D deficiency as well as a lack of production of cathelicidins. Efficiently activated monocytes will be capable of producing cathelicidin protein plus advancing 25(OH) vitamin D to the intracellular active form $1,25(OH)_2$ D [25]. Downstream from the Vitamin D receptor is the peptide, cathelicidin, with bactericidal capacity against *Mycobacterium tuberculosis* (Table 2). This suggests that hydroxylation of the cholecalciferol by the 1-alpha hydroxylase activates cathelicidin through toll like receptor activity. Raising the question as to whether correction of Vitamin D deficiency directly confers bactericidal potential. A question to be answered is the relationship between intracellular levels of calcium, vitamin D, and calcium channel blockers. The relationship of calcium channel blockers to expression of bactericidal proteins, cathelicidin and Beta-defensin in association with intracellular elaboration of $1,25(OH)_2$ vitamin D3 remains to be fully elucidated.

Since the demonstration of lower levels of 25(OH) vitamin D, vitamin D binding protein, and cathelicidin (LL37) in the serum of critically patients in the ICU compared to healthy controls, details of these observations have become of interest in trauma centers (Table 2). Eleven healthy controls were compared to 25 non-septic ICU control subjects versus 24 septic ICU study subjects. The individuals with sepsis had significantly higher levels of BUN and creatinine with lower levels of serum albumin, consistent with acute kidney injury. There was a positive correlation between 25(OH) vitamin D and cathelicidin (LL-37) levels [89]. Filtered 25(OH) vitamin D in complex with vitamin D binding protein can be resorbed from the glomerular filtrate with the assistance of proximal tubular protein, megalin [91] while the binding protein is degraded, the 25(OH) D is converted into 1,25(OH)2 D, and both 25(OH) D + $1,25(OH)_2$ D are resorbed into the circulation. Retention of phosphorus with lowering of calcium (despite reflex increases in parathyroid hormone) has been shown to be related to mortality [92]. Kidney failure with vitamin D deficiency can generate fibroblast growth factor to assist in excretion of retained phosphorus if vitamin D is replaced [93]. Klotho, the anti-aging gene seems to have a role in lowering the elevated level of phosphorus in kidney failure in cooperation with Fibroblast Growth Factor 23. Some insight into renal failure as a risk factor for infection might be gained (Table 2) by studying response to phosphorus retention in the setting of combined effects of parathyroid hormone, fibroblast growth factor 23, and Klotho [94] cooperating to exclude excess phosphorus at the sodium/phosphate co-transporter in the proximal tubule. But while PTH and FGF 23 are capable of cooperating at the proximal tubular site for phosphate transport to decrease phosphorus resorption, they are simultaneously capable of competing with each other at a different site where PTH assists in activation of 1-alpha hydroxylase (Table 2) for completion of synthesis if $1,25(OH)_2$ cholecalciferol (calcitriol), while inhibition of 1-alpha hydroxylase by Fibroblast Growth Factor is occurring simultaneously by way of inhibiting excess intestinal absorption of calcium in the setting of risk for deposition of calcium phosphorus in vascular tissues (calciphylaxis). In terms of infection risk, if the function of white blood cells is inhibited at low levels of both phosphorus and vitamin D, then the suggestion might be that kidney dysfunction and infection are more closely associated with Fibroblast Growth Factor 23 than with parathyroid hormone.

6. Conclusions

It is doubtful that sponsors can be found to support placebo-controlled trials for long-term calcium channel blockade in immunosuppressed individuals (such as solid organ transplant recipients) for the purpose of determining whether there is a beneficial effect on non-cardiovascular health care outcomes. However, there are transplant registries and multicenter trial databases from which additional information might be uncovered. An effort should be made to elucidate whether benefits or risks associated with calcium channel blockers in transplant populations as an initial example of immunosuppressed study groups with equally well-recorded information. Since long-term risk of infection in recipients of kidney transplant allografts is more important than cardiovascular complications [37] and outcome results for protection from sepsis by calcium channel blockers are in conflict [3,39] further studies need to be explored in terms of the multiple mechanisms reviewed herein. With respect to other populations immunosuppressed either iatrogenically or by virtue of underlying disease, we consider that further observations suggesting that this inexpensive class of medications might enhance treatment of infectious diseases should direct future trials. Another untested possibility is the hypothesis that use of statins to lower doses and cost of expensive antibiotics [95] may be effective since these drugs also exhibit some calcium channel blocking activity [96].

References

1. Dial, S.; Nessim, S.J.; Kezouh, A.; Benisty, J.; Suissa, S. Antihypertensive agents acting on the renin-angiotensin system, and the risk of sepsis. *Br. J. Clin. Pharmacol.* **2014**, *78*, 1151–1158. [CrossRef] [PubMed]
2. Zeng, L.; Hunter, K.; Gaughan, J.; Podder, S. Preadmission use of calcium channel blockers and outcomes after hospitalization with pneumonia: A retrospective propensity-matched cohort study. *Am. J. Ther.* **2017**, *24*, e30–e38. [CrossRef] [PubMed]
3. Weinrauch, L.A.; D'Elia, J.A.; Gleason, R.E.; Shaffer, D.; Monaco, A.P. Role of calcium channel blockers in diabetic renal transplant patients: Preliminary observations on protection from sepsis. *Clin. Nephrol.* **1995**, *44*, 185–192. [PubMed]
4. Meldrum, D.R.; Ayala, A.; Chaudry, I.H. Mechanism of diltiazem's immunomodulatory effects after hemorrhage and resuscitation. *Am. J. Physiol.* **1993**, *265*, C412–C421. [CrossRef] [PubMed]
5. Rose, S.; Baumann, H.; Tahques, G.P.; Sayeed, M.M. Diltiazem and superoxide dismutase modulate hepatic acute phase response in gram-negative sepsis. *Shock* **1994**, *1*, 87–93. [CrossRef] [PubMed]
6. Weir, M.R. Therapeutic benefits of calcium channel blockers in cyclosporine-treated organ transplant recipients: Blood pressure control and immunosuppression. *Am. J. Med.* **1991**, *90*, 32S–36S. [CrossRef]
7. Chrysostomou, A.; Walker, R.G.; Russ, G.R.; D'Apice, A.J.F.; Kincaid-Smith, I.; Mathew, T.H. Diltiazem in renal allograft recipients receiving cyclosporine. *Transplantation* **1993**, *55*, 300–304. [CrossRef] [PubMed]
8. Van Riemsdijk, I.C.; Mulder, P.; deFijter, J.W.; Bruijn, J.A.; van Hoof, J.P.; Hoitsma, A.J.; Tegzess, A.M.; Weimar, W. Addition of isradipine (Lomir) results in better renal function after renal transplantation: A double blind randomized placebo controlled multicenter study. *Transplantation* **2000**, *70*, 122–126. [PubMed]
9. Kuypers, D.R.; Neumayer, H.H.; Fritsche, K.; Rodicio, J.L.; Vanrenterghem, Y. Lacidipine Study Group. Calcium channel blockade and preservation of renal graft function in cyclosporine-treated recipients: A prospective randomized placebo-controlled 2-year study. *Transplantation* **2004**, *78*, 1204–1211. [CrossRef] [PubMed]
10. Ahmed, K.; Michael, B.; Burk, J.F. Effects of isradipine on renal hemodynamics in renal transplant patients treated with cyclosporine. *Clin. Nephrol.* **1997**, *48*, 307–310. [PubMed]
11. Neumayer, H.-H.; Kunzendorf, U.; Schreiber, M. Protective effects of calcium antagonists in human renal transplantation. *Kidney Int.* **1992**, *36*, S87–S93.
12. Varley, K.G.; Dhalla, N.S. Excitation-contraction coupling in heart. XII. Subcellular calcium transport in isoproterenol-induced myocardial necrosis. *Exp. Mol. Pathol.* **1973**, *19*, 94–105. [CrossRef]
13. Judah, J.D.; McLean, A.E.; McLean, E.K. Biochemical mechanisms of liver injury. *Am. J. Med.* **1970**, *49*, 609–616. [CrossRef]
14. Ellrodt, G.; Chew, C.Y.; Singh, B.N. Therapeutic implications of slow-channel blockade in cardio-circulatory disorders. *Circulation* **1980**, *62*, 669–679. [CrossRef] [PubMed]
15. Akizuki, O.; Inayoshi, A.; Kitayama, T.; Yao, K.; Shirakura, S.; Saski, K.; Kusaka, H.; Marshbara, M. Blockade of T-type voltage-dependent Ca^{2+} by benidipine, a dihydropyridine calcium channel blocker, inhibits aldosterone production in human adreno-cortical cell line NCI-H 295 R. *Eur. J. Pharmacol.* **2008**, *584*, 424–434. [CrossRef] [PubMed]
16. Dietz, J.D.; Du, S.; Bolten, C.W.; Payne, M.A.; Xia, C.; Blinn, J.R.; Funder, J.W.; Hu, X. A number of marketed dihydropyridine calcium channel blockers have mineralocorticoid receptor antagonist activity. *Hypertension* **2008**, *51*, 742–748. [CrossRef] [PubMed]

17. Bellien, J.; Joannides, R.; Iacob, M.; Anaud, P.; Thuillez, C. Calcium activated potassium channels and nitric oxide regulate human peripheral conduit artery mechanics. *Hypertension* **2005**, *46*, 210–216. [CrossRef] [PubMed]

18. Doerschug, K.C.; Delsing, A.S.; Schmidt, G.A.; Ashare, A. Renin-angiotensin system activation correlates with microvascular dysfunction in a prospective cohort study of clinical sepsis. *Crit. Care* **2010**, *14*, R24. [CrossRef] [PubMed]

19. Lund, D.D.; Brooks, R.M.; Faraci, F.M.; Heistad, D.D. Role of angiotensin II in endothelial dysfunction by lipopolysaccharide in mice. *Am. J. Physiol. Heart Circ. Physiol.* **2007**, *293*, H3726–H3731. [CrossRef] [PubMed]

20. Laesser, M.; Oi, Y.; Ewert, J.; Fandriks, L.; Aneman, A. The angiotensin II receptor blocker candesartan improves survival and mesenteric perfusion in an acute porcine endotoxic model. *Acta Anaethesiol. Scand.* **2004**, *48*, 198–204. [CrossRef]

21. Bukoski, R.D.; Xua, H. On the vascular inotropic action of 1,25(OH) vitamin D3. *Am. J. Hypertens.* **1993**, *6*, 388–396. [CrossRef] [PubMed]

22. Walthers, M.R.; Wicker, D.C.; Riggle, P.C. 1,25-dihydroxy D3 receptors identified in the rat heart. *J. Mol. Cell Cardiol.* **1986**, *18*, 67–72. [CrossRef]

23. Horiuchi, H.; Nagata, I.; Komorlya, K. Protective effect of vitamin D analogues on endotoxin shock in mice. *Agents Action* **1991**, *33*, 343–348. [CrossRef]

24. Asakura, H.; Aoshima, K.; Suga, Y.; Yamazaki, M.; Morishita, E.; Saito, M.; Miyamoto, K.-I.; Nakao, S. Beneficial effect of the active form of vitamin D3 against LPS-induced DIC, but not against tissue-factor-induced DIC in rats. *Thromb. Haemost.* **2001**, *85*, 287–290. [PubMed]

25. Liu, P.T.; Stenger, S.; Li, H.; Wenzel, L.; Tan, B.H.; Krutzik, S.R.; Ochoa, M.T.; Schauber, J.; Wu, K.; Meinken, C.; et al. Toll-like receptor triggering of a vitamin D-medicated human antimicrobial response. *Science* **2008**, *311*, 1770–1773. [CrossRef] [PubMed]

26. Moller, S.; Laigaard, F.; Olgaard, K.; Hemmingsen, C. Effect of 1,25-dihydroxy-vitaminD3 in experimental sepsis. *Int. J. Med. Sci.* **2007**, *4*, 190–195. [CrossRef] [PubMed]

27. Ramires, F.J.A.; Sun, Y.; Weber, K.T. Myocardial fibrosis associated with aldosterone or angiotensin ll administration: Attenuation by calcium channel blockade. *J. Mol. Cell. Cardiol.* **1998**, *30*, 475–483. [CrossRef] [PubMed]

28. Jacobsson, J.; Odlind, B.; Tufveson, G.; Wahlberg, J. Improvement of renal preservation by adding lidoflazine to University of Wisconsin solution. An experimental study in the rat. *Cryobiology* **1992**, *29*, 305–309. [CrossRef]

29. Grgic, H.; Wulff, I.; Eichler, C.; Flothmann, R.; Kohler, R.; Hoyer, J. Blockade of T-lymphocyte Kca 3.1 and Kv1.3 channels as novel immunosuppression strategy to prevent allograft rejection. *Transplant. Proc.* **2009**, *41*, 2601–2606. [CrossRef] [PubMed]

30. Roach, J.W.; Sublett, J.; Gao, G.; Wang, Y.-D.; Tuomanen, E.I. Calcium efflux is essential for bacterial survival in the eukaryotic host. *Mol. Microbiol.* **2008**, *70*, 435–444.

31. Weinrauch, L.A.; Kaldany, A.; Miller, D.G.; Yoburn, D.C.; Belok, S.; Healy, R.W.; Leland, O.S.; D'Elia, J.A. Cardio-renal failure: Treatment of refractory biventricular failure by peritoneal dialysis. *Uremia Investig.* **1984**, *8*, 1–8. [CrossRef]

32. Cooper, G.; White, J.; D'Elia, J.; DeGirolami, P.; Arkin, C.; Kaldany, A.; Platt, R. Lack of utility of routine screening tests for early detection of peritonitis in patients requiring intermittent peritoneal dialysis. *Infect. Control* **1984**, *5*, 321–325. [CrossRef] [PubMed]

33. D'Elia, J.A.; Weinrauch, L.A.; Paine, D.F.; Domey, P.E.; Smith-Ossman, S.; Williams, M.E.; Kaldany, A. Increased infection rate in diabetic dialysis patients exposed to cocaine. *Am. J. Kidney Dis.* **1991**, *17*, 349–352. [CrossRef]

34. Lindley, E.M.; Hall, A.K.; Hess, J.; Abraham, J.; Smith, B.; Hopkins, P.N.; Shihab, F.; Welt, F.; Owan, T.; Fang, J.C. Cardiovascular risk assessment and management in pre-renal transplant candidates. *Am. J. Cardiol.* **2016**, *117*, 146–150. [CrossRef] [PubMed]

35. Cecka, J.M.; Terasaki, P.I. The UNOS Scientific Renal Transplant Registry. *Clin. Transpl.* **1993**, *1*, 1–18.

36. Almond, P.S.; Matas, A.; Gillingham, K.; Dunn, D.L.; Payne, W.D.; Gores, P.; Gruessner, R.; Nagarian, J.S. Risk factors for chronic rejection in renal allograft recipients. *Transplantation* **1993**, *55*, 752–756. [CrossRef] [PubMed]

37. Weinrauch, L.A.; D'Elia, J.A.; Weir, M.R.; Bunnapradist, S.; Finn, P.; Liu, J.; Claggett, B.; Monaco, A.P. Infection and malignancy outweigh cardiovascular mortality in kidney transplant recipients: Post-hoc analysis of the FAVORIT trial. *Am. J. Med.* **2018**, *131*, 165–172. [CrossRef] [PubMed]

38. Hayer, M.K.; Ferrugia, D.; Begaj, I.; Ray, D.; Sharif, A. Infection-related mortality for kidney allograft recipients with pre-transplant diabetes mellitus. *Diabetologia* **2014**, *57*, 554–561. [CrossRef] [PubMed]

39. Weinrauch, L.A.; Liu, J.; Claggett, B.; Finn, P.V.; Weir, M.R.; D'Elia, J.A. Calcium channel blockade and survival in recipients of successful renal transplant: An analysis of the FAVORIT trial results. *Int. J. Nephrol. Renov. Dis.* **2018**, *11*, 1–7. [CrossRef] [PubMed]

40. Clunn, G.F.; Sever, P.S.; Hughes, A.D. Calcium channel regulation in vascular smooth muscle cells: Synergistic effects of statins and calcium channel blockers. *Int. J. Cardiol.* **2010**, *139*, 2–6. [CrossRef] [PubMed]

41. Alexiewicz, J.M.; Smogorzewski, M.; Klin, M.; Akmal, M.; Massry, S.G. Effective treatment of hemodialysis patients with nifedipine on metabolism and function of polymorphonuclear leukocytes. *Am. J. Kidney Dis.* **1995**, *25*, 440–444. [CrossRef]

42. Obialo, C.I.; Conner, A.C.; Lebon, L.F. Calcium blocking agents do not ameliorate hemodialysis catheter bacteremia. *Dial. Transplant.* **2002**, *31*, 848–854.

43. Nanni, G.; Pannochia, N.; Tacchino, R.; Foco, M.; Piccioni, E.; Castagneto, M. Increased incidence of Infection in verapamil-treated kidney transplant recipients. *Transpl. Proc.* **2000**, *32*, 551–553. [CrossRef]

44. Seyrek, N.; Markinkowski, W.; Smorgorzewski, M.; Demerdash, T.M.; Massry, S.G. Amlodipine prevents and reverses the elevation in [Ca2+] and the impaired phagocytosis of PMNL of diabetic rats. *Nephrol. Dial. Transpl.* **1997**, *12*, 265–272. [CrossRef]

45. Krol, E.; Agueel, R.; Smorgorzewski, M.; Kumar, D.; Massry, S.G. Amlodipine reverses the elevation in [Ca^{2+}] and the impairment in PMNLs of NIDDM patients. *Kidney Int.* **2003**, *64*, 2188–2195. [CrossRef] [PubMed]

46. Lee, C.; Xu, D.Z.; Feketova, E.; Nemeth, Z.; Kannan, K.B.; Hasko, G.; Deitch, E.A.; Hauser, C.J. Calcium entry inhibition during resuscitation from shock attenuates inflammatory lung injury. *Shock* **2008**, *30*, 29–35. [CrossRef] [PubMed]

47. Hotchkiss, R.S.; Karl, I.E. Ca2+, a regulator of the inflammatory response—The good, the bad, and the possibilities. *Shock* **1997**, *7*, 308–310. [CrossRef] [PubMed]

48. Pletz, M.W.; Mikhaylov, N.; Schumacher, U.; van der Liden, M.; Duesberg, C.B.; Fuehner, T.; Klugman, C.P.; Welte, T.; Makarewicz, O. Antihypertensives suppress the emergence of fluoroquinolone-resistant mutants in pneumococci: An in vitro study. *Int. J. Med. Microbiol.* **2013**, *303*, 176–181. [CrossRef] [PubMed]

49. Song, L.; Cui, R.; Yang, Y.; Wu, X. Role of calcium channels in cellular anti-tuberculosis effects: Potential of voltage-gated calcium-channel blockers in tuberculosis therapy. *J. Microbiol. Immunol. Infect.* **2015**, *48*, 471–476. [CrossRef] [PubMed]

50. Scheibel, L.W.; Colambani, P.M.; Hess, A.D.; Aikawa, M.; Atkinson, T.; Milhous, W.K. Calcium and calmodulin antagonists inhibit human malaria parasites (*Plasmodium falciparum*): Implications for drug design. *Proc. Natl. Acad. Sci. USA* **1987**, *84*, 7311–7314. [CrossRef]

51. Martins, Y.C.; Clemmer, L.; Orjuela-Sanchez, P.; Zanini, G.M.; Ong, P.K.; Frangos, J.A.; Carvalho, L.J. Slow and continuous delivery of a low dose of nimodipine improves survival and electrocardiogram parameters in rescue therapy of mice with experimental cerebral malaria. *Malar. J.* **2013**, *12*, 138–154. [CrossRef] [PubMed]

52. Gryseels, B.; Mbaye, A.; DeVias, S.J.; Stelma, F.F.; Guisse, F.; Van Lieshout, L.; Faye, D.; Diop, M.; Ly, A.; Tchuem-Tchuente, L.A.; et al. Are poor responses to praziquantel for treatment of Schistosoma mansoni infections in Senegal due to resistance? An overview of the evidence. *Trop. Med. Int. Health* **2001**, *6*, 864–873. [CrossRef] [PubMed]

53. Silva-Mores, V.; Couto, F.F.; Vasconcelos, M.M.; Araujo, N.; Coelho, P.M.; Katyz, N.; Grnfell, R.F. Anti-Schistosomal activity of a calcium channel antagonist on Schistosoma and adult Schistosoma Mansoni worms. *Memorias do Instituo Oswaldo Cruz* **2013**, *108*, 600–604. [CrossRef]

54. Kashif, M.; Manna, P.P.; Akhter, Y.; Alaidarous, M.; Rub, A. Screening of novel inhibitors against Leishmania donovani calcium channel ion channel to fight Leishmaniasis. *Infect. Disord. Drug Targets* **2017**, *17*, 120–129. [CrossRef] [PubMed]

55. Pollo, L.A.E.; deMoraes, M.H.; Cisilotto, J.; Creczynski-Pasa, T.B.; Biavatti, M.W.; Steindel, M.; Sandjo, L.P. Synthesis and in vitro evaluation of Ca^{2+} channel blockers 1,4-dihydropyridines analogues against Trypanosoma cruzi and Leishmania amazonensis: SAR analysis. *Parasitol. Int.* **2017**, *66*, 789–797. [CrossRef] [PubMed]

56. Kanatani, S.; Fuks, J.M.; Olafsson, E.B.; Westermark, L.; Chambers, B.; Varas-Godoy, M.; Ulen, P.; Barrigan, A. Voltage-dependent calcium channel signaling mediates GABA a receptor-induced migratory activation of dendritic cells infected by Toxoplasma gondii. *PLoS Pathog.* **2017**, *13*, e1006739. [CrossRef] [PubMed]

57. Liz, X.-Z.; Nikaido, H. Efflux-mediated drug resistance in bacteria: An update. *Drugs* **2009**, *69*, 1555–1623.

58. Adams, K.N.; Szumowski, J.D.; Ramakrishnan, L. Verapamil and its metabolite nor-verapamil inhibit macrophage-induced, bacterial efflux pump-mediated tolerance to multiple anti-tubercular drugs. *J. Infect. Dis.* **2014**, *210*, 456–466. [CrossRef] [PubMed]

59. Adams, K.N.; Takaki, K.; Connolly, L.E. Drug tolerance in replicating mycobacteria mediated by a macrophage-induced efflux mechanism. *Cell* **2011**, *145*, 39–53. [CrossRef] [PubMed]

60. Gupta, S.; Cohen, K.A.; Winglee, K.; Maiga, M.; Diarra BBishai, W.R. Efflux inhibition with verapamil potentiates bedaquiline in Mycobacterium tuberculosis. *Antimicrob. Agents Chemother.* **2014**, *58*, 574–576. [CrossRef] [PubMed]

61. Srikrishna, G.; Gupta, S.; Dooley, K.E.; Bishai, W.R. Can the addition of verapamil to bedaquiline-containing regimens improve tuberculosis treatment outcomes? A novel approach to optimizing TB treatment. *Future Microbiol.* **2015**, *10*, 1257–1260. [CrossRef] [PubMed]

62. Diacon, A.H.; Donald, P.R.; Pym, A.; Grobusch, M.; Patientia, R.F.; Mahanyele, R.; Bantubani, N.; Narasimooloo, R.; DeMarez, R.; van Heeswijk, R.; et al. Randomized pilot trial of eight weeks of bedaquiline (tme207) treatment for multidrug-resistant tuberculosis: Long-term outcome, tolerability, and effect on emergence of drug resistance. *Antimicrob. Agents Chemother.* **2012**, *56*, 3271–3276. [CrossRef] [PubMed]

63. Siroy, A.; Mailaender, C.; Harder, D.; Koerber, S.; Wolschendorph, F.; Danilchanka, O.; Wang, Y.; Heinz, C.; Niederweis, M. RV 1698 of microtubular proteins, a new class of channel-forming outer membrane proteins. *J. Biol. Chem.* **2008**, *283*, 17827–17837. [CrossRef] [PubMed]

64. Danichanka, O.; Sun, J.; Pavlemok, M.; Maueroder, C.; Speer, A.; Siroy, A.; Mayhew, N.; Doomlos, K.S.; Munez, L.E.; Herrmann, M.; et al. An outer membrane channel protein of Mycobacterium tuberculosis with exotoxin activity. *Proc. Natl. Acad. Sci. USA* **2014**, *111*, 6750–6755. [CrossRef] [PubMed]

65. Bryant, J.M.; Grogono, D.M.; Greaves, D.; Foweraker, J.; Roddick, I.; Inns, T.; Reacher, M.; Haworth, C.S.; Curran, M.D.; Harris, S.R.; et al. Whole-genome sequencing to identify transmission of Mycobacterium abscessus between patients with cystic fibrosis: A prospective cohort study. *Lancet* **2013**, *9877*, 1551–1560. [CrossRef]

66. Borahay, M.A.; Killic, G.S.; Yallampalli, C.; Snyder, R.R.; Hankins, G.D.V.; Al-Hendry, A.; Boehning, D. Simvastatin potentially induces calcium-dependent apoptosis of human leiomyoma cells. *J. Biol. Chem.* **2014**, *289*, 35075–35086. [CrossRef] [PubMed]

67. Gangola, P.; Rosen, B.P. Maintenance of intracellular calcium in *Escherichia coli*. *J. Biol. Chem.* **1987**, *262*, 12570–12574. [PubMed]

68. Orlandi, V.T.; Martegani, E.; Bolognese, F. Catalase A is involved in the response to photo-oxidative stress in *Pseudomonas aeruginosa*. *Photodiagn. Photodyn. Ther.* **2018**, *22*, 233–240. [CrossRef] [PubMed]

69. Khelef, N.; DeShager, D.; Friedman, R.L. In vitro and in vivo analogue of Bordetella pertussis catalase and Fe-superoxide dismutase mutants. *FEMS Lett.* **1996**, *142*, 231–235. [CrossRef]

70. DiMarco, T.M.; Freitas, F.Z.; Almeida, R.S.; Brown, N.A.; des Reis, T.F.; Zambelli-Ramalho, L.N.; Savoldi, M.; Goldman, M.H.S. Functional characterization of an Aspergillus fumigatus calcium transporter (PmcA) that is essential for fungal infection. *PLoS ONE* **2012**, *7*, e37591. [CrossRef]

71. Krauss, P.R.; Nichols, C.B.; Heitman, J. Calcium- and calcineurin-independent roles for calmodulin in Cryptococcus neoformans morphogenesis and high-temperature growth. *Eukaryot. Cell* **2005**, *4*, 1079–1087. [CrossRef] [PubMed]

72. Yu, Q.; Ding, X.; Bing, Z.; Xu, N.; Jia, C.; Mao, J.; Zhang, B.; Xing, L.; Li, M. Inhibitory effect of verapamil on Candida albicans hyphal development, adhesion, and gastrointestinal colonization. *FEMS Yeast Res.* **2014**, *14*, 633–641. [CrossRef] [PubMed]

73. Casciano, J.C.; Duchemin, N.J.; Lamontagne, J.; Steel, L.F.; Bouchard, M.J. Hepatitis B virus modulates store operated calcium entry to enhance viral replication in primary hepatocytes. *PLoS ONE* **2017**, *12*, e0168328. [CrossRef] [PubMed]

74. Casciano, J.C.; Bouchard, M.J. Hepatitis virus X protein modulates cytosolic Ca+ signaling in primary human hepatocytes. *Virus Res.* **2018**, *246*, 23–27. [CrossRef] [PubMed]

75. Dreyer, E.B.; Kaiser, P.K.; Offermann, J.T.; Lipton, S.A. HIV-1 coat protein neurotoxicity prevented by calcium channel antagonists. *Science* **1990**, *248*, 364–367. [CrossRef] [PubMed]

76. Reddish, F.N.; Miller, C.L.; Gorkhali, R.; Yang, J.J. Calcium dynamics mediated by the endoplasmic/sarcoplasmic reticulum and related diseases. *Int. J. Mol. Sci.* **2017**, *18*, 1024. [CrossRef] [PubMed]

77. Liu, S.; Rodriguez, A.V.; Tosteson, M.T. Role of simvastatin and beta cyclodextrin on inhibition of poliovirus infection. *Biochem. Biophys. Res. Commun.* **2006**, *347*, 51–59. [CrossRef] [PubMed]

78. Xie, J.; Pan, H.; Yao, J.; Han, W. SOCE and cancer: Recent progress and new perspectives. *Int. J. Cancer* **2016**, *138*, 2067–2077. [CrossRef] [PubMed]

79. Packer, M. Should we be combining GLP-1 receptor agonists and SGLT2 inhibitors in treating diabetes? *Am. J. Med.* **2018**, *131*, 461–463. [CrossRef] [PubMed]

80. Packer, M.; Anker, S.D.; Butler, J.; Filippatos, G.; Zannad, F. Effects of sodium-glucose cotransporter 2 inhibitors for the treatment of patients with heart failure: Proposal of a novel mechanism of action. *J. Am. Med. Assoc.* **2017**, *2*, 1025–1029. [CrossRef] [PubMed]

81. Byrne, N.J.; Parajuli, N.; Levasseur, J.L.; Boisvenue, J.; Becker, D.I.; Masson, G.; Fedak, P.W.M.; Verma, S.; Dyck, J.R.B. Empagliflozin prevents worsening of cardiac function in an experimental model of pressure overload-induced heart failure. *J. Am. Coll. Cardiol. Basic Transl. Sci.* **2017**, *2*, 347–354. [CrossRef] [PubMed]

82. Brault, M.; Ray, J.; Gomez, Y.H.; Mantzoros, C.S.; Daskalopoulou, S.S. Statin treatment and new onset diabetes: A review of proposed mechanisms. *Metabolism* **2014**, *63*, 735–745. [CrossRef] [PubMed]

83. Singh, K.; Kumar, M.; Pavodai, E.; Naran, K.; Warner, D.F.; Rominski, P.G.; Chibale, K. Synthesis of new verapamil analogues and their evaluation in combination with rifampicin analogues against Mycobacterium tuberculosis and molecular docking studies in the binding of efflux protein Rv1258c. *Bioorgan. Med. Chem. Lect.* **2014**, *14*, 2985–2990. [CrossRef] [PubMed]

84. Parikh, S.M.; Mammoto, T.; Schultz, A.; Yuan, H.T.; Christiani, D.; Karamuchi, S.A.; Sukhatme, V.P. Excess Angiopoietin-2 may contribute to pulmonary vascular leak in sepsis in humans. *PLoS Med.* **2006**, *3*, e46. [CrossRef] [PubMed]

85. David, S.; Kumpers, P.; van Slyke, P.; Parikh, S.M. Mending leaky blood vessels: The angiopoietin-Tie2 pathway in sepsis. *J. Pharmacol. Exp. Ther.* **2013**, *345*, 2–6. [CrossRef] [PubMed]

86. Retzlaff, J.; Thamm, K.; Ghosh, C.C.; Ziegler, W.; Haller, H.; Parikh, S.M.; David, S. Flunarizine suppresses endothelial angiopoietin-2 in a calcium-dependent fashion. *Science* **2017**, *7*, 44113. [CrossRef] [PubMed]

87. Grinde, A.A.; Mansbach, J.M.; Carmargo, C.A., Jr. Association between serum 25-hydroxyvitamin D Level and upper respiratory infection in the Third National Health and Nutrition Examination Survey. *Arch. Intern. Med.* **2009**, *169*, 384–390. [CrossRef] [PubMed]

88. Laaksi, I.; Ruohola, J.P.; Tuohimaa, P.; Auviven, A.; Haataja, R.; Pihlajamaki, H.; Yikomi, T. An association of serum vitamin D concentrations <40 nmol/L, with acute respiratory tract infection in young Finnish men. *Am. J. Clin. Nutr.* **2007**, *86*, 714–717. [PubMed]

89. Jengh, L.; Yamshchikov, A.V.; Judd, S.E.; Blumberg, H.M.; Martin, G.S.; Ziegler, T.R.; Tangpricha, V. Alterations in vitamin D status and anti-microbial peptide levels in patients in the intensive care unit with sepsis. *J. Transl. Med.* **2009**, *7*, 28–36. [CrossRef] [PubMed]

90. Chonchol, M.; Greene, T.; Zang, Y.; Hoofnagle, A.N.; Cheung, A.K. Low vitamin D and high fibroblast growth factor 23 serum levels associate with infectious and cardiac deaths in the HEMO study. *J. Am. Soc. Nephrol.* **2016**, *27*, 227–237. [CrossRef] [PubMed]

91. Nykjaer, A.; Dragun, D.; Walther, D.; Vorum, H.; Jacobsen, C.; Herz, J.; Mersen, F.; Christensen, E.I.; Willnow, T.E. An endocytic pathway essential for renal uptake of the steroid 25-(OH) vitamin D. *Cell* **1999**, *96*, 507–515. [CrossRef]

92. Kestenbaum, B.; Sampson, J.N.; Rudser, K.D.; Patterson, D.J.; Seliger, S.L.; Young, B.; Sherrard, D.J.; Andress, D.L. Serum phosphate levels and mortality risk among people with chronic kidney disease. *J. Am. Soc. Nephrol.* **2005**, *16*, 520–528. [CrossRef] [PubMed]

93. Fliser, D.; Kolleritis, B.; Neyer, U.; Ankerst, D.P.; Lhotta, K.; Lingenhel, A.; Rita, E.; Kronenberg, F.; Kuen, E.; Konig, P.; et al. Fibroblast growth factor (FGF 23) predicts progression of chronic kidney disease: The mile to moderate kidney disease (MMKD) study. *J. Am. Soc. Nephrol.* **2007**, *18*, 2600–2608. [CrossRef] [PubMed]

94. Razzaque, M.S. The FGF 23-Klotho axis: Endocrine regulation of phosphate homeostasis. *Nat. Rev. Endocrinol.* **2009**, *5*, 611–619. [CrossRef] [PubMed]

95. Lee, M.-Y.; Lin, K.-D.; Hsu, W.-H.; Chang, H.-L.; Yang, Y.-H.; Hsiao, P.-J.; Shin, S.-J. Statin, calcium channel blocker and beta blocker therapy may decrease the incidence of tuberculosis infection in elderly Taiwanese patients with type 2 diabetes. *Int. J. Mol. Sci.* **2015**, *16*, 11369–11384. [CrossRef] [PubMed]

96. Ali, N.; Begum RFaisal, M.S.; Khan, A.; Mabi, M.; Shehzadi, G.; Ullah, S.; Ali, W. Current statin show calcium channel blocking activity through voltage gated channels. *BioMed. Cent. Pharmacol. Toxicol.* **2016**, *17*, 43–50. [CrossRef] [PubMed]

11

Calcium and Nuclear Signaling in Prostate Cancer

Ivan V. Maly and Wilma A. Hofmann *

Department of Physiology and Biophysics, Jacobs School of Medicine and Biomedical Sciences,
University at Buffalo, 955 Main Street, Buffalo, NY 14203, USA; ivanmaly@buffalo.edu
* Correspondence: whofmann@buffalo.edu

Abstract: Recently, there have been a number of developments in the fields of calcium and nuclear signaling that point to new avenues for a more effective diagnosis and treatment of prostate cancer. An example is the discovery of new classes of molecules involved in calcium-regulated nuclear import and nuclear calcium signaling, from the G protein-coupled receptor (GPCR) and myosin families. This review surveys the new state of the calcium and nuclear signaling fields with the aim of identifying the unifying themes that hold out promise in the context of the problems presented by prostate cancer. Genomic perturbations, kinase cascades, developmental pathways, and channels and transporters are covered, with an emphasis on nuclear transport and functions. Special attention is paid to the molecular mechanisms behind prostate cancer progression to the malignant forms and the unfavorable response to anti-androgen treatment. The survey leads to some new hypotheses that connect heretofore disparate results and may present a translational interest.

Keywords: metastasis; nuclear import; myosin IC; calcium; prostate cancer

1. Introduction

Despite the recent progress in diagnosis and treatment, prostate cancer remains one of the most common and lethal malignancies [1,2], with approximately 3.3 million men living with the condition only in the United States. The disease is characterized by the comparative ease of the initial diagnosis, a long period of indolence relative to other common cancers, and an abrupt and less predictable progression to the lethal stages. When sharpened up, the questions facing the practitioners and researchers have been formulated as concerning the possibility of treatment for the patients that need it, and the actual need of those for whom treatment is possible [3]. In many cases, the critical events in disease progression from the stage that is believed to merit only watchful waiting to one that is lethal are the emergence of castration resistance and metastasis [4]. Mechanistically grounded prognostic biomarkers and potential drug targets associated with these steps toward poor survival are of particular interest as outcomes of the molecular biological research.

From the perspective of molecular sciences, the puzzle of prostate cancer is encapsulated in a complex perturbation of multiple regulatory pathways and mechanisms. In the face of this complexity, the biological signal integration by the calcium ion (Ca^{2+}), in addition to the central importance of gene expression regulation, presents a particular interest. The intracellular propagation of a calcium-mediated signal [5] and the import of downstream molecules into the nucleus [6] can be seen as two common physicochemical events in the dysregulation involving the cell and its environment. Moreover, calcium channels and GPCRs are among the favored "druggable" targets in therapeutics development [7–9], and metastasis is manifested primarily by the molecular-motor driven cell motility [10,11]. The unexpected results of the recent experiments implicate the GPCR and myosin molecular motor molecules in the intranuclear calcium signaling and calcium-regulated nuclear transport [12,13]. Motivated by the emergent role of these mechanisms in multiple aspects of prostate cancer cell regulation, below we review the recent advances in molecular biology of prostate cancer with a special focus on calcium signaling and nuclear transport.

2. Genomic Background

2.1. NKX3.1 Insufficiency

85% of high-grade prostate intraepithelial neoplasia (PIN) and prostate adenocarcinomas display a loss of heterozygosity in the 8p21.2 locus that includes the NKX3.1 homeobox gene [14]. The incidence correlates with disease grade [15], and there is evidence of an epigenetic downregulation at the locus as well [16]. NKX3.1 is an early marker of developing prostate epithelium during budding from the urogenital sinus, and its expression throughout the development of the gland is important for ductal branching and expression of secreted proteins [17–19]. It is also important in adulthood, when homo- and heterozygous mutants display hyperplasia and often PIN [20,21]. As shown by Lei et al. [22], NKX3.1 stabilizes the tumor suppressor p53, inhibits the activation of protein kinase B (PKB/AKT), and blocks prostate cancer initiation caused by the phosphatase and tensin homolog deleted on chromosome 10 (PTEN) loss in xenografts. At the same time, NKX3.1 inactivation in mice leads to a deficiency in the response to oxidative stress [23]. In human prostate cell lines, it has been shown that NKX3.1 promotes the response to DNA damage [24] but is ubiquitinated under the action of inflammatory cytokines [25]. Recent experiments have demonstrated, in addition, that the proteolytic control of NKX3.1 expression levels in prostate cancer cells is regulated by mitogens such as epithelial growth factor (EGF) in an intracellular calcium- and protein kinase C (PKC)-sensitive manner [26]. Thus, on the protein level, the effect of most common genomic abnormality that has been identified in prostate cancer converges with the effect of the posttranslational expression regulation by calcium signaling. Additional experiments are needed to establish the relative impact of these two factors at different stages of the disease development.

2.2. Amplification and Susceptibility in MYC

The MYC oncogene is somatically amplified (as part of the 8q24 region) in a subset of advanced prostate tumors [27,28]. Specifically nuclear MYC was found to be upregulated in many PIN lesions and most carcinomas in the absence of gene amplification [29]. Numerous single nucleotide polymorphisms have been identified in the gene-poor region next to MYC that contains MYC's long-range regulatory elements [30–33]. It has also been established that overexpression of MYC in transgenic mice drives formation of PIN as well as progression to invasive adenocarcinoma [34]. The tumors formed under these conditions are characterized by downregulation of NKX3.1 and a simultaneous upregulation of Pim1, MYC's partner in lymphomas. Coexpression of MYC and Pim1, on the other hand, leads to formation of carcinomas that exhibit the neuroendocrine phenotype [35]. Similarly to the recent results concerning NKX3.1, it has been known for some time that MYC in the prostate cancer cells is under the control of calcium-mediated signaling. Specifically, it has been shown that MYC is controlled by Notch, which involves the calcium/calmodulin-dependent kinase II (CAMKII)-regulated production of the nuclear import-competent cleaved form of Notch-1 [36]. The crosstalk of this calcium-regulated nuclear signaling pathway, which connects with one of the key regulators of prostate cancer progression, with other factors influencing the intracellular calcium homeostasis appears to be a promising direction for future research.

2.3. TMPRSS2-ERG Fusion

Most prostate carcinomas exhibit deletions (less commonly translocations) on chromosome 21q that activate ETS-family transcription factors, usually via an N-terminal fusion with the androgen receptor (AR)-activated gene TMPRSS2 [37–41]. TMPRSS2-ERG is found in approximately 15% of high-grade PINs and 50% of localized prostate cancer [42,43]. In agreement with these clinical sample analyses, AR binding in the lymph node cancer of prostate (LNCaP) cell line brings the TMPRSS2 and *ERG* loci in physical proximity [44,45]. This finding suggests a role for androgen signaling in the induction of fusions under conditions leading to DNA damage. In this connection, it is interesting that double-stranded breaks also occur under the action of topoisomerase II, which is recruited to the AR-responsive elements by androgen signaling [46]. At the same time, whole-genome chromatin immunoprecipitation analysis has shown that ERG binding can silence AR-responsive genes [47]. Conceivably, this effect may reinforce the physiological hormonal changes in the aging prostate, as well as those accompanying therapeutic castration. Furthermore, in cell culture and transgenic mice, activation of *ETS* expression contributes quantitatively to promotion of invasivity and epithelial-mesenchymal transition (EMT) [38,48,49]. By itself, expression of truncated human *ERG* in mice leads only to a weak PIN [48,49]. It however synergizes with a loss of *PTEN*, leading to a high-grade PIN and carcinoma [50,51]. Collectively, these results put the *TMPRSS2-ERG* fusion at the center of a web of mechanisms responsible for the prostate cancer progression.

Recent work has begun to clarify the relationship of this common genomic condition with perturbations of calcium signaling and nuclear import. Epigenomic profiling has identified the genes that are differentially methylated in the *TMPRSS2-ERG* fusion vs. non-fusion prostate tumors [52]. Among the top-ranked genes is the calcium-channel gene *CACNA1D*, a target of ERG. The most recent extension of this work employed retrospective analysis to demonstrate a negative correlation of the calcium channel blocker use by the prostate cancer patients with both high-grade disease and occurrence of fusion tumors [53]. The latter result can be interpreted as a selective disadvantage of a fusion in the absence of a functional reinforcement by means of the perturbation of calcium signaling. Prostate cancer and benign epithelial prostate cells overexpressing ERG display a characteristic invasivity in vitro [48], an effect that was most recently reversed using peptidomimetic inhibitors of ERG that were fused with a nuclear localization sequence (NLS) peptides to colocalize with ERG in the nucleus [54]. It is remarkable that of the two forms of the fusion protein found in prostate cancer cells, one lacks the NLS and does not enter the nucleus [55]. It binds the fully functional form of the fusion and results in its downregulation on the protein level as well as a reduced expression of MYC. The evidence points to the importance of further study of prostate oncoprotein isoforms that may be characterized by different nucleocytoplasmic partitioning.

The *TMPRSS2-ERG* fusion activity is also associated with Y-box binding protein 1 (YB-1), a genomic-instability response protein that enters the nucleus in response to genotoxic stress [56]. YB-1 contributes to the upregulation of AR during the development of castration resistance [57]. The nuclear localization of this protein has been most recently associated statistically with a poor prognosis for prostate cancer patients [58]. Its accumulation was at the same time shown to be linked to the *TMPRSS2-ERG* fusion, as well as *PTEN* deletion. It appears worth investigating whether the conditions for nuclear import, for example, the calcium-mediated depolarization of the nuclear envelope that is reviewed below, impact the YB-1 regulated downstream effects of the *TMPRSS2-ERG* fusion.

Additionally, the fusion drives expression of Toll-like receptors (TLR), including TLR4 [59]. This results in an increased phosphorylation of nuclear factor κB (NF-κB) on the serine residue that prevents binding of inhibitor of κB (IκB) that masks the NLS. The active fusion isoform and the phosphorylated factor both localize exclusively in the nucleus, as seen also in the orthotopic tumor samples. In vitro, the same experiments demonstrate, the cells' proliferation is controlled by the transcriptional activity of NF-κB, and the latter is increased with the expression of the fusion product. These findings have led to the speculation that the activation of TLR4 by bacterial lipopolysaccharides, as well as tumor microenvironment proteins, may impact the progression of prostate cancer. Adding to the hypothetical effects that may be induced by the *TMPRSS2-ERG* fusion, it should be noted that NF-κB is known to be able to drive an aberrant ligand-independent nuclear accumulation of AR by direct binding [60]. Conceivably, this may mean that the fusion can aid the cells in bypassing the critical androgen-regulated step during the establishment of androgen independence in the progressing tumor. One more gene activated downstream from *TMPRSS2-ERG* is the chemokine receptor *CXCR4* [61], whose role in intranuclear calcium signaling will be reviewed in Section 8.1.

2.4. PTEN Loss

The phosphatase gene *PTEN* undergoes homozygous deletion early in prostate carcinogenesis [62]. The loss has been found to be correlated with the development of aggressive and castration-resistant disease [63–65]. At the same time, it has been shown that conditional deletion of *PTEN* in prostate epithelium leads to PIN and adenocarcinoma [66]. Inactivation of *PTEN* in genetically engineered mice was found to cooperate with the loss of function of *NKX3.1* [67], upregulation of *c-MYC* [68], and expression of the human *TMPRSS2-ERG* fusion [50,51]. Among the molecular partners required for tumor formation on the *PTEN* loss background, the studies have identified mammalian target of rapamycin complex 2 (mTORC2) and phosphoinositide 3-kinase (PI3K) isoform p110β [69,70]. It is remarkable that prostate epithelial cells from *NKX3.1; PTEN* mutant mice display androgen independence even before development of the cancerous phenotype [71]. If this finding is translatable to the conditions in the human prostate, it indicates sufficiency of the combination of these common early and advanced genomic abnormalities for the critical step in the disease progression. This possibility raises the importance of further investigation, as suggested in Section 2.1, of the calcium-dependent regulation of NKX3.1, which appears to parallel the background genomic effect on the protein level.

In LNCaP cells, PTEN was found to interact with AR directly, resulting in an inhibition of the latter's nuclear localization and promotion of its degradation [72]. PTEN's own localization in heterozygous (+/−) prostate cancer cells was found to be predominantly nuclear as assessed by immunofluorescence in vitro [73]. The nuclear retention of PTEN was investigated also on model non-prostate-cancer cell lines [74] and found to be regulated by reactive oxygen species (ROS). The functional importance of the nuclear localization has been demonstrated in the same study in mouse xenograft models. Using human prostate cancer cells expressing mutant forms of PTEN, it was possible to demonstrate that nuclear PTEN acts in concert with p53 to suppress tumor formation. Most recently, the nuclear transport receptor importin 11 has been identified as a tumor suppressor that protects PTEN from proteolytic degradation [75]. This suggest the existence of a secondary function of the molecular interaction that presumably underpins the phosphatase's nuclear import. Generally, it appears to be an example of the nuclear import serving also as a regulatory mechanism impacting the protein-level expression of a critical tumor suppressor, akin to the chaperone-mediated sequestration mechanisms that will be reviewed in Section 7.

3. Developmental Pathways

Gene expression studies suggest that developmental signaling pathways may be reactivated during neoplastic processes in the adult prostate [76,77]. In the mouse, the wingless-integration adenomatous-polyposis-coli (WNT-APC)-β-catenin pathway is seen as contributing to carcinogenesis positively [78–80]. Consistent with this is the observation of an enhanced nuclear localization of β-catenin specifically in castration-resistant tumors [81]. In man, on the other hand, the nuclear localization of β-catenin has been seen as negatively correlated with tumor progression [82,83]. Further pointing to an explanation whereby the WNT-β-catenin signaling in prostate cancer development may be stage-specific is the finding that the WNT5A protein is upregulated specifically in the metastatic human samples [84]. The same work also demonstrated the existence of calcium waves that are induced by WNT5A in cultured metastatic prostate cancer cells, as well as this protein's stimulation of the actin cytoskeleton activity and cell motility (the latter being also under the control of CAMKII). Subsequent experiments showed that the calcium wave induced by this and other WNT proteins results in an intranuclear calcium mobilization and import of β-catenin into the nucleus, which is dependent on a thapsigargin-sensitive depolarization of the nuclear envelope [85]. The results support an interpretation whereby WNT signaling has a dual role as an anti-dedifferentiation regulator and a promoter of cell motility. It will be interesting to determine whether the intranuclear calcium mobilization by WNT impacts the functions of nuclear myosin (reviewed in Section 8.2), and what mechanisms or quantitative properties distinguish it from the one that has for some time been viewed as associated with apoptosis [86,87].

Among other developmental pathways, Hedgehog and fibroblast growth factor (FGF) signaling should be mentioned. Hedgehog signaling plays a significant role in prostate cancer progression [88]. Although experiments on tumor-derived primary cultures were seen initially as pointing toward an autocrine mechanism [89], targeted experimentation identified the mechanism as paracrine signaling to the epithelium from the prostate stroma [90], similarly to other epithelial cancers [91]. Nuclear accumulation of the downstream effector of Hedgehog signaling, transcription factor Gli1, as seen in clinical samples, is associated with metastasis and poor survival [92]. A breaking report [93] identifies the mode of the intracellular Hedgehog signal propagation in the prostate cancer cells as a novel non-canonical pathway involving AR binding-driven nuclear entry of the Gli transcription factors. The contribution of FGF signaling is analogous, in that both epithelial activation of FGF receptor 1 and stromal overexpression of FGF10 play a role [94,95], potentially leading downstream to the activation of extracellular signal-regulated kinase (ERK). Kwabi-Addo et al. [96] call attention to the alternative translation initiation isoforms of FGF2, which display an intranuclear accumulation and are expressed in transgenic adenocarcinoma of the mouse prostate (TRAMP) models, and posit that these isoforms may present a special interest in the context of paracrine signaling, because of the possibility of direct translation of the extracellular signal to the gene expression effects in the target cell. Overall, the ongoing work on the developmental pathways' dysregulation in prostate cancer emphasizes the need for the elucidation of the pivotal molecular mechanisms behind the calcium-regulated nuclear import, which may involve such novel biophysical principles as gating by the nuclear envelope depolarization.

4. Kinase Cascades

The *PTEN* loss of function, which may result, in particular, from the reviewed genomic abnormality, has been shown to cause an upregulation of the AKT/mTOR pathway, primarily via activation of the kinase AKT1 [97–99]. An alternative activating mechanism for this signaling pathway is the activating mutation in AKT1 itself, which has been found in human prostate cancer [100]. Experimental expression of the constitutively activated isoform p110β of PI3K is also sufficient for induction of neoplasia in mice [101]. In addition, activation of either AKT or ERK enables androgen-independent growth both in vitro and in vivo [102]. Collectively, these findings establish a broad molecular background among the intracellular signaling kinase classes, whose perturbation may support the development of prostate cancer.

Underscoring the functional importance of the activated kinases' localization in the cell, it has been shown that tumor suppressor PML recruits both phospho-AKT (pAKT) and its phosphatase PP2 to nuclear bodies [103]. Complicating the picture, loss of *PML* leads to an impairment of the PP2 activity toward pAKT and an accumulation of the latter in the nucleus. The kinase acts in the nucleus by inhibiting FOXO transcription factors, in particular the FOXO3A-mediated transcription of Bim and p27 (Kip1), suggesting a two-pronged mechanism that involves both the suppression of apoptosis and an acceleration of the cell cycle. Promisingly for the translation of this insight into the practice of prostate cancer drug development, the AKT inhibitor ML-9 has recently been found effective at inducing prostate cancer cell death in vitro in a calcium-dependent manner [104].

Adding to the in vitro and in vivo evidence, ERK is frequently found to be activated in the clinical prostate cancer, often at advanced stages and in combination with AKT. As reviewed above, the simultaneous activation promotes progression and androgen independence [102,105]. Moreover, simultaneous targeting of the two pathways was sufficient to inhibit hormone-refractory prostate cancer in the mouse model [106]. Mechanistically, the activation of mTOR and IκB kinase (IKK) by AKT leads to stimulation of NF-κB [107]. Supporting the significance of this regulation, the abnormal

NF-κB signaling correlates with androgen sensitivity, metastasis, and outcome in human prostate cancer and the mouse model. The studies implicate both the expression and the nuclear localization of NF-κB [108–111], and identify among the downstream mechanisms its ability to regulate expression of AR [112].

Phosphorylation of mitogen-activated protein kinases (MAPK) by expression of constitutively active RAF or RAS in the mouse prostate epithelium also leads to a nuclear accumulation of pERK, and this action promotes induction of tumorigenesis [79,113]. However, such mutations are rare in human prostate cancer. Notwithstanding, the collective perturbation of the RAS/RAF pathway via small but collectively significant alterations in the expression of the individual pathway members is prevalent in advanced human prostate cancer [114]. Phospho-MAPK, whether induced by EGF in vitro or prevalent in certain subsets of prostate cancer epithelial cells as assessed by immunofluorescence in vivo, preferentially partitions into the nucleus [115], which can be seen as indicative of the predominantly nuclear downstream functions. These functions remain to be investigated. Taken together, the data reviewed in the last sections demonstrate how a large number of fundamental molecular perturbations behind the development of prostate cancer are interlinked by calcium signaling and the calcium-regulated nuclear import mechanisms (Table 1).

Table 1. General regulatory factors involved in the calcium and nuclear signaling in prostate cancer. The molecular factors and their cell-physiological effects are categorized, and the corresponding calcium and nuclear events sketched out in the approximate order of their treatment in the text.

Category	Factor	Physiology	Partners	Calcium Regulation/Nuclear Signal
genomic	NKX3.1 insufficiency	development	p53, AKT, PTEN	Ca^{2+}-regulated proteolysis
	MYC amplification	cell cycle	NKX3.1, PIM1	Ca^{2+}-regulated Notch nuclear entry
	TMPRSS2-ERG fusion	EMT	PTEN, CACNA1D	NF-κB nuclear entry, Ca^{2+} channel expression
	PTEN loss	hormone independence	mTORC2, PI3K	AR nuclear localization
developmental	WNT	cell motility	APC	Ca^{2+} wave, β-catenin nuclear entry
	Hedgehog	signaling from stroma	AR	Gli1 nuclear entry
	FGF	signaling from stroma	ERK	nuclear entry of alternative translation isoforms
kinases and phosphatases	AKT-mTOR	apoptosis	FOXO3A, Bim, p27	pAKT accumulation in nuclear bodies, Ca^{2+}-dependent inhibition
	ERK, MAPK	hormone independence	RAS, RAF, AKT	pERK/pMAPK accumulation in nucleus
	calcineurin	anti-senescence	NFATc1, MYC, IL6, STAT3	NFATc1 nuclear import

5. Anti-Senescence Signaling

A limitless proliferation potential achieved through anti-senescence factors has been proposed to be one of the biological hallmarks of cancer cells [116,117]. The recent characterization of store-operated calcium entry (SOCE) channel components STIM1 and ORAI1 in prostate cancer sheds light on the involvement of calcium in these processes. It was found [118] that the expression of both STIM1 and ORAI1 correlated negatively with the Gleason score (the commonly used histopathological assessment of prostate cancer progression) and was lower in prostate carcinomas relative to the normal tissue. STIM1 was also significantly reduced in hyperplasia. In agreement with these results obtained by immunohistochemistry on patient samples, STIM1 expression was reduced in carcinoma cell lines relative to the benign prostate hyperplasia (BPH) cell line BPH-1, and additionally reduced in the malignant PC3 line relative to the less metastatic DU145. The endogenous expression level differences in these experiments had the expected functional consequences for SOCE activity. Overexpression of the two factors, on the other hand, induced growth inhibition via a cell cycle arrest and senescence, as detected morphologically and by β-galactosidase staining. The apoptosis markers were similarly affected. The complex phenotype was additionally characterized by an increased migration of the cultured cells and a decreased ability of their conditioned medium to induce recruitment of macrophages in vitro. These effects were correlated with a signature of downstream TGF-β signaling activation, manifested in an overexpression of Snail and WNT1 and a rise of p-Smads and nuclear β-catenin, as well as a perturbed secretion of cytokines. On the other hand, inoculation of the STIM1- and ORAI1-overespressing cells into non-obese diabetic, severe combined immunodeficiency (NOD/SCID) mice showed a comparative reduction of tumor growth and a relative loss of BrdU-positive cells and E-cadherin immunostaining. The latter can be seen as indicative of EMT, in agreement with the complex phenotype observed in vitro. An additional complicating finding both in vivo and in vitro has been an induction, in STIM1- and ORAI1-overexpressing cells and tumors, of a relative overexpression of the apoptosis-inhibiting [119,120] form of the tumor necrosis factor (TNF) receptor, decoy receptor 2 (DcR2), even though the anti-apoptotic Bcl-2 and XIAP in these experiments were reduced. A consistent interpretation may be possible, stemming from the fact that DcR2 is, at the same time, an established senescence marker [121,122].

Additional recent findings linking senescence in prostate cancer to calcium signaling pertain to the nuclear factor of activated T-cells c1 (NFATc1), a protein previously implicated in the regulation of expression of prostate specific membrane antigen in an ionomycin-responsive manner in LNCaP cells [123], as well as in the expression of the osteomimicry markers in PC3 and C4-2B cells [124]. NFATc subfamily proteins (the majority of the NFAT types) are activated via dephosphorylation by the calcium/calmodulin-dependent serine/threonine phosphatase calcineurin, upon which the NLS is exposed and the protein undergoes import into the nucleus, where it performs the functions of a transcription activator [125]. In a general agreement with previous work on NFATc1, Manda et al. [126] find the protein expressed in human adenocarcinoma samples (both in the neoplastic epithelium and in the stroma) as well as in tumorigenic prostate cancer cell lines, but not in the epithelium of the normal prostate or benign RWPE-1 cells. Targeted expression of the activated nuclear form of NFATc1 in a mouse model leads to PIN and subsequently adenocarcinoma marked by an elevated expression of proinflammatory cytokines, apparent proliferative influence of NFATc1[+] cells on the neighboring cells, and castration resistance. Revealingly, MYC was upregulated in both NFATc1[+] and—apparently via the elevated IL6 and pSTAT3—in the neighboring cells. A double transgenic model with a *PTEN* deletion displayed a synergy of the factors and an accelerated tumorigenesis. Compared with the common *PTEN* deletion model, the prostates were marked by a suppressed expression and nuclear localization of the senescence marker p21 and an insignificant β-galactosidase staining. The mechanism of the p21 suppression may involve the STAT3-SKP2 pathway demonstrated in the gastric and cervical cancer cells [127,128]. Altogether, the data suggest that the calcium-regulated nuclear import step may be critical for the establishment of anti-senescence signaling in advanced prostate cancer.

6. Channels and Transporters

6.1. Transient Receptor Potential (TRP) Channels

Channels and transporters are among the best-studied molecular classes orchestrating the calcium signaling. In the recent years, considerable attention has been given to the role of TRP channels in the development of prostate cancer. The function of TRPV6, a vanilloid subfamily plasma membrane calcium channel, in particular, is now comparatively well characterized. Earlier termed CaT-like, it is a marker of locally advanced, metastatic, and androgen-insensitive prostate cancer [129]. On mRNA level, in situ hybridization uncovered a correlation of its expression with a number of cancer progression characteristics, including the Gleason score [130]. Remarkably, while its expression is detectable in most androgen-insensitive lesions, the level is reduced relative to untreated tumors. In-vitro studies have shown that this channel, as well as the related TRPC6, increases prostate cancer cell proliferation, acting through NFAT [131,132]. A subsequent detailed immunohistochemical analysis established also the correlation of the TRPV6 expression with prostate-specific antigen (PSA) and TRPC1, and with a lack of the apoptosis marker caspase-3 cleaved fragment [133]. Intriguingly, the presence of TRPV6 in the more advanced tumors, where it is found in luminal cells, was correlated with the appearance of ORAI1 in the same cells, which normally localizes to the basal cells. It has been possible to establish the involvement of TRPV6 in SOCE in prostate cancer cells, as well as its physical association with ORAI1 via STIM1 and TRPC1 under the conditions of SOCE. The latter conditions also induced a translocation of TRPV6 to the plasma membrane, as observed using a confocal microscope. The same study showed that overexpression of TPRV6 in prostate cancer cells in vitro increases their resistance to apoptosis induced, for example, by cisplatin, while in heterotopic xenografts in mice its experimentally manipulated level of expression correlates with the resulting tumor mass. Most recently, an inhibitor of this channel was evaluated in a first-in-human phase I clinical trial with promising results [134]. The work on TRPV6, to date, can serve as a model of calcium-mediated mechanism elucidation leading to translationally relevant results in the prostate cancer field.

Among the other TRP channels, as reviewed most recently by Cui et al. [8], the nonselective cationic channel vanilloid 2 (TRPV2) has been associated specifically with metastatic prostate cancer. Experiments in an earlier TRPV2 study [135] demonstrated the channel's involvement in the promotion of prostate cancer cells' migration and invasion both in vitro and in vivo. The effect, accompanied by elevated expression levels of the invasion enzymes MMP2 and MMP9, as well as cathepsin B, appeared to be mediated by an elevated basal cytosolic concentration of Ca^{2+} that was dependent on the channel's expression level. For progress in system-level understanding of the mechanism behind the emergence of prostate cancer invasivity, it will be of interest to characterize the functional interplay of these effects with the myosin-driven secretion of metalloproteases [136], a process that is likely open to calcium regulation in the light of the molecular dynamics investigations that will be reviewed in the last section.

Another TRP channel, melastatin family 8 (TRPM8, initially termed TRP-P8), was first found to be overexpressed in prostate cancer by Tsavaler et al. [137]. In androgen-responsive prostate cancer cells in vitro, it is expressed on the plasma membrane and in the endoplasmic reticulum (ER), and the level of expression is positively regulated by androgen [138]. The channel is capable of initiating an agonist-induced elevation of the intracellular calcium ion concentration. In keeping with the general theme of fine-tuned homeostasis, which is widely accepted to characterize the calcium regulation [139], both the agonist and antagonist action on TRPM8 in in-vitro experiments was able to induce apoptosis. In the same experiments, the channel was found to be expressed at a low and unregulated level in androgen-insensitive cells. This finding agrees with the loss of TRPM8 in transition to androgen independence that was detected in a genome-wide expression profiling study [140]. Subsequent work [141] found that experimental re-expression of this channel in androgen-insensitive cells reduced their migration in vitro. At the same time, in non-prostate-cancer model cell culture, PSA was able to potentiate the TRPM8 agonist-induced calcium current by enhancing trafficking of the channel to the plasma membrane via its phosphorylation downstream of the bradykinin 2 receptor signaling pathway. In resection samples and primary cultures, the plasma membrane localization and activity of this channel was found to be characteristic of the luminal cells of normal tissue, BPH, and especially in situ tumors, whereas the dedifferentiated transit amplifying and intermediate phenotype retained primarily the endoplasmic isoform [142]. As pointed out by the authors of this investigation, one functional consequence of the endoplasmic localization of the channel may be the apoptosis-resistant state characterized by a lowered intraendoplasmic calcium concentration, which had been seen in other studies of advanced prostate cancer [143,144]. Thus, the channel's role appears to be different, depending on the stage of prostate cancer, and change from its expression maintaining the pro-survival calcium homeostasis to its partial loss potentially contributing to the development of metastatic potential in parallel with the acquisition of androgen independence.

A recent study [145] has implicated a second melastatin family TRP channel, TRPM7, in the positive regulation of prostate cancer cells' migration, invasion, and expression of MMPs, as well as in downregulation of E-cadherin that accompanies EMT. Furthermore, the related TRPM4 has a function in the proliferation of prostate cancer cells, according to a breaking report [146]. In this instance, the downstream cascade includes a relative dephosphorylation of β-catenin and an elevation of this regulator's nuclear localization. Completing the present picture of a broad and mechanistically diverse involvement of the melastatin subfamily channels in the prostate cancer development, TRPM2 is one more known member that controls proliferation of prostate cancer cells by inhibiting the nuclear ADP-ribosylation [147]. The intracellular localization of TRPM2, which is associated with the plasmalemma of benign prostate epithelial cells, is altered in tumorigenic prostate cell lines, where it appears in intranuclear clusters. It will be interesting to investigate if the channels of this broad class are similar to the GPCRs to be reviewed in Section 8.1 in exhibiting nuclear translocation and influencing the trans-nuclear envelope potential and intranuclear calcium concentration in prostate cancer cells. Their downstream targets, hypothetically, could include the physical nuclear organization and transcription-related functions of nuclear myosins (see Section 8.2).

6.2. Other Channels and Transporters

The other types of channels and transporters that have been implicated in the development of prostate cancer include molecules responsible for a remarkably diverse set of calcium homeostasis mechanisms, such as voltage-gated and calcium-activated channels, as well as intracellular pumps. In particular, there is pharmacological evidence of a contribution from T-type calcium channels, including, specifically, $Ca_v3.2$, to the proliferative activity of prostate cancer cells in vitro and to cancer progression in xenograft experiments [148,149]. A similar involvement of this type of channels has been documented in other tumor types [150]. The $Ca_v3.2$ isoform may also be involved in neuroendocrine differentiation in prostate cancer, which is associated with castration resistance, distant metastasis, and poor prognosis [151]. Revealingly, induction of neuroendocrine differentiation of LNCaP epithelial prostate cancer cells by androgen deprivation in vitro was found to be accompanied by an upregulation of $Ca_v3.2$ and an increase in the associated calcium current [152,153]. These results suggest that a calcium-mediated reinforcement loop similar to the one already discussed in connection with the CACNA1D gene (which encodes $Ca_v1.3$, see Section 2.3) may be contributing to the neuroendocrine differentiation in the androgen-deprived prostate.

An intriguing connection to the cytoskeleton signaling has recently been uncovered that involves a plasma membrane channel that is calcium-activated but conducts potassium ions. The big potassium calcium-sensitive (BKCa) channels are expressed in excitable cells [154] and aberrantly in cancers of various origin [155]. They are overexpressed in a significant subset of advanced prostate cancers, and inhibiting their expression can suppress PC3 cell proliferation in vitro [156]. A recent study [157] has extended these results, using a xenograft model, and additionally demonstrated a stimulating effect of these channels' overexpression on cell migration and invasion. Interestingly, the effects depended only slightly on the channels' conductance, but were exerted instead in an integrin- and phosphorylated focal adhesion kinase (FAK)-dependent manner. The immunofluorescence colocalization and immunoprecipitation showed a formation of a BKCa-αVβ3 integrin complex, accompanied by recruitment and phosphorylation of FAK. The possible interplay of this mechanism with the tumor stroma's influence on the focal adhesion signaling, as well as the potential feedback from the extracellular matrix modification by the invasive epithelial cells themselves, present interest for future research.

The sarcoendoplasmic calcium pump SERCA's expression in epithelial prostate cancer cells is upregulated by the action of EGF, dihydrotestosterone (DHT), or serum, according to the in-vitro studies on LNCaP cells [158]. Stimulation of proliferation caused by these agents is suppressed by the SERCA inhibitor thapsigargin. Thus, the pump appears to be both an intermediary of the response to mitogens and subject to the downstream regulation caused by these factors, which may sensitize the tumor to the subsequent stimulation. Although the SERCA isoform implicated by the cited study was the non-tissue specific isoform (2B), a prostate-selective thapsigargin analog has been designed [159] and, most recently, reached phase II in the clinical trials for prostate cancer [8].

In addition to the reviewed roles of calcium-release activated calcium channel protein ORAI1 in senescence and SOCE, the channel appears to be involved also in the capacitative entry mediating apoptosis in androgen-sensitive prostate cancer cells in culture [160]. The androgen dependence of expression of ORAI1 itself, which was demonstrated in this study, raises the possibility of its downregulation being part of the mechanism of the emergence of aggressive cancer following the androgen deprivation therapy. Extending these results, the store-independent entry was recently associated with prostate cancer-specific enhanced expression of ORAI3 and formation of a heteromeric channel with ORAI1 [161]. Interestingly, the channel is gated by arachidonic acid, opening a new link between calcium signaling and the regulation by metabolism and inflammation.

The mitochondrial calcium uniporter, which contributes to the proapoptotic accumulation of calcium in mitochondria, is a comparatively new pump that has been studied in relation to cancer [8]. In-silico screens pointed to a negative regulation of its expression by miR-25, and this prediction was borne out by the experimental measurements [162]. Established prostate cancer cell lines express large quantities of this micro-RNA and are, at the same time, characterized by suppressed levels of the uniporter. Experimental restoration of its expression has been shown to sensitize prostate cancer cells in vitro to apoptosis. This is a promising direction for a further investigation in vivo.

An important apoptotic pathway is mediated by a sustained Ca^{2+} flux from the ER to mitochondria through the endoplasmic inositol 1,4,5-trisphosphate receptors, InsP3R [163]. The regulation of this mechanism has been linked to the deficiency of the *PTEN* gene, which, as reviewed above, is common in prostate cancer. The role of InsP3R has been elucidated in a series of experiments on established prostate cancer cell lines, which similarly exhibit the *PTEN* inactivation [164]. LNCaP cells, in particular, are characterized by a frameshift mutation in *PTEN*, and PKB/ACT in these cells was found to be constitutively phosphorylated and activated in the PI3K-dependent manner [165]. The phosphorylation in these experiments has proved determinative of the cells' resistance to apoptosis upon withdrawal of the serum factors. Subsequent work [166] has traced the mechanism of the resistance to phosphorylation of InsP3R by AKT. Most recently, a similar effect has been described in PC3 cells, where it is mediated by the lack of competition from the deficient *PTEN* product for binding to InsP3R3, which opens the latter to binding the F-box protein FBXL2 [167]. FBXL2 has the function of the receptor subunit of the ubiquitin ligase complexes, and can target InsP3R3 for degradation. The different mechanisms in the lymph node- and bone-metastatic cell lines may be indicative of diverging functions that can be discovered even for the comparatively well-characterized molecular classes at the different stages of prostate cancer progression. While the conditions-dependent multifunctionality may play the role of a complicating factor in drug development, it can also present an opportunity to more selectively target the tumor development at its specific crucial steps.

Finally, in the discussion of the channels involved in calcium signaling in prostate cancer, ryanodine receptors (RyR) have also been found to be expressed in prostate cancer cell lines [168]. The isoforms RyR1-3 vary in quantity depending on the cell line, a finding that should be further investigated, as it may be indicative of modification of the response mediated by these channels, depending on the stage in the tumor development or its environment. Calcium release is observed under the action of the RyR agonist caffeine and has been found to mildly promote apoptosis of the lymph-node metastatic LNCaP cells in vitro [169]. A synthetic agonist 4-chloro-*m*-cresol was also effective at eliciting this calcium response in these experiments, pointing to a possibility of targeting the receptor through a prostate-selective drug design in the future. On the whole, the calcium and calcium-operated channels and transporters remain one of the most systematically studied molecular classes in relation to their roles in prostate cancer (Table 2).

Table 2. Calcium and nuclear signaling in prostate cancer that is associated with specific channels and transporters.

Category	Channel/Transporter	Related Cell Physiology	Partners	Calcium Regulation/Nuclear Signal
TRP channels	TRPM2	proliferation	nuclear ADP-ribosylase	channel accumulation in nuclear clusters
	TRPM4	proliferation	β-catenin	β-catenin nuclear accumulation
	TRPM7	EMT	E-cadherin, MMPs	PM Ca^{2+} channel
	TRPM8	apoptosis	AR, PSA	lowered ER Ca^{2+}
	TRPV2	invasion	MMPs	elevated basal Ca^{2+}
	TRPV6	proliferation	NFAT, ORAI1, STIM1	SOCE
other channels, transporters	BKCa	invasion	integrin	Ca^{2+}-activated phosphorylation of FAK
	SERCA	proliferation	EGF, DHT	ER Ca^{2+}
	ORAI1	apoptosis	ORAI3, arachidonic acid	capacitative entry via the gated heteromer
	mitochondrial uniporter	apoptosis	miR-25	mitochondrial Ca^{2+}
	InsP3R	apoptosis	AKT, PTEN, FBXL2	ER-to-mitochondria current

7. Additional Topics

7.1. Calpain Proteolysis

Upon elevation of the intracellular calcium concentration in prostate cancer cells, as demonstrated in experiments in vitro [170], the calcium-regulated protease calpain cleaves β-catenin, producing a stable, transcriptionally active fragment that enters the nucleus. At the same time, microarray experiments demonstrated an elevation of the calpain expression in metastatic prostate cancer, which may suggest sensitization of this nuclear import pathway to calcium signaling in cells undergoing the transition to malignancy. Alongside the calcineurin-NFAT (Section 5) and calmodulin-myosin (Section 8.2) mechanisms, the calpain-β-catenin mechanism can be seen as part of an emerging paradigmatic triad representative of the mechanistically diverse calcium-regulated nuclear import pathways, many of which probably remain to be elucidated.

Another target for the calcium-induced cleavage by calpain in prostate cancer cells is E-cadherin, whose inactivation is characteristic of adenocarcinomas and is often seen as a marker for EMT [171]. The intracellular calcium-induced calpain proteolysis of FAK [172], at the same time, is a contributing factor to the modulation of cell adhesion in developing prostate cancer. The likely BKCa-mediated crosstalk with the mechanism reviewed in the last section is a promising avenue for future research. On the whole, the data on calpain demonstrate a remarkably multifaceted role of the calcium-regulated proteolysis in prostate cancer development and emphasize the simultaneous challenge and opportunity for the systems-based design of novel therapeutics that could reverse the dysregulation of this process.

7.2. 14-3-3 Mediated Nucleocytoplasmic Redistribution

As reviewed above (Section 4), AKT is one of the key signaling kinases involved in prostate cancer development, which exerts its effects, in particular, by inhibiting the FOXO3A transcription activator of the proapoptotic factor Bim through a mechanism that involves nucleocytoplasmic partitioning. Additional light has been shed on this issue by the work that has implicated the chaperone protein 14-3-3 [173]. The serine phosphorylation of FOXO3A by AKT creates a binding site for 14-3-3 and leads to retention of the complex in the cytoplasm, sequestering the transcription factor. The cytoplasmic localization of FOXO3A and 14-3-3 is correlated with the Gleason score in human samples and with disease progression in TRAMP mice [174]. In the reviewed work by Trotman et al. [103], it was shown, in particular, that pAKT in the mouse model prostate lesions accumulates in the nuclei. The mechanistic significance of this fact has been illuminated by the results that were obtained on non-prostate-cancer cells in vitro [175]. In the latter experiments, the nuclear pAKT was found to phosphorylate FOXO3A/FKHL1 immediately prior to the 14-3-3 binding and nuclear export of this transcription factor. The chaperone-mediated sequestration and nuclear export induced by the intranuclear activated kinase is an intriguing element of the nuclear transport-related mechanistic repertoire of prostate cancer, which merits further study.

7.3. Autonomic-System Regulation

In addition to the *PTEN* inactivation and constitutive activation of AKT, an acetylcholine-induced and calcium-dependent pathway leading to this kinase's activation has been uncovered recently [176]. Following the in-vivo indications that the autonomic, including cholinergic, nerve infiltration promoted prostate cancer development [177], the in-vitro work by Wang et al. [176] revealed the existence of autocrine acetylcholine signaling in epithelial prostate cancer cells, which proceeds through the muscarinic GPCR CHRM3 and a Ca^{2+} influx. Phosphorylation of AKT under these conditions was found to be mediated by calmodulin and CAMK kinase (CAMKK). CHRM3 is known to be upregulated in human prostate cancer samples, and experiments in vivo and in vitro have implicated the autocrine mechanism in the prostate cancer cells' proliferative and migratory potential. An antagonist treatment in vivo also increased the percentage of apoptotic cells and reduced the castration-resistant tumor growth. The most recent extension of this work has implicated the autocrine calcium-dependent mechanism in BPH and maintaining the prostate epithelial progenitor cells in the proliferative state [178]. It has also been demonstrated in LNCaP cells that carbachol-induced CAMKK-dependent blockage of apoptosis is mediated by AKT phosphorylation of the Bcl-2 associated death promoter protein BAD [179]. The effect on BAD is linked to its 14-3-3 binding and sequestration in the cytosol from the mitochondria, as shown in non-prostate-cancer cells [180]. Similar to the cytoplasmic sequestration of FOXO3A from the nucleus (last section), this mechanism adds complexity to the web of the spatially-restricted interactions of the molecules involved in the intracellular transport downstream of calcium signaling.

It is interesting to note that although the sympathetic, adrenergic effects in the Magnon et al. experiments [177] were found to be mediated by the stromal $\beta 1$ and $\beta 2$ receptors, a similar adrenergic effect on proliferation of primary prostate epithelial cells was demonstrated in the reviewed work of Thebault et al. [131], where it was mediated by $\alpha 1$ receptors. The downstream adrenergic signaling seen in the epithelial cell culture proceeded through a store-independent calcium entry via the TRPC6 channels, calcineurin, and NFAT-dependent transcription, thereby implicating the reviewed (Section 5) calcium-dependent nuclear import pathway.

8. New Molecular Classes in Calcium Signaling to the Nucleus

8.1. GPCRs

Alongside the adrenergic receptors, whose role was reviewed above, the chemokine cysteine (C)-X-C receptor 4 (CXCR4) is another GPCR that is remarkable for its involvement in the nuclear calcium signaling. It has been implicated in bone tropism of prostate cancer cells [181], in which it participates through its interaction with its bone-associated agonist stromal cell-derived factor 1 (SDF-1). It has also been found to promote migration and invasion of prostate cancer cells under the conditions when PTEN is inactivated by ROS [182]. It resides on the plasma membrane but also localizes specifically to the nuclei in human prostate cancer samples, as revealed by means of immunohistochemistry [13]. It similarly accumulates in the nuclear fractions of cultured prostate cancer cells, possesses an NLS, and associates with the nuclear import receptor transportin-$\beta 1$. Exposure to SDF-1 results in dissociation of the α subunit from the CXCR4 in the nuclear fraction, leading to intranuclear calcium release. In addition to the identification of a new class of nuclear calcium signaling molecules associated with prostate cancer, this recent study points to a novel mechanism of escape from any therapeutic receptor antagonist action that is limited to the cell surface.

8.2. Myosins

Non-muscle myosins have recently been shown to have multiple functions in prostate cancer. The longest-known functions of the myosin family concern cell motility, and recent work on prostate cancer cells has finely delineated this class of functions among the myosin family members including IB, IXB, X, and XVIIIA [183]. The contributions of these molecules to the shaping of the actin cytoskeleton and motile morphology of prostate cancer cells underpin mechanically the acquisition of a metastatic phenotype by the originally non-motile epithelial cells [184]. Myosin IC too has long been known as a contributor to cell migration and to intracellular transport in various cell types [185]. The most recent results have implicated it in the prostate cancer cell migration and invasion [136]. The three isoforms of this protein, including the recently discovered one (A) that is associated with metastatic prostate cancer [186,187]; and the one (B) originally described as the nuclear myosin [188], differ kinetically, according to the latest work [189].

Following the discovery of the NLS [190] within the IQ domain of myosin IC that mediates the interaction with apo-calmodulin, it was demonstrated that in prostate cancer cells, elevation of the intracellular calcium concentration drives the nuclear import of this myosin [12]. This import depends on importin-β1 and is presumed to be mediated by the unmasking of the NLS when, as established for myosin IC in vitro [191], the calcium-calmodulin dissociates from the IQ domain. The most recent work [192] has demonstrated that binding to phosphoinositides, mediated in part by the NLS region, contributes to the nuclear import of myosin IC. It will be of interest to determine whether similar mechanisms regulate the nucleocytoplasmic partitioning of the other myosins known to localize to the nucleus, such as subfamilies II, V, VI, X, XVI, and XVIII [193].

The nuclear functions have to date been best characterized for myosin IC isoform B. They include promotion of transcription, which is effected through an association with RNA polymerases I and II [194–200]. This myosin isoform also associates with RNA transcripts and pre-ribosomal units, participating in the export of the ribosome subunits from the nucleus [201,202]. The movements of chromosomes inside the nucleus similarly depend on nuclear myosin and actin [203,204]. Myosin IC isoform C—long thought to be cytoplasmic but now recognized as the most responsive to the calcium-regulated nuclear import in prostate cancer cells [12]—displays a capacity for interaction with RNA polymerase II and for maintaining the level of polymerase I activity [205]. Since calmodulin, acting as the light chain, modulates the motor properties of myosin IC [206] and itself undergoes nuclear import via the facilitated pathway [207], it appears promising to investigate next the intranuclear calcium regulation of the myosin functions in prostate cancer.

The tumor-suppressor qualities of myosin IC have recently been characterized in non-prostate-cancer cells [208]. Partially similar results involving suppression of anchorage-independent growth had previously been obtained in relation to myosin XVIII in other cancers, as reviewed by Oudekirk and Krendel [10], and most recently in a clinical association study for myosin VC in colorectal cancer [209]. At the same time, in breast and prostate cancer *MYO18A* (which encodes the non-muscle myosin XVIIIA) appeared to be an oncogene [210]. Future work should be able to shed light on the possible reciprocal regulation by means of the calcium-controlled switch between the cytoplasmic and nuclear functions of the non-muscle myosins, leading to a unified picture of these phenomena.

9. Summary and Outlook

The intracellular calcium signal and the nuclear entry of the activated factors can be seen as dual, linked integrating events that undergird the molecular complexity behind the critical stages of prostate cancer development (Tables 1–3). As these multiple examples in the reviewed literature from the recent years show, the nuclear vs. cytoplasmic localization frequently presents a unique differentiating feature that is associated with a prognosis for progression of the disease. Although localized calcium signals have the capacity to channel the information transmitted by the specific pathways, the influences impinging on the intracellular calcium homeostasis establish a network for a physical integration of the conditions of the tumor microenvironment. Of special interest from the molecular biological viewpoint are the events linking the calcium signaling to the regulation of the nuclear entry, as well as the dynamic steady-state nucleocytoplasmic distribution of the transcription regulators and cytoskeletal effector molecules. The polyfunctionality and novel intracellular localizations of the previously known factors in prostate cancer development that have been uncovered in the most recent work point, on the one hand, at a greater complexity that lies ahead in the molecular exploration of the disease mechanisms. On the other hand, however, they hold out a promise for a mechanistically grounded validation of new prognostic markers and potential targets for drug development, which would be able to disrupt the critical nodes in the prostate cancer network.

Table 3. Additional molecular classes involved in the calcium and nuclear signaling in prostate cancer.

Category	Factor	Physiology	Partners	Calcium Regulation/ Nuclear Signal
proteases	calpain	metastasis, EMT	AR, β-catenin, E-cadherin	nuclear entry of cleaved β-catenin
chaperones	14-3-3	apoptosis	AKT, FOXO3A, Bim	nuclear export of FOXO3A
autonomic regulation	CHRM3	castration resistance	CAMKK, AKT, BAD	acetylcholine-induced Ca^{2+} influx
	adrenergic receptors	proliferation	TPRC6, calcineurin, NFAT	store-independent entry
GPCRs	CXCR4	bone tropism	SDF-1, PTEN, ROS, importin-β1	nuclear entry, intranuclear Ca^{2+} release
molecular motors	myosin IC	migration, invasion, metastasis	calmodulin, importin-β1, RNA polymerase I and II, nuclear actin	Ca^{2+}-induced nuclear entry

Acknowledgments: Supported in part by the NCI of the National Institutes of Health under award number R21CA220155 to Wilma A. Hofmann.

Author Contributions: Ivan V. Maly—conceptualization and writing; Wilma A. Hofmann—conceptualization, review and editing.

Abbreviations

AR	Androgen receptor
BKCa	Big potassium calcium-sensitive (channel)
BPH	Benign prostate hyperplasia
CAMK	Calcium/calmodulin-dependent kinase
CAMKK	Calcium/calmodulin-dependent kinase kinase
DcR2	Decoy receptor 2
DHT	Dihydrotestosterone
EGF	Epithelial growth factor
EMT	Epithelial-mesenchymal transition
ER	Endoplasmic reticulum
ERK	Extracellular signal-regulated kinase
FAK	Focal adhesion kinase
FGF	Fibroblast growth factor
GPCR	G protein-coupled receptor
IκB	Inhibitor of κB
IKK	Inhibitor of κB kinase
InsP3R	Inositol 1,4,5-trisphosphate receptor

LNCaP   Lymph node cancer of prostate (cell line)
MAPK   Mitogen-activated protein kinase
mTORC2   Mammalian target of rapamycin complex 2
NFAT   Nuclear factor of activated T-cells
NF-κB   Nuclear factor κB
NLS   Nuclear localization signal (sequence)
PI3K   Phosphoinositide 3-kinase
PIN   Prostate intraepithelial neoplasia
PKB   Protein kinase B
PKC   Protein kinase C
PSA   Prostate-specific antigen
PTEN   Phosphatase and tensin homolog deleted on chromosome 10
ROS   Reactive oxygen species
RyR   Ryanodine receptor
SDF-1   Stromal cell-derived factor 1
SOCE   Store-operated calcium entry
TLR   Toll-like receptor
TRAMP   Transgenic adenocarcinoma of the mouse prostate
TRP   Transient receptor potential (channel)
YB-1   Y-box binding protein 1

References

1. Miller, K.D.; Siegel, R.L.; Lin, C.C.; Mariotto, A.B.; Kramer, J.L.; Rowland, J.H.; Stein, K.D.; Alteri, R.; Jemal, A. Cancer treatment and survivorship statistics, 2016. *CA Cancer J. Clin.* **2016**, *66*, 271–289. [CrossRef] [PubMed]

2. Siegel, R.L.; Miller, K.D.; Jemal, A. Cancer statistics, 2016. *CA Cancer J. Clin.* **2016**, *66*, 7–30. [CrossRef] [PubMed]

3. Carlsson, S.; Vickers, A. Spotlight on prostate cancer: The latest evidence and current controversies. *BMC Med.* **2015**, *13*, 60. [CrossRef] [PubMed]

4. Shen, M.M.; Abate-Shen, C. Molecular genetics of prostate cancer: New prospects for old challenges. *Genes Dev.* **2010**, *24*, 1967–2000. [CrossRef] [PubMed]

5. Bootman, M.D. Calcium signaling. *Cold Spring Harb. Perspect. Biol.* **2012**, *4*, a011171. [CrossRef] [PubMed]

6. Cautain, B.; Hill, R.; de Pedro, N.; Link, W. Components and regulation of nuclear transport processes. *FEBS J.* **2015**, *282*, 445–462. [CrossRef] [PubMed]

7. Godfraind, T. Discovery and Development of Calcium Channel Blockers. *Front. Pharmacol.* **2017**, *8*, 286. [CrossRef] [PubMed]

8. Cui, C.; Merritt, R.; Fu, L.; Pan, Z. Targeting calcium signaling in cancer therapy. *Acta Pharm. Sin. B* **2017**, *7*, 3–17. [CrossRef] [PubMed]

9. Hauser, A.S.; Attwood, M.M.; Rask-Andersen, M.; Schioth, H.B.; Gloriam, D.E. Trends in GPCR drug discovery: New agents, targets and indications. *Nat. Rev. Drug Discov.* **2017**, *16*, 829–842. [CrossRef] [PubMed]

10. Ouderkirk, J.L.; Krendel, M. Non-muscle myosins in tumor progression, cancer cell invasion, and metastasis. *Cytoskeleton* **2014**, *71*, 447–463. [CrossRef] [PubMed]

11. Li, Y.R.; Yang, W.X. Myosins as fundamental components during tumorigenesis: Diverse and indispensable. *Oncotarget* **2016**, *7*, 46785–46812. [CrossRef] [PubMed]

12. Maly, I.V.; Hofmann, W.A. Calcium-regulated import of myosin IC into the nucleus. *Cytoskeleton* **2016**, *73*, 341–350. [CrossRef] [PubMed]

13. Don-Salu-Hewage, A.S.; Chan, S.Y.; McAndrews, K.M.; Chetram, M.A.; Dawson, M.R.; Bethea, D.A.; Hinton, C.V. Cysteine (C)-X-C receptor 4 undergoes transportin 1-dependent nuclear localization and remains functional at the nucleus of metastatic prostate cancer cells. *PLoS ONE* **2013**, *8*, e57194. [CrossRef] [PubMed]

14. Emmert-Buck, M.R.; Vocke, C.D.; Pozzatti, R.O.; Duray, P.H.; Jennings, S.B.; Florence, C.D.; Zhuang, Z.; Bostwick, D.G.; Liotta, L.A.; Linehan, W.M. Allelic loss on chromosome 8p12-21 in microdissected prostatic intraepithelial neoplasia. *Cancer Res.* **1995**, *55*, 2959–2962. [PubMed]

15. Bethel, C.R.; Faith, D.; Li, X.; Guan, B.; Hicks, J.L.; Lan, F.; Jenkins, R.B.; Bieberich, C.J.; De Marzo, A.M. Decreased NKX3.1 protein expression in focal prostatic atrophy, prostatic intraepithelial neoplasia, and adenocarcinoma: Association with gleason score and chromosome 8p deletion. *Cancer Res.* **2006**, *66*, 10683–10690. [CrossRef] [PubMed]

16. Asatiani, E.; Huang, W.X.; Wang, A.; Rodriguez Ortner, E.; Cavalli, L.R.; Haddad, B.R.; Gelmann, E.P. Deletion, methylation, and expression of the NKX3.1 suppressor gene in primary human prostate cancer. *Cancer Res.* **2005**, *65*, 1164–1173. [CrossRef] [PubMed]

17. Bhatia-Gaur, R.; Donjacour, A.A.; Sciavolino, P.J.; Kim, M.; Desai, N.; Young, P.; Norton, C.R.; Gridley, T.; Cardiff, R.D.; Cunha, G.R.; et al. Roles for Nkx3.1 in prostate development and cancer. *Genes Dev.* **1999**, *13*, 966–977. [CrossRef] [PubMed]

18. Schneider, A.; Brand, T.; Zweigerdt, R.; Arnold, H. Targeted disruption of the Nkx3.1 gene in mice results in morphogenetic defects of minor salivary glands: Parallels to glandular duct morphogenesis in prostate. *Mech. Dev.* **2000**, *95*, 163–174. [CrossRef]

19. Tanaka, M.; Komuro, I.; Inagaki, H.; Jenkins, N.A.; Copeland, N.G.; Izumo, S. Nkx3.1, a murine homolog of Ddrosophila bagpipe, regulates epithelial ductal branching and proliferation of the prostate and palatine glands. *Dev. Dyn.* **2000**, *219*, 248–260. [CrossRef]

20. Abdulkadir, S.A.; Magee, J.A.; Peters, T.J.; Kaleem, Z.; Naughton, C.K.; Humphrey, P.A.; Milbrandt, J. Conditional loss of Nkx3.1 in adult mice induces prostatic intraepithelial neoplasia. *Mol. Cell. Biol.* **2002**, *22*, 1495–1503. [CrossRef] [PubMed]

21. Kim, M.J.; Bhatia-Gaur, R.; Banach-Petrosky, W.A.; Desai, N.; Wang, Y.; Hayward, S.W.; Cunha, G.R.; Cardiff, R.D.; Shen, M.M.; Abate-Shen, C. Nkx3.1 mutant mice recapitulate early stages of prostate carcinogenesis. *Cancer Res.* **2002**, *62*, 2999–3004. [PubMed]

22. Lei, Q.; Jiao, J.; Xin, L.; Chang, C.J.; Wang, S.; Gao, J.; Gleave, M.E.; Witte, O.N.; Liu, X.; Wu, H. NKX3.1 stabilizes p53, inhibits AKT activation, and blocks prostate cancer initiation caused by PTEN loss. *Cancer Cell* **2006**, *9*, 367–378. [CrossRef] [PubMed]

23. Ouyang, X.; DeWeese, T.L.; Nelson, W.G.; Abate-Shen, C. Loss-of-function of Nkx3.1 promotes increased oxidative damage in prostate carcinogenesis. *Cancer Res.* **2005**, *65*, 6773–6779. [CrossRef] [PubMed]

24. Bowen, C.; Gelmann, E.P. NKX3.1 activates cellular response to DNA damage. *Cancer Res.* **2010**, *70*, 3089–3097. [CrossRef] [PubMed]

25. Markowski, M.C.; Bowen, C.; Gelmann, E.P. Inflammatory cytokines induce phosphorylation and ubiquitination of prostate suppressor protein NKX3.1. *Cancer Res.* **2008**, *68*, 6896–6901. [CrossRef] [PubMed]

26. Decker, J.; Jain, G.; Kießling, T.; Philip, S.; Rid, M.; Barth, T.T.; Möller, P.; Cronauer, M.V.; Marienfeld, R.B. Loss of the Tumor Suppressor NKX3.1 in Prostate Cancer Cells is Induced by Prostatitis Related Mitogens. *J. Clin. Exp. Oncol.* **2016**, *5*. [CrossRef]

27. Sato, K.; Qian, J.; Slezak, J.M.; Lieber, M.M.; Bostwick, D.G.; Bergstralh, E.J.; Jenkins, R.B. Clinical significance of alterations of chromosome 8 in high-grade, advanced, nonmetastatic prostate carcinoma. *J. Natl. Cancer Inst.* **1999**, *91*, 1574–1580. [CrossRef] [PubMed]

28. Jenkins, R.B.; Qian, J.; Lieber, M.M.; Bostwick, D.G. Detection of c-myc oncogene amplification and chromosomal anomalies in metastatic prostatic carcinoma by fluorescence in situ hybridization. *Cancer Res.* **1997**, *57*, 524–531. [PubMed]

29. Gurel, B.; Iwata, T.; Koh, C.M.; Jenkins, R.B.; Lan, F.; Van Dang, C.; Hicks, J.L.; Morgan, J.; Cornish, T.C.; Sutcliffe, S.; et al. Nuclear MYC protein overexpression is an early alteration in human prostate carcinogenesis. *Mod. Pathol.* **2008**, *21*, 1156–1167. [CrossRef] [PubMed]

30. Sotelo, J.; Esposito, D.; Duhagon, M.A.; Banfield, K.; Mehalko, J.; Liao, H.; Stephens, R.M.; Harris, T.J.; Munroe, D.J.; Wu, X. Long-range enhancers on 8q24 regulate c-Myc. *Proc. Natl. Acad. Sci. USA* **2010**, *107*, 3001–3005. [CrossRef] [PubMed]

31. Jia, L.; Landan, G.; Pomerantz, M.; Jaschek, R.; Herman, P.; Reich, D.; Yan, C.; Khalid, O.; Kantoff, P.; Oh, W.; et al. Functional enhancers at the gene-poor 8q24 cancer-linked locus. *PLoS Genet.* **2009**, *5*, e1000597. [CrossRef] [PubMed]

32. Al Olama, A.A.; Kote-Jarai, Z.; Giles, G.G.; Guy, M.; Morrison, J.; Severi, G.; Leongamornlert, D.A.; Tymrakiewicz, M.; Jhavar, S.; Saunders, E.; et al. Multiple loci on 8q24 associated with prostate cancer susceptibility. *Nat. Genet.* **2009**, *41*, 1058–1060. [CrossRef] [PubMed]

33. Gudmundsson, J.; Sulem, P.; Gudbjartsson, D.F.; Blondal, T.; Gylfason, A.; Agnarsson, B.A.; Benediktsdottir, K.R.; Magnusdottir, D.N.; Orlygsdottir, G.; Jakobsdottir, M.; et al. Genome-wide association and replication studies identify four variants associated with prostate cancer susceptibility. *Nat. Genet.* **2009**, *41*, 1122–1126. [CrossRef] [PubMed]

34. Ellwood-Yen, K.; Graeber, T.G.; Wongvipat, J.; Iruela-Arispe, M.L.; Zhang, J.; Matusik, R.; Thomas, G.V.; Sawyers, C.L. Myc-driven murine prostate cancer shares molecular features with human prostate tumors. *Cancer Cell* **2003**, *4*, 223–238. [CrossRef]

35. Wang, J.; Kim, J.; Roh, M.; Franco, O.E.; Hayward, S.W.; Wills, M.L.; Abdulkadir, S.A. Pim1 kinase synergizes with c-MYC to induce advanced prostate carcinoma. *Oncogene* **2010**, *29*, 2477–2487. [CrossRef] [PubMed]

36. Mamaeva, O.A.; Kim, J.; Feng, G.; McDonald, J.M. Calcium/calmodulin-dependent kinase II regulates notch-1 signaling in prostate cancer cells. *J. Cell. Biochem.* **2009**, *106*, 25–32. [CrossRef] [PubMed]

37. Mehra, R.; Tomlins, S.A.; Shen, R.; Nadeem, O.; Wang, L.; Wei, J.T.; Pienta, K.J.; Ghosh, D.; Rubin, M.A.; Chinnaiyan, A.M.; et al. Comprehensive assessment of TMPRSS2 and ETS family gene aberrations in clinically localized prostate cancer. *Mod. Pathol.* **2007**, *20*, 538–544. [CrossRef] [PubMed]

38. Tomlins, S.A.; Laxman, B.; Dhanasekaran, S.M.; Helgeson, B.E.; Cao, X.; Morris, D.S.; Menon, A.; Jing, X.; Cao, Q.; Han, B.; et al. Distinct classes of chromosomal rearrangements create oncogenic ETS gene fusions in prostate cancer. *Nature* **2007**, *448*, 595–599. [CrossRef] [PubMed]

39. Tomlins, S.A.; Rhodes, D.R.; Perner, S.; Dhanasekaran, S.M.; Mehra, R.; Sun, X.W.; Varambally, S.; Cao, X.; Tchinda, J.; Kuefer, R.; et al. Recurrent fusion of TMPRSS2 and ETS transcription factor genes in prostate cancer. *Science* **2005**, *310*, 644–648. [CrossRef] [PubMed]

40. Iljin, K.; Wolf, M.; Edgren, H.; Gupta, S.; Kilpinen, S.; Skotheim, R.I.; Peltola, M.; Smit, F.; Verhaegh, G.; Schalken, J.; et al. TMPRSS2 fusions with oncogenic ETS factors in prostate cancer involve unbalanced genomic rearrangements and are associated with HDAC1 and epigenetic reprogramming. *Cancer Res.* **2006**, *66*, 10242–10246. [CrossRef] [PubMed]

41. Rouzier, C.; Haudebourg, J.; Carpentier, X.; Valerio, L.; Amiel, J.; Michiels, J.F.; Pedeutour, F. Detection of the TMPRSS2-ETS fusion gene in prostate carcinomas: Retrospective analysis of 55 formalin-fixed and paraffin-embedded samples with clinical data. *Cancer Genet. Cytogenet.* **2008**, *183*, 21–27. [CrossRef] [PubMed]

42. Mosquera, J.M.; Perner, S.; Genega, E.M.; Sanda, M.; Hofer, M.D.; Mertz, K.D.; Paris, P.L.; Simko, J.; Bismar, T.A.; Ayala, G.; et al. Characterization of TMPRSS2-ERG fusion high-grade prostatic intraepithelial neoplasia and potential clinical implications. *Clin. Cancer Res.* **2008**, *14*, 3380–3385. [CrossRef] [PubMed]

43. Albadine, R.; Latour, M.; Toubaji, A.; Haffner, M.; Isaacs, W.B.; E, A.P.; Meeker, A.K.; Demarzo, A.M.; Epstein, J.I.; Netto, G.J. TMPRSS2-ERG gene fusion status in minute (minimal) prostatic adenocarcinoma. *Mod. Pathol.* **2009**, *22*, 1415–1422. [CrossRef] [PubMed]

44. Lin, C.; Yang, L.; Tanasa, B.; Hutt, K.; Ju, B.G.; Ohgi, K.; Zhang, J.; Rose, D.W.; Fu, X.D.; Glass, C.K.; et al. Nuclear receptor-induced chromosomal proximity and DNA breaks underlie specific translocations in cancer. *Cell* **2009**, *139*, 1069–1083. [CrossRef] [PubMed]

45. Mani, R.S.; Tomlins, S.A.; Callahan, K.; Ghosh, A.; Nyati, M.K.; Varambally, S.; Palanisamy, N.; Chinnaiyan, A.M. Induced chromosomal proximity and gene fusions in prostate cancer. *Science* **2009**, *326*, 1230. [CrossRef] [PubMed]

46. Haffner, M.C.; Aryee, M.J.; Toubaji, A.; Esopi, D.M.; Albadine, R.; Gurel, B.; Isaacs, W.B.; Bova, G.S.; Liu, W.; Xu, J.; et al. Androgen-induced TOP2B-mediated double-strand breaks and prostate cancer gene rearrangements. *Nat. Genet.* **2010**, *42*, 668–675. [CrossRef] [PubMed]

47. Yu, J.; Yu, J.; Mani, R.S.; Cao, Q.; Brenner, C.J.; Cao, X.; Wang, X.; Wu, L.; Li, J.; Hu, M.; et al. An integrated network of androgen receptor, polycomb, and TMPRSS2-ERG gene fusions in prostate cancer progression. *Cancer Cell.* **2010**, *17*, 443–454. [CrossRef] [PubMed]

48. Tomlins, S.A.; Laxman, B.; Varambally, S.; Cao, X.; Yu, J.; Helgeson, B.E.; Cao, Q.; Prensner, J.R.; Rubin, M.A.; Shah, R.B.; et al. Role of the TMPRSS2-ERG gene fusion in prostate cancer. *Neoplasia* **2008**, *10*, 177–188. [CrossRef] [PubMed]

49. Klezovitch, O.; Risk, M.; Coleman, I.; Lucas, J.M.; Null, M.; True, L.D.; Nelson, P.S.; Vasioukhin, V. A causal role for ERG in neoplastic transformation of prostate epithelium. *Proc. Natl. Acad. Sci. USA* **2008**, *105*, 2105–2110. [CrossRef] [PubMed]

50. Carver, B.S.; Tran, J.; Gopalan, A.; Chen, Z.; Shaikh, S.; Carracedo, A.; Alimonti, A.; Nardella, C.; Varmeh, S.; Scardino, P.T.; et al. Aberrant ERG expression cooperates with loss of PTEN to promote cancer progression in the prostate. *Nat. Genet.* **2009**, *41*, 619–624. [CrossRef] [PubMed]

51. King, J.C.; Xu, J.; Wongvipat, J.; Hieronymus, H.; Carver, B.S.; Leung, D.H.; Taylor, B.S.; Sander, C.; Cardiff, R.D.; Couto, S.S.; et al. Cooperativity of TMPRSS2-ERG with PI3-kinase pathway activation in prostate oncogenesis. *Nat. Genet.* **2009**, *41*, 524–526. [CrossRef] [PubMed]

52. Geybels, M.S.; Alumkal, J.J.; Luedeke, M.; Rinckleb, A.; Zhao, S.; Shui, I.M.; Bibikova, M.; Klotzle, B.; van den Brandt, P.A.; Ostrander, E.A.; et al. Epigenomic profiling of prostate cancer identifies differentially methylated genes in TMPRSS2:ERG fusion-positive versus fusion-negative tumors. *Clin. Epigenet.* **2015**, *7*, 128. [CrossRef] [PubMed]

53. Geybels, M.S.; McCloskey, K.D.; Mills, I.G.; Stanford, J.L. Calcium Channel Blocker Use and Risk of Prostate Cancer by TMPRSS2:ERG Gene Fusion Status. *Prostate* **2017**, *77*, 282–290. [CrossRef] [PubMed]

54. Wang, X.; Qiao, Y.; Asangani, I.A.; Ateeq, B.; Poliakov, A.; Cieslik, M.; Pitchiaya, S.; Chakravarthi, B.; Cao, X.; Jing, X.; et al. Development of Peptidomimetic Inhibitors of the ERG Gene Fusion Product in Prostate Cancer. *Cancer Cell* **2017**, *31*, 532–548. [CrossRef] [PubMed]

55. Rastogi, A.; Tan, S.H.; Mohamed, A.A.; Chen, Y.; Hu, Y.; Petrovics, G.; Sreenath, T.; Kagan, J.; Srivastava, S.; McLeod, D.G.; et al. Functional antagonism of TMPRSS2-ERG splice variants in prostate cancer. *Genes Cancer* **2014**, *5*, 273–284. [PubMed]

56. Gimenez-Bonafe, P.; Fedoruk, M.N.; Whitmore, T.G.; Akbari, M.; Ralph, J.L.; Ettinger, S.; Gleave, M.E.; Nelson, C.C. YB-1 is upregulated during prostate cancer tumor progression and increases P-glycoprotein activity. *Prostate* **2004**, *59*, 337–349. [CrossRef] [PubMed]

57. Shiota, M.; Takeuchi, A.; Song, Y.; Yokomizo, A.; Kashiwagi, E.; Uchiumi, T.; Kuroiwa, K.; Tatsugami, K.; Fujimoto, N.; Oda, Y.; et al. Y-box binding protein-1 promotes castration-resistant prostate cancer growth via androgen receptor expression. *Endocr. Relat. Cancer* **2011**, *18*, 505–517. [CrossRef] [PubMed]

58. Heumann, A.; Kaya, O.; Burdelski, C.; Hube-Magg, C.; Kluth, M.; Lang, D.S.; Simon, R.; Beyer, B.; Thederan, I.; Sauter, G.; et al. Up regulation and nuclear translocation of Y-box binding protein 1 (YB-1) is linked to poor prognosis in ERG-negative prostate cancer. *Sci. Rep.* **2017**, *7*, 2056. [CrossRef] [PubMed]

59. Wang, J.; Cai, Y.; Shao, L.J.; Siddiqui, J.; Palanisamy, N.; Li, R.; Ren, C.; Ayala, G.; Ittmann, M. Activation of NF-{kappa}B by TMPRSS2/ERG Fusion Isoforms through Toll-Like Receptor-4. *Cancer Res.* **2011**, *71*, 1325–1333. [CrossRef] [PubMed]

60. Nadiminty, N.; Lou, W.; Sun, M.; Chen, J.; Yue, J.; Kung, H.J.; Evans, C.P.; Zhou, Q.; Gao, A.C. Aberrant activation of the androgen receptor by NF-kappaB2/p52 in prostate cancer cells. *Cancer Res.* **2010**, *70*, 3309–3319. [CrossRef] [PubMed]

61. Singareddy, R.; Semaan, L.; Conley-Lacomb, M.K.; St John, J.; Powell, K.; Iyer, M.; Smith, D.; Heilbrun, L.K.; Shi, D.; Sakr, W.; et al. Transcriptional regulation of CXCR4 in prostate cancer: Significance of TMPRSS2-ERG fusions. *Mol. Cancer Res.* **2013**, *11*, 1349–1361. [CrossRef] [PubMed]

62. Wang, S.I.; Parsons, R.; Ittmann, M. Homozygous deletion of the PTEN tumor suppressor gene in a subset of prostate adenocarcinomas. *Clin. Cancer Res.* **1998**, *4*, 811–815. [PubMed]

63. Schmitz, M.; Grignard, G.; Margue, C.; Dippel, W.; Capesius, C.; Mossong, J.; Nathan, M.; Giacchi, S.; Scheiden, R.; Kieffer, N. Complete loss of PTEN expression as a possible early prognostic marker for prostate cancer metastasis. *Int. J. Cancer* **2007**, *120*, 1284–1292. [CrossRef] [PubMed]

64. Sircar, K.; Yoshimoto, M.; Monzon, F.A.; Koumakpayi, I.H.; Katz, R.L.; Khanna, A.; Alvarez, K.; Chen, G.; Darnel, A.D.; Aprikian, A.G.; et al. PTEN genomic deletion is associated with p-Akt and AR signalling in poorer outcome, hormone refractory prostate cancer. *J. Pathol.* **2009**, *218*, 505–513. [CrossRef] [PubMed]

65. McMenamin, M.E.; Soung, P.; Perera, S.; Kaplan, I.; Loda, M.; Sellers, W.R. Loss of PTEN expression in paraffin-embedded primary prostate cancer correlates with high Gleason score and advanced stage. *Cancer Res.* **1999**, *59*, 4291–4296. [PubMed]

66. Wang, S.; Gao, J.; Lei, Q.; Rozengurt, N.; Pritchard, C.; Jiao, J.; Thomas, G.V.; Li, G.; Roy-Burman, P.; Nelson, P.S.; et al. Prostate-specific deletion of the murine Pten tumor suppressor gene leads to metastatic prostate cancer. *Cancer Cell* **2003**, *4*, 209–221. [CrossRef]

67. Kim, M.J.; Cardiff, R.D.; Desai, N.; Banach-Petrosky, W.A.; Parsons, R.; Shen, M.M.; Abate-Shen, C. Cooperativity of Nkx3.1 and Pten loss of function in a mouse model of prostate carcinogenesis. *Proc. Natl. Acad. Sci. USA* **2002**, *99*, 2884–2889. [CrossRef] [PubMed]

68. Kim, J.; Eltoum, I.E.; Roh, M.; Wang, J.; Abdulkadir, S.A. Interactions between cells with distinct mutations in c-MYC and Pten in prostate cancer. *PLoS Genet.* **2009**, *5*, e1000542. [CrossRef] [PubMed]

69. Jia, S.; Liu, Z.; Zhang, S.; Liu, P.; Zhang, L.; Lee, S.H.; Zhang, J.; Signoretti, S.; Loda, M.; Roberts, T.M.; et al. Essential roles of PI(3)K-p110β in cell growth, metabolism and tumorigenesis. *Nature* **2008**, *454*, 776–779. [CrossRef] [PubMed]

70. Guertin, D.A.; Stevens, D.M.; Saitoh, M.; Kinkel, S.; Crosby, K.; Sheen, J.H.; Mullholland, D.J.; Magnuson, M.A.; Wu, H.; Sabatini, D.M. mTOR complex 2 is required for the development of prostate cancer induced by Pten loss in mice. *Cancer Cell* **2009**, *15*, 148–159. [CrossRef] [PubMed]

71. Gao, H.; Ouyang, X.; Banach-Petrosky, W.A.; Shen, M.M.; Abate-Shen, C. Emergence of androgen independence at early stages of prostate cancer progression in Nkx3.1; Pten mice. *Cancer Res.* **2006**, *66*, 7929–7933. [CrossRef] [PubMed]

72. Lin, H.K.; Hu, Y.C.; Lee, D.K.; Chang, C. Regulation of androgen receptor signaling by PTEN (phosphatase and tensin homolog deleted on chromosome 10) tumor suppressor through distinct mechanisms in prostate cancer cells. *Mol. Endocrinol.* **2004**, *18*, 2409–2423. [CrossRef] [PubMed]

73. Trotman, L.C.; Wang, X.; Alimonti, A.; Chen, Z.; Teruya-Feldstein, J.; Yang, H.; Pavletich, N.P.; Carver, B.S.; Cordon-Cardo, C.; Erdjument-Bromage, H.; et al. Ubiquitination regulates PTEN nuclear import and tumor suppression. *Cell* **2007**, *128*, 141–156. [CrossRef] [PubMed]

74. Chang, C.J.; Mulholland, D.J.; Valamehr, B.; Mosessian, S.; Sellers, W.R.; Wu, H. PTEN nuclear localization is regulated by oxidative stress and mediates p53-dependent tumor suppression. *Mol. Cell. Biol.* **2008**, *28*, 3281–3289. [CrossRef] [PubMed]

75. Chen, M.; Nowak, D.G.; Narula, N.; Robinson, B.; Watrud, K.; Ambrico, A.; Herzka, T.M.; Zeeman, M.E.; Minderer, M.; Zheng, W.; et al. The nuclear transport receptor Importin-11 is a tumor suppressor that maintains PTEN protein. *J. Cell Biol.* **2017**, *216*, 641–656. [CrossRef] [PubMed]

76. Schaeffer, E.M.; Marchionni, L.; Huang, Z.; Simons, B.; Blackman, A.; Yu, W.; Parmigiani, G.; Berman, D.M. Androgen-induced programs for prostate epithelial growth and invasion arise in embryogenesis and are reactivated in cancer. *Oncogene* **2008**, *27*, 7180–7191. [CrossRef] [PubMed]

77. Pritchard, C.; Mecham, B.; Dumpit, R.; Coleman, I.; Bhattacharjee, M.; Chen, Q.; Sikes, R.A.; Nelson, P.S. Conserved gene expression programs integrate mammalian prostate development and tumorigenesis. *Cancer Res.* **2009**, *69*, 1739–1747. [CrossRef] [PubMed]

78. Bruxvoort, K.J.; Charbonneau, H.M.; Giambernardi, T.A.; Goolsby, J.C.; Qian, C.N.; Zylstra, C.R.; Robinson, D.R.; Roy-Burman, P.; Shaw, A.K.; Buckner-Berghuis, B.D.; et al. Inactivation of Apc in the mouse prostate causes prostate carcinoma. *Cancer Res.* **2007**, *67*, 2490–2496. [CrossRef] [PubMed]

79. Pearson, H.B.; Phesse, T.J.; Clarke, A.R. K-ras and Wnt signaling synergize to accelerate prostate tumorigenesis in the mouse. *Cancer Res.* **2009**, *69*, 94–101. [CrossRef] [PubMed]

80. Yu, X.; Wang, Y.; Jiang, M.; Bierie, B.; Roy-Burman, P.; Shen, M.M.; Taketo, M.M.; Wills, M.; Matusik, R.J. Activation of β-Catenin in mouse prostate causes HGPIN and continuous prostate growth after castration. *Prostate* **2009**, *69*, 249–262. [CrossRef] [PubMed]

81. Wang, G.; Wang, J.; Sadar, M.D. Crosstalk between the androgen receptor and β-catenin in castrate-resistant prostate cancer. *Cancer Res.* **2008**, *68*, 9918–9927. [CrossRef] [PubMed]

82. Horvath, L.G.; Henshall, S.M.; Lee, C.S.; Kench, J.G.; Golovsky, D.; Brenner, P.C.; O'Neill, G.F.; Kooner, R.; Stricker, P.D.; Grygiel, J.J.; et al. Lower levels of nuclear β-catenin predict for a poorer prognosis in localized prostate cancer. *Int. J. Cancer* **2005**, *113*, 415–422. [CrossRef] [PubMed]

83. Whitaker, H.C.; Girling, J.; Warren, A.Y.; Leung, H.; Mills, I.G.; Neal, D.E. Alterations in β-catenin expression and localization in prostate cancer. *Prostate* **2008**, *68*, 1196–1205. [CrossRef] [PubMed]

84. Wang, Q.; Symes, A.J.; Kane, C.A.; Freeman, A.; Nariculam, J.; Munson, P.; Thrasivoulou, C.; Masters, J.R.; Ahmed, A. A novel role for Wnt/Ca^{2+} signaling in actin cytoskeleton remodeling and cell motility in prostate cancer. *PLoS ONE* **2010**, *5*, e10456. [CrossRef] [PubMed]

85. Thrasivoulou, C.; Millar, M.; Ahmed, A. Activation of intracellular calcium by multiple Wnt ligands and translocation of β-catenin into the nucleus: A convergent model of Wnt/Ca^{2+} and Wnt/β-catenin pathways. *J. Biol. Chem.* **2013**, *288*, 35651–35659. [CrossRef] [PubMed]

86. Nicotera, P.; Zhivotovsky, B.; Orrenius, S. Nuclear calcium transport and the role of calcium in apoptosis. *Cell Calcium* **1994**, *16*, 279–288. [CrossRef]

87. Hsu, J.L.; Pan, S.L.; Ho, Y.F.; Hwang, T.L.; Kung, F.L.; Guh, J.H. Costunolide induces apoptosis through nuclear Ca^{2+} overload and DNA damage response in human prostate cancer. *J. Urol.* **2011**, *185*, 1967–1974. [CrossRef] [PubMed]
88. Karhadkar, S.S.; Bova, G.S.; Abdallah, N.; Dhara, S.; Gardner, D.; Maitra, A.; Isaacs, J.T.; Berman, D.M.; Beachy, P.A. Hedgehog signalling in prostate regeneration, neoplasia and metastasis. *Nature* **2004**, *431*, 707–712. [CrossRef] [PubMed]
89. Sanchez, P.; Hernandez, A.M.; Stecca, B.; Kahler, A.J.; DeGueme, A.M.; Barrett, A.; Beyna, M.; Datta, M.W.; Datta, S.; Ruiz i Altaba, A. Inhibition of prostate cancer proliferation by interference with SONIC HEDGEHOG-GLI1 signaling. *Proc. Natl. Acad. Sci. USA* **2004**, *101*, 12561–12566. [CrossRef] [PubMed]
90. Shaw, A.; Gipp, J.; Bushman, W. The Sonic Hedgehog pathway stimulates prostate tumor growth by paracrine signaling and recapitulates embryonic gene expression in tumor myofibroblasts. *Oncogene* **2009**, *28*, 4480–4490. [CrossRef] [PubMed]
91. Yauch, R.L.; Gould, S.E.; Scales, S.J.; Tang, T.; Tian, H.; Ahn, C.P.; Marshall, D.; Fu, L.; Januario, T.; Kallop, D.; et al. A paracrine requirement for hedgehog signalling in cancer. *Nature* **2008**, *455*, 406–410. [CrossRef] [PubMed]
92. Gonnissen, A.; Isebaert, S.; Haustermans, K. Hedgehog signaling in prostate cancer and its therapeutic implication. *Int. J. Mol. Sci.* **2013**, *14*, 13979–14007. [CrossRef] [PubMed]
93. Li, N.; Truong, S.; Nouri, M.; Moore, J.; Al Nakouzi, N.; Lubik, A.A.; Buttyan, R. Non-canonical activation of hedgehog in prostate cancer cells mediated by the interaction of transcriptionally active androgen receptor proteins with Gli3. *Oncogene* **2018**. [CrossRef] [PubMed]
94. Acevedo, V.D.; Gangula, R.D.; Freeman, K.W.; Li, R.; Zhang, Y.; Wang, F.; Ayala, G.E.; Peterson, L.E.; Ittmann, M.; Spencer, D.M. Inducible FGFR-1 activation leads to irreversible prostate adenocarcinoma and an epithelial-to-mesenchymal transition. *Cancer Cell* **2007**, *12*, 559–571. [CrossRef] [PubMed]
95. Memarzadeh, S.; Xin, L.; Mulholland, D.J.; Mansukhani, A.; Wu, H.; Teitell, M.A.; Witte, O.N. Enhanced paracrine FGF10 expression promotes formation of multifocal prostate adenocarcinoma and an increase in epithelial androgen receptor. *Cancer Cell* **2007**, *12*, 572–585. [CrossRef] [PubMed]
96. Kwabi-Addo, B.; Ozen, M.; Ittmann, M. The role of fibroblast growth factors and their receptors in prostate cancer. *Endocr. Relat. Cancer* **2004**, *11*, 709–724. [CrossRef] [PubMed]
97. Chen, M.L.; Xu, P.Z.; Peng, X.D.; Chen, W.S.; Guzman, G.; Yang, X.; Di Cristofano, A.; Pandolfi, P.P.; Hay, N. The deficiency of *Akt1* is sufficient to suppress tumor development in *Pten+/−* mice. *Genes Dev.* **2006**, *20*, 1569–1574. [CrossRef] [PubMed]
98. Thomas, G.V.; Horvath, S.; Smith, B.L.; Crosby, K.; Lebel, L.A.; Schrage, M.; Said, J.; De Kernion, J.; Reiter, R.E.; Sawyers, C.L. Antibody-based profiling of the phosphoinositide 3-kinase pathway in clinical prostate cancer. *Clin. Cancer Res.* **2004**, *10*, 8351–8356. [CrossRef] [PubMed]
99. Shen, M.M.; Abate-Shen, C. Pten inactivation and the emergence of androgen-independent prostate cancer. *Cancer Res.* **2007**, *67*, 6535–6538. [CrossRef] [PubMed]
100. Boormans, J.L.; Hermans, K.G.; van Leenders, G.J.; Trapman, J.; Verhagen, P.C. An activating mutation in AKT1 in human prostate cancer. *Int. J. Cancer* **2008**, *123*, 2725–2726. [CrossRef] [PubMed]
101. Lee, S.H.; Poulogiannis, G.; Pyne, S.; Jia, S.; Zou, L.; Signoretti, S.; Loda, M.; Cantley, L.C.; Roberts, T.M. A constitutively activated form of the p110β isoform of PI3-kinase induces prostatic intraepithelial neoplasia in mice. *Proc. Natl. Acad. Sci. USA* **2010**, *107*, 11002–11007. [CrossRef] [PubMed]
102. Gao, H.; Ouyang, X.; Banach-Petrosky, W.A.; Gerald, W.L.; Shen, M.M.; Abate-Shen, C. Combinatorial activities of Akt and B-Raf/Erk signaling in a mouse model of androgen-independent prostate cancer. *Proc. Natl. Acad. Sci. USA* **2006**, *103*, 14477–14482. [CrossRef] [PubMed]
103. Trotman, L.C.; Alimonti, A.; Scaglioni, P.P.; Koutcher, J.A.; Cordon-Cardo, C.; Pandolfi, P.P. Identification of a tumour suppressor network opposing nuclear Akt function. *Nature* **2006**, *441*, 523–527. [CrossRef] [PubMed]
104. Kondratskyi, A.; Yassine, M.; Slomianny, C.; Kondratska, K.; Gordienko, D.; Dewailly, E.; Lehen'kyi, V.; Skryma, R.; Prevarskaya, N. Identification of ML-9 as a lysosomotropic agent targeting autophagy and cell death. *Cell Death Dis.* **2014**, *5*, e1193. [CrossRef] [PubMed]
105. Uzgare, A.R.; Isaacs, J.T. Enhanced redundancy in Akt and mitogen-activated protein kinase-induced survival of malignant versus normal prostate epithelial cells. *Cancer Res.* **2004**, *64*, 6190–6199. [CrossRef] [PubMed]

106. Kinkade, C.W.; Castillo-Martin, M.; Puzio-Kuter, A.; Yan, J.; Foster, T.H.; Gao, H.; Sun, Y.; Ouyang, X.; Gerald, W.L.; Cordon-Cardo, C.; et al. Targeting AKT/mTOR and ERK MAPK signaling inhibits hormone-refractory prostate cancer in a preclinical mouse model. *J. Clin. Investig.* **2008**, *118*, 3051–3064. [CrossRef] [PubMed]

107. Dan, H.C.; Cooper, M.J.; Cogswell, P.C.; Duncan, J.A.; Ting, J.P.; Baldwin, A.S. Akt-dependent regulation of NF-{kappa}B is controlled by mTOR and Raptor in association with IKK. *Genes Dev.* **2008**, *22*, 1490–1500. [CrossRef] [PubMed]

108. Fradet, V.; Lessard, L.; Begin, L.R.; Karakiewicz, P.; Masson, A.M.; Saad, F. Nuclear factor-kappaB nuclear localization is predictive of biochemical recurrence in patients with positive margin prostate cancer. *Clin. Cancer Res.* **2004**, *10*, 8460–8464. [CrossRef] [PubMed]

109. Ismail, H.A.; Lessard, L.; Mes-Masson, A.M.; Saad, F. Expression of NF-kappaB in prostate cancer lymph node metastases. *Prostate* **2004**, *58*, 308–313. [CrossRef] [PubMed]

110. Lessard, L.; Karakiewicz, P.I.; Bellon-Gagnon, P.; Alam-Fahmy, M.; Ismail, H.A.; Mes-Masson, A.M.; Saad, F. Nuclear localization of nuclear factor-kappaB p65 in primary prostate tumors is highly predictive of pelvic lymph node metastases. *Clin. Cancer Res.* **2006**, *12*, 5741–5745. [CrossRef] [PubMed]

111. Luo, J.L.; Tan, W.; Ricono, J.M.; Korchynskyi, O.; Zhang, M.; Gonias, S.L.; Cheresh, D.A.; Karin, M. Nuclear cytokine-activated IKKα controls prostate cancer metastasis by repressing Maspin. *Nature* **2007**, *446*, 690–694. [CrossRef] [PubMed]

112. Zhang, L.; Altuwaijri, S.; Deng, F.; Chen, L.; Lal, P.; Bhanot, U.K.; Korets, R.; Wenske, S.; Lilja, H.G.; Chang, C.; et al. NF-kappaB regulates androgen receptor expression and prostate cancer growth. *Am. J. Pathol.* **2009**, *175*, 489–499. [CrossRef] [PubMed]

113. Jeong, J.H.; Wang, Z.; Guimaraes, A.S.; Ouyang, X.; Figueiredo, J.L.; Ding, Z.; Jiang, S.; Guney, I.; Kang, G.H.; Shin, E.; et al. BRAF activation initiates but does not maintain invasive prostate adenocarcinoma. *PLoS ONE* **2008**, *3*, e3949. [CrossRef] [PubMed]

114. Taylor, B.S.; Schultz, N.; Hieronymus, H.; Gopalan, A.; Xiao, Y.; Carver, B.S.; Arora, V.K.; Kaushik, P.; Cerami, E.; Reva, B.; et al. Integrative genomic profiling of human prostate cancer. *Cancer Cell* **2010**, *18*, 11–22. [CrossRef] [PubMed]

115. Gioeli, D.; Mandell, J.W.; Petroni, G.R.; Frierson, H.F., Jr.; Weber, M.J. Activation of mitogen-activated protein kinase associated with prostate cancer progression. *Cancer Res.* **1999**, *59*, 279–284. [PubMed]

116. Roderick, H.L.; Cook, S.J. Ca^{2+} signalling checkpoints in cancer: Remodelling Ca^{2+} for cancer cell proliferation and survival. *Nat. Rev. Cancer* **2008**, *8*, 361–375. [CrossRef] [PubMed]

117. Hanahan, D.; Weinberg, R.A. Hallmarks of cancer: The next generation. *Cell* **2011**, *144*, 646–674. [CrossRef] [PubMed]

118. Xu, Y.; Zhang, S.; Niu, H.; Ye, Y.; Hu, F.; Chen, S.; Li, X.; Luo, X.; Jiang, S.; Liu, Y.; et al. STIM1 accelerates cell senescence in a remodeled microenvironment but enhances the epithelial-to-mesenchymal transition in prostate cancer. *Sci. Rep.* **2015**, *5*, 11754. [CrossRef] [PubMed]

119. van Noesel, M.M.; van Bezouw, S.; Salomons, G.S.; Voute, P.A.; Pieters, R.; Baylin, S.B.; Herman, J.G.; Versteeg, R. Tumor-specific down-regulation of the tumor necrosis factor-related apoptosis-inducing ligand decoy receptors DcR1 and DcR2 is associated with dense promoter hypermethylation. *Cancer Res.* **2002**, *62*, 2157–2161. [PubMed]

120. Sheikh, M.S.; Fornace, A.J., Jr. Death and decoy receptors and p53-mediated apoptosis. *Leukemia* **2000**, *14*, 1509–1513. [CrossRef] [PubMed]

121. Collado, M.; Gil, J.; Efeyan, A.; Guerra, C.; Schuhmacher, A.J.; Barradas, M.; Benguria, A.; Zaballos, A.; Flores, J.M.; Barbacid, M.; et al. Tumour biology: Senescence in premalignant tumours. *Nature* **2005**, *436*, 642. [CrossRef] [PubMed]

122. Althubiti, M.; Lezina, L.; Carrera, S.; Jukes-Jones, R.; Giblett, S.M.; Antonov, A.; Barlev, N.; Saldanha, G.S.; Pritchard, C.A.; Cain, K.; et al. Characterization of novel markers of senescence and their prognostic potential in cancer. *Cell Death Dis.* **2014**, *5*, e1528. [CrossRef] [PubMed]

123. Lee, S.J.; Lee, K.; Yang, X.; Jung, C.; Gardner, T.; Kim, H.S.; Jeng, M.H.; Kao, C. NFATc1 with AP-3 site binding specificity mediates gene expression of prostate-specific-membrane-antigen. *J. Mol. Biol.* **2003**, *330*, 749–760. [CrossRef]

124. Kavitha, C.V.; Deep, G.; Gangar, S.C.; Jain, A.K.; Agarwal, C.; Agarwal, R. Silibinin inhibits prostate cancer cells- and RANKL-induced osteoclastogenesis by targeting NFATc1, NF-kappaB, and AP-1 activation in RAW264.7 cells. *Mol. Carcinog.* **2014**, *53*, 169–180. [CrossRef] [PubMed]

125. Mognol, G.P.; Carneiro, F.R.; Robbs, B.K.; Faget, D.V.; Viola, J.P. Cell cycle and apoptosis regulation by NFAT transcription factors: New roles for an old player. *Cell Death Dis.* **2016**, *7*, e2199. [CrossRef] [PubMed]

126. Manda, K.R.; Tripathi, P.; Hsi, A.C.; Ning, J.; Ruzinova, M.B.; Liapis, H.; Bailey, M.; Zhang, H.; Maher, C.A.; Humphrey, P.A.; et al. NFATc1 promotes prostate tumorigenesis and overcomes PTEN loss-induced senescence. *Oncogene* **2016**, *35*, 3282–3292. [CrossRef] [PubMed]

127. Wei, Z.; Jiang, X.; Qiao, H.; Zhai, B.; Zhang, L.; Zhang, Q.; Wu, Y.; Jiang, H.; Sun, X. STAT3 interacts with Skp2/p27/p21 pathway to regulate the motility and invasion of gastric cancer cells. *Cell. Signal.* **2013**, *25*, 931–938. [CrossRef] [PubMed]

128. Huang, H.; Zhao, W.; Yang, D. Stat3 induces oncogenic Skp2 expression in human cervical carcinoma cells. *Biochem. Biophys. Res. Commun.* **2012**, *418*, 186–190. [CrossRef] [PubMed]

129. Wissenbach, U.; Niemeyer, B.A.; Fixemer, T.; Schneidewind, A.; Trost, C.; Cavalie, A.; Reus, K.; Meese, E.; Bonkhoff, H.; Flockerzi, V. Expression of CaT-like, a novel calcium-selective channel, correlates with the malignancy of prostate cancer. *J. Biol. Chem.* **2001**, *276*, 19461–19468. [CrossRef] [PubMed]

130. Fixemer, T.; Wissenbach, U.; Flockerzi, V.; Bonkhoff, H. Expression of the Ca^{2+}-selective cation channel TRPV6 in human prostate cancer: A novel prognostic marker for tumor progression. *Oncogene* **2003**, *22*, 7858–7861. [CrossRef] [PubMed]

131. Thebault, S.; Flourakis, M.; Vanoverberghe, K.; Vandermoere, F.; Roudbaraki, M.; Lehen'kyi, V.; Slomianny, C.; Beck, B.; Mariot, P.; Bonnal, J.L.; et al. Differential role of transient receptor potential channels in Ca^{2+} entry and proliferation of prostate cancer epithelial cells. *Cancer Res.* **2006**, *66*, 2038–2047. [CrossRef] [PubMed]

132. Lehen'kyi, V.; Flourakis, M.; Skryma, R.; Prevarskaya, N. TRPV6 channel controls prostate cancer cell proliferation via Ca^{2+}/NFAT-dependent pathways. *Oncogene* **2007**, *26*, 7380–7385. [CrossRef] [PubMed]

133. Raphael, M.; Lehen'kyi, V.; Vandenberghe, M.; Beck, B.; Khalimonchyk, S.; Vanden Abeele, F.; Farsetti, L.; Germain, E.; Bokhobza, A.; Mihalache, A.; et al. TRPV6 calcium channel translocates to the plasma membrane via Orai1-mediated mechanism and controls cancer cell survival. *Proc. Natl. Acad. Sci. USA* **2014**, *111*, E3870–E3879. [CrossRef] [PubMed]

134. Fu, S.; Hirte, H.; Welch, S.; Ilenchuk, T.T.; Lutes, T.; Rice, C.; Fields, N.; Nemet, A.; Dugourd, D.; Piha-Paul, S.; et al. First-in-human phase I study of SOR-C13, a TRPV6 calcium channel inhibitor, in patients with advanced solid tumors. *Investig. New Drugs* **2017**, *35*, 324–333. [CrossRef] [PubMed]

135. Monet, M.; Lehen'kyi, V.; Gackiere, F.; Firlej, V.; Vandenberghe, M.; Roudbaraki, M.; Gkika, D.; Pourtier, A.; Bidaux, G.; Slomianny, C.; et al. Role of cationic channel TRPV2 in promoting prostate cancer migration and progression to androgen resistance. *Cancer Res.* **2010**, *70*, 1225–1235. [CrossRef] [PubMed]

136. Maly, I.V.; Domaradzki, T.M.; Gosy, V.A.; Hofmann, W.A. Myosin isoform expressed in metastatic prostate cancer stimulates cell invasion. *Sci. Rep.* **2017**, *7*, 8476. [CrossRef] [PubMed]

137. Tsavaler, L.; Shapero, M.H.; Morkowski, S.; Laus, R. Trp-p8, a novel prostate-specific gene, is up-regulated in prostate cancer and other malignancies and shares high homology with transient receptor potential calcium channel proteins. *Cancer Res.* **2001**, *61*, 3760–3769. [PubMed]

138. Zhang, L.; Barritt, G.J. Evidence that TRPM8 is an androgen-dependent Ca^{2+} channel required for the survival of prostate cancer cells. *Cancer Res.* **2004**, *64*, 8365–8373. [CrossRef] [PubMed]

139. Monteith, G.R.; Davis, F.M.; Roberts-Thomson, S.J. Calcium channels and pumps in cancer: Changes and consequences. *J. Biol. Chem.* **2012**, *287*, 31666–31673. [CrossRef] [PubMed]

140. Henshall, S.M.; Afar, D.E.; Hiller, J.; Horvath, L.G.; Quinn, D.I.; Rasiah, K.K.; Gish, K.; Willhite, D.; Kench, J.G.; Gardiner-Garden, M.; et al. Survival analysis of genome-wide gene expression profiles of prostate cancers identifies new prognostic targets of disease relapse. *Cancer Res.* **2003**, *63*, 4196–4203. [PubMed]

141. Gkika, D.; Flourakis, M.; Lemonnier, L.; Prevarskaya, N. PSA reduces prostate cancer cell motility by stimulating TRPM8 activity and plasma membrane expression. *Oncogene* **2010**, *29*, 4611–4616. [CrossRef] [PubMed]

142. Bidaux, G.; Flourakis, M.; Thebault, S.; Zholos, A.; Beck, B.; Gkika, D.; Roudbaraki, M.; Bonnal, J.L.; Mauroy, B.; Shuba, Y.; et al. Prostate cell differentiation status determines transient receptor potential melastatin member 8 channel subcellular localization and function. *J. Clin. Investig.* **2007**, *117*, 1647–1657. [CrossRef] [PubMed]

143. Vanoverberghe, K.; Vanden Abeele, F.; Mariot, P.; Lepage, G.; Roudbaraki, M.; Bonnal, J.L.; Mauroy, B.; Shuba, Y.; Skryma, R.; Prevarskaya, N. Ca^{2+} homeostasis and apoptotic resistance of neuroendocrine-differentiated prostate cancer cells. *Cell Death Differ.* **2004**, *11*, 321–330. [CrossRef] [PubMed]

144. Vanden Abeele, F.; Skryma, R.; Shuba, Y.; Van Coppenolle, F.; Slomianny, C.; Roudbaraki, M.; Mauroy, B.; Wuytack, F.; Prevarskaya, N. Bcl-2-dependent modulation of Ca^{2+} homeostasis and store-operated channels in prostate cancer cells. *Cancer Cell* **2002**, *1*, 169–179. [CrossRef]

145. Chen, L.; Cao, R.; Wang, G.; Yuan, L.; Qian, G.; Guo, Z.; Wu, C.L.; Wang, X.; Xiao, Y. Downregulation of TRPM7 suppressed migration and invasion by regulating epithelial-mesenchymal transition in prostate cancer cells. *Med. Oncol.* **2017**, *34*, 127. [CrossRef] [PubMed]

146. Sagredo, A.I.; Sagredo, E.A.; Cappelli, C.; Baez, P.; Andaur, R.E.; Blanco, C.; Tapia, J.C.; Echeverria, C.; Cerda, O.; Stutzin, A.; et al. TRPM4 regulates Akt/GSK3-β activity and enhances β-catenin signaling and cell proliferation in prostate cancer cells. *Mol. Oncol.* **2018**, *12*, 151–165. [CrossRef] [PubMed]

147. Zeng, X.; Sikka, S.C.; Huang, L.; Sun, C.; Xu, C.; Jia, D.; Abdel-Mageed, A.B.; Pottle, J.E.; Taylor, J.T.; Li, M. Novel role for the transient receptor potential channel TRPM2 in prostate cancer cell proliferation. *Prostate Cancer Prostatic Dis.* **2010**, *13*, 195–201. [CrossRef] [PubMed]

148. Haverstick, D.M.; Heady, T.N.; Macdonald, T.L.; Gray, L.S. Inhibition of human prostate cancer proliferation in vitro and in a mouse model by a compound synthesized to block Ca^{2+} entry. *Cancer Res.* **2000**, *60*, 1002–1008. [PubMed]

149. Gray, L.S.; Perez-Reyes, E.; Gomora, J.C.; Haverstick, D.M.; Shattock, M.; McLatchie, L.; Harper, J.; Brooks, G.; Heady, T.; Macdonald, T.L. The role of voltage gated T-type Ca^{2+} channel isoforms in mediating "capacitative" Ca^{2+} entry in cancer cells. *Cell Calcium* **2004**, *36*, 489–497. [CrossRef] [PubMed]

150. Panner, A.; Wurster, R.D. T-type calcium channels and tumor proliferation. *Cell Calcium* **2006**, *40*, 253–259. [CrossRef] [PubMed]

151. Parimi, V.; Goyal, R.; Poropatich, K.; Yang, X.J. Neuroendocrine differentiation of prostate cancer: A review. *Am. J. Clin. Exp. Urol.* **2014**, *2*, 273–285. [PubMed]

152. Mariot, P.; Vanoverberghe, K.; Lalevee, N.; Rossier, M.F.; Prevarskaya, N. Overexpression of an α1H (Cav3.2) T-type calcium channel during neuroendocrine differentiation of human prostate cancer cells. *J. Biol. Chem.* **2002**, *277*, 10824–10833. [CrossRef] [PubMed]

153. Rossier, M.F.; Lesouhaitier, O.; Perrier, E.; Bockhorn, L.; Chiappe, A.; Lalevee, N. Aldosterone regulation of T-type calcium channels. *J. Steroid Biochem. Mol. Biol.* **2003**, *85*, 383–388. [CrossRef]

154. Contreras, G.F.; Castillo, K.; Enrique, N.; Carrasquel-Ursulaez, W.; Castillo, J.P.; Milesi, V.; Neely, A.; Alvarez, O.; Ferreira, G.; Gonzalez, C.; et al. A BK (Slo1) channel journey from molecule to physiology. *Channels* **2013**, *7*, 442–458. [CrossRef] [PubMed]

155. Ge, L.; Hoa, N.T.; Wilson, Z.; Arismendi-Morillo, G.; Kong, X.T.; Tajhya, R.B.; Beeton, C.; Jadus, M.R. Big Potassium (BK) ion channels in biology, disease and possible targets for cancer immunotherapy. *Int. Immunopharmacol.* **2014**, *22*, 427–443. [CrossRef] [PubMed]

156. Bloch, M.; Ousingsawat, J.; Simon, R.; Schraml, P.; Gasser, T.C.; Mihatsch, M.J.; Kunzelmann, K.; Bubendorf, L. KCNMA1 gene amplification promotes tumor cell proliferation in human prostate cancer. *Oncogene* **2007**, *26*, 2525–2534. [CrossRef] [PubMed]

157. Du, C.; Zheng, Z.; Li, D.; Chen, L.; Li, N.; Yi, X.; Yang, Y.; Guo, F.; Liu, W.; Xie, X.; et al. BKCa promotes growth and metastasis of prostate cancer through facilitating the coupling between αvβ3 integrin and FAK. *Oncotarget* **2016**, *7*, 40174–40188. [CrossRef] [PubMed]

158. Legrand, G.; Humez, S.; Slomianny, C.; Dewailly, E.; Vanden Abeele, F.; Mariot, P.; Wuytack, F.; Prevarskaya, N. Ca^{2+} pools and cell growth. Evidence for sarcoendoplasmic Ca^{2+}-ATPases 2B involvement in human prostate cancer cell growth control. *J. Biol. Chem.* **2001**, *276*, 47608–47614. [CrossRef] [PubMed]

159. Denmeade, S.R.; Mhaka, A.M.; Rosen, D.M.; Brennen, W.N.; Dalrymple, S.; Dach, I.; Olesen, C.; Gurel, B.; Demarzo, A.M.; Wilding, G.; et al. Engineering a prostate-specific membrane antigen-activated tumor endothelial cell prodrug for cancer therapy. *Sci. Transl. Med.* **2012**, *4*, 140ra186. [CrossRef] [PubMed]

160. Flourakis, M.; Lehen'kyi, V.; Beck, B.; Raphael, M.; Vandenberghe, M.; Abeele, F.V.; Roudbaraki, M.; Lepage, G.; Mauroy, B.; Romanin, C.; et al. Orai1 contributes to the establishment of an apoptosis-resistant phenotype in prostate cancer cells. *Cell Death Dis.* **2010**, *1*, e75. [CrossRef] [PubMed]

161. Dubois, C.; Vanden Abeele, F.; Lehen'kyi, V.; Gkika, D.; Guarmit, B.; Lepage, G.; Slomianny, C.; Borowiec, A.S.; Bidaux, G.; Benahmed, M.; et al. Remodeling of channel-forming ORAI proteins determines an oncogenic switch in prostate cancer. *Cancer Cell* **2014**, *26*, 19–32. [CrossRef] [PubMed]

162. Marchi, S.; Lupini, L.; Patergnani, S.; Rimessi, A.; Missiroli, S.; Bonora, M.; Bononi, A.; Corra, F.; Giorgi, C.; De Marchi, E.; et al. Downregulation of the mitochondrial calcium uniporter by cancer-related miR-25. *Curr. Biol.* **2013**, *23*, 58–63. [CrossRef] [PubMed]

163. Orrenius, S.; Zhivotovsky, B.; Nicotera, P. Regulation of cell death: The calcium-apoptosis link. *Nat. Rev. Mol. Cell Biol.* **2003**, *4*, 552–565. [CrossRef] [PubMed]

164. Vlietstra, R.J.; van Alewijk, D.C.; Hermans, K.G.; van Steenbrugge, G.J.; Trapman, J. Frequent inactivation of PTEN in prostate cancer cell lines and xenografts. *Cancer Res.* **1998**, *58*, 2720–2723. [PubMed]

165. Carson, J.P.; Kulik, G.; Weber, M.J. Antiapoptotic signaling in LNCaP prostate cancer cells: A survival signaling pathway independent of phosphatidylinositol 3'-kinase and Akt/protein kinase B. *Cancer Res.* **1999**, *59*, 1449–1453. [PubMed]

166. Khan, M.T.; Wagner, L., 2nd; Yule, D.I.; Bhanumathy, C.; Joseph, S.K. Akt kinase phosphorylation of inositol 1,4,5-trisphosphate receptors. *J. Biol. Chem.* **2006**, *281*, 3731–3737. [CrossRef] [PubMed]

167. Kuchay, S.; Giorgi, C.; Simoneschi, D.; Pagan, J.; Missiroli, S.; Saraf, A.; Florens, L.; Washburn, M.P.; Collazo-Lorduy, A.; Castillo-Martin, M.; et al. PTEN counteracts FBXL2 to promote IP3R3- and Ca²⁺-mediated apoptosis limiting tumour growth. *Nature* **2017**, *546*, 554–558. [CrossRef] [PubMed]

168. Kobylewski, S.E.; Henderson, K.A.; Eckhert, C.D. Identification of ryanodine receptor isoforms in prostate DU-145, LNCaP, and PWR-1E cells. *Biochem. Biophys. Res. Commun.* **2012**, *425*, 431–435. [CrossRef] [PubMed]

169. Mariot, P.; Prevarskaya, N.; Roudbaraki, M.M.; Le Bourhis, X.; Van Coppenolle, F.; Vanoverberghe, K.; Skryma, R. Evidence of functional ryanodine receptor involved in apoptosis of prostate cancer (LNCaP) cells. *Prostate* **2000**, *43*, 205–214. [CrossRef]

170. Rios-Doria, J.; Kuefer, R.; Ethier, S.P.; Day, M.L. Cleavage of β-catenin by calpain in prostate and mammary tumor cells. *Cancer Res.* **2004**, *64*, 7237–7240. [CrossRef] [PubMed]

171. Rios-Doria, J.; Day, K.C.; Kuefer, R.; Rashid, M.G.; Chinnaiyan, A.M.; Rubin, M.A.; Day, M.L. The role of calpain in the proteolytic cleavage of E-cadherin in prostate and mammary epithelial cells. *J. Biol. Chem.* **2003**, *278*, 1372–1379. [CrossRef] [PubMed]

172. Park, J.J.; Rubio, M.V.; Zhang, Z.; Um, T.; Xie, Y.; Knoepp, S.M.; Snider, A.J.; Gibbs, T.C.; Meier, K.E. Effects of lysophosphatidic acid on calpain-mediated proteolysis of focal adhesion kinase in human prostate cancer cells. *Prostate* **2012**, *72*, 1595–1610. [CrossRef] [PubMed]

173. Shukla, S.; Shukla, M.; Maclennan, G.T.; Fu, P.; Gupta, S. Deregulation of FOXO3A during prostate cancer progression. *Int. J. Oncol.* **2009**, *34*, 1613–1620. [PubMed]

174. Shukla, S.; Bhaskaran, N.; Maclennan, G.T.; Gupta, S. Deregulation of FoxO3a accelerates prostate cancer progression in TRAMP mice. *Prostate* **2013**, *73*, 1507–1517. [CrossRef] [PubMed]

175. Brunet, A.; Kanai, F.; Stehn, J.; Xu, J.; Sarbassova, D.; Frangioni, J.V.; Dalal, S.N.; DeCaprio, J.A.; Greenberg, M.E.; Yaffe, M.B. 14-3-3 transits to the nucleus and participates in dynamic nucleocytoplasmic transport. *J. Cell Biol.* **2002**, *156*, 817–828. [CrossRef] [PubMed]

176. Wang, N.; Yao, M.; Xu, J.; Quan, Y.; Zhang, K.; Yang, R.; Gao, W.Q. Autocrine Activation of CHRM3 Promotes Prostate Cancer Growth and Castration Resistance via CaM/CaMKK-Mediated Phosphorylation of Akt. *Clin. Cancer Res.* **2015**, *21*, 4676–4685. [CrossRef] [PubMed]

177. Magnon, C.; Hall, S.J.; Lin, J.; Xue, X.; Gerber, L.; Freedland, S.J.; Frenette, P.S. Autonomic nerve development contributes to prostate cancer progression. *Science* **2013**, *341*, 1236361. [CrossRef] [PubMed]

178. Wang, N.; Dong, B.J.; Quan, Y.; Chen, Q.; Chu, M.; Xu, J.; Xue, W.; Huang, Y.R.; Yang, R.; Gao, W.Q. Regulation of Prostate Development and Benign Prostatic Hyperplasia by Autocrine Cholinergic Signaling via Maintaining the Epithelial Progenitor Cells in Proliferating Status. *Stem Cell Rep.* **2016**, *6*, 668–678. [CrossRef] [PubMed]

179. Schmitt, J.M.; Smith, S.; Hart, B.; Fletcher, L. CaM kinase control of AKT and LNCaP cell survival. *J. Cell. Biochem.* **2012**, *113*, 1514–1526. [CrossRef] [PubMed]

180. Hekman, M.; Albert, S.; Galmiche, A.; Rennefahrt, U.E.; Fueller, J.; Fischer, A.; Puehringer, D.; Wiese, S.; Rapp, U.R. Reversible membrane interaction of BAD requires two C-terminal lipid binding domains in conjunction with 14-3-3 protein binding. *J. Biol. Chem.* **2006**, *281*, 17321–17336. [CrossRef] [PubMed]

181. Taichman, R.S.; Cooper, C.; Keller, E.T.; Pienta, K.J.; Taichman, N.S.; McCauley, L.K. Use of the stromal cell-derived factor-1/CXCR4 pathway in prostate cancer metastasis to bone. *Cancer Res.* **2002**, *62*, 1832–1837. [PubMed]

182. Chetram, M.A.; Don-Salu-Hewage, A.S.; Hinton, C.V. ROS enhances CXCR4-mediated functions through inactivation of PTEN in prostate cancer cells. *Biochem. Biophys. Res. Commun.* **2011**, *410*, 195–200. [CrossRef] [PubMed]

183. Makowska, K.A.; Hughes, R.E.; White, K.J.; Wells, C.M.; Peckham, M. Specific Myosins Control Actin Organization, Cell Morphology, and Migration in Prostate Cancer Cells. *Cell Rep.* **2015**, *13*, 2118–2125. [CrossRef] [PubMed]

184. Peckham, M. How myosin organization of the actin cytoskeleton contributes to the cancer phenotype. *Biochem. Soc. Trans.* **2016**, *44*, 1026–1034. [CrossRef] [PubMed]

185. Barylko, B.; Jung, G.; Albanesi, J.P. Structure, function, and regulation of myosin 1C. *Acta Biochim. Pol.* **2005**, *52*, 373–380. [PubMed]

186. Ihnatovych, I.; Sielski, N.L.; Hofmann, W.A. Selective expression of myosin IC Isoform A in mouse and human cell lines and mouse prostate cancer tissues. *PLoS ONE* **2014**, *9*, e108609. [CrossRef] [PubMed]

187. Ihnatovych, I.; Migocka-Patrzalek, M.; Dukh, M.; Hofmann, W.A. Identification and characterization of a novel myosin Ic isoform that localizes to the nucleus. *Cytoskeleton* **2012**, *69*, 555–565. [CrossRef] [PubMed]

188. Nowak, G.; Pestic-Dragovich, L.; Hozak, P.; Philimonenko, A.; Simerly, C.; Schatten, G.; de Lanerolle, P. Evidence for the presence of myosin I in the nucleus. *J. Biol. Chem.* **1997**, *272*, 17176–17181. [CrossRef] [PubMed]

189. Zattelman, L.; Regev, R.; Usaj, M.; Reinke, P.Y.A.; Giese, S.; Samson, A.O.; Taft, M.H.; Manstein, D.J.; Henn, A. N-terminal splicing extensions of the human MYO1C gene fine-tune the kinetics of the three full-length myosin IC isoforms. *J. Biol. Chem.* **2017**, *292*, 17804–17818. [CrossRef] [PubMed]

190. Dzijak, R.; Yildirim, S.; Kahle, M.; Novak, P.; Hnilicova, J.; Venit, T.; Hozak, P. Specific nuclear localizing sequence directs two myosin isoforms to the cell nucleus in calmodulin-sensitive manner. *PLoS ONE* **2012**, *7*, e30529. [CrossRef] [PubMed]

191. Gillespie, P.G.; Cyr, J.L. Calmodulin binding to recombinant myosin-1c and myosin-1c IQ peptides. *BMC Biochem.* **2002**, *3*, 31. [CrossRef]

192. Nevzorov, I.; Sidorenko, E.; Wang, W.; Zhao, H.; Vartiainen, M.K. Myosin-1C uses a novel phosphoinositide-dependent pathway for nuclear localization. *EMBO Rep.* **2018**, *19*, 290–304. [CrossRef] [PubMed]

193. De Lanerolle, P. Nuclear actin and myosins at a glance. *J. Cell Sci.* **2012**, *125*, 4945–4949. [CrossRef] [PubMed]

194. Ye, J.; Zhao, J.; Hoffmann-Rohrer, U.; Grummt, I. Nuclear myosin I acts in concert with polymeric actin to drive RNA polymerase I transcription. *Genes Dev.* **2008**, *22*, 322–330. [CrossRef] [PubMed]

195. Percipalle, P.; Fomproix, N.; Cavellan, E.; Voit, R.; Reimer, G.; Kruger, T.; Thyberg, J.; Scheer, U.; Grummt, I.; Farrants, A.K. The chromatin remodelling complex WSTF-SNF2h interacts with nuclear myosin 1 and has a role in RNA polymerase I transcription. *EMBO Rep.* **2006**, *7*, 525–530. [CrossRef] [PubMed]

196. Hofmann, W.A.; Vargas, G.M.; Ramchandran, R.; Stojiljkovic, L.; Goodrich, J.A.; de Lanerolle, P. Nuclear myosin I is necessary for the formation of the first phosphodiester bond during transcription initiation by RNA polymerase II. *J. Cell. Biochem.* **2006**, *99*, 1001–1009. [CrossRef] [PubMed]

197. Kysela, K.; Philimonenko, A.A.; Philimonenko, V.V.; Janacek, J.; Kahle, M.; Hozak, P. Nuclear distribution of actin and myosin I depends on transcriptional activity of the cell. *Histochem. Cell Biol.* **2005**, *124*, 347–358. [CrossRef] [PubMed]

198. Philimonenko, V.V.; Zhao, J.; Iben, S.; Dingova, H.; Kysela, K.; Kahle, M.; Zentgraf, H.; Hofmann, W.A.; de Lanerolle, P.; Hozak, P.; et al. Nuclear actin and myosin I are required for RNA polymerase I transcription. *Nat. Cell Biol.* **2004**, *6*, 1165–1172. [CrossRef] [PubMed]

199. Fomproix, N.; Percipalle, P. An actin-myosin complex on actively transcribing genes. *Exp. Cell Res.* **2004**, *294*, 140–148. [CrossRef] [PubMed]

200. Pestic-Dragovich, L.; Stojiljkovic, L.; Philimonenko, A.A.; Nowak, G.; Ke, Y.; Settlage, R.E.; Shabanowitz, J.; Hunt, D.F.; Hozak, P.; de Lanerolle, P. A myosin I isoform in the nucleus. *Science* **2000**, *290*, 337–341. [CrossRef] [PubMed]

201. Obrdlik, A.; Louvet, E.; Kukalev, A.; Naschekin, D.; Kiseleva, E.; Fahrenkrog, B.; Percipalle, P. Nuclear myosin 1 is in complex with mature rRNA transcripts and associates with the nuclear pore basket. *FASEB J.* **2010**, *24*, 146–157. [CrossRef] [PubMed]

202. Cisterna, B.; Necchi, D.; Prosperi, E.; Biggiogera, M. Small ribosomal subunits associate with nuclear myosin and actin in transit to the nuclear pores. *FASEB J.* **2006**, *20*, 1901–1903. [CrossRef] [PubMed]

203. Mehta, I.S.; Elcock, L.S.; Amira, M.; Kill, I.R.; Bridger, J.M. Nuclear motors and nuclear structures containing A-type lamins and emerin: Is there a functional link? *Biochem. Soc. Trans.* **2008**, *36*, 1384–1388. [CrossRef] [PubMed]

204. Chuang, C.H.; Carpenter, A.E.; Fuchsova, B.; Johnson, T.; de Lanerolle, P.; Belmont, A.S. Long-range directional movement of an interphase chromosome site. *Curr. Biol.* **2006**, *16*, 825–831. [CrossRef] [PubMed]

205. Venit, T.; Dzijak, R.; Kalendova, A.; Kahle, M.; Rohozkova, J.; Schmidt, V.; Rulicke, T.; Rathkolb, B.; Hans, W.; Bohla, A.; et al. Mouse nuclear myosin I knock-out shows interchangeability and redundancy of myosin isoforms in the cell nucleus. *PLoS ONE* **2013**, *8*, e61406. [CrossRef] [PubMed]

206. Lu, Q.; Li, J.; Ye, F.; Zhang, M. Structure of myosin-1c tail bound to calmodulin provides insights into calcium-mediated conformational coupling. *Nat. Struct. Mol. Biol.* **2015**, *22*, 81–88. [CrossRef] [PubMed]

207. Pruschy, M.; Ju, Y.; Spitz, L.; Carafoli, E.; Goldfarb, D.S. Facilitated nuclear transport of calmodulin in tissue culture cells. *J. Cell Biol.* **1994**, *127*, 1527–1536. [CrossRef] [PubMed]

208. Visuttijai, K.; Pettersson, J.; Mehrbani Azar, Y.; van den Bout, I.; Orndal, C.; Marcickiewicz, J.; Nilsson, S.; Hornquist, M.; Olsson, B.; Ejeskar, K.; et al. Lowered Expression of Tumor Suppressor Candidate MYO1C Stimulates Cell Proliferation, Suppresses Cell Adhesion and Activates AKT. *PLoS ONE* **2016**, *11*, e0164063. [CrossRef] [PubMed]

209. Letellier, E.; Schmitz, M.; Ginolhac, A.; Rodriguez, F.; Ullmann, P.; Qureshi-Baig, K.; Frasquilho, S.; Antunes, L.; Haan, S. Loss of Myosin Vb in colorectal cancer is a strong prognostic factor for disease recurrence. *Br. J. Cancer* **2017**, *117*, 1689–1701. [CrossRef] [PubMed]

210. Buschman, M.D.; Field, S.J. MYO18A: An unusual myosin. *Adv. Biol. Regul.* **2018**, *67*, 84–92. [CrossRef] [PubMed]

Permissions

All chapters in this book were first published by MDPI; hereby published with permission under the Creative Commons Attribution License or equivalent. Every chapter published in this book has been scrutinized by our experts. Their significance has been extensively debated. The topics covered herein carry significant findings which will fuel the growth of the discipline. They may even be implemented as practical applications or may be referred to as a beginning point for another development.

The contributors of this book come from diverse backgrounds, making this book a truly international effort. This book will bring forth new frontiers with its revolutionizing research information and detailed analysis of the nascent developments around the world.

We would like to thank all the contributing authors for lending their expertise to make the book truly unique. They have played a crucial role in the development of this book. Without their invaluable contributions this book wouldn't have been possible. They have made vital efforts to compile up to date information on the varied aspects of this subject to make this book a valuable addition to the collection of many professionals and students.

This book was conceptualized with the vision of imparting up-to-date information and advanced data in this field. To ensure the same, a matchless editorial board was set up. Every individual on the board went through rigorous rounds of assessment to prove their worth. After which they invested a large part of their time researching and compiling the most relevant data for our readers.

The editorial board has been involved in producing this book since its inception. They have spent rigorous hours researching and exploring the diverse topics which have resulted in the successful publishing of this book. They have passed on their knowledge of decades through this book. To expedite this challenging task, the publisher supported the team at every step. A small team of assistant editors was also appointed to further simplify the editing procedure and attain best results for the readers.

Apart from the editorial board, the designing team has also invested a significant amount of their time in understanding the subject and creating the most relevant covers. They scrutinized every image to scout for the most suitable representation of the subject and create an appropriate cover for the book.

The publishing team has been an ardent support to the editorial, designing and production team. Their endless efforts to recruit the best for this project, has resulted in the accomplishment of this book. They are a veteran in the field of academics and their pool of knowledge is as vast as their experience in printing. Their expertise and guidance has proved useful at every step. Their uncompromising quality standards have made this book an exceptional effort. Their encouragement from time to time has been an inspiration for everyone.

The publisher and the editorial board hope that this book will prove to be a valuable piece of knowledge for researchers, students, practitioners and scholars across the globe.

List of Contributors

Simone Patergnani, Alberto Danese, Esmaa Bouhamida, Paolo Pinton and Carlotta Giorgi
Department of Medical Sciences, Laboratory for Technologies of Advanced Therapies, University of Ferrara, 44121 Ferrara, Italy

Gianluca Aguiari
Department of Biomedical and Surgical Specialty Sciences, University of Ferrara, 44121 Ferrara, Italy

Maurizio Previati
Department of Morphology, Surgery and Experimental Medicine, Section of Human Anatomy and Histology, Laboratory for Technologies of Advanced Therapies (LTTA), University of Ferrara, 44121 Ferrara Italy

Marcial Sanchez-Tecuatl and Juan Manuel Ramirez-Cortes
Electronics Department, National Institute of Astrophysics, Optics and Electronics, 72840 Puebla, Mexico

Ajelet Vargaz-Guadarrama and Roberto Berra-Romani
Biomedicine School, Benemerita Universidad Autonoma de Puebla, 72410 Puebla, Mexico

Pilar Gomez-Gil
Computer Science Department, National Institute of Astrophysics, Optics and Electronics, 72840 Puebla, Mexico

Claudia M. Tellez Freitas, Deborah K. Johnson and K. Scott Weber
Department of Microbiology and Molecular Biology, Brigham Young University, Provo, UT 84604, USA

Germano Guerra and Angelica Perna
Department of Medicine and Health Sciences "Vincenzo Tiberio", University of Molise, via F. De Santis, 86100 Campobasso, Italy

Angela Lucariello and Antonio De Luca
Department of Mental Health and Preventive Medicine, Section of Human Anatomy, University of Campania "L. Vanvitelli", 81100 Naples, Italy

Laura Botta and Francesco Moccia
Laboratory of General Physiology, Department of Biology and Biotechnology "L. Spallanzani", University of Pavia, via Forlanini 6, 27100 Pavia, Italy

Viviana Meraviglia, Maria Cristina Florio, Benedetta M. Motta, Corrado Corti, Christian X. Weichenberger, Yuri D'Elia, Marcelo D. Rosato-Siri, Silvia Suffredini, Chiara Piubelli, Peter P. Pramstaller, Francisco S. Domingues and Alessandra Rossini
Institute for Biomedicine, Eurac Research, 39100 Bolzano, Italy

Leonardo Bocchi, Monia Savi and Donatella Stilli
Department of Chemistry, Life Sciences and Environmental Sustainability, University of Parma, 43124 Parma, Italy

Roberta Sacchetto
Department of Comparative Biomedicine and Food Science, University of Padova, 35020 Legnaro (Padova), Italy

Giulio Pompilio
Vascular Biology and Regenerative Medicine Unit, Centro Cardiologico Monzino, IRCCS, 20138 Milano, Italy
Dipartimento di Scienze Cliniche e di Comunita, Universita degli Studi di Milano, 20122 Milano, Italy

Daniel Dumitru Banciu
Department of Anatomy, Animal Physiology and Biophysics, Faculty of Biology, University of Bucharest, Splaiul Independentei 91-95, 050095 Bucharest, Romania

Adela Banciu
Department of Anatomy, Animal Physiology and Biophysics, Faculty of Biology, University of Bucharest, Splaiul Independentei 91-95, 050095 Bucharest, Romania
Faculty of Medical Engineering, University Politehnica of Bucharest, Gheorge Polizu Street 1-7, 011061 Bucharest, Romania

Mihai Radu
Department of Life and Environmental Physics, Horia Hulubei National Institute of Physics and Nuclear Engineering, Reactorului 30, 077125 Magurele, Romania

Sanda Maria Cretoiu
Department of Cell and Molecular Biology and Histology, Carol Davila University of Medicine and Pharmacy, 050474 Bucharest, Romania

Cosmin Catalin Mustaciosu
Department of Life and Environmental Physics, Horia Hulubei National Institute of Physics and Nuclear Engineering, Reactorului 30, 077125 Magurele, Romania
Faculty of Applied Chemistry and Materials Science, University Politehnica of Bucharest, 011061 Bucharest, Romania

Dragos Cretoiu
Department of Cell and Molecular Biology and Histology, Carol Davila University of Medicine and Pharmacy, 050474 Bucharest, Romania
Alessandrescu-Rusescu National Institute of Mother and Child Health, Fetal Medicine Excellence Research Center, 020395 Bucharest, Romania

Junjie Xiao
Cardiac Regeneration and Ageing Lab, Experimental Center of Life Sciences, School of Life Science, Shanghai University, Shanghai 200444, China

Nicolae Suciu
Alessandrescu-Rusescu National Institute of Mother and Child Health, Fetal Medicine Excellence Research Center, 020395 Bucharest, Romania
Department of Obstetrics and Gynecology, Polizu Clinical Hospital, 011062 Bucharest, Romania

Beatrice Mihaela Radu
Department of Anatomy, Animal Physiology and Biophysics, Faculty of Biology, University of Bucharest, Splaiul Independentei 91-95, 050095 Bucharest, Romania
Life, Environmental and Earth Sciences Division, Research Institute of the University of Bucharest (ICUB), 91-95 Splaiul Independenței, 050095 Bucharest, Romania

Daniela Sorriento and Bruno Trimarco
Dipartmento di "Scienze Biomediche Avanzate", Universita "Federico II" di Napoli, Via Pansini 5, 80131 Napoli, Italy

Gaetano Santulli
Dipartmento di "Scienze Biomediche Avanzate", Universita "Federico II" di Napoli, Via Pansini 5, 80131 Napoli, Italy
Department of Medicine, Albert Einstein College of Medicine, Montefiore University Hospital, 1300 Morris Park Avenue, Bronx, NY 10461, USA

Michele Ciccarelli and Guido Iaccarino
Dipartmento di Medicina, Chirurgia e Odontoiatria "Scuola Medica Salernitana"/DIPMED, Universita degli Studi di Salerno, Via S. Allende, 84081 Baronissi (SA), Italy

Angela Serena Maione and Maddalena Illario
Dipartmento di "Scienze Mediche Traslazionali", Universita "Federico II" di Napoli, Via Pansini 5, 80131 Napoli, Italy

Luigi Catacuzzeno and Fabio Franciolini
Department of Chemistry, Biology and Biotechnology, University of Perugia, 06134 Perugia, Italy

Francesco Moccia
Laboratory of General Physiology, Department of Biology and Biotechnology "L. Spallanzani", University of Pavia, I-27100 Pavia, Italy

John A. D'Elia and Larry A. Weinrauch
E P Joslin Research Laboratory, Kidney and Hypertension Section, Joslin Diabetes Center, Department of Medicine, Mount Auburn Hospital, Harvard Medical School, Boston and Cambridge, 521 Mount Auburn Street Watertown, MA 02472, USA

Ivan V. Maly and Wilma A. Hofmann
Department of Physiology and Biophysics, Jacobs School of Medicine and Biomedical Sciences, University at Buffalo, 955 Main Street, Buffalo, NY 14203, USA

Index